TERRESTRIAL VERTEBRATES OF TIDAL MARSHES: EVOLUTION, ECOLOGY, AND CONSERVATION

Russell Greenberg, Jesús E. Maldonado, Sam Droege, and M. Victoria McDonald
Associate Editors

Studies in Avian Biology No. 32
A PUBLICATION OF THE COOPER ORNITHOLOGICAL SOCIETY

Cover painting (Saltmarsh Sharp-tailed Sparrow) and black-and-white drawings by Julie Zickefoose
Painting on back cover "Cloudy Day, Rhode Island" by Martin Johnson Heade
Painting © 2006 Museum of Fine Arts Boston

STUDIES IN AVIAN BIOLOGY

Edited by

Carl D. Marti
1310 East Jefferson Street
Boise, ID 83712

Spanish translation by
Cecilia Valencia

Studies in Avian Biology is a series of works too long for *The Condor*, published at irregular intervals by the Cooper Ornithological Society. Manuscripts for consideration should be submitted to the editor. Style and format should follow those of previous issues.

Price $24.00 including postage and handling. All orders cash in advance; make checks payable to Cooper Ornithological Society. Send orders to Cooper Ornithological Society, % Western Foundation of Vertebrate Zoology, 439 Calle San Pablo, Camarillo, CA 93010

Permission to Copy

The Cooper Ornithological Society hereby grants permission to copy chapters (in whole or in part) appearing in *Studies in Avian Biology* for personal use, or educational use within one's home institution, without payment, provided that the copied material bears the statement "©2006 The Cooper Ornithological Society" and the full citation, including names of all authors. Authors may post copies of their chapters on their personal or institutional website, except that whole issues of *Studies in Avian Biology* may not be posted on websites. Any use not specifically granted here, and any use of *Studies in Avian Biology* articles or portions thereof for advertising, republication, or commercial uses, requires prior consent from the editor.

ISBN: 0-943610-70-2

Library of Congress Control Number: 2006933990
Printed at Cadmus Professional Communications, Ephrata, Pennsylvania 17522
Issued: 15 November 2006

Copyright © by the Cooper Ornithological Society 2006

Painting on back cover: *Cloudy Day, Rhode Island*, 1861.
Oil on canvas. 29.53 × 64.45 cm (11⅝ × 25⅜ in). Museum of Fine Arts, Boston.
Gift of Maxim Karolik for the M. and M. Karolik Collection of American Paintings, 1815–1865
47.1158

CONTENTS

LIST OF AUTHORS v

FOREWORD .... David Challinor   1

INTRODUCTION

Tidal marshes: Home for the few and the highly selected .... Russell Greenberg   2

BIOGEOGRAPHY AND EVOLUTION OF TIDAL-MARSH FAUNAS

The quaternary geography and biogeography of tidal saltmarshes ....
  .... Karl P. Malamud-Roam, Frances P. Malamud-Roam, Elizabeth B. Watson,
  Joshua N. Collins, and B. Lynn Ingram 11

Diversity and endemism in tidal-marsh vertebrates ....
  .... Russell Greenberg and Jesús E. Maldonado   32

Evolution and conservation of tidal-marsh vertebrates: molecular approaches ....
  .... Yvonne L. Chan, Christopher E. Hill, Jesús E. Maldonado,
  and Robert C. Fleischer   54

ADAPTATION TO TIDAL MARSHES

Avian nesting response to tidal-marsh flooding: literature review and a case for adaptation in the Red-winged Blackbird .... Steven E. Reinert   77

Flooding and predation: trade-offs in the nesting ecology of tidal-marsh sparrows ....
  .... Russell Greenberg, Christopher Elphick, J. Cully Nordby,
  Carina Gjerdrum, Hildie Spautz, Gregory Shriver, Barbara Schmeling,
  Brian Olsen, Peter Marra, Nadav Nur, and Maiken Winter   96

Osmoregulatory biology of saltmarsh passerines .... David L. Goldstein 110

Social behavior of North American tidal-marsh vertebrates ....
  .... M. Victoria McDonald and Russell Greenberg 119

Trophic adaptations in sparrows and other vertebrates of tidal marshes ....
  .... J. Letitia Grenier and Russell Greenberg 130

REGIONAL STUDIES

Breeding birds of northeast saltmarshes: habitat use and conservation ....
  .... Alan R. Hanson and W. Gregory Shriver 141

Impacts of marsh management on coastal-marsh birds habitats ....
  .... Laura R. Mitchell, Steven Gabrey, Peter P. Marra, and R. Michael Erwin 155

Environmental threats to tidal-marsh vertebrates of the San Francisco Bay estuary ......
  .... John Y. Takekawa, Isa Woo, Hildie Spautz, Nadav Nur, J. Letitia Grenier,
  Karl Malamud-Roam, J. Cully Nordby, Andrew N. Cohen,
  Frances Malamud-Roam, and Susan E. Wainwright-De La Cruz 176

Are southern California's fragmented salt marshes capable of sustaining endemic bird populations?..Abby N. Powell 198

CONSERVATION BIOLOGY

The diamondback terrapin: the biology, ecology, cultural history, and conservation status of an obligate estuarine turtle .... Kristen M. Hart and David S. Lee 206

High tides and rising seas: potential effects on estuarine waterbirds.........................
......R. Michael Erwin, Geoffrey M. Sanders, Diann J. Prosser, and Donald R. Cahoon 214

The impact of invasive plants on tidal-marsh vertebrate species: common reed (*Phragmites australis*) and smooth cordgrass (*Spartina alterniflora*) as case studies.......
.... Glenn R. Guntenspergen and J. Cully Nordby 229

Tidal saltmarsh fragmentation and persistence of San Pablo Song Sparrows (*Melospiza melodia samuelis*): assessing benefits of wetland restoration in San Francisco Bay.................................John Y. Takekawa, Benjamin N. Sacks, Isa Woo, Michael L. Johnson, and Glenn D. Wylie 238

Multiple-scale habitat relationships of tidal-marsh breeding birds in the San Francisco Bay estuary .... Hildie Spautz, Nadav Nur, Diana Stralberg, and Yvonne Chan 247

The Clapper Rail as an indicator species of estuarine-marsh health............................
.... James M. Novak, Karen F. Gaines, James C. Cumbee, Jr., Gary L. Mills, Alejandro Rodriguez-Navarro, and Christopher S. Romanek 270

A unified strategy for monitoring changes in abundance of birds associated with North American tidal marshes..................... Courtney J. Conway and Sam Droege 282

An agenda for research on the ecology, evolution, and conservation of tidal-marsh vertebrates ....The Symposium Contributors 298

LITERATURE CITED 300

LIST OF AUTHORS

DONALD R. CAHOON
BARC-East, Building 308
10300 Baltimore Avenue
Beltsville, MD 20705

YVONNE CHAN
Department of Biological Sciences
Stanford University
371 Serra Mall
Stanford, CA 94305
(Current address: PRBO Conservation Science,
3820 Cypress Drive #11,
Petaluma, CA 94954)

ANDREW N. COHEN
San Francisco Estuary Institute,
7770 Pardee Lane,
Oakland, CA 94621

JOSHUA N. COLLINS
San Francisco Estuary Institute
7770 Pardee Lane
Oakland, CA 94621

JAMES C. CUMBEE, JR.
Savannah River Ecology Laboratory
P.O. Drawer E
Aiken, SC 29802
and
Institute of Ecology
University of Georgia
Athens, GA 30602

COURTNEY J. CONWAY
U.S. Geological Survey
Arizona Cooperative Fish and Wildlife Research Unit
104 Biological Sciences East, University of Arizona
Tucson, AZ 85721

SAM DROEGE
USGS Patuxent Wildlife Research Center
12100 Beech Forest Drive
Laurel, MD 20774

CHRISTOPHER ELPHICK
Ecology and Evolutionary Biology
University of Connecticut
75 N. Eagleville Road
Storrs, CT 06269-3043

R. MICHAEL ERWIN
USGS Patuxent Wildlife Research Center
Department of Environmental Sciences
University of Virginia
Charlottesville VA 22904

ROBERT FLEISCHER
Genetics Program
National Zoological Park/
National Museum of Natural History
Smithsonian Institution
3001 Connecticut Avenue, NW
Washington, DC 20008

STEVEN GABREY
Biology Department
Northwestern Louisiana State University
Natchitoches, LA 71497

KAREN F. GAINES
Department of Biology
University of South Dakota
Vermillion, SD 57069
(Current address: Department of Biological Sciences,
Eastern Illinois University, Charleston IL 61920)

CARINA GJERDRUM
Ecology and Evolutionary Biology
University of Connecticut
75 N. Eagleville Road
Storrs, CT 06269-3043

DAVID L. GOLDSTEIN
Department of Biological Sciences
Wright State University
Dayton, OH 45435

RUSSELL GREENBERG
Smithsonian Migratory Bird Center
National Zoological Park
Washington, DC 20008

J. LETITIA GRENIER,
Department of Environmental Science, Policy and
Management
University of California
151 Hilgard Hall #3110
Berkeley, CA 94720-3110
(Current address: San Francisco Estuary Institute,
7770 Pardee Lane, Oakland, CA 94621)

GLENN R. GUNTENSPERGEN
U.S. Geological Survey,
Patuxent Wildlife Research Center,
Laurel, MD 20708

ALAN R. HANSON
Canadian Wildlife Service
P.O. Box 6227
Sackville, NB
E4L 1G6 Canada

KRISTEN M. HART
Duke University
Nicholas School of the Environment and Earth
Sciences
Marine Laboratory,
135 Duke Marine Lab Road
Beaufort, NC 28516-9721
(Current address: U.S. Geological Survey, Center for
Coastal and Watershed Studies, 600 Fourth Street
South, St. Petersburg, FL 33701)

CHRIS HILL
Department of Biology
Coastal Carolina University
Conway, SC 29528-1954

LYNN B. INGRAM
Departments of Geography and Earth
and Planetary Sciences
University of California
Berkeley, CA 94720

MICHAEL L. JOHNSON
John Muir Institute of the Environment
University of California
Davis, CA 95616

DAVID LEE
The Tortoise Reserve
P.O. Box 7082
White Lake, NC 28337

FRANCES MALAMUD-ROAM
University of California
Department of Geography
Berkeley, CA 94720

KARL MALAMUD-ROAM
Contra Costa Mosquito and
Vector Control District
55 Mason Circle
Concord, CA 94520

JESÚS E. MALDONADO
Genetics Program
National Zoological Park/
National Museum of Natural History
Smithsonian Institution
3001 Connecticut Ave., NW
Washington, DC 20008

PETER P. MARRA
Smithsonian Environmental Research Center
P.O. Box 28
647 Contees Wharf Road
Edgewater, MD 21037

M. VICTORIA MCDONALD
Deprtment of Biology
University of Central Arkansas
Conway, AR 72035

GARY L. MILLS
Savannah River Ecology Laboratory
P.O. Drawer E
Aiken, SC 29802

LAURA R. MITCHELL
Prime Hook National Wildlife Refuge
11978 Turkle Pond Road
Milton, DE 19968
(Current address: Eastern Massachusetts
NWR Complex, 73 Weir Hill Road,
Sudbury, MA 01776)

J. CULLY NORDBY
Department of Environmental Science, Policy, and
Management
University of California
Berkeley, CA 94720

JAMES M. NOVAK
Savannah River Ecology Laboratory
P.O. Drawer E
Aiken, SC 29802
and
Institute of Ecology
University of Georgia
Athens, GA 30602
(Current address: Department of Biological Sciences,
Eastern Illinois University, Charleston IL 61920)

NADAV NUR
PRBO Conservation Science
3820 Cypress Drive #11
Petaluma, CA 94954

ABBY N. POWELL
USGS, Alaska Cooperative Fish and Wildlife
Research Unit
University of Alaska, Fairbanks
Fairbanks, AK 99775-7020

DIANN J. PROSSER
USGS Patuxent Wildlife Research Center
BARC-East, Building 308
10300 Baltimore Avenue
Beltsville, MD 20705

STEVEN E. REINERT
11 Talcott Street
Barrington, RI 02806

ALEJANDRO RODRIGUEZ-NAVARRO
Savannah River Ecology Laboratory
P.O. Drawer E
Aiken, SC C 29802
(Current address: Instituto Andaluz de Ciencias
de la Tierra. CSIC, Universidad de Granada, 18002
Granada, Spain)

CHRISTOPHER S. ROMANEK
Savannah River Ecology Laboratory
P.O. Drawer E
Aiken, SC 29802
and
Department of Geology
University of Georgia
Athens, GA 30602

BENJAMIN N. SACKS
John Muir Institute of the Environment
University of California
Davis, CA 95616

GEOFFREY M. SANDERS
National Park Service
4598 MacArthur Boulevard, NW
Washington, DC 20007

BARBARA SCHMELING
Smithsonian Environmental Research Center
P.O. Box 28
647 Contees Wharf Road
Edgewater, MD 21037

W. GREGORY SHRIVER
National Park Service
Marsh-Billings-Rockefeller NHP
54 Elm Street
Woodstock, VT 05091
(Current address: 257 Townsend Hall,
Department of Entomology and Wildlife Ecology,
University of Delaware, Newark, DE 19716-2160)

HILDIE SPAUTZ
PRBO Conservation Science
3820 Cypress Drive #11
Petaluma, CA 94954
(Current address: Wetland Wildlife Associates, P.O.
Box 2330, El Cerrito, CA 94530)

DIANA STRALBERG
PRBO Conservation Science
4990 Shoreline Highway
Stinson Beach, CA 94970

JOHN Y. TAKEKAWA
U. S. Geological Survey
Western Ecological Research Center
San Francisco Bay Estuary Field Station
Vallejo, CA 94592
(Current address: U. S. Geological Survey,
505 Azuar Drive, P. O. Box 2012,
Vallejo, CA 94592)

SUSAN E. WAINWRIGHT-DE LA CRUZ
U.S. Geological Survey, Western Ecological Research
Center
San Francisco Bay Estuary Field Station
505 Azuar Drive
Vallejo, CA 94592

ELIZABETH B. WATSON
Department of Geography
University of California
Berkeley, CA 94720

MAIKEN WINTER
State University of New York
College of Environmental Sciences and Forestry
Syracuse, NY 13210
(Current address: Laboratory of Ornithology, Ithaca,
NY 14850)

ISA WOO
Humboldt State University Foundation
Arcata, CA 95521

GLENN D. WYLIE
U. S. Geological Survey
Western Ecological Research Center
Dixon Field Station
Dixon, CA 95620

FOREWORD

DAVID CHALLINOR

With unremitting pressure on both North American coasts to satisfy the demands for new marinas and other shore developments, the extent of tidal marshes is continually shrinking. Having grown up and lived adjacent to Connecticut tidal marshes for more than 80 yr, I have watched both their alteration and demise. Despite the relatively small space occupied by tidal marshes, their value as a crucial habitat for a disproportionate number of vertebrate species is attracting increasing attention. How birds, mammals, and reptiles have adapted to exploit this relatively impoverished floral habitat was the focus of a symposium held in October 2002 at the Patuxent National Wildlife Research Center, Patuxent, Maryland.

The collection of twenty papers presented at this gathering is assembled in this volume. The section devoted to avian adaptation to tidal marshes contains a wealth of new research results on how marsh denizens differ from their dry-land interior congeners. We learn how, long ago, they may have split from their more common relatives in order to live in such a dynamic habitat where, twice daily, salty water floods and flows from their territories. A larger part of this volume focuses on the conservation biology of tidal marshes and calls attention to such immediate threats as invading exotic plants, water pollution, drainage and a host of other habitat-modifying forces. A less immediate but still real menace to current tidal marshes is the rising ocean, but if the pace is slow enough, the marshes can retreat to higher ground. Such advances and retreats have been well recorded in the geological record.

This volume fills a crucial gap in our understanding of the dynamics of tidal-marsh vertebrate fauna and, furthermore, devotes a thoughtful concluding paper to an agenda for future research on marsh fauna. The Smithsonian's Migratory Bird Center, The U.S. Geological Survey, and the USDI Fish and Wildlife Service deserve great credit for sponsoring this symposium; its resulting volume assures not only the permanent record of the proceedings but a clear recommendation for future research on the fauna of tidal marshes.

TIDAL MARSHES: HOME FOR THE FEW AND THE HIGHLY SELECTED

RUSSELL GREENBERG

WHY STUDY TIDAL MARSHES?

Tidal marshes consist of grass or small shrub-dominated wetlands that experience regular tidal inundation. In subtropical and tropical regions, marshes give way to mangrove swamps dominated by a small number of salt-tolerant tree species. Tidal marshes can be fresh, brackish, saline, or hyper-saline with respect to salt concentrations in sea water. In this volume we focus on marshes (not mangroves [*Rhizophora, Avicennia,* and *Laguncularia*]) that are brackish to saline (5–35 ppt salt concentration). Tidal saltmarshes are widely distributed along most continental coastlines (Chapman 1977). Although found along thousands of kilometers of shorelines, the aerial extent of tidal marsh is quite small. We estimate that, excluding arctic marshes and tropical salt flats, tidal marshes cover ≈45,000 km^2 which, to put this in perspective, would cover a land area merely twice the size of the state of New Jersey. To place this figure further in an ecological context, the total area of another threatened ecosystem, tropical rain forest, is approximately 14,000,000 km^2 or >300 times greater than the amount of tidal marsh even after deforestation). Although the area covered by tidal marsh is small, this ecosystem forms a true ecotone between the ocean and land, and therefore plays a key role in both marine and terrestrial ecological processes. In the parlance of modern conservation biology, the tidal-marsh ecosystem provides numerous critical ecological services, including protecting shorelines from erosion, providing nursery areas for fish, crabs and other marine organisms, and improving water quality for estuaries.

Tidal saltmarshes are primarily associated with the large estuaries of mid-latitudes, in North America, Eurasia, and southern South America, with some in Australia and South Africa. Tidal marshes are highly productive yet, in some ways, inhospitable to birds and other vertebrates. Surrounded by a highly diverse source fauna from the interior of the continental land mass, relatively few species cross the threshold of the maximum high-tide line and colonize intertidal wetlands. In this volume, we discuss myriad approaches to understanding which species have colonized the landward side, how they have evolved to meet the adaptive challenges of tidal marsh ecosystems, and in what ways we can act to conserve these small but unique tidal marsh faunas.

Studies of tidal-marsh faunas have significance far beyond understanding the vagaries of this particular habitat. Tidal marshes, with their abrupt selective gradients and relatively simple biotic assemblages, provide a living laboratory for the study of evolutionary processes. The following are just a few of the major conceptually defined fields within biology that have focused on tidal marshes as a model system: (1) evolutionary biologists seeking to investigate systems where morphological changes may have evolved in the face of recent colonization and current gene flow between saltmarsh and inland populations, (2) ecologists interested in how life history and behavior may shift in the face of a local, but strongly divergent environment, (3) physiological ecologists, wishing to see how different organisms cope with the abiotic factors governing successful colonization of saltmarshes, (4) biogeographers interested in patterns of diversity in endemism in this habitat along different coasts and in different continents, and (5) conservation biologists, because of the disproportionately high frequency of endangered and threatened taxa that are endemic to tidal marshes.

Many of us have spent years in tidal marshes in pursuit of our particular study species. We came together for this project because we began to think beyond our particular study species and study marsh, slough, or estuary. It became apparent to us that tidal marsh vertebrates face a number of severe environmental threats that might best be understood by gaining a more global and less local estuary-centric perspective. Furthermore, although tidal marshes provide a laboratory for studying local ecological differentiation, the mechanisms and ultimate factors shaping this local divergence can best be understood by studying common adaptive challenges and their solutions in a more comparative manner. As we contacted vertebrate zoologists working around the globe, it became apparent that few tidal-marsh researchers think beyond their particular coastline. We believed that if we could provide the catalyst for a more holistic and global thinking about tidal marsh vertebrates, that would be an important step forward.

In October 2002, we held a symposium at Patuxent National Wildlife Research Center to bring researchers together from different coasts and marshes. But we took one step further. Both during the organization of the symposium and the subsequent preparation of this volume, we made a concerted effort to go beyond our ornithological roots and to pull together research from other vertebrate groups, as well as more process-oriented tidal-marsh ecologists. Including other classes of terrestrial vertebrates has opened our collective eyes and we appreciate the cooperation of the editors of *Studies in Avian Biology* to allow so much non-avian material in our publication.

Tidal marshes are among the most productive ecosystems in the world, with high levels of primary production created by vascular plants, phytoplankton, and algal mats on the substrate (Adam 1990, Mitsch and Gosselink 2000). Abundant plant and animal food resources are available through both the terrestrial vegetation and the marine food chains associated with tidal channels. It is small wonder that saltmarshes often support high abundances of the species that live there.

On the other hand, the fauna and flora associated with salt and brackish marshes are depauperate. Our attention is drawn to tidal-marsh systems not primarily for the diversity of birds and other terrestrial vertebrates, but for the high proportion of endemic taxa (subspecies or species with endemic subspecies). In the course of preparing this volume, we have identified 25 species of mammals, reptiles, and breeding birds that are either wholly restricted or have recognized subspecies that are restricted to tidal marshes (Table 1).

Tidal marshes present enormous adaptive challenges to animals attempting to colonize them. The vegetation is often quite distinct from adjacent upland or freshwater marsh habitats. Perhaps more severe are the challenges from the physical environment (Dunson and Travis 1994). In particular, animals must cope with the salinity of the water, the retained salinity in the food supply, the regular ebb and flow of tides, and the less predictable storm surges. Less obvious differences include basic geochemical processes, which, among other things can alter the dominant coloration of the substrate. How these challenges shape individual physiological, morphological and behavioral adaptations has often been the focus of excellent research, but efforts to integrate the effect of these environmental factors are far fewer.

The availability of tidal-marsh habitat as a setting for evolution and adaptation by colonizing terrestrial vertebrate species has varied greatly throughout the Pleistocene (Malamud-Roam et al., *this volume*). Perhaps because of this, the current fauna is a mosaic of species with old and very recent associations with this habitat (Chan et al., *this volume*). In North America, the fauna consists of repeated invasions from species in a few select genera of which sparrows (*Ammodramus* and *Melospiza*), shrews (*Sorex*), voles (*Microtus*), and water snakes (*Nerodia*) are the most frequently involved. On the other hand, tidal marshes are inhabited by a few ancient taxa, such as the diamondback terrapin (*Maloclemys terrapin*), that have evolved in estuarine habitats since the Tertiary. A plethora of recent work on molecular phylogenies of these species allows us to examine the pattern and time of invasions by new taxa. Furthermore, we can examine the nature of adaptation of taxa with older and more recent associations with tidal marshes (Grenier and Greenberg, *this volume*).

Because of this high level of differentiation of tidal marsh taxa, the restricted distribution of this habitat, and its location in some of the most heavily settled areas of the world, it is not surprising that many populations are very small and have shown rapid declines. Tidal marsh vertebrates face the continuing challenges of fragmentation, ditching and impoundment, reduction in area, pollution, and the establishment of invasive species (Daiber 1982). In addition, sea-level rise will not only influence the extent and zonation of tidal marshes (Erwin et al. 1994, *this volume*), but the salinity and perhaps the frequency of storm surges as well.

Given the enormous pressures on delicate coastal ecosystems, it should not be a surprise that the 25 species and the close to 50 subspecies that they represent are disproportionately endangered, threatened, or otherwise of heightened conservation concern (Table 1). One saltmarsh subspecies of ornate shrew from Baja California (*Sorex ornatus juncensis*) may already be extinct. Federally endangered taxa include the salt marsh harvest mouse (*Reithrodontomys raviventris*), three western subspecies of the Clapper Rail (*Rallus longirostris*), and the Florida meadow vole (*Microtus pennsylvanicus dukecampbelli*). The Atlantic Coast subspecies of the salt marsh water snake (*Nerodia clarkia taeniatus*) is listed as threatened by the USDI Fish and Wildlife Service. Although only seasonally associated with saltmarshes, the Orange-bellied Parrot (*Neophema chrysogaster*) of Australia and the Saunder's Gull (*Larus saunderi*) of Asia, may be added to the global list of species that may depend upon saltmarshes. Many of the other subspecies listed in Table 1 are on various state and regional lists for threatened or vulnerable species.

TABLE 1. Vertebrate taxa restricted to tidal marshes.

Species	Subspecies	Distribution	Status
Diamondback terrapin (*Malaclemys terrapin*)	terrapin centrata tequesta rhizophorarum macrospilota pileata littoralis	Atlantic coast of North America	Endangered in Massachusetts, threatened in Rhode Island, species of special concern in six other states.
Gulf saltmarsh snake (*Nerodia clarkii*)	clarkii taeniata	Gulf of Mexico and Atlantic coast of Florida	*taeniata* is threatened.
Carolina water snake (*Natrix sipedon*)	williamengelsi	Carolina coast of North America	State species of concern.
Northern brown snake (*Storeria dekayi*)	limnetes	Gulf of Mexico, North America	
Black Rail (*Laterallus jamaicensis*)[a]	jamaicensis coturniculus	Atlantic, Gulf of Mexico, and Pacific coasts of North America	Species of conservation concern (USDI Fish and Wildlife Service 2002).
Clapper Rail (*Rallus longirostrus*)	*obsoletus* group *crepitans* group	Atlantic, Gulf of Mexico, and Pacific coasts of North America	Populations in California are endangered.
Willet (*Catoptrophorus semipalmatus*)	semipalmatus	Atlantic coast of North America	None
Common Yellowthroat (*Geothlypis trichas*)	sinuosa	San Francisco Bay	State species of concern.
Marsh Wren (*Cistothorus palustris*)	palustris waynei griseus marianae	Atlantic coast of North America	*C. p. griseus* and *C. p. marianae* subspecies of conservation concern in Florida.
Song Sparrow (*Melospiza melodia*)	samuelis pusillula maxillaris	San Francisco Bay	State of California subspecies of concern.
Swamp Sparrow (*Melospiza georgiana*)	nigrescens	Mid-Atlantic North American coast	Maryland subspecies of concern.
Savanna Sparrow (*Passerculus sandwichensis*)	*rostrata* group *beldingi* group	Western Mexico and Southern and Baja California	Threatened in California.
Seaside Sparrow (*Ammodramus maritimus*)	Atlantic Coast group Gulf Coast group	Atlantic and Gulf of Mexico coasts	One subspecies endangered (*A. m. mirabilis*), one subspecies extinct (*A. m. nigrescens*). Species of national conservation concern (USDI Fish and Wildlife Service 2002).
Salt Marsh Sharp-tailed Sparrow (*Ammodramus caudacutus*)	caudacutus diversus	Atlantic coast of North America (non-breeding)	Species of national conservation concern (USDI Fish and Wildlife Service 2002).
Nelson's Sharp-tailed Sparrow (*Ammodramus nelsoni*)	subvirgatus alterus	Atlantic and Gulf of Mexico coast of North America (non-breeding)	Species of national conservation concern (USDI Fish and Wildlife Service 2002).

TABLE 1. CONTINUED.

Species	Subspecies	Distribution	Status
Slender-billed Thornbill (*Acanthiza iradelei*)	rosinae	South coast of Australia	None
Masked shrew (*Sorex cinereus*)	nigriculus	Tidal marshes at mouth of Tuckahoe river, Cape May, New Jersey	None
Ornate shrew (*Sorex ornatus*)	sinuosus salarius salicornicus juncensis	San Pablo Bay, Monterey Bay, Los Angeles Bay, El Socorro marsh, Baja California.	State of California subspecies of concern. Extinct?
Wandering shrew (*Sorex vagrans*)	halicoetes	South arm of San Francisco Bay	State of California subspecies of concern.
Louisiana swamp rabbit (*Sylvilagus aquaticus*)	littoralis	Gulf coast	
Salt marsh harvest mouse (*Reithrodontomys raviventris*)	raviventris halicoetes	San Francisco Bay	Both California and federal endangered species.
Western harvest mouse (*Reithrodontomys megalotis*)	distichlis limicola	Monterey Bay, Los Angeles Bay	No status. State of California subspecies of concern.
California vole (*Microtus californicus*)	paludicola sanpabloenis halophilus stephensi	San Francisco Bay, San Pablo Bay, Monterey Bay, Los Angeles coast	Subspecies sanpabloenis and stephensi are California subspecies of concern.
Meadow vole (*Microtus pennsylvanicus*)	dukecampbelli nigrans	Gulf Coast, Waccasassa Bay in Levy County, and Suwannee National Wildlife Refuge, Florida; East coast Chesapeake Bay Area	Federally endangered.
White-tailed deer (*Odocoileus virginianus*)	mcilhennyi	Gulf coast	None

a Black Rail is included, although small populations of both North American subspecies can be found in inland freshwater marshes (Eddleman et al. 1994).

THREATS TO TIDAL SALTMARSHES

As we have suggested, the threats to the already local and restricted saltmarsh taxa are a bellwether of the overall threats to the integrity of salt marsh ecosystems. The following represents some of the major environmental issues facing the small amount of remaining tidal marsh.

DEVELOPMENT

Coastal areas along protected temperate shorelines are prime areas for human habitation. By the end of the last century, 37% of the world's population was found within 100 km of the coast (Cohen et al. 1997). At the same time, 42% of the U.S. population lived in coastal counties along the Pacific, Atlantic, and Gulf of Mexico (NOAA http://spo.nos.noaa.gov/projects/population/population.html). The impact of human populations around major navigable estuaries where most tidal marsh is found is undoubtedly higher than random sections of coastline. In particular, the filling and development of the shoreline of tidal estuaries such as the San Francisco and Chesapeake bays and the Rio Plata has led to the direct loss of large areas of saltmarsh. The loss of >80% of the original wetlands around San Francisco Bay is of particular concern (Takekawa et al., chapter 11, *this volume*), since its three major embayments support more endemic tidal marsh taxa than any other single coastal locality.

GRAZING AND AGRICULTURE

Marshes are often populated by palatable and nutritious forage plants and hence have been directly grazed or grasses have been harvested for hay. Harvesting salt hay for forage and mulch was an important industry in marshes along the east coast of North America in the 18th and 19th centuries (Dreyer and Niering 1995). Although no longer a common practice in North American tidal marshes, the use of coastal wetlands to support livestock still occurs in the maritime provinces of Canada and is common in Europe and parts of South America.

Apart from grazing and haying over the course of human history, large and unknown areas of tidal marsh have been diked and converted to agricultural use, such as the low countries of Northern Europe (Bos et al. 2002), areas of rice farming in Korea and China, and salt production.

A more profound change than the addition of grazing livestock to many marsh systems is the loss of large grazing animals towards the end of the Pleistocene (Levin et al. 2002). We know from studies of reintroduced horses, that tidal marsh grasses—particularly smooth cordgrass (*Spartina alterniflora*)—are highly palatable and preferred forage (Furbish and Albano 1994). In many marshes the largest vertebrate herbivores have shifted from ungulates to microtine and cricitid rodents. Nowadays, the most important herbivores in some marshes may be snails and snail populations are controlled by crabs (Sillman and Bertness 2002). But in the Tertiary and Pleistocene, large mammals might have been keystone herbivores in tidal marsh systems. It would be fair to say that the ecological and evolutionary impact of the loss of such herbivores is not fully understood (G. Chmura, pers. comm.)

DITCHING, CHANNEL DEVELOPMENT, AND CHANGES IN HYDROLOGY

Tidal marshes have borne the brunt of an array of management activities that either directly or indirectly affect their functioning. Barriers to or canalization of tidal flow can disrupt natural cycles of inundation. The reduction of tidal flow has been implicated in major vegetation changes in tidal marshes in Southern California (Zedler et al. 2001). Water management projects for creating shipping navigation channels have had a particularly large impact on the coastal marshes of the Mississippi Delta (Mitsch and Gosselink (2000). On the other hand, upstream impoundment of water may reduce the input of freshwater and induce salt water incursions into freshwater systems. Shifts towards higher salinity over the past 150 yr have been documented for the marshes of the Meadowlands in the Hudson River estuary (Sipple 1971). On an even larger scale, the balance between fresh-water flow and salt-water intrusion has been the subject of considerable interest in the estuaries of the Suisun Bay and lower Sacramento-San Joaquin deltas of the San Francisco Bay area (Goman 2001). The California Water Project has doubtlessly influenced this, but early Holocene shifts in plant composition suggest natural variation in the pattern of salt water incursion has been profound.

On a micro-scale, saltmarshes have been variously ditched for insect control (Daiber 1986) and opened with large water impoundments to provide habitat for insect control and to provide habitat for waterfowl (Erwin et al. 1994, Wolfe 1996). In some areas, human engineering of water distribution and vegetation in marshes has all but replaced the natural engineering of wildlife—particularly the muskrat (*Ondatra zibethicus*; Errington 1961).

MARSH BURNING

Lightning fires can be an important source of natural disturbance to coastal marshes, occurring at particularly high frequencies along the southern Atlantic and Gulf coasts (Nyman and Chabreck 1995). The frequency of marsh burning has increased due to human activities, including the purposeful use of fire as a management tool to increase food for waterfowl and trappable wildlife. However, the effect of such management on non-target organisms and ecosystem function is just beginning to be evaluated (Mitchell et al., *this volume*).

INVASIVE SPECIES

Coastal ecosystems have been on the receiving end of human-caused introductions that have resulted in species invading and changing tidal marshes. The most critical invasions have consisted of dominant tidal-marsh plants, because as they take over marshlands, they change the face of the habitat. Species of *Spartina* have been prone to establishing themselves on foreign shores (West Coast of the US, China, parts of Northern Europe, New Zealand, and Tasmania). Even along its native shoreline, smooth cordgrass is spreading as a result of nitrification and other environmental changes (Bertness et al 2002). The common reed *(Phragmites australis)*, a native species, has spread in the high marshes of eastern North America, often creating large barren monocultures (Benoit and Askins 1999).

We have focused on how invasions of dominant plant species change the basic habitat structure and productivity in many, as yet poorly understood, ways. Major changes have occurred in the benthic fauna of major North American estuaries (Cohen and Carlton 1998) and the effect this has had on the feeding ecology of tidal marsh vertebrates has not been well documented. Vertebrate species themselves are often invasive, and the tidal-marsh fauna itself has been dramatically changed through human introductions. Species of *Rattus* and the house mouse (*Mus musculus*) are now distributed in marshes around the world. The rats, in particular, are known to be important nest predators and are hypothesized to have a negative impact on endangered taxa, such as the Clapper Rail. Other predator populations, including red fox (*Vulpes vulpes*) and Virginia opossum (*Didelphis virginiana*), have spread through human introductions and activities. The nutria (*Myocastor coypus*) has spread throughout the southeastern US resulting in severe levels of grazing damage. Although we know of no introduced breeding bird species, a variety of reptiles have colonized mangroves and subtropical saltmarshes of Florida.

TOXINS, POLLUTANTS, AND AGRICULTURAL RUN-OFF

Estuaries receive run-off from agricultural fields and urban development spread over large watersheds. Tidal marshes are often sprayed directly with pesticides, a practice that will probably increase under the threat of emergent mosquito-borne diseases, such as West Nile virus. In addition, tidal marshes that fringe estuaries also bear the brunt of any oil or chemical spills into the marine environment that drift into the shores. The effects of pollution are both acute and long term; the latter including the effects of increased nutrient loads into the tidal-marsh ecosystem and the former comprised of the toxic effects of chemicals to the vegetation and wildlife (Clark et al. 1992). The impact on dominant vegetation of increased nitrogen inputs into tidal marshes has been documented, at least for marshes along the Atlantic Coast of North America (Bertness et al. 2002).

INCREASE IN CARBON DIOXIDE, SEA-LEVEL RISE, CHANGES IN SALINITY, AND GLOBAL WARMING

Sea level is rising in response to global increases in atmospheric temperatures. If, on a local scale, coastline accretion does not keep pace with this rise, then the leading edge of coastal marshes will become permanently inundated and lost as wildlife habitat. Over time, high marsh becomes middle and then low marsh with increasing sea levels. New high marsh forms after major disturbance of upland communities allows marsh invasion. Depending upon the shape of the estuarine basin and the land use on the lands above the maximum high-tide line, the possibility of upland expansion may be curtailed along many coastlines. Estimates for coastal wetland loss as a result of sea-level rise range from 0.5–1.5% per year.

Global warming may result in other, less obvious impacts on coastal marsh systems. Perhaps of equal concern as the loss of marshland is the change in salinity resulting from salt-water intrusion into brackish-marsh systems. The actual warming itself may favor the spread of lower latitude species into higher latitude coastlines. Warmer conditions may also favor the increase in the seasonal activity of mosquitoes and other disease-transmitting insects and help the spread of associated diseases. Finally, increases in atmospheric carbon dioxide (CO_2) have a demonstrable impact on the productivity and transpiration of salt-marsh plants. These effects vary between species and may shift the mix of tidal marsh dominants. Already it has been demonstrated that increases in CO_2 favor the spread of C^3 versus C^4 plants (Arp et al. 1993).

WHAT THIS VOLUME IS ABOUT

In this volume, the authors collectively provide a sweeping view of what we know about vertebrates—primarily terrestrial vertebrates—in the highly threatened tidal-marsh systems. The contents provide a broad view of tidal-marsh biogeography, more focused discussions of adaptations of different taxa to the challenges of tidal-marsh life, and a comprehensive account of the major conservation and management issues facing marshes and their wildlife. The following provides a brief guide to the narrative trail we explore.

BIOGEOGRAPHY

We examine what is known—from both direct evidence and inference—about the changes in the quantity and distribution of tidal marshes from the Tertiary to recent times, with a focus on the San Francisco Bay estuaries, home of the greatest single concentration of endemic vertebrate species and subspecies. Having set the historical stage, we examine the distribution of tidal marshes and their vertebrate biota throughout the world. The disparate distributional literature for mammals and birds, and as much as possible, reptiles and amphibians has been sifted through to determine which species of these taxa occupy tidal marshes along different coasts and on different continents. Emphasis is placed on the distribution of differentiated taxa (subspecies and species) that occupy tidal marshes in different regions. Distributional patterns are synthesized and some preliminary hypotheses to explain the distributions are proposed. In addition, some of the features that characterize successful colonists of tidal marshes are explored.

In recent years, molecular phylogenies of groups that feature tidal-marsh taxa have been developed and the genetic structure of tidal marsh taxa has been detailed as well. This new information allows us to begin to estimate the length of historical association of various taxa and how this has affected adaptation to tidal marshes.

ADAPTATION TO TIDAL MARSHES

Tidal marshes present myriad adaptive opportunities and challenges to the few species that colonize them. In a series of chapters, adaptation to tidal marsh life is explored from a variety of perspectives. Focusing on nesting biology of birds, we explore the role of tidal cycles and flooding events in shaping this central feature of avian ecology. Adaptations to saline environments are examined by focusing on the physiology of salinity tolerance in sparrows, a group that is not generally known for its maritime distribution. In the course of focusing in on sparrow adaptations, we review the different behavioral, physiological and morphological adaptations of vertebrates in brackish to salty environments. The volume further explores shared adaptations to the tropic opportunities with emphasis on the bill morphology of sparrows and background matching coloration of a suite of terrestrial species. Finally, we examine shifts in communication, demography and social organization that accompany successful occupation of tidal marshes.

CONSERVATION BIOLOGY: ANTHROPOGENIC ENVIRONMENTAL IMPACTS ON TIDAL MARSHES OF THE PREVIOUS AND NEXT CENTURY

Tidal marshes have already been reduced in area, fragmented, ditched, and altered by the damming of streams and rerouting of water sources. To place the environmental issues facing saltmarsh vertebrates in context, we will provide regional reviews of four North American tidal-marsh areas—Northeast, Southeast, San Francisco Bay, and southern California—that together present the range of conservation issues. Two chapters address species specific approaches to evaluating both local- and landscape-level effects of habitat change. We finally turn to more synthetic treatments of environmental issues outlined above with chapters focusing on sea-level rise, invasive species, toxins (focusing on Clapper Rails), and the effect of active salt-marsh management, including burning, open-water management, and mosquito-control efforts.

If nothing else is accomplished, we hope that we will bring greater attention to the conservation of the tidal-marsh endemics. The first step towards a more concerted conservation effort is a systematic source of information on the population status and long-term trends of saltmarsh vertebrate populations. To catalyze this, we provide a collaborative chapter outlining approaches to the long-term monitoring of tidal-marsh birds. Future collaborations should focus on establishing similar systems for mammals and, in some areas, snakes and turtles. Such monitoring programs are only a first step. We hope they will provide the backbone to an active research program on tidal-marsh vertebrates.

We end the volume with a menu of exciting and important areas for both applied and basic research. By following these research leads, we will achieve the ability to better manage and protect the healthy, restore the degraded, and reestablish the lost marshlands, while achieving a greater understanding of how animals adapt to this unique environment.

ACKNOWLEDGMENTS

This publication grew from a symposium held in October 2002 at the Patuxent Wildlife Visitors Center which brought together scientists from throughout North America to focus on the scientific and conservation issues facing vertebrates in tidal marshes. We thank J. Taylor and the USDI Fish and Wildlife Service and the Smithsonian Migratory Bird Center for providing financial support to the symposium. We also would like to extend our appreciation to the Friends of Patuxent and the staff of the visitor center and the Smithsonian Migratory Bird center for logistical support. We received incisive reviews of all of the manuscripts from 36 subject-matter experts and this has greatly improved the quality of the publication. The authors of papers in the volume were encouraged to revise their contributions to make them as inductive as possible. This involved a good deal of time and patience over and beyond what is normally expected contributors and we (the editors) appreciate this extra effort. I thank S. Droege, M.V. McDonald, and M. Deinlein for comments on a draft of the introduction. The following provided funds to support publication of this volume: Canadian Wildlife Service; Migratory Bird Center, Smithsonian Institution; The National Museum of Natural History, Smithsonian Institution; Biology Department, Northwestern State University; USGS Patuxent Wildlife Research Center; Department of Geography, University of California, Berkeley; University of South Dakota; USDI Fish and Wildlife Service; USGS, Alaska Cooperative Fish and Wildlife Research Unit; USGS, Western Ecology Research Center; Department of Biology, University of South Dakota; and Department of Biological Sciences, Wright State University.

BIOGEOGRAPHY AND EVOLUTION OF TIDAL-MARSH FAUNAS

Bay-capped Wren-spinetail (*Spartonoica maluroides*)
Drawing by Julie Zickefoose

THE QUATERNARY GEOGRAPHY AND BIOGEOGRAPHY OF TIDAL SALTMARSHES

Karl P. Malamud-Roam, Frances P. Malamud-Roam, Elizabeth B. Watson, Joshua N. Collins, and B. Lynn Ingram

Abstract. Climate change and sea-level change largely explain the changing distribution and structure of tidal saltmarshes over time, and these geographic attributes, in turn, are primarily responsible for the biogeography of tidal-saltmarsh organisms. This paper presents a general model of these relationships, and uses the San Francisco Bay-delta estuary (California) to demonstrate some of the model's implications and limitations. Throughout the Quaternary period, global cycles of glaciation and deglaciation have resulted in ca. 100-m variations in global mean sea level, which have been accompanied by large changes in the location of the intertidal coastal zone, and hence of potential sites for tidal marshes. Other climate-related variables (e.g., temperature and exposure to storms) have in turn substantially controlled both the location and size of marshes within the coastal zone and of specific physical environments (i.e., potential habitats) within marshes at any time. Since the most recent deglaciation resulted in a global rise in sea level of 100–130 m between about 21,000 and 7,000 yr BP, and a slower rise of about 10 m over the last 7,000 yr, modern tidal saltmarshes are relatively young geomorphic and ecological phenomena, and most continue to evolve in elevation and geomorphology. Therefore, the distribution of taxa between and within marshes reflects not only salinity and wetness at the time, the dominant controls on marsh zonation, but also antecedent conditions at present marsh sites and the extent and connectedness of habitat refugia during and since the glacial maximum. Unfortunately, direct stratigraphic evidence of paleomarsh extent and distribution is almost nonexistent for the Late Glacial-Early Holocene, and is incomplete for the late Holocene.

Key Words: biogeography, glacial-deglacial cycles, global climate change, Quaternary, San Francisco Bay, sea-level change, spatial patterns, tidal saltmarsh.

LA GEOGRAFÍA Y BIOGRAFÍA CUATERNARIA DE MARISMAS SALADAS DE MAREA

Resumen. Tanto el cambio climático como el cambio en el nivel del mar explican ampliamente el cambio en la distribución y la estructura de marismas saladas de marea en el transcurso del tiempo; y estos atributos geográficos a su vez, son los principales responsables de la biogeografía de los organismos de marismas saladas de marea. Este artículo presenta un modelo general de estas relaciones y utiliza el estuario Bahía-delta de San Francisco (California) para demostrar algunas de las implicaciones y limitaciones del modelo. A lo largo del período cuaternario, ciclos globales de glaciación y deglaciación han resultado en variaciones ca. 100-m en la media global del nivel del mar, lo cual ha sido acompañado por un gran número de cambios en la ubicación de la zona costera intermareal y por ende, de sitios potenciales para marismas de marea. Otras variables relacionadas al clima (ej. temperatura y exposición a tormentas) han hecho que se controle substancialmente tanto la ubicación, como el tamaño de marismas a lo largo de la zona costera asi como de ambientes físicos (ej. habitats potenciales) entre los marismas en cualquier tiempo. A partir de la más reciente deglaciación que resultó en un incremento en el nivel del mar de 100–130 m entre 21,000 y 7,000 años AP, y un incremento más lento de cerca de 10 m en los últimos 7,000 años, las marismas saladas de marea modernas son un fenómeno relativamente joven morfológica y ecológicamente, que deberá seguir evolucionando en elevación y geomorfología. Es por esto que la distribución del taxa entre y dentro de los marismas no solo refleja salinidad y humedad en el tiempo, los controles dominantes de la zona de marisma, sino que también condiciones anteriores en sitios presentes de marisma y el alcance y conectividad del hábitat de refugio durante y a partir del máximo glacial. Desafortunadamente, es casi inexistente la evidencia directa estratigráfica del alcance y distribución del paleo marisma, para el Heleoceno Tardío Glacial-Temprano.

Two related but distinct phenomena—climate change and sea-level change—largely explain the changing distribution and structure of tidal saltmarshes over time, and this historical geography, in turn, is primarily responsible for the present biogeography of the organisms that inhabit them. Marsh biogeography, the distribution of tidal-saltmarsh organisms at all spatial scales, has become a significant research question in recent years, and the conservation of these organisms a major priority for natural resource managers (Estuary Restoration Act

2000, Zedler 2001), but the limited extent of these ecosystems and the limited distribution of their fauna have made it difficult to formulate useful general conceptual models of marsh distribution, structure, and function (Daiber 1986, Goals Project 1999, Zedler 2001). This is reflected in the literature on marshes and marsh organisms, which has historically focused heavily on the attributes of specific sites (Zedler 1982, Stout 1984, Teal 1986, Goals Project 1999), and on generalities which emphasize the significance of local conditions as controls on marsh form and function (Chapman 1974, Adam 1990, Mitsch and Gosselink 2000).

One general principle widely recognized is that tidal saltmarshes are very young landscapes in geologic time and young ecosystems in evolutionary time, having existed in their present locations for no more than a few thousand years due to the transition from a glacially dominated global climate to warmer conditions with higher sea levels over the last 20,000 yr (Zedler 1982, Josselyn 1983, Teal 1986, Mitsch and Gosselink 2000). Although the youth of tidal saltmarshes can further serve to emphasize their uniqueness in time as well as in space, the primary aim of this paper is to explore how climate change and sea-level change can instead serve as organizing principles of a supplemental general conceptual model of tidal-saltmarsh geography and biogeography. We accomplish this by first articulating a standard model of tidal-saltmarsh geography and biogeography that is implicit in most of the literature, and then by proposing the supplemental model. Then to justify and expand the model, we present sections on the mechanisms, patterns, and consequences of global climate change; on the distribution of marshes and marsh types at multiple spatial scales; and on the distribution of taxa between and within marshes. Finally, although the underlying causes we review are essentially global, their local effects can vary dramatically, and the San Francisco Bay-delta estuary (California) is used to illustrate the complex interplay of global processes and local settings.

THE STANDARD MODEL OF TIDAL SALTMARSH GEOGRAPHY AND BIOGEOGRAPHY

Tidal saltmarshes, by definition, are coastal areas characterized by (1) tidal flooding and drying, (2) salinity in sufficient quantity to influence the biotic community, and (3) non-woody vascular vegetation (Mitsch and Gosselink 2000), although some authors have emphasized the role of tides (Daiber 1986, Zedler 2001), others of salt (Chapman 1974, Adam 1990), and others of the specialized flora of these areas (Eleuterius 1990). Because climate change and other global-scale or long-term phenomena can influence water level and salinity patterns independently, it is important to carefully distinguish between marshes that are tidal, those that are salty, and those that are both.

In addition to their defining characteristics and their relative youth, tidal saltmarshes share relatively few attributes on a global scale, although some generalities have been noted. Tidal saltmarshes typically have high biotic productivity and food webs dominated by detritus rather than herbivory (Mitsch and Gosselink 2000). They frequently, although not inevitably, provide habitat for taxa that are only found in this type of environment, that are limited in geographic range, and/or that are rare (Zedler 2001). Tidal saltmarshes sometimes have high biodiversity at some taxonomic levels, but this varies considerably depending on the metric used, e.g., whether periodic visitors or only obligate residents are counted, marsh size and shape, the size and distribution of other marshes in the region, the elevation and distribution of landforms on the marsh, the degree of spatial variation in physical conditions within the marsh, the proximity and quality of adjacent refugia during high tides or other stressors, and the extent of anthropogenic disturbance. Although small, isolated, disturbed, and highly salty and/or highly tidal marshes can provide significant habitat for some taxa, they generally have low biodiversity at most taxonomic levels (Goals Project 1999, Zedler 2001).

Although the phrase is not commonly used, it is clear that a standard model of tidal-saltmarsh geography and biogeography (Malamud-Roam 2000) is implicit in the literature and is used to explain both the similarities and differences between marshlands (Daiber 1986, Adam 1990, Mitsch and Gosselink 2000, Goals Project 1999, Zedler 2001). This standard model includes several basic elements spanning a range of spatial and temporal scales: (1) distribution of marshes—tidal saltmarshes exist where favorable local conditions (protection from waves and storms, relatively gradual bedrock slope, and sediment accumulation faster than local coastal submergence) exist within latitudinal zones warm enough for vegetation but too cold for mangroves, (2) distribution of landforms—although geomorphic features of marshes are relatively stable, marshes are depositional environments and become higher and drier over time unless local sediment supplies are limiting, (3) distribution of marsh organisms between marshes—salinity gradients along estuaries dominate distribution of habitat types and hence of taxa, and (4) distribution of marsh organisms within marshes—plants and animals are found in zones primarily reflecting elevation and hence wetness or hydroperiod. Local hydroperiod is modified by channel and pond configuration. As sediments accumulate, plants and animals adapted to drier conditions replace those more adapted to frequent or prolonged flooding.

In this standard model, long-term temporal changes in the distribution of marshes, marsh habitats, and marsh organisms are generally recognized to be consequences of climate change and, in particular, of deglaciation. Many authors recognize that modern tidal saltmarshes are young features, reflecting global sea-level rise during the late Pleistocene and early Holocene (ca. the last 21,000 yr), that this rise has been due to glacial melting and thermal expansion of ocean water, and that the rate of rise dropped dramatically about 7,000–5,000 yr BP (to 1–2 mm/yr), leading to relatively stable coastlines since that time (Chapman 1974, Mitsch and Gosselink 2000). Climate change, deglaciation, and global sea-level change are almost always presented as past phenomena, significant primarily for controlling the timing of marsh establishment and for setting in motion processes of landscape evolution and/or ecosystem succession (Zedler 1982, Josselyn 1983, Teal 1986, Mitsch and Gosselink 2000). Spatial differences in rates of relative sea-level rise, due to local crustal movements, have been described primarily where they have been large enough to result in marsh drowning (Atwater and Hemphill-Haley 1997) or dessication (Price and Woo 1988).

On shorter time scales—decades to centuries—the preferred explanations for changes in the distribution of marsh types and organisms have varied greatly, apparently reflecting trends in environmental sciences in general, as well as disciplinary differences and individual interests. Although relatively fixed successional pathways, emphasizing biotic, especially plant, roles in modifying the marsh environment, were commonly discussed in previous decades (Chapman 1974), explanations of progressive changes in marshes then shifted primarily to landscape evolution with an emphasis on geomorphic responses to local sediment supplies and coastal submergence rates (Josselyn 1983, Mitsch and Gosselink 2000). More recently, at least five trends are apparent in the literature: (1) a recognition that dynamic equilibrium can occur at relatively long time scales, and that change is rarely continuous in one direction for long (Mitsch and Gosselink 2000), (2) an increasing focus on the patterns and consequences of disturbance, and in particular human disturbance (Daiber 1986, Zedler 2001), (3) a shift in emphasis from fixed pathways to thresholds and bifurcation points between possible paths or trajectories of change (Zedler 2001, Williams and Orr 2002), (4) an explicit integration of geomorphic and biotic processes and interactions between them (American Geophysical Union 2004), and (5) a burgeoning concern that anthropogenic climate change might substantially increase the rate of sea-level change, with perhaps dramatic consequences for tidal saltmarshes (Keldsen 1997).

A HISTORICALLY FOCUSED SUPPLEMENTAL MODE

Although all of the elements and variations of the standard model are useful, they do not appear to adequately explain biodiversity, adaptive radiations, endemism, rarity, colonization-invasion patterns, historic marsh distribution, or many other qualities critical to conservation biology. Classical biogeography theory argues that these are most likely controlled by the historical distribution of habitats (e.g., islands, and refugia; MacArthur and Wilson 1967, Lomolino 2000, Walter 2004), and recent global-change research indicates that this historical geography has been largely controlled by large-scale climate dynamics. We therefore suggest that the standard model be supplemented by the conceptual model of tidal saltmarsh geography and biogeography shown in Fig. 1, which emphasizes climate change and sea-level change as organizing principles, and which sets local phenomena explicitly in the context of global and millennial scales of space and time than is typical.

The flow chart shown in Fig. 1 expands the standard model largely by emphasizing distinctions between related causes for observed phenomena. First, although global mean (eustatic) sea-level rise associated with the most recent deglaciation is still the primary causal factor in marsh history, climate change and sea-level change are distinct, with climate change influencing marsh form and function through many mechanisms. Second, climate and sea level determine not only the current locations and extent of marshes, but also their past distribution, extent, and connectedness; these antecedent conditions, especially the amount and location of habitat refugia, have probably strongly influenced the large-scale distribution of taxa. Third, the history of the coastal zone, which can be mapped with some precision, is distinct from the actual extent and distribution of marshes at any time, which has responded to many global and local variables, and which is, hence, much less definite. Fourth, the distribution of physical

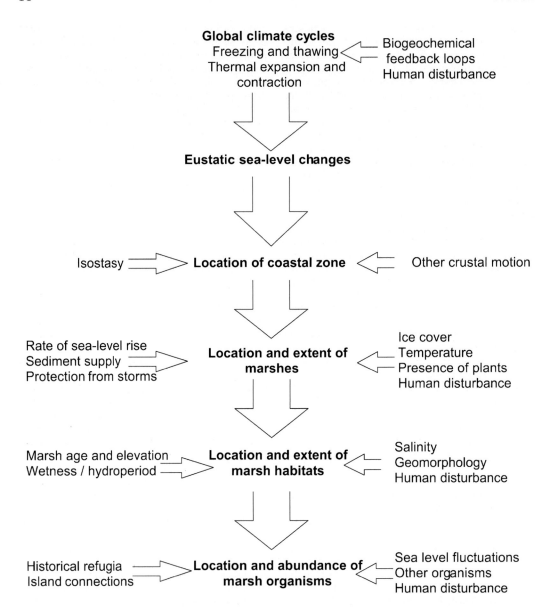

FIGURE 1. Conceptual model of historical geography and biogeography of tidal salt marshes. Major causal pathways are shown as large vertical arrows, and secondary causes are horizontal arrows. In the interest of simplicity and clarity, indirect or feedback influences are omitted from the figure, but discussed in the text.

environments within marshes, which is analogous to the distribution of potential habitats, is influenced both by external parameters and by antecedent internal feedback mechanisms. Fifth, climatic and oceanographic phenomena continue to cause fluctuations in both marsh elevation and sea level on many time scales, heavily influencing marsh hydrology, and thus the distribution of taxa within them. Details of and evidence for the model are discussed in the sections that follow.

Calibration of any historical geography model requires preserved evidence, generally buried in sediment, but the direct sedimentary evidence for past marshes is very limited (Goman 1996, Malamud-Roam 2002). Although tidal saltmarshes do provide good depositional environments for plant material, they represent a small proportion of the land surface at any time and their locations have changed significantly over time; therefore intertidal depositional environments will make up only a small portion of sediment formations potentially spanning millions of years. In addition, the response of intertidal sediments to exposure or drowning ensures that preservation of the intertidal marsh sedimentary record is not good before the last few thousand years (Bradley 1985). As relative sea level drops, intertidal areas become exposed and the peat sediments can be lost to erosion and oxidation. Conversely, as relative sea-level rises, intertidal areas can become flooded if the change in sea level is greater than the ability of the marshes to accumulate sediments vertically. Thus former marshes can become buried both by the rising sea and by estuarine sediments (as in the case of the San Francisco Bay; Ruddiman 2001). These processes have resulted in the scarcity of marsh deposits from pre-Holocene periods. The best sedimentary records from tidal marshes cover no more than the past 5,000–10,000 yr, a period in which the deposits are both close to the surface and generally accessible beneath present tidal marshes. Although Holocene tidal-marsh deposits are especially valuable because they often contain abundant, well-preserved modern macro and microfossil assemblages that can be interpreted with regard to paleo-environmental conditions and because they can be dated very precisely using radiocarbon dating (Goman 1996, Malamud-Roam 2002), they do not provide direct records of the extent or locations of habitat during the last glacial maximum or during the years of rapid sea level rise that followed it.

QUATERNARY CLIMATE CHANGE AND SEA-LEVEL CHANGE

The primary causal factor in our model is spatio-temporal variation in climate, because climatic and oceanographic conditions of the world have varied dramatically over the last 2,000,000 yr, and in particular, because the world's coastlines were very different places just 21,000 yr ago. Understanding the present biogeography of tidal saltmarshes thus requires awareness of previous conditions when they were most different from the present; an understanding of how and when variables changed to their current states; and awareness of the terminology used to characterize these changes. In this section we first introduce the Quaternary period and its divisions to facilitate understanding of the climate literature. We then describe the world climate, and conditions along temperate coastlines in particular, during the peak of the most recent glacial maximum and during the years that followed. Changes in sea level are the primary mechanisms through which climate change impacts coastal zones, and the next sub-sections address eustatic and relative local sea-level variation. We conclude the section with an introduction to other consequences of climate change, and in particular latitudinal shifts in temperature, that can influence tidal saltmarshes.

THE QUATERNARY, THE PLEISTOCENE, AND THE HOLOCENE

The global climate system of the last 2,000,000 yr or so has been characterized by large and relatively regular oscillations between glacial phases when large portions of the continental surfaces are covered by ice sheets, and when mean sea level is low, and interglacial phases when retreat of the ice sheets results in higher global sea levels (Hays et al. 1977, Ruddiman 2001). This time of alternating glacial and interglacial phases is known as the Quaternary Period, and its initiation is generally dated at about 1.8–2,600,000 yr BP, but various authors have focused on periods ranging from the last 3,000,000 yr (Ruddiman 2001) to the last 750,000 yr for which good paleoclimate records exist (Bradley 1985). Like all geological time periods, the Quaternary Period is formally delineated by rock strata, and the Quaternary was named in 1829 by the French geologist Jules Desnoyers to describe certain sedimentary and volcanic deposits in the Seine Basin in northern France which contained few fossils but were in positions above the previously described third or Tertiary series of rocks. The Scottish geologist Charles Lyell recognized that Quaternary deposits were primarily deposited by glaciers but that the most recent deposits did not appear of glacial origin. Thus, in 1839 he divided the Quaternary into an older Pleistocene Series, comprising the great majority of the deposits and popularly known as the time of Ice Ages, and a younger Recent Series which is now associated with the Holocene Epoch (Bradley 1985). Later, the Quaternary became popularly known as the Age of Man, but the paleotological and climatic records do not coincide well enough for this phrase to have any specific meaning (Bradley 1985). These terms are generally important for interpreting the climate change literature, and more specifically because the period of maximum difference from present coastal conditions—during the last glacial maximum (LGM; ca. 21,000 yr BP)—does not coincide with a transition between geological time periods; in fact glacial materials continued to be deposited for some 10,000 yr after the LGM while the glaciers retreated. Thus, the most recent low sea stand, which does coincide with LGM, occurred

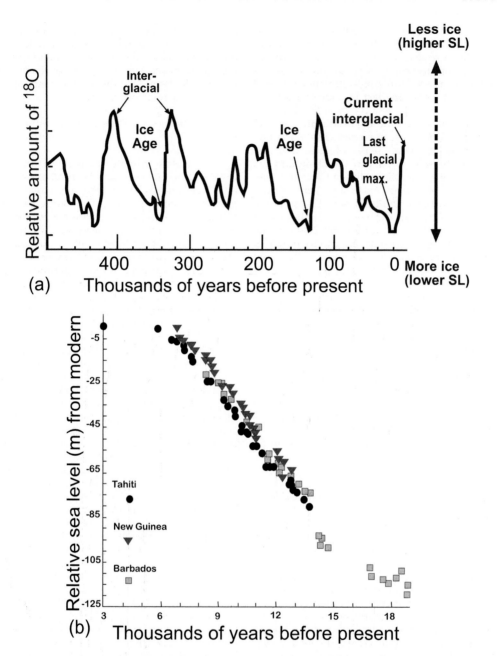

FIGURE 2. (a) Generalized oxygen isotope curve (after Bassinot et al. 1994) showing the cyclical changes in global climate. Negative oxygen isotope ratios indicate warmer climatic periods (less water stored as ice on land) and positive ratios indicate generally cooler conditions (more water as ice). (b) Sea-level curve since the Last Glacial Maximum. Adapted from Quinn (2000) with source data from Fairbanks (1989), Chappell and Polach (1991), Edwards et al. (1993), and Bard et al. (1996).

during the late Pleistocene and sea level has been rising through both the latest Pleistocene and throughout the Holocene (Ruddiman 2001).

During the Quaternary, glacial and interglacial conditions have oscillated on roughly 100,000 yr cycles, with periods of slow cooling to glacial conditions over some 90,000 yr punctuated by relatively rapid warming to interglacial conditions lasting about 10,000 yr (Fig. 2a; Shackleton and Opdyke 1976, Bassinot et al. 1994). Periods when water was locked in glaciers are always associated with lowered sea level and colder mean temperatures, and generally with dryer conditions, but regional climate patterns varied substantially (Ruddiman 2001). Although the changing climate patterns are clearly seen in numerous sediment cores and other climate proxy records, explanations for the large-scale oscillations are still controversial (Ruddiman 2001) and beyond the scope of this paper. It is important to remember during the following discussion on the recent glacial maximum and deglaciation that this is that this is only the latest of at least four such cycles (Fig. 2), which almost certainly had major impacts on the evolutionary and dispersal histories of coastal taxa.

Conditions during and since the LGM that could have impacted tidal marshes and other coastal ecosystems have been inferred from many proxy records (Bradley 1985, Kutzbach et al. 1998, Ruddiman 2001). The exact date of the LGM has been somewhat inconsistent in the literature, primarily because of measurement and dating problems (Ruddiman 2001), and also possibly because the ice reached its maximum extent at somewhat different times in different places (McCabe and Clark 1998), but it is clear that the global maximum extent of ice was about 21,000 yr BP (Fairbanks 1989, Kutzbach et al. 1998, Ruddiman 2001). At this time, sea level was 110–140 m lower, ice covered the coasts year-round in many areas now seasonally or permanently free of ice (McCabe and Clark 1998), the world ocean was colder by about 4 C, varying from 8 C colder in the North Atlantic (Kutzbach et al.1998) to perhaps 2 C warmer in some tropical areas (CLIMAP 1981, etc. in Ruddiman 2001), atmospheric CO_2 was considerably lower than at present (Kutzbach et al. 1998), precipitation and runoff were lower world-wide although with potentially large regional variations (Kutzbach et al. 1998), fluvial supplies of sediment to the coastal zone were lower in some places than at present because of reduced runoff but were higher in others both because of the intense erosive impacts of glaciers and because of locally intense season runoff (Collier et al. 2000), and some coastal regions experienced more oceanic storms because they were not in the geological setting that now protects them.

GLACIAL DYNAMICS AND SEA-LEVEL CHANGE

The most significant aspects of Quaternary climate dynamics for tidal saltmarshes are: (1) the dramatic changes in local relative sea levels resulting from the advance and retreat of the world's ice sheets, (2) the dramatically varying rates of change during any period of rise or fall, and (3) the repetition of these cycles. The total change in mean global eustatic sea level associated with glacial melting and thermal expansion of the oceans during the most recent deglaciation has traditionally been reported at between 110 m (Ruddiman 2001) and 120 m (Fairbanks 1989), as measured on relatively stable coasts, although more recent work (Issar 2003, Clark et al. 2004) now consistently report 130–140 m as more likely. These values are similar to those from earlier Quaternary cycles (Ruddiman 2001), although the previous high stand (the Sangamon) was higher than at present by 6 m (Chen et al. 1991) to about 16 m (Bradley 1985). Although the mean rate of rise has been about 5–7 mm/yr during the 21,000 yr since the LGM, the rate has varied substantially, and clearly has been much slower than the mean (ca. 1–2 mm/yr) during the most recent 5,000–7,000 yr (Atwater et al. 1979, Nikitina et al. 2000). However, the relatively rapid rise of the late Pleistocene and early Holocene was not uniform either, instead consisting of at least two melt water pulses characterized by rapid rise and a period of slow rise (ca. 14,000–12,000 yr BP) in between them, before the current period of slow average rise (Ruddiman 2001). Recent data by Clark et al. (2004) strongly support the idea that the first melt water pulse (at 19,119 ± 180 yr ago), was truly catastrophic, raising global sea levels by about 10 m over a period of time too short to be measured in dated sediments. A second period of rapid rise was described by Raban and Galili (1985, in Issar 2003) of 5.2 mm/yr rise between 8,000 and 6,000 yr BP. These same authors also used archaeological evidence to infer a high stand of almost a meter above present mean sea level in the Mediterranean Sea about 1,500 yr BP, despite evidence of local tectonic stability for the last 8,000 yr; although few other authors have claimed that eustatic, as opposed to local, sea levels have been higher earlier in the Holocene than at present, it appears that, given the magnitude of uncertainties in dating, surveying, and land stability (Atwater et al. 1979) previous Holocene eustatic high stands are possible.

Rates of sea-level rise are critical to understanding marsh history because marsh formation depends on sediment accumulation exceeding the rate of relative sea-level rise (Mitsch and Gosselink 2000), and only when the rate of rise slowed to about the modern rate (1–2 mm/yr) did modern marshes form in their current locations. However, when local sea level drops, marshes can rapidly experience loss of peat soils to oxidation and/or can be colonized by upland plants, losing their marsh character (Zedler 2001). In either case, it is important to note that marshes do not respond directly to global influences such as eustatic sea-level changes, but instead to their local manifestations, and superimposed on the global eustatic patterns have been a range of local vertical crustal movements which have modified local relative sea-level curves. Thus, the rates of rise or fall in the sea relative to the land have differed substantially, especially during the late Holocene when the eustatic rate of change was relatively small (Nikitina et al. 2000).

A particularly significant form of local vertical land movement during this period is isostatic movements of the crust in response to its elastic response to the weight of the accumulated glacial ice. In high latitudes during glacial epochs, the accumulation of hundreds of meters of ice on the continents caused isostatic downwarping of the crust by hundreds of meters and to compensate for the crustal downwarping, adjacent areas were pushed up, creating a forebulge that was usually low and broad; in some settings the paired down-warp and uplift were of large amplitudes over a short distance (Peltier. 1994, Peltier et al. 2002). For example, the Pacific Northwest of North America was isostatically depressed by the weight of glacial ice on the continent to such a degree that relative sea level on the coast at British Columbia, Canada, was actually higher during the last glacial maximum than today (Barrie and Conway 2002, Clauge et al. 2002), and some sites in the British Isles experienced isostatic movements over 170 m during this time (Clark et al. 2004). On the Atlantic Coast of North America as well, isostatic rebound and simultaneous lowering of the forebulge land surfaces as the ice sheets receded have led to complex patterns of sea-level changes over time, including episodic reversals of sea-level change (Peltier 1994, Nikitina et al. 2000). These complex patterns result in part from isostatic adjustments of the crust lagging behind the ice retreat by differing amounts in different times and places (Barrie and Conway 2002), so that the crustal responses to glaciation and deglaciation have in many places modified the eustatic curve caused by glacial melting and thermal expansion long after the eustatic curve had flattened. This complex interaction of direct and indirect influences of climate on sea level have resulted both in coastlines with more modest (Mason and Jordan 2001) and/or more extreme (Barrie and Conway 2002) changes in height than predicted by eustatic changes alone. This has apparently been true throughout the period since the LGM, but would have had its greatest impacts on coastal processes during period of slow eustatic change, including the last 5,000–7,000 yr, when the rate of crustal movements in many areas have been greater than eustatic changes in sea level.

Relative sea level has also been impacted by non-glacial factors. Along the western coastline of North America, relative sea level of local coastlines has been affected by tectonic movement of the lithospheric plates on which the continent and the ocean rest. The abrupt changes in land surface of marshes relative to sea level that can result from underlying active faults have been clearly shown along the Washington coast (Atwater and Hemphill-Haley 1997). In other tectonically active areas such as the San Francisco Bay region, it is likely that local relative sea level may also have been affected by vertical activity along the faults, though the evidence for this in marsh sediments is ambiguous (Goman 1996). Finally, many authors have expressed concerns about the potential impact on marshes of accelerated sea-level rise due to anthropogenic global warming (Keldsen 1997, Goals Project 1999). Although this is a very significant threat to marsh species, anthropogenic influences on sea level have been of such recent origin that they seem unlikely to have had a significant impact yet on marsh biogeography compared to natural variations in sea level and to other human disturbances (Daiber 1986, Zedler 2001).

OTHER ATTRIBUTES OF GLOBAL CLIMATE CHANGE

In addition to relative changes in sea level, global scale changes in climate during the late Quaternary had other major impacts on areas where marshes are currently located. First, and most dramatically, many areas along the shores of modern Canada and northern Europe were covered with thick ice, meaning that no vegetated ecosystem of any sort existed in these areas until the ice melted and retreated (McCabe and Clark 1998, Ruddiman 2001). Recolonization by all species after ice retreat must have occurred from outside the ice-covered areas. Second, the oceans were considerably colder, meaning that temperature-dependent organisms would have been displaced towards the equator, although the specific locations of tolerable water temperature would also have been influenced by changes in ocean currents (Ruddiman 2001). Third, the large quantity of water locked up in glaciers could have led to an increase in oceanic salinity, changing the distribution of marsh organisms, although ocean salinity at the glacial maximum probably did not exceed the tolerances of truly halophitic plants and animals, the distribution of more brackish species could have been influenced by this phenomenon.

In addition to these primarily marine changes, the global-scale changes associated with glacial expansion and retreat were primarily climatic, and even though oceanic influences would have buffered the effects of these on tidal saltmarshes, biotic communities throughout the temperate zones were influenced by dramatic changes in temperature and precipitation during the Quaternary. Both proxy records (Bradley 1985) and numerical models (CLIMAP 1981 and COHMAP 1998 in Ruddiman 2001 and Kutzbach et al. 1998) have been used to discern the climate and associated biotic changes since the LGM. Similar to relative sea-level changes, a global story exists with significant variations over time and space. In general, the most recent comprehensive review (Kutzbach et al. 1998) concludes that the global climate was both cold and dry, and that the period between 14,000 and 6,000 yr ago had relatively strong northern summer monsoons and warm mid-latitude continental interiors. The models are too coarse to show detailed latitudinal changes along coast lines, but clearly show large southward shifts in northern tundra and forest biomes at LGM, and contraction of subtropical deserts in mid-Holocene. Of particular interest to coastal researchers is the conclusion by Kutzbach et al. (1998) that the exposed continental shelves during the low sea stand would have been vegetated to the extent that they compensate for the areas covered with ice, resulting in the total area of vegetated land remaining nearly constant through time. How much of this vegetated shelf might have been marshlands is not discussed in Kutzbach et al. (1998).

At a finer scale, climate since the LGM includes a number of apparently global periods or events, although local variations could be extreme (Fletcher et al 1993, Diffenbaugh and Sloan 2004). An aridity maximum apparently lasted from around the LGM to about 13,000 yr BP, when conditions quickly became warmer and moister and similar to the present (Adams and Faure 1997), though with a strong cold dry event around 11,000 yr BP (the Younger Dryas). Early Holocene conditions seem to have been slightly warmer than at present, peaking around 8,000–5,000 yr ago, at least across central and northern Europe. Evidence for other strong cold events is seen about 8,200 and 2,600 yr ago (Adams and Faure 1997), and more recently, a medieval warm period occurred between about AD 1110 and 1250 (ca. 810–750 yrs ago), followed by the well-known Little Ice Age of ca. AD 1300–1700 (Bradley 1985, Ruddiman 2001). Although many of these global changes and their local manifestations would presumably have been moderated close to coasts, a detailed review of their potential impacts on marshes and marsh organisms is beyond the scope of this paper.

DISTRIBUTION OF MARSHES, MARSH HABITATS, AND MARSH ORGANISMS

The extent and distribution of tidal marshes, and therefore the amount and connectedness of habitat for tidal marsh organisms, cannot be measured directly or even precisely estimated for the late Pleistocene or early Holocene, as rapid sea-level rise and coastal sediment accumulation have buried most, if not all, of these marshes from around the world (Bradley 1985, Malamud-Roam 2002). Therefore, fundamental parameters for interpreting tide-marsh biogeography, such as the number, size, and location of habitat areas must all be inferred indirectly for the period before, during, and after the LGM until about 5,000 yr ago. This is particularly challenging because this period includes not only the very different world of the glacial maximum, but also includes a time of slow cooling and dropping sea level before the LGM; at least two melt-water pulses, when the sea was rising very rapidly; a period of relative coastal stability between the melt water pulses; and the period after the rate of rise slowed, but before marshes were established enough to leave sedimentary records. Thus, the specific causes of specific biogeographic patterns in tidal marshes will inevitably remain somewhat ambiguous. However, the conceptual model in Fig. 1 allows for a structured approach to making these inferences, and for relating the possible or probable paleogeography of tidal marshes with the current distributions of specific marsh habitats and organisms.

The model shown in Fig. 1 is based on a series of strong causal relationships, primarily driven by global climate cycles leading to patterns of sea-level change, which determine the location of the intertidal coastal zone over time, which in turn sets the stage for the possibility of tidal marshes, habitat types, and specific organisms. Secondary influences on the location of the coastal zone, marshes within the coastal zone, marsh habitats, and taxa are shown as horizontal arrows. Indirect effects—global climate cycles causing glacially mediated isostatic rebound—and feedback loops—marshes require plants, just as many plants require marshes—are not shown with arrows in the interest of simplicity and clarity, but are discussed in the text that follows, and some can be inferred from the parameters in the side columns. One possible feedback mechanism that is unlikely to be significant is a role for tidal marsh extent or structure on global climate cycles. Although the role of tidal marshes and other wetlands on global carbon cycles, and hence on climate, has been investigated (Bartlett et al. 1990), tidal marshes cover such a very small fraction of the land's surface area (Chapman 1974) that they probably have had little effect on global atmospheric and oceanographic phenomena. In contrast, the extent of marshes is essentially defined by the extent of marsh vegetation, which not only has a major role in defining the habitat value of a marsh for fauna (Adam 1990, Zedler 2001), but also in determining the distribution of sedimentation and other physical processes which help maintain the marsh surface (Zedler 2001, American Geophysical Union 2004). Thus, the lower three parameters in Fig. 1 for specific marshes result from constantly interacting physical and biotic processes (American Geophysical Union 2004), resulting in local spatial and temporal variation in these parameters that is even more pronounced than with climate or sea level. In this section, we use a global climate and sea-level change perspective to explore these variations, reviewing first the distribution of the intertidal coastal zone and of marshes within it, and then the distribution of marsh habitats, and finally the mechanisms governing the distribution of specific organisms.

GLOBAL CLIMATE CHANGE, THE COASTAL ZONE, AND POTENTIAL MARSH LOCATIONS

Rising sea levels since the LGM drowned the marshes that existed at that time, and forced their flora and fauna to migrate, to evolve, or to perish. Although the mean vertical rise of sea level, and hence of the entire intertidal zone, was globally around 110–140 m over the last 21,000 yr, with up to about 170 m of additional local crustal movement during this period (Clark et al. 2004), this has been accompanied by a much more variable pattern of horizontal movement of the coastal zone during this time. This horizontal movement is determined not only by the local rate of relative sea-level rise, but also by the slope of the underlying bedrock at a site, and by the abundance and character of the sediments. Even on very steep coastlines, the horizontal movement of the coastal zone associated with deglaciation, and the rate of movement, were far greater than the vertical change. For example, in areas with a mean surface slope of 1%, the late Pleistocene–early Holocene eustatic rise would have resulted in a horizontal movement of the shoreline of about 11 km, and along flatter areas this movement could have covered scores of kilometers. Atwater (1979) estimated that the intertidal coastal zone expanded into south San Francisco Bay at a rate of about 30 m/yr horizontally during the early Holocene, a rate that could challenge the dispersal abilities of many marsh plants, especially those that reproduce primarily asexually, although it may be tolerable to most animals.

Changes in the location of the intertidal coastal zone control the potential distribution over time of tidal marshes, which can only occur along this narrow band, but the actual distribution of marshes at any time would have only reflected a subset of this potential distribution. Even at times and places where vegetation could migrate as fast as the shoreline was moving, a number of other factors preclude marsh formation in many coastal areas now (Chapman 1974, Adam 1990, Mitsch and Gosselink 2000), and presumably would have in the past. Thus, even if paleo-coastlines could be precisely mapped, these maps would not define the extent of marshlands along them.

THE DISTRIBUTION OF CONTEMPORARY AND PALEO-MARSHES

Tidal saltmarshes are found at sites along the fringes of most of the continents. Because of the lack of fossil or sedimentary evidence of late Pleistocene or early Holocene tidal marshes, the best guidance we have to their probable location within the paleo-coastal zone is their present distribution, which has been mapped by many authors on scales from local to global (Chapman 1974, Frey and Basan 1985, Daiber 1986, Adam 1990, Trenhaile 1997, Mitsch and Gosselink 2000). These authors and others consistently, if generally implicitly, attribute the distribution of marshes within the intertidal zone, on all spatial scales, to a common set of favorable regional and local conditions: (1) air and water temperatures warm enough for marsh plant growth and for freedom from permanent ice, but cool enough to preclude mangrove growth, (2) adequate protection from storms and destructive waves, (3) bedrock slope and sediment supply sufficient to allow net sediment accumulation (after resuspension and erosion) faster than local coastal submergence, (4) the presence of pioneer plants within dispersal distance of the incipient marsh, and (5) freedom from destructive human manipulation.

In addition, though this has been less frequently discussed, it is clear that some marsh and mudflat animals can significantly restrict marsh plant growth through herbivory and/or sediment disturbance, and must be considered potential constraints on marsh formation or stability (Collins and Resh 1989, Philippart 1994, Miller et al. 1996).

Traditionally, authors have used the existence of marshes as proof of where conditions are favorable, rather than to test theoretical models of potential marsh formation and stability against independently mapped physical attributes of sites. Although geomorphologists and ecologists have recently begun to rigorously model and quantify the needed inputs for marsh formation and maintenance (Temmerman et al. 2003), we know of no publications yet using these tools to estimate the extent of paleomarshes over any large areas or long time periods.

In comparison with sea-level changes, latitudinal temperature shifts associated with glaciation and deglaciation and their potential impacts on tidal marshes have received scant attention in the marsh literature, despite being a prominent feature of large-scale paleoclimate models (Kutzbach et al. 1998). In contrast to vertical fluctuations in sea level (110–140 m) and horizontal changes in the location of the coastal zone (ca. 10–40 km), zones of mean or extreme temperature and of major biomes can move toward the equator during glacial phases and toward the poles during interglacials by hundreds of kilometers. Although the extent of these shifts may have been smaller in the coastal zones than in the continental interiors because of a temperature-dampening effect of the sea, the extent of coastal mangroves was depressed during the full glacial, apparently due to the colder climate (Bhattacharyya and Chaudhany 1997, Wang et al. 1999), and shifts in the line between marshes and mangroves continues today, although perhaps due to other reasons (Saintilan and Williams 1999). The line of year-round ice, and hence the high-latitude limits of arctic-type tidal marshes, shifted by hundreds of kilometers toward the equator during the LGM (McCabe and Clark 1998), and the transition between arctic- and temperate-type tidal-marsh ecosystems also likely shifted towards the equator.

Another requirement of tidal marshes is protection from waves and storms above some critical threshold (Mitsch and Gosselink 2000, Zedler 2001); however, it is not clear what these thresholds are or how they vary between marsh types. One particular consequence of the horizontal movement of the coastal zone associated with sea-level changes is a perhaps substantial change in the degree of protection from storms and waves that can be provided by structural embayments. For example, the margins of both the San Francisco and Chesapeake bays are largely protected now from intense oceanic events, while at lower sea stands the intertidal zone would have been seaward of the structural basins, and would not have had the bedrock protection. In light of the long gradual continental shelf off the Atlantic Coast of North America, it is likely that barrier islands or barrier spits could have protected Atlantic marshes as they do now over large areas without rocky natural breakwaters (Odum et al. 1995), but it is not clear that equivalent geomorphology would have developed on the California coast. Nor is it clear how extensive or how protective barrier island-marsh systems may have been off any coasts during and since the LGM, as climate change can influence both fluvial sediment supplies and river mouth form (Finkelstein and Hardaway 1988).

Protection from storm and wave energy is critical for tidal-marsh formation and persistence because the geomorphic dynamic basic to marshes is net sediment accumulation equal to or slightly greater than local relative sea-level change. Thus, a key element in all explanations and numerical models of tidal-marsh formation and stability is sediment supply, and change in the sediment budgets of marshes is another potentially significant impact of climate change. The literature on tidal saltmarsh sediment dynamics is extensive (Frey and Basan 1985, Stoddart et al. 1989, Pethick 1992, Trenhaile 1997), and a comprehensive review is beyond the scope of this paper, but some key processes have clear relationships to climate, sea level, and runoff. Patterns of sedimentation on tidal saltmarshes depend partly on factors extrinsic to the marsh itself but also heavily upon dynamics within the marsh (Frey and Basan 1985, Trenhaile 1997, Malamud-Roam 2000), which has made large-scale or long-term mapping difficult. Generally, tidal marshes are maintained over time by a null to slightly positive sediment balance, with the more frequently inundated parts of the marsh surface often accreting more rapidly than the areas of the marsh less frequently inundated (Trenhaile 1997). Significant changes in climate can alter these patterns by changing the availability of both mineral and organic sediment. For example, lake core and coastal records indicate that sediment supplies during the last glacial maximum were lower in some places than during the Holocene (Grosjean et al. 2001, Wanket, 2002), changes that may be attributed to shorter growing seasons and, in the higher latitudes, a reduction in land area exposed to erosion, although, as previously noted, these patterns vary substantially from place to place.

An exhaustive comparative review of saltmarsh development on different coasts is beyond the scope of this paper, but a brief comparison of the geologic setting and modern distribution tidal saltmarshes along the Atlantic and Pacific coasts of the US helps explain differences between these regions and indicates possible causal relationships elsewhere. The Pacific and Atlantic coasts (and the Gulf of Mexico coast, although this region is not discussed here; see Stout 1984) of North America differ in their geomorphic and tectonic settings and this has probably had a significant impact on saltmarsh development. In contrast to the small and isolated tidal saltmarshes found along the Pacific coast, tidal marshes along the Atlantic Coast are presently larger and better connected (Josselyn 1983, Goals Project 1999, Zedler 2001). The effects of post-glacial isostacy has resulted in a complex north-south gradient in the relative rates of sea-level rise along the Atlantic Coast throughout the Holocene, and the rates of sea-level rise have changed over time (Fairbanks 1992; Peltier 1994, 1996). The two major estuaries on the U.S. Atlantic coast, the Delaware Bay and the Chesapeake Bay, are both subsiding, but at different rates. Tidal marshes surrounding these bay systems have been influenced by changing rates of relative sea level rise both between the two systems and within each system as they both have long north–south axes (Fletcher et al. 1990, Kearney 1996). The Chesapeake Bay system has had a slower rate of relative sea-level rise in the last 1,000 yr, and may have experienced a regression in sea level (Kearney 1996).

Marshes cannot form without the presence of pioneer marsh plants within dispersal range (Adam 1990, Malamud-Roam 2002). In addition, plants that can both colonize and tolerate wet and salty conditions are not only required for the establishment of tidal saltmarshes, but their presence is often critical to transformations of marsh type (Chapman 1974). Plant species do not generally disperse as well as many animal taxa, which initially implies that plant migration rates could be the major limiting factor on marsh establishment following deglaciation, but many of the plant species found in tidal marshes share a suite of evolutionary adaptations to the intertidal environment that may pre-adapt them to surviving during, and re-colonizing following, climate or sea-level changes. These adaptations include a high degree of phenotypic plasticity allowing the plants to respond quickly to rapidly changing conditions (Allison 1992, Dunton et al. 2001), asexual reproduction that can be an advantage for rapid establishment (Daehler 1998), increased chances of survivorship through clones (Pan and Price 2002), and specific physiological adaptations allowing exploitation of limited nutrients, tolerance of anoxia, absorption of water against osmotic pressure, and excretion of excess salts (Adam 1990, Eleuterius 1990).

Finally, marshes cannot form or persist in the presence of excessive disturbance by humans or other animals. People have caused a significant decrease in tidal-marsh extent in recent centuries, and in some places an increase in extent and habitat values through intentional restoration activies (Daiber 1986, Goals Project 1999, Zedler 2001). These impacts have been well reviewed elsewhere, and will not be further discussed here. Although other animals do not have the same capacity for short-term impacts as humans with heavy equipment, it is clear that herbivory or faunal disturbance of the substrate can be sufficient to preclude marsh formation or to limit the extent of marsh plant spread (Collins and Resh 1989, Philippart 1994, Miller et al. 1996). We know of no published research on the potential impacts of animals on the extent or distribution of paleomarshes.

MACRO-SCALE BIOGEOGRAPHY – BIOTIC DISTRIBUTION BETWEEN REGIONS OR ESTUARIES

Distributional patterns of tidal-marsh organisms, as with other organisms, occurs on multiple scales, and a convenient delineation with coastal or estuarine species is macro-scale or between regions, meso-scale or within regions, and micro-scale or within specific sites. The primary controls on macroscale biogeography of all taxa are the sites of origin or adaptive radiation of taxa, the presence or absence of dispersal routes to other areas, and the presence and extent of refugia habitat during periods when conditions are stressful and populations have been vulnerable to extirpations (Arbogast and Kenagy 2001, Smith et al. 2001). In the case of tidal marshes, macro-scale biogeographic differences are seen between continents, between oceanic coasts, and along latitudinal gradients, and all of these patterns were significantly shaped by Quaternary climatic dynamics. In particular, the variables that control the distribution of marshes can also independently affect the global- to regional-scale distributions of the organisms that inhabit them.

Different authors have categorized tidal saltmarshes into different numbers of regional types on the basis of their dominant vegetation (Chapman 1974, Frey and Basan 1985), and these largely correlate with latitude (particularly arctic-semi-arctic versus temperate) and ocean basin, but a relatively small number of plant genera and species dominate most the temperate tidal saltmarshes world-wide. Species of marsh rosemary (*Limonium*), *Suaeda* spp., pickleweed (*Salicornia*) and fat hen (*Atriplex*), as well as arrowgrass (*Triglochin maritima*), saltgrass (*Distichlis spicata*), and jaumea (*Jaumea carnosa*) are common tidal-marsh species in the temperate latitudes, while cordgrass (*Spartina* spp.) is common both on mudflats and higher in the intertidal zone.

Modern high-latitude tidal saltmarshes are distinct in many ways from temperate saltmarshes (Earle and Kershaw 1989, Gray and Mogg 2001), and may indicate the probable structure of marshes near the LGM ice margin. Although some of the present differences may be due to seasonality of day length or other variables that are functions of latitude rather than ice proximity or temperature, other attributes apparently could have been translated farther from the poles. For example, the alkali grass (*Puccinellia phryganodes*) out-competes species of *Spartina* at low temperatures (Gray and Mogg 2001). These authors also suggested that greater generic diversity occurs in Arctic than in temperate saltmarshes because the high latitude coastal waters are relatively low in salinity; if this is generally true, then it could indicate a significant impact of climate change on marsh biogeography, because deglaciation led to dramatic changes in the distribution of near-shore salinity near rivers draining the melting glaciers (Ruddiman 2001).

Latitudinal shifts in temperature not only result in specific places becoming colder or warmer, but organisms adapted to specific temperature ranges may have had to survive in suitable refugia at a great distance from their present distribution, and potentially in areas without suitable settings for the formation of extensive marshlands. One example of an estuarine species apparently strongly influenced by glacial temperature shifts is the coho salmon (*Oncorhynchus kisutch*); genetic analysis of this species in estuaries in the northern hemisphere has shown increasing genetic diversity from north to south, indicating that previous glaciations eliminated coho salmon from the northern part of its range and led to adaptive radiation as it recolonized suitable habitats (Smith et al. 2001).

One particularly well-studied group of coastal-zone dwellers that apparently re-colonized temperate regions during the deglaciation are the varieties of the brown alga (*Fucus serratus*), which is potentially a good model of the biogeographic processes underlying Holocene re-colonization of coastlines impacted directly by ice cover or indirectly by cold climate in the previous glacial maximum. Coyer et al. (2003) hypothesize that brown alga originally evolved in the North Atlantic and that present populations reflect re-colonization from a southern refugium since the LGM. The authors examined genetic structure across multiple spatial scales using micro-satellite loci in populations collected throughout the species' range. At the smallest scale (ca. 100 m) no evidence shows spatial clustering of alleles despite limited gamete dispersal (ca. 2 m from parent plants); instead, the minimal panmictic distance for this plant was estimated at between 0.5 and 2 km. At greater distances, even along contiguous coastlines, genetic isolation is significant, and population differentiation was strong within the Skagerrak-Kattegat-Baltic seas (SKB) region, even though the plant only (re)entered this area some 7,500 yr BP. On the largest scale, the genetic data suggest a central assemblage of populations with high allelic diversity on the Brittany Peninsula surrounded by four distinct clusters—SKB, the North Sea, and two from the northern Spanish coast—with lower diversity; plants from Iceland were most similar to those from northwest Sweden, and plants from Nova Scotia were most similar to those from Brittany. The authors were not sure if Brittany represents a refugium or a re-colonized area, but interpreted the low allelic diversity in the Spanish populations as evidence of present-day edge populations having undergone repeated bottlenecks as a consequence of thermally induced cycles of re-colonization and extinction.

In addition to re-colonization from extant areas of similar habitat, current occupants of tidal saltmarshes and other coastal areas may have evolved or found refuge in other types of environments and then colonized tidal saltmarsh habitats when they can come into contact with them. In addition to a number of specific marsh taxa which are discussed in other chapters in this volume, the possibility for broad groups of taxa is suggested by patterns of movement into the marine realm by previously terrestrial species not found on tidal marshes. For example, the non-halacarid marine mites apparently went through two distinct migration events in the past, based on their adaptive radiation (Proche and Marshall 2001). Another possibility is colonization from non-tidal freshwater marshes, such as those that have persisted continuously in the California inland delta for at least the last 35,000 yr, and which came into contact with oceanic tides and low levels of salt only about 4,000 yr BP (Atwater and Belknap 1980).

Meso- and Micro-scale Biogeography—Biotic Distribution within Regions and Marshes

As on the global scale, marsh types at smaller spatial scales are often distinguished by their dominant vegetation (Chapman 1974), but increasingly classifications of marshes have focused more on the distribution of physical parameters such as salinity, wetness, elevation, and geomorphic pattern, and on the potential habitat values these provide (Goals Project 1999, Malamud-Roam 2000). In particular, high marsh and low marsh are very commonly used divisions (Chapman 1974, Teal 1986, Goals Project 1999), reflecting the significance of elevation as a control on wetness and hydroperiod (Malamud-Roam 2000). Following an old geomorphic convention, the apparent age of the marsh, primarily as inferred from its elevation and landforms, is often used as well as a descriptive tool (Goals Project 1999).

Bioregions are conventionally defined as areas with essentially similar species composition, although the actual presence or absence and abundance of specific taxa between sites within the region can vary dramatically (Goals Project 1999). Thus meso-scale biogeographic variability presumably reflects habitat suitability and local patterns of migration, extirpation, dispersal, and recolonization more than large-scale historical isolation or long-term barriers to migration (MacArthur and Wilson 1967). As noted in the description of the standard model, the most obvious region in which tidal marshes share potential species is specific estuaries, and the most significant cause for differences in biotic composition of marsh communities within estuaries is gradient in salinity (Josselyn 1983, Adam 1990, Goals Project 1999). In addition, marsh size and the distribution of marshes within estuaries have also been widely investigated as examples of landscape-level variables controlling biotic-community structure (Goals Project 1999), and these variables have occasionally been used to analyze tidal marshes as habitat islands in a theoretical biogeography sense (Bell et al. 1997, Lafferty et al. 1999, Micheli and Peterson 1999).

Finally, the distribution of specific marsh taxa or biotic communities within marshes is usually seen as a consequence of modern physical variables, with the frequency and duration of tidal flooding and drying given the most emphasis, and soil salinity and nutrient limitation also attracting research (Zedler 1982, Stout 1984, Teal 1986, Mitsch and Gosselink 2000, Zedler 2001). Elevational zonation is the conventional characterization of plant distribution with the explicit recognition that marsh plants do not directly respond to elevation, but instead to wetness and hydro period, for which elevation serves as a reasonably useful proxy (Frey and Basan 1985, Malamud-Roam 2000). Although animals also respond to physical parameters, their distribution is also clearly influenced by the distribution of flora as well.

TIDAL SALTMARSHES OF THE SAN FRANCISCO BAY-DELTA ESTUARY

The tidal saltmarshes of the San Francisco Bay-delta estuary cover a large area, have many rare and endemic plant and animal taxa, have been intensively researched, and are the subject of intense current debate about how best to achieve protection and restoration of habitat values (Atwater et al. 1979, Josselyn 1983, Goals Project 1999; Malamud-Roam 2000, 2002). Therefore, these marshes are used to illustrate some the elements of the conceptual model, some significant site-specific patterns which may help explain the high rates of endemism found in the tidal saltmarshes there, and some associated conservation challenges. This estuary has been referred to in many ways in the literature (Malamud-Roam 2000), but hereafter will be referred to as the San Francisco estuary.

The basic configuration of the San Francisco estuary today is a series of bedrock basins linked by narrows or straits (Goals Project 1999, Malamud-Roam 2000). Inland of the Golden Gate, the only opening from the estuary to the Pacific Ocean, is Central Bay, followed in order upriver by San Pablo Bay, Carquinez Strait, Suisun Bay, and the delta of the Sacramento and San Joaquin rivers. An additional basin attached to Central Bay, known prosaically as South Bay, has little freshwater input, but the other basins form a classic estuarine gradient of decreasing salt and generally decreasing tidal character with distance upstream. Thus, although Central Bay has essentially oceanic salinity (~35 ppt) and tidal range (~2 m), the delta is a freshwater environment with tidal range ~1m, and Suisun Bay is an extensive brackish zone, the conditions of which vary substantially with the season and the year. All of the basins had extensive tidal marshes at the beginning of European contact with the site (ca. 1776), but some 90% or more of these have been diked, filled, or otherwise removed from the tides (Goals Project 1999).

In our model, we have treated the distribution of the intertidal coastal zone, the distribution of marshes, and the distribution of specific marsh habitats or communities as separate parameters; in practice, however, much of the evidence for each in the San Francisco estuary and elsewhere is provided by sediment cores collected at multiple sites (Bradley 1985, Malamud-Roam 2002). Dated sediments collected from below current marshes or estuaries can potentially provide evidence of sub-tidal estuarine and inter-tidal marsh history back to the LGM and of riverine and non-tidal marsh settings even further back. In particular, the basic elements of the formation and evolution of tidal marshes within the San Francisco estuary, which had been articulated by Atwater and his colleagues (Atwater et al. 1977, Atwater 1979, Atwater and Belknap 1980), have been elaborated in recent years using a range of methodologies including stable isotopes (Malamud-Roam and Ingram 2001, 2004; Malamud-Roam 2006), fossil pollen (May 1999, Byrne et al. 2001, Watson 2002), fossil seeds and metals (Goman 1996, Goman 2001, Goman and Wells 2000), and diatoms (Starratt 2004). Although the site specificity of each core means that a complete paleo-mapping has not been completed, the history of some areas is well known, and sufficient information on causal variables has been collected that interpolations of areas between the cored sites are being developed. In addition, these studies have begun to show how the physical environment and biotic communities of these sites have responded to changes in inputs such as runoff or sea level. Information on LGM refugial intertidal habitats outside the Golden Gate, however, is not available, and inferences about these areas are tentative.

Paleo-shoreline maps can be developed not only from sediment cores, but also from current bathymetric maps where relative sea-level curves are known, although these maps will be imprecise if either sediment accumulation is significant or if regional crustal motions are non-uniform (Atwater 1979, Nikitina et al. 2000). Mapped former shorelines for the San Francisco Bay, based on calculated sea-level rise for the south San Francisco Bay, show that ocean waters entered through the Golden Gate approximately 10,000 yr BP (Atwater 1979). Although the Golden Gate is currently >100 m deep, the 50 m bathymetric contour lies some 30 km offshore now (NOAA 2003), and the current estuary was certainly non-tidal during the LGM and for thousands of years after (Atwater 1979, Atwater and Belknap 1980, Goman 1996, Malamud-Roam 2002). Therefore, to estimate the LGM shoreline as a first step in modeling late-glacial-phase marsh refugia, modern bathymetric maps of the California coast were used to produce an approximation for the shoreline which existed ca. 21,000 yr BP along the California coast (Fig. 3) and outside the Golden Gate (Fig. 4). This paleo-shoreline is based on a LGM sea level 120 m lower than today, and does not account for sediment accumulation or local variations in crustal stability.

The paleo-shoreline maps indicate potential tidal-marsh sites, but neither they nor the many sediment cores that have been collected in the San Francisco estuary allow definitive maps of late Pleistocene or early Holocene tidal-marsh distribution; however, together with some observations of modern marshes, they do allow for some estimates and some conjectures. The San Francisco estuary clearly contains all the necessary conditions for tidal saltmarsh development and maintenance currently, and all of these can be estimated for at least some time into the past, although with varying degrees of precision. Evidence of ice or mangroves is lacking during the Quaternary in any of the environmental histories of the area (Goman 1996). Mineral sediments are supplied in large quantities by the Sacramento River and the San Joaquin River and smaller local rivers, which together drain a combined watershed region of ~40% of the state of California, and which have done so throughout the Quaternary (Goals Project 1999). Although tidal-marsh studies in the estuary reveal a pattern of incipient marsh formation and submergence in some sites until balance was achieved between sediment supply and sea-level rise (Malamud-Roam 2006), no clear evidence shows local crustal movement resulting in relative sea-level rise drowning marshes (Atwater et al. 1979, Atwater and Belknap 1980, Goman 1996). Plant and animal genetic material for the San Francisco estuary tidal marshes may have come from two sources: local invasions from adjacent uplands and fresh-water marshes, especially in the Sacramento-San Joaquin delta, and from small coastal saltmarshes that may have occupied the exposed coastline outside the Golden Gate. Finally, although human disturbance of the marshes has been substantial over the last 150 yr, no published evidence exists of extensive human disturbance prior to that time or of significant limitations on marsh formation by other animals (Goals Project 1999).

A major question is the extent to which the geological setting would have provided adequate protection from storms and wave energy for marsh establishment or persistence. The geologic constriction forming the Golden Gate now creates a buffer to the high-energy conditions that exist along the California coastline (NOAA 2003), but this would not have protected coastal environments during and for some 11,000 yr after the LGM (Atwater and Belknap 1980). Outside the Golden Gate, the principle feature that stands out in the paleo-shoreline maps (Figs. 3 and 4) is the absence of a large fully protected inlet or bay anywhere along the north and central coastline that could provide conditions similar to those inside the Golden Gate for extensive saltmarsh development during and shortly after the LGM, although a large (~5,000–10,000 km^2)

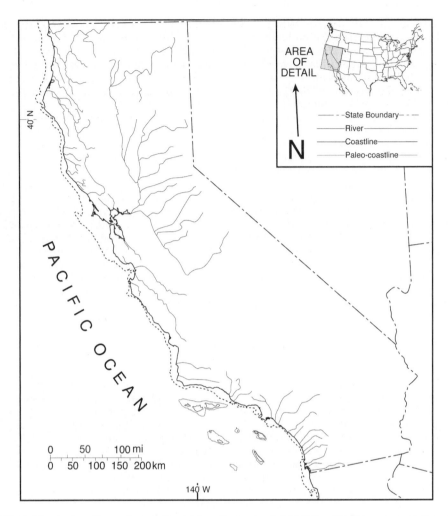

FIGURE 3. California shoreline and approximate shoreline present at 20,000 yr BP. This representation of paleo-shoreline assumes a drop of 120 m in sea level and does not account for local variations in geologic stability. This map was adapted from public domain bathymetric maps (U.S. Coast and Geodetic Survey 1967a, b, c, d; U.S. Coast and Geodetic Survey 1969, National Ocean Service 1974a, b.).

semi-enclosed basin—the Gulf of the Farallons and Cordell Bank—lies between the Farallon Ridge and the Golden Gate (NOAA 2003), and may have provided substantial protection for some of this period. Although Atlantic Coast marshes are extensive in many areas without bedrock protection, the lack of large marshlands along the central and northern California coastlines at present (NOAA 2003) and the structural-tectonic setting of this area, with steep bathymetry and a history of rapid vertical tectonic motion (Atwater and Hemphill-Haley 1997), suggests that LGM refugial tidal marshes outside the Golden Gate were very small and isolated, and may have been quite limited in size and possibly separated at times by large distances throughout the late Pleistocene and early Holocene.

Modern tidal marshes along the northern California coast outside the Golden Gate are currently associated primarily with river mouths (NOAA 2003), and several of these potential marsh sites can be seen in Figs. 3 and 4, such as at the mouth of the Eel River (Fig. 5), where a delta with seasonally variable sandy barrier spits and beaches currently creates some protected opportunities for salt-marsh development. In addition, sandy oceanic sediments have formed barriers and protected small marshes at Point Reyes and Tomales Bay, where structural barriers provide some protection to the sediments and marshes (NOAA 2003). Although direct evidence is lacking for tidal saltmarshes of the late glacial period in this region, it appears most likely that they also

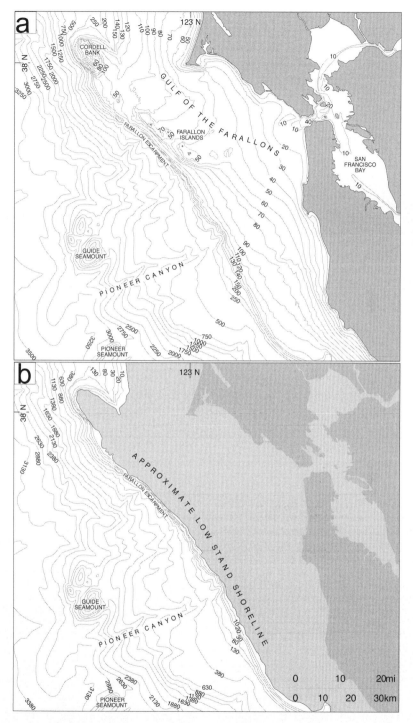

FIGURE 4. Near-shore bathymetry of north-central California during high and low stands of sea level. (a) During high stands, a large estuary is located east of the Golden Gate. (b) During low stands, shorelines are located east of the Farallon Islands. This representation of paleo-shoreline assumes a drop of approximately 120 m and does not account for changes in elevation as a result of tectonic uplift or subsidence. Bathymetry is in meters and reported relative to mean lower low water (MLLW). This map is adapted from a National Ocean Service (1974) bathymetric map.

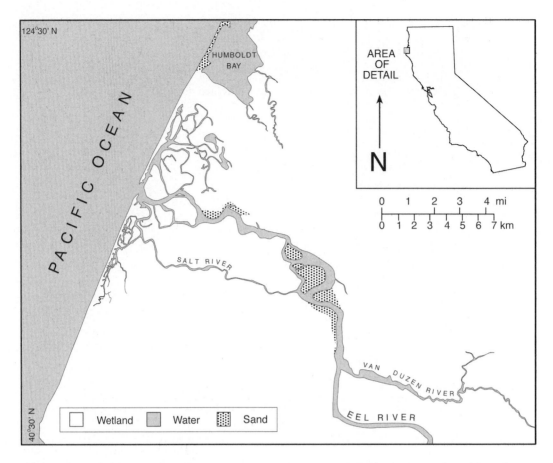

FIGURE 5. The Eel River delta before major coastal development occurred. This map was adapted from the U.S. Army Corps of Engineers (1916a, b).

developed as relatively small fringing coastal marshes where barrier spits and islands created by the build up of river-borne and coastal sediments provided some limited protection. Barrier features similar to those at the Eel River and Point Reyes may have existed throughout the glacial periods of the Quaternary where the Sacramento and San Joaquin rivers reached the paleo-shoreline, although the bathymetric maps indicate a significant drop in elevation just beyond the Farallon Islands, where the 21,000 yr BP shoreline would have been. At some point after the first melt-water pulse, the topographic ridge containing the Farallon Islands and the Cordel Bank west of Point Reyes (Fig. 4a) would have formed a semi-enclosed basin at the site of the current Gulf of the Farallons, which presumably provided some protection from storms during the latest Pleistocene and early Holocene. However, although some tidal marshes probably formed in the Gulf of the Farallons, the lack of evidence for extensive Atlantic or Gulf Coast marshes during this time argues that rapidly rising sea level probably kept them small.

It is unclear when tidal marshes first formed inside the Golden Gate after the LGM. Evidence from other coasts as well as from the San Francisco estuary indicates that tidal marshes were able to colonize the mudflats in the bay only after the rate of sea level slowed to less than 2 mm/yr, roughly 6,000 yr BP (Atwater 1979, Fairbanks 1989), and numerous sediment cores in the lower and middle San Francisco estuary (east through Suisun Bay) have not found evidence for tidal marshes before about 4,000–5,500 yr BP (Atwater 1979, Atwater and Belknap 1980, Goman 1996, Goman and Wells 2000, Malamud-Roam 2002, Malamud-Roam and Ingram 2004). In salty parts of the estuary, the mudflats were often first colonized by the pioneer plant, California cordgrass (*Spartina foliosa*; Malamud-Roam 2002), a California endemic that can withstand prolonged periods of inundation. This grass does best in fresh conditions (Cuneo 1987), but can tolerate high salinity and is therefore more commonly found in salt tidal marshes today. As the surface elevation of the mudflats rose, a result of the increased mineral and organic sediments accumulating due to the stands of California cordgrass, other marsh species became established, such as pickelweed (*Salicornia virginica*) and salt grass (*Distichlis spicata*), or sedge species (*Schoenplectus californica* and *S. acutus*) in the case of the brackish marshes.

In contrast to tidal marshes, there is clear sedimentary evidence for continuous non-tidal freshwater marshlands in the delta of the Central Valley dating back over 30,000 yr, reflecting drainage impeded by tectonic-structural barriers at the transition from the Central Valley to Suisun Bay (Schlemon and Begg 1973, Atwater and Belknap 1980). Precisely when the delta marshes began to experience tidal influence has been controversial, and the complex geologic history of the Suisun Basin and the western delta has precluded precise estimates of tidal introduction to the delta based solely on bathymetry. Schlemon and Begg (1973) interpreted 12,000-yr-old sediments at Sherman Island, in the western delta, as intertidal, but this was disputed by Atwater et al. (1979) and Atwater and Belknap (1980), who believed that the site was non-tidal freshwater marshes until perhaps 7,600 yr ago. More recent sediment cores show clear evidence of perhaps 7,000 yr of fresh-water marshes and considerable taxonomic diversity at Browns Island (Goman and Wells 2000, May 1999, Malamud-Roam 2002) and a similar history at several sites that are now sub-tidal (Watson, Chin, and Orzech, unpubl. data) in Suisun Bay near the delta, but the degree of tidal action in these sites is ambiguous.

The occurrence of high endemism in tidal-marsh plants and animals in the San Francisco estuary (Greenberg and Maldonado, *this volume*) likely has many causes, and in addition to the rapid expansion of habitats over a physically diverse estuary spanning over 100 km, and dispersal and possibly adaptive radiation in an estuary that is largely isolated from other tidal-marsh gene pools, colonization or recolonization also apparently took place from multiple directions (tidal saltmarshes, non-tidal freshwater marshes, non-tidal salty or alkaline marsh, and uplands). Fresh-water marshes have occupied the area adjacent to the confluence of the Sacramento and San Joaquin rivers for approximately 7,000 yr (Goman, 1996). Animal species that may have stopped in the delta during their annual migrations may have taken advantage of the newly available niches provided by the development of salty and freshwater tidal marshes in the San Francisco estuary. In addition to the delta, other wet and frequently salty or alkaline environments exist inland of the San Francisco Bay, including shallow seasonal lakes, pools and marshes. Resulting from the combination of California's mediterranean climate, soils which produce a subsurface hardpan and largely flat, but hummocky topography, vernal wetlands are common throughout the state of California, particularly in the Central Valley and along its adjacent coastal terraces and range from <1 ha to >20 ha in size (Holland and Jain 1977). Today vernal pools provide temporary habitats for many ducks, shorebirds, and passerines (Baker et al. 1992), and species richness is significantly correlated with the size of the vernal pool (Holland and Jain 1984). During the late Pleistocene and early Holocene, much of the Central Valley was covered by large vernal pools and lakes (Baskin 1994) and the marshy habitats that were associated with them may have provided some habitat for some of the vertebrate organisms occupying present day saltmarshes around the San Francisco Bay.

CONCLUSIONS AND IMPLICATIONS

The climate and sea-level variations seen since the last glacial maximum have had significant direct and indirect impacts on the location of the coastal zone, on the extent and distribution of saltmarshes worldwide, on the distribution of physical conditions and thus potential habitats within marshes, and ultimately on the biogeography at all spatial scales of the species associated with saltmarshes. The global-scale climate changes that led to rapid sea-level rise also influenced the distribution of marshes and their inhabitants through other, more subtle, mechanisms, including shifts in the distribution of sea-surface temperature, ice, rainfall and runoff, and sediments. Major consequences to tidal marshes of these global-scale changes and their local manifestations include frequent, periodic losses of habitat with associated consequences for population and genetic processes, sequential expansions from habitat refugia, and communities predisposed to invasion.

Some of the aspects of the historical geography and biogeography of tidal saltmarshes discussed in this paper are known conclusively while others, because of limitations in preserved data, are known only indirectly, inferentially, and/or imprecisely. There is no doubt that the global ocean rose everywhere relative to the land over the last 21,000 or so years, and that on a global scale the scale of this was about 110–140m, but there is incomplete knowledge of the precise extent of rise relative to local land surfaces, because of complex local crustal movements due both to glacial rebound and other geologic processes. It is clear that the rate of rise varied dramatically during this period, and that the most recent 6,000 yr or so have been characterized by relatively slow rise on a global scale, but the precise rate and timing of phases of faster and slower rise is unclear both globally and locally. Tidal saltmarshes have no doubt existed in ephemeral settings, and their current locations and forms have existed for no more than a few thousand years, but there are significant challenges in mapping their extent and connectedness during the last glacial maximum and during the following 15,000 yr. It is almost certain that the extent and connectedness of marshlands along all coasts increased and decreased in several phases during the late Pleistocene and early Holocene, potentially allowing for phases of adaptive radiation and dispersal, but the precise distribution of antecedent tidal marshes is not known and probably never will be. It is certain that both the air and sea water were colder during and shortly after the LGM at all current tidal marsh sites, but it is not yet clear how far from the poles coastal biota were pushed by these temperature shifts and the associated expansion of year-round ice cover.

Some other general principles are certain—as sea level rises, aquatic environments invade the terrestrial realm, and tidal marshes persist either by accreting vertically, or by migrating landward. A result of the rise and fall of global sea level on glacial timescales is the burial and/or erasure of former saltmarsh sedimentary records. Glacial cycles have led to north-south gradients on all coasts because of isostasy, changes in ice cover, and other causes unrelated to current latitudinal variations in physical conditions. Glacial cycles may have contributed to cases, like in the San Francisco Bay, where tidal marshes have developed largely in isolation from other coastal saltmarshes, with a consequently high rate of endemism in tidal marsh plants and animals. As tidal marshes have developed in their current locations, their inhabitants have colonized them not only from refugial tidal marshes, but for some taxa at least, from other wetlands or upland areas with very different natural histories. Tidal saltmarshes and their flora and fauna have suffered significant losses due to human development and today face potentially serious threats related to invasive species and global.

Because direct stratigraphic evidence is missing, the specific underlying mechanisms leading to some modern biogeographic patterns are not completely clear, but are strongly suggested both by the biotic distributions themselves and by the coastal environments implied from our model. For example, evidence is presented in Greenberg and Maldonado (*this volume*) that sparrows and other groups of the U.S. Atlantic Coast vary dramatically in the length of time that they have been genetically isolated from congeners, with a trend towards genetic longer isolation in the south than north. Although it is not possible to exactly map the low stand Atlantic coastline or its tidal marshes, it is clear that the sites of current northern marshes were under thick pack ice during the LGM and during earlier Pleistocene glacial advances, and that coasts near the ice front could have experienced significant storms associated with the pronounced temperature gradients. In contrast, more southern coasts, while kilometers east of their present location during low sea stands, would probably not have differed greatly in physical conditions from the present—gentle bedrock slope, sediment fluxes down the rivers and along the coasts, moderate tides, air and water temperatures within the current ranges of tidal marshes. Although droughts associated with glacial conditions would have reduced freshwater supplies and probably sediment fluxes, it seems likely that barrier islands and spits would have provided adequate protection from storms for significant marshes. Thus, the genetics of northern taxa may well represent recent colonization of tidal marshes and differentiation from upland types, while the southern taxa have had substantial time for specialization to tidal marsh conditions. This is in stark contrast to the California examples, where a lack of storm protection could have limited the extent of tidal marshlands along the length of the coast during low sea stands, with expansion and colonization of tidal marshes more determined by basin configuration and sea level than by latitude.

Some final questions remain unanswered despite the supplemental model:
1. Given the changing distribution of physical conditions in estuaries, especially in light of anthropogenic influences, where can marshes be effectively protected and restored for the long term?
2. Why are biodiversity, rarity, and endemism higher in some estuaries than in others?
3. In addition to maintaining marshes along salinity gradients, are other landscape-level attributes of patch size and distribution important for protection and restoration of rare, endemic, and/or native species?
4. How should restoration projects be planned to maximize the likelihood of producing desired taxa and minimize the abundance of pests?

5. What are the risk factors associated with invasive and/or non-native species in marshes?
6. Can marshes be designed to minimize invasion risk?
7. When temporal changes are noted in the distribution or abundance of marshes or marsh taxa, are these due to natural succession or landscape evolution, to natural periodicities in forcing functions, to unintentional human influences, and/or to intentional restoration activities? Although these questions have not been comprehensively answered in this review, it is hoped that the framework provided can suggest new interpretations and fruitful lines of research.

ACKNOWLEDGMENTS

We would like to thank the reviewers, Michelle Goman and Daniel Belknap, for their careful reading and extensive and useful suggestions. We would like to thank our funders The Smithsonian Institution, the USDA Fish and Wildlife Service, and Contra Costa Mosquito and Vector Control District. Finally, we thank Roger Byrne and Doris Sloan of the University of California at Berkeley and Dorothy Peteet of Columbia University for insights on Pleistocene conditions in the San Francisco Bay Area and the Atlantic Coast.

DIVERSITY AND ENDEMISM IN TIDAL-MARSH VERTEBRATES

RUSSELL GREENBERG AND JESÚS E. MALDONADO

Abstract. Tidal marshes are distributed patchily, predominantly along the mid- to high-latitude coasts of the major continents. The greatest extensions of non-arctic tidal marshes are found along the Atlantic and Gulf coasts of North America, but local concentrations can be found in Great Britain, northern Europe, northern Japan, northern China, and northern Korea, Argentina-Uruguay-Brazil, Australia, and New Zealand. We tallied the number of terrestrial vertebrate species that regularly occupy tidal marshes in each of these regions, as well as species or subspecies that are largely restricted to tidal marshes. In each of the major coastal areas we found 8–21 species of breeding birds and 13–25 species of terrestrial mammals. The diversity of tidal-marsh birds and mammals is highly inter-correlated, as is the diversity of species restricted to saltmarshes. These values are, in turn, correlated with tidal-marsh area along a coastline. We estimate approximately seven species of turtles occur in brackish or saltmarshes worldwide, but only one species is endemic and it is found in eastern North America. A large number of frogs and snakes occur opportunistically in tidal marshes, primarily in southeastern United States, particularly Florida. Three endemic snake taxa are restricted to tidal marshes of eastern North America as well. Overall, only in North America were we able to find documentation for multiple taxa of terrestrial vertebrates associated with tidal marshes. These include one species of mammal and two species of birds, one species of snake, and one species of turtle. However, an additional 11 species of birds, seven species of mammals, and at least one snake have morphologically distinct subspecies associated with tidal marshes. Not surprisingly, species not restricted entirely to tidal marshes are shared predominantly with freshwater marshes and to a lesser degree with grasslands. The prevalence of endemic subspecies in North American marshes can either be a real biogeographical phenomenon or be attributable to how finely species are divided into subspecies in different regions. The difference between North America and Eurasia is almost certainly a biological reality. Additional taxonomic and ecological work needs to be undertaken on South American marsh vertebrates to confirm the lack of endemism and specialization there. Assuming that the pattern of greater degree of differentiation in North American tidal-marsh vertebrates is accurate, we propose that the extension and stability of North American marshes and the existence of connected southern refugia along the Gulf Coast during the Pleistocene contributed to the diversification there. The relatively large number of endemics found along the west coast of North America seems anomalous considering the overall low diversity of tidal marsh species and the limited areas of marsh which are mostly concentrated around the San Francisco Bay area.

Key Words: biogeography, habitat specialization, saltmarsh, wetland vertebrates.

DIVERSIDAD Y ENDEMISMO EN VERTEBRADOS DE MARISMA DE MAREA

Resumen. Los marismas de marea se distribuyen en parches, predominantemente por las costas de media- a alta latitud de los grandes continentes. Las extensiones mayores de marismas de marea no-árticos son encontradas a lo largo de las costas del Atlántico y del Golfo de Norte América, pero concentraciones locales pueden ser encontradas en Gran Bretaña, el norte de Europa, el norte de Japón, el norte de China y el norte de Corea, Argentina-Uruguay-Brasil, Australia y Nueva Zelanda. Enumeramos el numero de especies de vertebrados terrestres que regularmente ocupan las marismas de marea en cada una de estas regiones, así como especies o subespecies que son ampliamente restringidas a marismas de marea. En cada una de las áreas costeras principales encontramos 8–21 especies de aves reproductoras y 13–25 especies de mamíferos terrestres. La diversidad de aves y mamíferos de marismas de marea se encuentra altamente inter-correlacionada, así como la diversidad de especies restringidas a marismas saladas. Estos valores son por lo tanto, correlacionados con el área marisma-marea a lo largo de la línea costera. Estimamos que aproximadamente siete especies de tortugas aparecen en aguas salobres o marismas saladas en todo el mundo, pero solo una especie es endémica, y es encontrada en el este de Norte América. Un gran numero de ranas y culebras aparecen oportunísticamente en marismas de marea, principalmente en el sureste de Estados Unidos, particularmente en Florida. Tres taxa de culebras endémicas son restringidas a marismas de marea también del este de Norte América. Sobre todo, solo en Norte América fuimos capaces de encontrar documentación de múltiples taxa de vertebrados terrestres asociados a marismas de marea. Esto incluye una especie de mamífero y dos especies de aves, una especie de culebra y una de tortuga. Sin embargo, 11 especies adicionales de aves, siete de mamíferos, y al menos una culebra, tienen subspecies que son morfológicamente distintas y que estan asociadas con marismas de marea. No es de sorprenderse, pero las especies no son restringidas completamente a marismas de marea son compartidas predominantemente con marismas de agua fresca y no a menor grado con pastizales. El predominio de subespecies endémicas en marismas de Norte América se debe ya sea a un fenómeno biogeográfico real, o puede ser atribuido a que especies son divididas finamente en subespecies en diferentes regiones. La diferencia entre Norte América y Eurasia es casi ciertamente una realidad biológica. Trabajo taxonómico y ecológico adicional debe de ser llevado a cabo en vertebrados de marisma en Sudamérica para confirmar la falta de endemismo y especialización ahí. Asumiendo que el patrón de mayor grado de diferenciación de vertebrados de marisma en Norte América es correcto, proponemos que la extensión y la estabilidad de marismas en Norte América, y la existencia de refugios sureños conectados a lo largo de la costa del Golfo durante el Pleistoceno, contribuyó a la diversificación ahí. El relativamente gran número de endemismos encontrados a lo largo de la costa oeste de Norte América, parece anormal considerando el total de diversidad baja de especies de marismas de marea y las áreas limitadas de marisma, las cuales están principalmente concentradas alrededor del área de la bahía de San Francisco.

Tidal marshes associated with well-protected shorelines of low relief at mid- to high latitudes, are found at the margins of all major continents except Antarctica (Chapman 1977). Along with tropical and subtropical mangrove swamps, tidal marshes are a true ecotone between marine and terrestrial systems, and as such present a number of adaptive challenges to vertebrate species. Perhaps the most obvious and critical feature of tidal marshes is salinity, with salt concentration ranging from zero parts per thousand (ppt) to concentrations >35 ppt. The regular influx of tidal waters causes water levels to be variable and in the intertidal zone creates regular fluctuations between flooded and exposed muddy substrates. Tidal levels vary throughout the year and in conjunction with storm systems can cause both seasonal and unpredictable flooding. Saltmarshes are dominated by a few species of salt-tolerant (halophytic) plants. Where the dominant plants are grasses and shrubs, the above-ground strata of the marsh come to resemble freshwater marshes and grasslands. In more arid regions, lower portions of tidal marsh are dominated by other halophytic plants, such as *Salicornia* spp. that render the marshes structurally quite distinct from most interior habitats. Species diversity of plants increases with latitude (up to a point) and decreases with salinity (Chapman 1977).

Marine invertebrates, such as amphipods, decapods, and gastropods, dominate the fauna of the muddy substrate of tidal marshes (Daiber 1982). The low species diversity of plants, their high reliance on vegetative means of reproduction, and the regular washing of the substrate by tidal waters decrease the availability of seeds and fruits to saltmarsh vertebrates. These adaptive challenges, and others, combine to form a selective environment that should lead to local adaptive modifications of terrestrial vertebrates colonizing the marshes. However, several biogeographical features may act to reduce the ability of genetically based divergence to evolve in local tidal marsh populations. Overall, tidal marshes are limited in extent and with the exception of a few large estuarine systems, e.g., San Francisco and Chesapeake bays in North America, and the estuary of the Rio de la Plata in South America, tidal marshes are often linearly distributed along coastlines or found in small pockets associated with river mouths, deltas, and the inland shores of barrier islands. This distribution results in a large edge effect that may reduce the isolation of tidal marsh and other habitats. Furthermore, as documented by Malamud-Roam et al. (*this volume*), tidal marshes have been highly unstable in their location and extension throughout recent geological history. Many of the largest areas of estuarine tidal marsh located at mid-latitudes are a consequence of flooding resulting from the sea-level rise associated with the melting of Pleistocene glaciers. The ice sheets covered arctic tidal marshes, among the most extensive in the world (Mitsch and Gosselink 2000). However, lower sea levels may have exposed more coastal plain and created a greater area of tidal marsh during the glacial maxima (G. Chmura, pers. comm.).

As we examine the species richness and endemism associated with tidal marshes in different regions, we will consider the following regional factors that might affect global patterns: the extent of tidal marsh habitat and its spatial distribution and the historical stability of tidal marsh and related habitats. Because information on these factors is either unavailable or difficult to synthesize for a number of important regions, the discussion will remain speculative and qualitative in nature. We begin with a brief discussion of marshes of different coastlines throughout the world based in large part on the classification system of Chapman (1977).

MAJOR AREAS OF TIDAL MARSHES

Global Overview

The lower marsh zone (between marsh edge and the mean high tide line) is dominated by one or two plant species whose physiognomy determines the overall structure of the simple habitat in terms of vegetation. Throughout the world, marshes differ in whether this lower zone is covered predominantly by cord grasses (*Spartina*) or, in areas with more arid climates or microclimates, by succulent halophytic plants, e.g., *Salicornia*, *Batis*, *and Suaeda*. Low marshes along the Pacific Rim, such as in Korea-Japan, California, Western Mexico, and those in southern Australia are generally covered by the succulent types of plants but eastern US and eastern South America are covered primarily with cord grasses. Lower marshes in Europe are often un-vegetated or have a sparse cover (Lefeuvre and Dame 1994).

Tidal marshes have patchy distributions (Fig. 1). Marshes form along low-energy shorelines that are associated with barrier islands along the outer coast or river mouths and estuaries. The extent and pattern of zonation in marshes is affected by the tidal patterns of their particular coastline. For most coastlines along open oceans, tidal ranges average between 1-2 m (NOAA 2004a, b, c, d). Marshes along the Atlantic and Pacific coasts of North America, southern coast of Australia, the North Sea, and the coast of Argentina generally experience these magnitudes of tidal flux. Along North American coastlines, tides increase as one moves northward. Certain isolated and shallow bodies of water—Baltic Sea of Europe, Laguna de Patos in Brazil, and Mediterranean and Caspian seas—experience very low tidal flux (NOAA 2004a, c). The shoreline of the Gulf of Mexico generally has small tidal amplitudes (0.5-1 m) compared to the Atlantic Coast (NOAA 2004a, b).

A few major estuaries are noteworthy because of the large areas of tidal salt and brackish marsh they support. San Francisco Bay in central California, which has three biologically relevant subdivisions (San Pablo, Suisun, and lower San Francisco bays), the Chesapeake and Delaware bays which flank the Delmarva Peninsula, and the estuary of the Río de la Plata along the eastern South American coast. These estuary systems represent the largest and most diverse tidal marshes in the world today. For example, approximately 90% of the original Pacific Coast wetlands from Cabo San Lucas to the Canadian border was found in the San Francisco Bay estuaries. By contrast, tectonically older coastlines with greater barrier island development usually have more continuous distribution of tidal marshes. The prime example of this can be found along the mid-Atlantic Coast of eastern North America. Along these coasts, marshes are concentrated in a few major estuaries. Delaware Bay, Chesapeake Bay, and Pamlico Sound estuaries contain 45% of the saltmarshes along the Atlantic Coast of North America, and the first two estuaries contain over three-quarters of the mid-Atlantic marshes (Field et al. 1991).

The major estuaries of temperate coasts are recent formations with similar histories. The history of the San Francisco Bay estuaries is explored fully in Malamud-Roam et al. (*this*

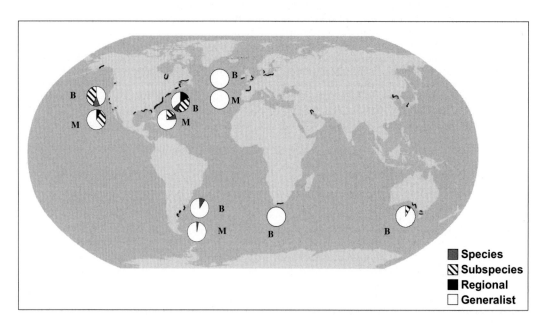

FIGURE 1. The world distribution of tidal marshes. The map is based on Chapman (1977), with revisions using more recent information on the distribution and quantity of tidal marsh along different coastlines. Circles indicate the proportion of native species of birds and mammals along different major coastlines with different levels of differentiation for tidal-marsh habitats. Tidal marsh includes brackish to salt and excludes tidal fresh water marsh.

volume). The Chesapeake Bay experienced a similar rapid development between 12,000 and 8,000 yr before present (BP), stabilizing into its present form at 3,000 yr BP (Colman and Mixon 1988, Bratton et al. 2003). Approximately 14,000 yr BP, river channels in the deltas of the Rio de la Plata and Rio Grande do Sul in South America drained to the edge of the continental shelf. At about 11,000–6,000 yr BP the sea shore moved westward across the coastal plain, shallow brackish bays developed in what would become the current estuarine valley. In the last 4,000 yr, as the sea levels became stable, coastlines became accretional, and areas of shallow coastal plain bordered the major portions of the estuary and adjacent Atlantic coastline accompanied the formation of protective sandbars (Urien et al.1980, Lopez-Laborde 1997).

These recently formed estuaries not only provide long protected shorelines adjacent to shallow coastal plains ideal for the formation of large areas of tidal marsh, but also tidal flux is generally greater in estuaries, and this, together with the mixing of salt and fresh waters over a large area, leads to the formation of more heterogeneous marsh communities and increased marsh zonation.

NORTH AMERICAN EAST AND GULF COASTS

Temperate-zone tidal marshes are by far the most extensive along the eastern and southern coasts of North America where they currently cover ca. 15,000 km^2, primarily along the southeast Atlantic and Gulf coasts (Field et al. 1991). The Gulf Coast supports ca. 9,880 km^2 of tidal marsh, with concentrations in the Mississippi delta. The remaining 5,000–6,000 km^2 are found along the Atlantic Coast, half along the south coast, 42% in the mid-Atlantic region, and the 8% in New England and the Canadian Maritimes (Field et al. 1991; Hanson and Shriver, *this volume*). Because of the lower tidal influx and the outflow from the Mississippi River, >40% of the Gulf Coast tidal marshes are brackish. The proportion of brackish marshes is well below 10% for other North American coastlines. Unlike most other regions, the lower tidal zones of eastern North America are naturally dominated by grasses of the genus *Spartina*, particularly smooth cordgrass (*Spartina alterniflora*), and not by members of the Chenopodiaceae, such as *Salicornia*. Upper zones of marshes are often a mix of several saltmeadow cordgrass (*Spartina patens*) with big cordgrass (*S. cynosuroides*) in the south and Townsend's cordgrass (*S. townsendii*) in the north, salt grass (*Distichlis spicata*) with marsh elder (*Iva frutescens*), and *Baccharis* shrubs. Brackish marshes along upper estuaries also support black needlerush (*Juncus roemerianus*), cattails (*Typha*), bulrushes (*Schoenoplectus*), and other rush species (*Juncus*). along with a number of forbs. East Coast marshes have been altered by changes in hydrology and nutrient influx and the overall result is the increased dominance of high marsh by the common reed (*Phragmites australis*) and, in low to mid-marsh, the spread of smooth cordgrass at the expense of a diversity of other high-marsh grasses and forbs (Bertness et al. 2002).

Along the southern Atlantic and Gulf coasts, the lower-marsh zone is dominated by smooth cordgrass which occurs patchily through the Caribbean and along the tropical South American coast to Argentina, where it, once again, dominates low, saline marshes. The upper, more brackish zones are dominated by black needlerush. Some of the largest concentrations of tidal marsh in the world are found along the mid- to south Atlantic and Gulf Coasts of North America.

NORTH AMERICAN WEST COAST

Tidal marshes of the West Coast of the US are limited in extent (ca. 440 km^2) of which ca. 70% are associated with the San Francisco Bay. Other small pockets are associated with a few other major estuaries such as Willapa Bay, Puget Sound, and river mouths (Field et al. 1991). The San Francisco Bay and delta estuary encompass <7% (4,140 km^2) of the land area in California but drain more than 40% (155,400 km^2) of its surface (Nichols et al. 1986). A band of *Salicornia* with a narrow outer zone of California cordgrass (*Spartina foliosa*) dominates the lower tidal zones, with a diversity of other plant species found in the upper zones (*Grindelia, Atriplex,* and *Baccharis*, among others). Brackish marshes, such as those of Suisun Bay and the lower Sacramento delta, are dominated by species of rush (*Schoenoplectus* and *Bolboschoenus* spp. and *Juncus* spp.). This represents one of the largest remaining areas of habitat for tidal-marsh vertebrates, yet the tidal marshes have been dramatically altered since the middle of the 19th century. Although efforts to restore ecological functions are underway, numerous threats to both endemic and widespread marsh organisms are still present, including habitat loss (Takekawa et al., chapter 11, *this volume*). Furthermore, invasive *Spartina* species (common cordgrass [*Spartina alterniflora*] and dense-flowered cordgrass [*Spartina densiflora*]) are encroaching on tidal flats and lower marsh edge (Gutensbergen and Nordby, *this volume*).

Tiny pockets of coastal marsh dominated by California cordgrass (with ample tidal flushing) and *Salicornia* spp. can be found along the southern and Baja California coasts (Field et al. 1991, Baja California Wetland Inventory 2004) and along the Mexican mainland from the mouth of the Colorado River south through Sinaloa. The isolated and localized marshes of southern California have suffered from urban development (Zedler et al. 2001; A. Powell, *this volume*), fragmentation, and changes in hydrology, thereby favoring monocultural patches of pickleweed.

BRITAIN AND NORTHERN EUROPE

Tidal marshes in Europe are found around estuaries in the British Isles as well as along the coasts of the Waddell and Baltic Seas. The area of European coastal marshes is ca. 450 km^2 for the British Isles and 950 km^2 for the rest of western Europe (Dijkema 1990). These are high estimates as the areas of adjacent water bodies are included. Additional pockets of coastal marsh are found along the Mediterranean Sea and Persian Gulf (Chapman 1977). Large marshes (here defined as >5 km^2) are relatively uncommon, ca. 25 in Great Britain and 50 along the European mainland. Marshes in northern Europe respond to a sharp salinity gradient, from the coasts of the eastern Atlantic with salinity levels equivalent to full sea water to the barely brackish (0–5 ppt) marshes of the Baltic Sea.

European marshes have been diked, grazed, and harvested for hay since their most recent post-glacial development began (Hazelden and Boorman 2001). It has been estimated that 70% of the remaining European saltmarsh are thus exploited (Dijkema 1990). The original extent of tidal marsh may have been on the order of 1,000 km^2 for Great Britain and 3,000 km^2 along the northern European mainland.

European marshes show patterns of zonation that differ from other regions. Low areas of marsh are often devoid of vegetation, unlike North American marshes (Lefeuvre and Dame 1994). Mid-marsh zones are often now dominated by *Spartina* which has spread considerably, particularly along the British Isles, with the advent of Townsend's cordgrass and the hybrid common cordgrass or in more sandy areas by *Salicornia*, with upper zones covered with *Puccinellia, Juncus, Schoenoplectus, Carex,* and *Festuca* with patches of *Phragmites* reeds along the upper edges. Because of variation in soil type, tidal action, salinity, and long histories of human use, European marshes show considerable geographic variation in composition and zonation. For example, marshes in Scandinavia are dominated by grasses (*Puccinellia maritima, Festuca rubra,* and *Agrostris stolonifera*); they are referred to as salt meadows and have been heavily grazed. By contrast, Mediterranean marshes have a greater representation by *Salicornia, Salsola,* and *Suaeda* spp. (Chapman 1977) and are rarely grazed.

EASTERN SOUTH AMERICA

Salt and brackish coastal marshes are largely restricted to the large shoreline of protected lagoons in Rio Grande do Sul, Brazil, and along estuaries in the Rio de la Plata of Argentina and Uruguay. Although we have an incomplete assessment of the quantity of such coastal marsh, recent estimates based on remote sensing and geographic information services suggest that Argentina alone has >2,000 km^2 (Isacch et al. 2006). The area of saltmarsh in Uruguay and Brazil is considerably smaller than this, perhaps on the order of 250 km^2 of which 150 km^2 is found at subtropical latitudes (C. S. B. Costa, unpubl. data). Floristically, these marshes resemble the marshes of southeastern US in their domination by smooth cordgrass in low marsh zones and a mixture of dense-flowered cordgrass, saltmeadow cordgrass, and *Distichlis* in higher zones. *Juncus* and *Bolboschoenus* dominate the interior zones with low salinity. The marshes along protected lagoons experience little tidal flux and their flooding is the result of less predictable wind (Costa et al. 2003). One of the largest South American tidal marshes is located in Bahía Samborombón Argentina. Extending from Punta Piedras to the northern point of Cabo San Antonio, Punta Rosa, it is included in the Depresión Del Salado region, which occupies practically the entire east-central portion of Buenos Aires Province. It encompasses about 150 km of coastline. Many channels and creeks as well as two main rivers—the Salado and the Samborombón—feed into this coastal area. The coastline of Bahía Samborombón has not yet felt the impact by human activity in the nearby pampas and grassland areas, because humans deem this area as suboptimal for cattle and other agricultural activities. Consequently, the tidal marsh areas have become refuges for many species that used to inhabit the large pampas grasslands and that are now not available for wildlife. In Bahía de Samborombón, salt grass, dense-flowered cordgrass, and *Sarcoconia ambigua* are the dominant plant species, with *Sporobolus indicus, Puccinellia glaucescens, Sida leprosa, Lepidium parodii, Spergularia villosa, Sisyrinchium platense* and *Paspalum vaginatum* as sub-dominant. This community also has large patches of pickleweed mixed with salt grass. At the mouths of rivers in flooded areas, smooth cordgrass is the dominant species and in partially flooded areas, smooth cordgrass is the common species.

Australia

Saltmarshes can be found locally along all Australian coastlines. The estimate for coastal saltmarsh area is 6,020 km^2 (N. Montgomery, pers. comm.), with the greatest extent of Australian saltmarshes found along the tropical coast in an adjacent zone to mangrove vegetation (Adam 1990). For example, only 660 km^2 (11%) of coastal marsh is found in New South Wales, Victoria, Tasmania, and South Australia combined (N. Montgomery, pers. comm.). Australia is unique among the continents in having the vast majority of its saltmarshes along tropical coastlines. Tropical saltmarshes are high level flats above mangroves—often hypersaline and with large salt flats and sparse vegetation (P. Adam, pers. comm.). In more favorable areas, flats are covered with grasses, such as *Sporobolus virginicus*. In general, tropical saltmarshes are restricted to sites that are too saline or are otherwise inhospitable for mangrove vegetation. Overall, Australia is sufficiently arid that streams are small and estuaries are few. In addition, much of the coastline of south Australia experiences small tidal ranges and much of the coastal marsh is found around coastal lagoons.

Mediterranean areas support sparse vegetation dominated by chenopod shrubs in the marshes along warm-temperate coastlines with more consistently moist conditions are limited in extent, but resemble marshes from other warm-temperate regions. Low marshes are characterized by *Sarcocornia quinquefolia* (Chenopodiacea) and the upper marshes by sedges and tall rushes. As in other regions, alien species are colonizing marshes with common cordgrass being the most invasive but thus far restricted primarily to Tasmania and Victoria (Adam 1990).

Asia

Tidal saltmarsh is found primarily in southwest Japan, along the Korean Peninsula and the shores of the Yellow Sea and shows a restricted distribution. We have not been able to obtain much quantitative information, but although Korea has extensive tidal flats, the total amount of saltmarsh may be <100 km^2, mostly dominated by flats of *Suaeda japonica* backed by *Phragmites* beds (N. Moore, pers. comm.). Despite a 50% decline in marsh area due to rice and salt production since 1950, the China coast still supports substantial areas of *Sueda*-dominated marsh, mostly along the yellow-sea. (Dachang 1996). Estimates for coverage are as high as 22,000 km^2. Native salt meadow is being invaded by *Spartina* from Europe and North America with there being now about 1,100 km^2 of this new habitat (Shuqing 2003). Lower marshes are locally dominated by *Suaeda, Zoysia,* and *Salicornia,* and upper marshes have a diversity of reeds (*Phragmites, Scirpus*) and shrubs (e.g., *Artemisia capillars*).

Africa

Most temperate saltmarsh in Africa is concentrated along the South African coast. This coast is generally exposed and only 18% of the 250 estuaries are permanently open to the sea (Colloty et al. 2000). The remaining estuaries are intermittently closed off as a result of reduced freshwater inflow and the development of a sandbar at the mouth. True inter-tidal saltmarshes, found in the permanently open estuaries comprise about 27 km^2 and tidal marsh in the closed (supra-tidal) systems account for 51 km^2. The total area in saltmarsh is, therefore, only 71 km^2. The inter-tidal marshes are dominated by small cordgrass *Spartina maritima, Sarcocornia perennis, Salicornia meyeriana, Triglochin bulbosa, Triglochin striata, Chenolea diffusa,* and *Limonium linifolium*. The dominant supratidal species are *Sporobolus virginicus, Sarcocornia pillansii,* and *Stenotaphrum secundatum* (Colloty et al. 2000).

METHODS FOR THE ANALYSIS OF DIVERSITY AND ENDEMISM

General Approach

The focus of the paper is terrestrial vertebrates. These include species that depend upon the marsh vegetation and the underlying substrate for a substantial part of their use of the tidal marsh habitats. Thus we exclude primarily aquatic species that may enter or feed in the channels and lagoons within tidal marshes, including fish, fish-eating and other birds feeding in aquatic habitats, mammals, and turtles. We also exclude strictly aerial-feeding taxa such as swallows, swifts, and bats. The line between aquatic and terrestrial species sometimes requires a subjective judgment. Among birds, for example, we have generally excluded herons, egrets, ibises, and other wading species, as well as waterfowl and most gulls and terns. We include songbirds, rails, and some species of shorebirds, such as Willets (*Catoptrophorus semipalmatus*), which have a substantial dependence upon the marsh vegetation itself for something more than roosting. The inclusion or exclusion of resident shorebirds from the list of tidal-marsh species is probably the most problematic of any group. When we turn to analysis we have restricted most to species that breed in tidal marshes. Among mammals, we exclude seals, sea otters, and cetaceans which feed in tidal channels and may rest upon tidal mud bars, but are primarily aquatic species.

Because this paper represents the first systematic compendium and classification of terrestrial vertebrates in tidal marshes, we used a diversity of sources as background information for our ecological classification. We relied primarily on the natural history literature for the major groups and supplemented this with information of knowledgeable informants and the gray literature as well. We were unable to obtain information from some areas, such as New Zealand, and the Black Sea, and therefore do not include faunas from such areas with small amounts of saltmarsh.

Species were classified into four groups, based on their demonstrable evolutionary specialization to tidal marsh habitats (Fig. 1): 1 = species that are largely or wholly restricted to tidal marshes; 2 = species that have recognized subspecies that are largely or wholly restricted to tidal marshes; 3 = species that have populations that are largely or wholly restricted to tidal marshes (these populations are not known to be differentiated); and 4 = species that occur in tidal marshes and other habitats as well. For species in the last two groups, we classified the dominant alternate habitats to tidal marsh: FM = fresh-water marsh and other fresh-water aquatic habitats, grass = grassland, agricultural fields and pastures and other open habitats, scrub = shrubby second-growth habitats, or varied = other non-tidal habitats. We designate which tidal-marsh species are known to regularly occur in salt as opposed to brackish marsh. This categorization is based largely upon the designation in the literature and not any controlled measurement or consistent criteria. However, we consider classifying species this way as providing a first approximation of which species occur in the most saline marshes. The following sections provide our major sources for the different higher taxa of vertebrates considered.

AMPHIBIANS AND REPTILES

Nomenclature is based on (Uetz et al. 2005) and Frost (2004). We developed our habitat classification for amphibians and reptiles from several major natural history sources (Stebbins 1954, Conant 1969, Dunson and Mazzotti 1989, Ernst and Barbour 1989, Conant et al. 1998) as well as a lengthy review paper on the distribution of amphibians and reptiles in saline habitats throughout the world (Neill 1958). The review is the most comprehensive on this topic to date and is global in scope, but clearly focused more on North American and Europe than other saltmarsh regions although it has considerable information on mangrove swamp herpetofaunas. We therefore restrict our analysis for these classes to North America and the western Palearctic. In terms of a global assessment, the information on turtles is probably the most comprehensive.

BIRDS

Bird nomenclature is based on Sibley and Monroe (1990). Information on North American avifauna was obtained from the Birds of North America series (Poole 2006) and the field notes of the senior author, as well as more specialized accounts (Gill 1973, Benoit and Askins 1999). Habitat information on rails was based largely on Taylor (1998). Australian bird distributions were obtained from Higgins (1999), Higgins et al. (2001), and personal communications with several Australian ornithologists. Information on European and British birds was obtained from communications with John Marchant and Phil Atkinson, and found in Williamson (1967), Glue (1971) Greenhaugh (1971), Møller (1975), Larssen (1976), Spaans (1994), and more specialized references on particular taxa (Taillandier 1993, Allano et al. 1994). Information on Asian birds was based on Brazil (1991) as well as personal communications from Hisashi Nijati and Nick Moores. Information on the birds of South America was based on Wetmore (1926), Ridgely and Tudor (1989, 1994), Stotz et al. (1996), Martinez et al. (1997), Dias and Maurío (1998), Isacch et al. (2004) and communications from several South American ornithologists (Rafael Dias, Pablo Petracci, Santiago Claramunt, and Rosendo Fraga). Classification of birds of South Africa estuarine marshes was based on Hockey and Turpie (1999). Subspecific designations were based on Cramp (1988, 1992) and Cramp and Perrins (1993, 1994), American Ornithologists' Union (1957), Hellmayr (1932, 1938), Cory and Hellmayr (1927), and Hellmayr and Conover (1942), and Higgins (1999) and Higgins et al. (2001).

MAMMALS

Information on North American mammals was obtained by surveying reference materials containing range maps and varying details of habitat information, primarily Hall (1981), Wilson and Ruff (1999) and the mammalian species accounts of the American Society of

Mammalogy (<http://www.science.smith.edu/departments/Biology/VHAYSSEN/msi/> [26 July 2006]). We also surveyed books dedicated to mammals of states where coastal tidal marshes were present: Álvarez-Castañeda and Patton (1999), Ingles (1967), Linzey (1998), Lowery (1974), Webster et al. (1985), and Williams (1979, 1986).

Published information for mammals that live in tidal-marsh ecosystems in South America is sparse. For this reason, we surveyed general field guides on Neotropical mammals (Emmons 1990) and books that had general distribution and habitat information for Neotropical mammals (Redford and Eisenberg 1992, Eisenberg and Redford 1999). Information was also obtained from field researchers with knowledge of the marsh ecosystems of the area (M. L. Merino and S. Gonzalez) and from reports and other unpublished literature (Milovich et al.1992, Merino et al. 1993, Yorio 1998, Bó et al. 2002). We also surveyed over 700 mammalian species accounts (<http://www.science.smith.edu/departments/Biology/VHAYSSEN/msi/> [26 July 2006]) and searched for information regarding distribution and habitat use for mammals reported in or around the major marsh ecosystems of Argentina, Uruguay, and Brazil.

Information on European mammals was obtained from general reference materials such as Nowak (1999) as well as references that dealt exclusively with European and British mammals (Bjärvall and Ulström 1986, Mitchell-Jones et al. 1999). We also received anecdotal information from researchers that had experience surveying mammals in European marshes such as M. Delibes, J. Flowerdew, A. Grogan, R. Strachan, R. Trout, and D. W. Yalden.

Compared to North America and Europe, little is known about tidal-marsh mammals in Asia, Africa, and Australia. Therefore we do not include information in this review from these areas.

RESULTS

Tidal-Marsh Faunas

Amphibians and reptiles

We were able to locate references to 43 species of amphibians and reptiles regularly found in tidal marshes in North America (Table 1). However, the only saltmarsh taxa characteristic of saltmarshes outside of North America we were able to document were two species of saltmarsh inhabiting skinks (*Egernia*) in Australia (Chapple 2003). Within North America, 37 of the 41 North American species are restricted to the Atlantic or Gulf coasts. In his review, Neill (1958) emphasized that Florida was a particularly important area for finding salt-marsh populations of reptiles and amphibians. Despite the large number of species that have been found, at least locally, to inhabit tidal marshes, only colubrid snakes and emydid turtles have species or subspecies that were restricted to tidal marsh and adjacent estuarine habitats. Of the 13 species of snakes that have been reported from tidal marshes, three have subspecies restricted to tidal marshes or mangroves—northern water snake (*Nerodia sipedon*) and saltmarsh snake (*N. clarkia* [Myers 1988, Lawson et al. 1991, Gaul 1996]) and brown snake (*Storeria dekayi* [Anderson 1961]); the subspecies involved are found in the southeastern US. Similarly, one of the ten species of turtles that have been found in tidal marshes is restricted to estuarine habitats, diamondback terrapin (*Malaclemys terrapin*). American crocodiles (*Crocodylus acutus*) are generally restricted to brackish or salt-water estuarine habitats, but predominantly occupy subtropical to tropical mangrove swamps.

Birds

The number of breeding tidal marsh birds varied between eight and 21 for the different continents with the highest number of species in the North and South American tidal marshes (Tables 2–6; Fig. 2a). However, the number of species found in saltmarsh was highest for North America (11), with other continents ranging between six–eight species. In the case of South Africa, the data were not available to categorize species as brackish or salt marsh and this region is not included in analyses for which this distinction is involved. The number of endemic species (two) and the number of species with at least one endemic subspecies (11) was highest for North America. In fact, we found only one endemic species or species with endemic subspecies outside of North America involving a single subspecies of the Slender-billed Thornbird (*Acanthiza iredalei rosinae*; Mathew 1994). Research is underway to determine if two species of South American saltmarshes (Bay-capped Wren-spinetail [*Spartonoica maluroides*] and the Dot-winged crake [*Porzana spiloptera*]) might have locally differentiated populations. The Zitting Cisticola (*Cisticola juncidis*) is found in, but not restricted to, tidal marshes in southern Europe, Africa, Asia, and Australia and should be examined, as well, for local differentiation. In North America, endemism is somewhat greater along the East Coast where two endemic species and six species with at least one endemic

TABLE 1. Species of tidal-marsh amphibians and reptiles of North America.

Species	Family	Class[a]	Coast[b]	Saltmarsh[c]	Alternate habitat[d]
Southern chorus frog (*Pseudacris nigrita*)	Hylidae	4	ENA	+	Varied
Spotted chorus frog (*Pseudacris clarkii*)	Hylidae	4	ENA	+	Grass
Little glass frog (*Pseudacris ocularis*)	Hylidae	4	ENA	+	FM
Green tree frog (*Hyla cinerea*)	Hylidae	4	ENA		FM
Gray tree frog (*Hyla versicolor*)	Hylidae	4	ENA		FM
Pine woods tree frog (*Hyla femoralis*)	Hylidae	4	ENA		Varied
Eastern narrow-mouthed toad (*Gastrophryne carolinensis*)	Microhylidae	4	ENA		Varied
Southern leopard frog (*Rana sphenocephala*)	Ranidae	4	ENA		FM
Pickerel frog (*Rana palustris*)	Ranidae	4	ENA		FM
Pig frog (*Rana grylio*)	Ranidae	4	ENA		FM
Common snapping turtle (*Chelydra serpentine*)	Chelydridae	4	ENA	+	FM
Spotted turtle (*Clemmys guttata*)	Emydidae	4	ENA		FM
Pacific pond turtle (*Emys marmorata*)	Emydidae	4	WNA		FM
Painted turtle (*Chrysemys picta*)	Emydidae	4	ENA		FM
Florida cooter (*Pseudemys concinna*)	Emydidae	4	ENA		FM
Florida redbelly turtle (*Pseudemys nelsoni*)	Emydidae	4	ENA		FM
Striped mud turtle (*Kinosternon baurii*)	Kinosternidae	4	ENA		FM
Eastern mud turtle (*Kinosternon subrubrum*)	Kinosternidae	4	ENA	+	FM
Green anole (*Anolis carolinensis*)	Iguanidae	4	ENA		Varied
Western mourning skink (*Egernia luctuosa*)	Scincidae	4	AUS	+	FM
Swamp skink (*Egernia coventryi*)	Scincidae	4	AUS	+	FM
Slender glass lizard (*Ophisaurus attenuatus*)	Anguidae	4	ENA	+	Grass
Saltmarsh snake (*Nerodia clarkia*)	Colubridae	1	ENA	+	
Northern water snake (*Nerodia sipedon*)	Colubridae	2	ENA	+	FM
West Mexican water snake (*Nerodia valida*)	Colubridae	4	WNA		FM
Mississippi green water snake (*Nerodia cyclopion*)	Colubridae	4	ENA		FM
Graham's crayfish snake (*Regina grahami*)	Colubridae	4	ENA		FM
Black swamp snake (*Seminatrix pygeae*)	Colubridae	4	ENA		FM
Common garter snake (*Thamnophis sirtalis*)	Colubridae	4	WNA		FM
Ring-necked snake (*Diadophis punctatus*)	Colubridae	4	WNA		Varied
Mud snake (*Farancia abacura*)	Colubridae	3	ENA	+	FM
Eastern indigo snake (*Drymarchon corais*)	Colubridae	4	ENA		Varied
Eastern racer (*Coluber constrictor*)	Colubridae	4	ENA		Varied
Rough green snake (*Opheodrys aestivus*)	Colubridae	4	ENA		FM
Eastern rat snake (*Elaphe obsoleta*)	Colubridae	4	ENA		Varied
Common king snake (*Lampropeltis getula*)	Colubridae	4	ENA		Varied
Eastern diamond-backed rattlesnake (*Crotalus adamanteus*)	Viperidae	4	ENA		Varied
Timber rattlesnake (*Crotalus horridus*)	Viperidae	4	ENA	+	FM
American crocodile (*Crocodylus acutus*)	Crocodylidae	3	ENA	+	FM
American alligator (*Alligator mississippiensis*)	Crocodylidae	4	ENA	+	FM

[a] Class: 1 = endemic species; 2 = species with endemic subspecies; 3 = species with population locally restricted to tidal marsh; 4 = generalist.
[b] Coast: W = west (Pacific) coast; E = east (Atlantic or Gulf) coast.
[c] + = regularly found in salt marsh (salinity ≈ sea water).
[d] Alternate habitat (for generalists) FM = freshwater marsh or other aquatic habitat; grass = grasslands or fields; scrub = scrub habitats; varied = different habitats.

subspecies can be found, compared to the West Coast where no endemic species occur and five species have at least one endemic subspecies. Along the Gulf Coast, we found only one endemic species and four species with at least one endemic subspecies.

We focus our analysis on breeding birds, recognizing that a number of birds will appear in tidal marshes opportunistically during migration or winter periods. However, some species are relatively specialized on tidal marshes during the non-breeding season. For example, the endangered Orange-bellied Parrot (*Neophema chrysogaster*) over-winters largely in saltmarshes in southern Australia, where it feeds extensively on the seeds of *Halosarcia* and other chenopods (Loyn et al. 1986). In Britain and northern Europe, the

TABLE 2. Species of tidal-marsh birds of North America.

Species	Family	Class[a]	Coast[b]	Saltmarsh[c]	Alternate habitat[d]
Short-eared Owl (*Asio flammeus*)	Strigidae	4	W, E	+	FM, grass
Yellow Rail (*Coturnicops noveboracensis*)	Rallidae	4 NB[e]	E		FM, grass
Black Rail (*Laterallus jamaicensis*)	Rallidae	3[f]	W, E	+	FM, grass
Clapper Rail (*Rallus longirostris*)	Rallidae	2	W, E	+	FM
King Rail (*Rallus elegans*)	Rallidae	4 NB[e]	E		FM
Virginia Rail (*Rallus limicola*)	Rallidae	4 NB[e]	W, E		FM
Willet (*Catoptrophorus semipalmatus*)	Scolopacidae	2	E	+	FM, grass
Northern Harrier (*Circus cyaneus*)	Accipitridae	4	W, E	+	FM, grass
Willow Flycatcher (*Empidonax traillii*)	Tyrannidae	4	E		FM, scrub
Song Sparrow (*Melospiza melodia*)	Fringillidae	2	W	+	FM, scrub
Swamp Sparrow (*Melospiza georgiana*)	Fringillidae	2	E		FM
Savannah Sparrow (*Passerculus sandwichensis*)	Fringillidae	2	W	+	Grass
Seaside Sparrow (*Ammodramus maritimus*)	Fringillidae	1	E	+	FM
Saltmarsh Sharp-tailed Sparrow (*Ammodramus caudacutus*)	Fringillidae	1	E	+	
Nelson's Sharp-tailed Sparrow (*Ammodramus nelsoni*)	Fringillidae	2	E	+	FM
Yellow Warbler (*Dendroica petechia*)	Fringillidae	4	E		FM, scrub
Common Yellowthroat (*Geothlypis trichas*)	Fringillidae	2	W, E	+ NB[e]	FM, scrub
Red-winged Blackbird (*Agelaius phoeniceus*)	Fringillidae	4	W, E	+	FM, varied
Boat-tailed Grackle (*Quiscalus major*)	Fringillidae	2	E	+	FM, varied

[a] Class: 1 = endemic species; 2 = species with endemic subspecies; 3 = species with population locally restricted to tidal marsh; 4 = generalist.
[b] Coast: W = west (Pacific) coast; E = east (Atlantic or Gulf) coast.
[c] + = regularly found in salt marsh (salinity ≈ sea water).
[d] Alternate habitat (for generalists) FM = freshwater marsh or other aquatic habitat; grass = grasslands or fields; scrub = scrub habitats; varied = different habitats.
[e] NB = found primarily in non-breeding season.
[f] Black Rails might be considered tidal-marsh endemics, particularly the eastern subspecies, where inland breeding populations are very sporadic and small (Eddleman et al. 1994).

TABLE 3. Species of tidal-marsh birds of Europe and Asia.

Species	Family	Class[a]	Saltmarsh[b]	Alternate habitat[c]
Short-eared Owl (*Asio flammeus*)			+	FM, grass
Water Rail (*Rallus aquaticus*)	Rallidae	4		FM
Redshank (*Tringa totanus*)	Scolopacidae	4	+	FM
Marsh Harrier (*Circus aeruginosus*)	Accipitridae	4		FM, grass
Bluethroat (*Luscinia svecica*)	Muscicapidae	4		Varied
Eurasian Penduline-Tit (*Remiz pendulinus*)	Paridae	4		FM
Zitting Cisticola (*Cisticola juncidis*)	Sylviidae	4		FM, grass
Middendorff's Grasshopper Warbler (*Locustella ochotensis*)	Sylviidae	4		FM
Japanese Marsh Warbler (*Locustella pryeri*)	Sylviidae	4		FM
Sedge Warbler (*Acrocephalus schoenobaenus*)	Sylviidae	4		FM
Eurasian Reed-Warbler (*Acrocephalus scirpaceus*)	Sylviidae	4		FM
Clamorous Reed-Warbler (*Acrocephalus stentoreus*)	Sylviidae	4		FM
Great Reed-Warbler (*Acrocephalus arundinaceus*)	Sylviidae	4		FM
Bearded Parrotbill (*Panurus biarmicus*)	Sylviidae	4		FM
Reed Parrotbill (*Panurus heuderi*)	Sylviidae	4		FM
Eurasian Skylark (*Alauda arvensis*)	Alaudidae	4	+	Grass
Horned Lark (*Eremophila alpestris*)	Alaudidae	4		Grass
Yellow Wagtail (*Motacilla flava*)	Motacillidae	4	+	Grass
Twite (*Carduelis flavirostris*)	Fringillidae	4 NB[d]		FM, grass
Pallas's Bunting (*Emberiza pallasi*)	Fringillidae	4		FM
Reed Bunting (*Emberiza schoeniclus*)	Fringillidae	4		FM
Lapland Longspur (*Calcarius lapponicus*)	Fringillidae	4 NB[d]	+	
Snow Bunting (*Plectrophenax nivalis*)	Fringillidae	4 NB[d]	+	

[a] Class: 1 = endemic species; 2 = species with endemic subspecies; 3 = species with population locally restricted to tidal marsh; 4 = generalist.
[b] + = regularly found in salt marsh (salinity ≈ sea water).
[c] Alternate habitat (for generalists) FM = freshwater marsh or other aquatic habitat; grass = grasslands or fields; scrub = scrub habitats; varied = different habitats.
[d] NB = found primarily in non-breeding season.

TABLE 4. Species of birds of eastern South American tidal marsh.

Species	Family	Class[a]	Saltmarsh[b]	Alternate habitat[c]
Short-eared Owl (*Asio flammeus*)	Strigidae	4		Grass
Speckled Rail (*Coturnicops notatus*)	Rallidae	4		FM
Rufous-sided Crake (*Laterallus melanophaius*)	Rallidae	4		FM
Dot-winged Crake (*Porzana spiloptera*)	Rallidae	3	+	FM
Blackish Rail (*Pardirallus nigricans*)	Rallidae	4		FM
Plumbeous Rail (*Pardirallus sanguinolentus*)	Rallidae	4	+	FM
South American Painted-Snipe (*Nycticryphes semicollaris*)	Rostratulidae	4	+	FM
Long-winged Harrier (*Circus buffoni*)	Accipitridae	4		FM
Cinereous Harrier (*Circus cinereus*)	Accipitridae	4	+	FM
Spectacled Tyrant (*Hymenops perspicillatus*)	Tyrannidae	4		Grass
Sulphur-bearded Spintail (*Cranioleuca sulphurifera*)	Furnariidae	4		FM
Yellow-chinned Spintail (*Certhiaxis cinnamomea*)	Furnariidae	4		FM, mangroves
Hudson's Canastero (*Asthenes hudsoni*)	Furnariidae	4	+	FM, grass
Freckle-bearded Thornbird (*Phacellodomus striaticollis*)	Furnariidae	4		Scrub
Bay-capped Wren-Spintail (*Spartonoica maluroides*)	Furnariidae	3	+	FM
Wren-like Rushbird (*Phleopryptes melanops*)	Furnariidae	4		FM
Sedge Wren (*Cistothorus platensis*)	Trogolodytidae	4	+	FM, grass
Correndera's Pipit (*Anthus correndera*)	Motacillidae	4		Grass
Grassland Yellow Finch (*Sicalis luteola*)	Fringillidae	4		Grass
Great Pampa-Finch (*Embernagra platensis*)	Fringillidae	4		Grass
Yellow-winged Blackbird (*Agelaius thilius*)	Fringillidae	4	+	FM

[a] Class: 1 = endemic species; 2 = species with endemic subspecies; 3 = species with population locally restricted to tidal marsh; 4 = generalist.
[b] + = regularly found in salt marsh (salinity ≈ sea water).
[c] Alternate habitat (for generalists) FM = freshwater marsh or other aquatic habitat; grass = grasslands or fields; scrub = scrub habitats; varied = different habitats.

TABLE 5. Bird species of Australian tidal marshes.

Species	Family	Class[a]	Coast[b]	Saltmarsh[c]	Alternate habitat[d]
Grass Owl (*Tyto capensis*)	Tytonidae	4	N, E		Grass
Buff-banded Rail (*Gallirallus phillipensis*)	Rallidae	4	All		FM
Lewin's Rail (*Lewinia pectoralis*)	Rallidae	4	S, E		FM
Australian Crake (*Porzana fluminea*)	Rallidae	4	S, E	+	FM
Blue-winged Parrot (*Neophema chrysostoma*)	Psittacidae	4 NB			Scrub
Rock Parrot (*Neophema petrophila*)	Psittacidae	4 NB			Scrub
Orange-bellied Parrot (*Neophema chrysogaster*)	Psittacidae	4 NB			Scrub
Swamp Harrier (*Circus approximan*)	Accipitridae	4	All		FM, grass
White-winged Fairywren (*Melurus leucopterus*)	Maluridae		S, W	+	Scrub
Red-backed Fairywren (*Malurus melanocephalus*)	Maluridae		N, E		Scrub
Singing Honeycreeper (*Lichenostomus virescens*)	Meliphagidae	4	N, E, S		Scrub
Orange Chat (*Epithianura aurifrons*)	Meliphagidae		W	+	Scrub
White-fronted Chat (*Epthianura albifrons*)	Meliphagidae	4	S	+	Scrub
Slender-billed Thornbill (*Acanthiza iredalei*)	Pardalotidae	2	S	+	Scrub
Zitting Cisticola (*Cisticola juncidis*)	Sylviidae	4	N, W		Grass
Brown Songlark (*Cincloramphus cruralis*)	Alaudidae	4	E, W, S		Grass
Australasian Pipit (*Anthus novaseseelandiae*)	Passeridae	4	All		Grass

[a] Class: 1 = endemic species; 2 = species with endemic subspecies; 3 = species with population locally restricted to tidal marsh; 4 = generalist.
[b] Coast: N = north; S = south; W = west (Pacific) coast; E = east (Atlantic) coast.
[c] + = regularly found in salt marsh (salinity ≈ sea water).
[d] Alternate habitat (for generalists) FM = freshwater marsh or other aquatic habitat; grass = grasslands or fields; scrub = scrub habitats; varied = different habitats.

Twite (*Acanthis flavirostris*), Snow Bunting (*Plectrophenax nivalis*), Lapland Longspur (*Calcarius lapponicus*), and Meadow Pipit (*Anthus pratensis*) are particularly noted for their dependence on saltmarshes in the winter; the Twite and other finches feed on the seeds of *Salicornia* (Brown and Atkinson 1996; J. Marchant, pers. comm.).

The endemic bird taxa include two species and the subspecies of four other species of sparrows and they all occur in North America. The endemic sparrow taxa are generally grayer with more

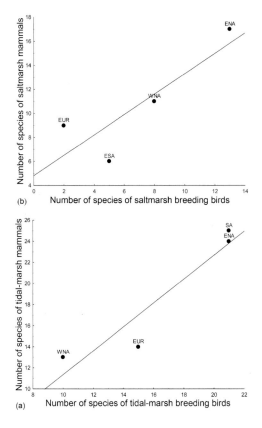

FIGURE 2. The number of species of native mammals plotted against the number of breeding birds for a single coastline associated with (a) tidal marshes and (b) salt marshes.

distinct black markings than the upland subspecies, and are also larger with relatively larger bills (Grenier and Greenberg 2005; Grenier and Greenberg, *this volume*). The bird species most restricted to saltmarshes is the Saltmarsh Sharp-tailed Sparrow (*Ammodramus caudacutus*). A closely related species, the Nelson's Sharp-tailed Sparrow (*A. nelsoni*) occupies prairie wetlands, but has two isolated subspecies found in tidal marshes. Both occupy brackish and saltmarshes, as well as nearby freshwater meadows (Peters 1942, Greenlaw and Rising 1994) during the breeding season and saltmarshes in the winter. Seaside Sparrows (*Ammodramus maritima*) are also almost entirely restricted to saltmarsh, particularly low marsh dominated by smooth cordgrass. Two distinct Florida subspecies (*A. m. nigrescens* and *A. m. mirabilis*) are or were found in inland flooded prairie habitats in addition to *Spartina* marshes. Seaside Sparrows are characterized by their generally grayish coloration and a larger bill compared to related congeners.

Two species of *Melospiza* have distinct tidal-marsh subspecies. The Coastal Plain Swamp Sparrow (*M. georgiana nigrescens*) is found in brackish tidal marshes of the mid-Atlantic coast and is distinctly grayer, blacker, and has a larger bill than conspecific inland populations (Greenberg and Droege 1990). Three subspecies of Song Sparrows (*Melospiza melodia*) have been described for the San Francisco Bay estuaries (Marshall 1948b). These subspecies tend to be grayer, or in one case more yellow, than local upland Song Sparrows. The subspecies in the brackish Suisun Bay marshes (*M. m maxillaris*) has particularly black markings and a larger bill. Other endemic subspecies of birds have been described as darker than other populations, including the Salt-Marsh Yellowthroat (*Geothlypis trichas sinuosus*; Grinnell 1913) and the Slender-billed Thornbill (*Acanthiza iradelei rosinae*; Mathew 1994). Atlantic Coast subspecies of Marsh Wrens (*Cistothorus palustris*) have been reported to be grayer than inland subspecies (Kroodsma and Verner 1997). The saltmarsh subspecies of Savannah Sparrows (*Passerculus sandwichensis*) have been included with the Pacific coast *P. s. beldingi* group and the Large-billed Sparrow of the coast of the Gulf of California (*P. s. rostrata*; Rising 2001). The former is relatively grayish with heavy black markings; bill size varies clinally but is relatively slender compared to other Savannah Sparrows. Members of the *rostrata* group have large bills and are grayish with reduced streaking on the back.

Finally, the Clapper Rails (*Rallus longirostris*) tend to be small and gray compared to King Rails (*Rallus elegans*), with the West Coast Clapper Rails having a warmer rust coloration ventrally (more similar to King Rails) (Eddleman and Conway 1998). Clapper Rails are largely restricted to saltmarshes, except tropical populations along the west coast of Mexico, the Caribbean, and the Atlantic Coast of South America, which occupy mangrove swamps. The Black Rail (*Laterallus jamaicensis*) is often considered a salt-marsh-restricted species and occurs primarily in tidal areas, but small populations can be found in inland freshwater marshes (Eddleman et al. 1994). As a primarily saltmarsh bird, it shows the tendency towards melanism described for other taxa in this habitat.

Mammals

The number of native tidal marsh species ranges from 15 (Europe) to 25 (South America) to 35 (North America) for the different continents. When individual coastlines are considered, the east coast of South America has

TABLE 6. BIRD SPECIES OF SOUTH AFRICAN TIDAL MARSHES.

Species	Family	Class[a]	Alternate habitat[b]
Marsh Owl (*Asio capensis*)	Strigidae	4	Grass
Kaffir Rail (*Rallus caerulescens*)	Rallidae	4	FM
Red-chested Flufftail (*Sarothrura rufa*)	Rallidae	4	FM
Red-knobbed Coot (*Fulica cristata*)	Rallidae	4	FM
Black-winged Stilt (*Himantopus himantopus*)	Charadriidae	4	FM
Water Thick-knee (*Burhinus vermiculatus*)	Burhinidae	4	Grass
Yellow-billed Kite (*Milvus migrans*)	Accipitridae	4	Varied
African Bush-Warbler (*Bradypterus boboecala*)	Sylviidae	4	FM
Zitting Cisticola (*Cisticola juncidis*)	Sylviidae	4	Grass
Tinkling Cisticola (*Cisticola tinniens*)	Sylviidae	4	FM
Cape Wagtail (*Motacilla capensis*)	Motacillidae	4	Varied
Cape Longclaw (*Macronyx capensis*)	Motacillidae	4	FM

[a] Class: 1 = endemic species; 2 = species with endemic subspecies; 3 = species with population locally restricted to tidal marsh; 4 = generalist.
[b] Alternate habitat (for generalists) FM = freshwater marsh or other aquatic habitat; grass = grasslands or fields; scrub = scrub habitats; varied = different habitats.

the highest species richness (25) followed by eastern North America (24) with lower values for Europe and western North America (15 and 14, respectively). In contrast, eastern North America has substantially more salt marsh species (excluding brackish marsh taxa) with 17; western North America has 10, Europe nine, and South America six. Endemic taxa are restricted to North America with the only entirely endemic species, the salt marsh harvest mouse (*Reithrodontomys raviventris*) occurring along the West Coast, and the West and East coasts each supporting four species that have at least one endemic subspecies.

North America has the most species of tidal-marsh mammals, yet these species comprise a small portion of the total fauna. Of the 90 species that have been reported in various habitats in coastal states from the eastern US and 167 from the Pacific coastal states, only 26% and 8%, respectively, are commonly present in brackish and/or saltwater tidal marshes (Table 7). These statistics apply to native species. In addition, exotic species such as the Virginia opossum (*Didelphis virginiana*; introduced to the West Coast), house mouse (*Mus musculus*), black rat (*Rattus rattus*), brown rat (*Rattus norvegicus*), house cat (*Felis catus*), and red fox (*Vulpes vulpes*; introduced in coastal southern California) are commonly found in tidal marshes. As far as we know, only in mammals do exotic species commonly inhabit tidal marshes and then only in North America. These species display a strong tendency to replace native species in tidal-marsh areas that have been altered and surrounded by human development (Takekawa et al. chapter 11, *this volume*). This case is most dramatic in disturbed tidal marshes in southern and central California where house mice and black rats have become the dominant small mammal species and harvest mice, meadow voles (*Microtus pennsylvanicus*) and shrews are either absent or occur in low densities (J. Maldonado, pers. obs.).

Endemism appears to be restricted to both coasts of North America. Despite the large number of tidal-marsh species in South American marshes, only one species, the pampas cat (*Felis colocolo*), is regionally restricted to tidal marshes (Table 8). In Europe we found a similar pattern, namely no endemic mammalian species or subspecies endemic to tidal marshes or species with regional populations restricted to tidal marshes. All species recorded as occurring in tidal marshes were species that inhabit both upland habitats and tidal marshes and their alternative habitat is freshwater marsh or grassland or else they are generalist species that used various habitats (Table 9).

On the Pacific Coast of California, the ornate shrew (*Sorex ornatus*) and vagrant shrew (*S. vagrans*) as well as western harvest mouse (*Reithrodontomys megalotis*) and California vole (*Microtus californicus*) have subspecies endemic to tidal marshes and most of these are restricted to tidal marshes in the San Francisco Bay, the Monterey Bay, and the southern California area (Rudd 1955, Thaeler 1961).

On the East Coast, the cinereus shrew (*Sorex cinereus*) has a subspecies (Tuckahoe masked shrew [*S. c. nigriculus*]) that is probably restricted to the salt-water littoral marshes of southern New Jersey (Green 1932). Two subspecies of the meadow vole, are associated with coastal saltmarshes. The Florida saltmarsh vole (*Microtus pennsylvanicus dukecampbelli*) is currently listed as endangered in the state of Florida (Florida Fish and Wildlife Conservation Commission 2004) and was originally reported

TABLE 7. Species of tidal-marsh mammals of North America.

Species	Family	Class[a]	Coast[b]	Saltmarsh[c]	Alternate habitat[d]
Virginia opossum (*Didelphis virginiana*)[e]	Didelphidae	4	E, W	+	Varied
Cinereus shrew (*Sorex cinereus*)	Soricidae	2	E	+	
Ornate shrew (*Sorex ornatus*)[f]	Soricidae	2	W	+	
Vagrant shrew (*Sorex vagrans*)[f]	Soricidae	2	W	+	
Southeastern shrew (*Sorex longirostris*)	Soricidae	4	E		FM
Marsh shrew (*Sorex bendirii*)	Soricidae	4	W	+	Varied
Northern short-tailed shrew (*Blarina brevicauda*)	Soricidae	4	E	+	FM
Least shrew (*Cryptotis parva*)	Soricidae	4	E	+	FM
Desert shrew (*Notiosorex crawfordi*)	Soricidae	4	W	+	Scrub
Eastern mole (*Scalopus aquaticus*)	Talpidae	4	E		Varied
Broad-footed mole (*Scapanus latimanus*)	Talpidae	4	W		Varied
Swamp rabbit (*Sylvilagus aquaticus*)	Leporidae	2	E	+	
Marsh rabbit (*Sylvilagus palustris*)	Leporidae	3	E	+	FM, grass
Brush rabbit (*Sylvilagus bachmani*)	Leporidae	4	W		Varied
Salt marsh harvest mouse (*Reithrodontomys raviventris*)	Muridae	1	W	+	
Western harvest mouse (*Reithrodontomys megalotis*)	Muridae	2	W	+	
Eastern harvest mouse (*Reithrodontomys humulis*)	Muridae	4	E	+	Varied
California vole (*Microtus californicus*)	Muridae	2	W	+	
Meadow vole (*Microtus pennsylvanicus*)	Muridae	2	E	+	
Long-tailed vole (*Microtus longicaudus*)	Muridae	4	E		FM
Townsend's vole (*Microtus townsendii*)	Muridae	4	E	+	FM, grass
Muskrat (*Ondatra zibethicus*)	Muridae	4	E	+	FM
Marsh rice rat (*Oryzomys palustris*)	Muridae	4	E	+	FM
Round-tailed muskrat (*Neofiber alleni*)	Muridae	4	E	+	Grass
Eastern woodrat (*Neotoma floridana*)	Muridae	4	E	+	Varied
White-footed mouse (*Peromyscus leucopus*)	Muridae	4	E		Varied
House mouse (*Mus musculus*)[e]	Muridae	4	E, W	+	Varied
Black rat (*Rattus rattus*)[e]	Muridae	4	E, W	+	Varied
Brown rat (*Rattus norvegicus*)[e]	Muridae		E, W	+	Varied
Nutria (*Myocastor coypus*)[e]	Muridae	4	E	+	FM, grass
Gray fox (*Urocyon cinereoargenteus*)	Canidae	4	E		Varied
Red fox (*Vulpes vulpes*)[e]	Canidae	4	E, W		Varied
Coyote (*Canis latrans*)	Canidae	4	E, W		Varied
Northern raccoon (*Procyon lotor*)	Procyonidae	4	E, W	+	FM
American mink (*Mustela vison*)	Mustelidae	3	E	+	Varied
Northern river otter (*Lontra canadensis*)	Mustelidae	4	E		Varied
Striped skunk (*Mephitis mephitis*)	Mustelidae	4	E, W	+	Varied
Domestic cat (*Felis catus*)[e]	Felidae	4	E, W	+	Varied
Horse (*Equus caballus*)[d]	Equidae	4	E	+	Varied
White-tailed deer (*Odocoileus virginianus*)	Cervidae	2	E	+	
Elk (*Cervus elaphus*)	Cervidae	4	W	+	Varied
Sika deer (*Cervus nippon*)[e]	Cervidae	4	E	+	Varied[c]

Note: Taxonomic arrangement of mammals in table as in Nowak (1999) and common and scientific names are based on Wilson and Reeder (1993).
[a] Class: 1 = endemic species; 2 = species with endemic subspecies; 3 = species with population locally restricted to tidal marsh; 4 = generalist.
[b] Coast: W = west (Pacific) coast; E = east (Atlantic or Gulf) coast.
[c] + = regularly found in salt marsh (salinity ≈ sea water).
[d] Alternate habitat (for generalists) FM = freshwater marsh or other aquatic habitat; grass = grasslands or fields; scrub = scrub habitats; varied = different habitats.
[e] Non-native species: Virginia opossum and red fox native on the east coast.
[f] Taxonomy based on Hall (1981) but its designation is controversial (Chan et al., *this volume*).

to be restricted to a single population in a saltmarsh near Cedar Key, Florida (Wood et al. 1982). More recently, an additional population was discovered from a location 19 km north of the first population at the Lower Suwannee National Wildlife refuge in Levy County Florida (USDI Fish and Wildlife Service 2004). The other subspecies of meadow vole (*Microtus pennsylvanicus nigrans*) is common and has been reported from the uplands as well as tidal marshes in eastern Virginia and Maryland. Both subspecies have been described as being darker than conspecifics. A third isolated saltmarsh population of meadow vole occurs in the Santee delta of South Carolina, but its taxonomic status has not yet been studied (W. Post, pers. comm.).

TABLE 8. Species of tidal-marsh mammals of eastern South America.

Species	Family	Class[a]	Saltmarsh[b]	Alternate habitat[c]
Lutrine opossum (*Lutreolina crassicaudata*)	Didelphidae	4		Varied
White-eared opossum (*Didelphis albiventris*)	Didelphidae	4		Varied
Southern long-nosed armadillo (*Dasypus hybridus*)	Dasypodidae	4		Varied
Screaming hairy armadillo (*Chaetophractus vellerosus*)	Dasypodidae	4		Varied
Large hairy armadillo (*Chaetophractus villosus*)	Dasypodidae	4		Varied
Torres's crimson-nosed mouse (*Bibimys torresi*)	Muridae	4		FM
Web-footed marsh rat (*Holochilus brasiliensis*)	Muridae	4		FM
House mouse (*Mus musculus*)[d]	Muridae	4	+	Varied
Black rat (*Rattus rattus*)[d]	Muridae	4	+	Varied
Brown rat (*Rattus norvegicus*)[d]	Muridae	4	+	Varied
Nutria (*Myocastor coypus*)	Muridae	4	+	FM
Swamp rat (*Scapteromys tumidus*)	Muridae	4	+	FM
Drylands vesper mouse (*Calomys musculinus*)	Muridae	4		FM, grass
Red hocicudo (*Oxymycterus rufus*)	Muridae	4		FM, grass
Azara's grass mouse (*Akodon azarae*)	Muridae	4	+	Grass
Bunny rat (*Reithrodon auritus*)	Muridae	4	?	Grass
Yellow pygmy rice rat (*Oligoryzomys flavescens*)	Muridae	4	+	Varied
Small vesper mouse (*Calomys laucha*)	Muridae	4		Varied
Brazilian guinea pig (*Cavia aperea*)	Caviidae	4		Varied
Capybara (*Hydrochaeris hydrochaeris*)	Hydrochaeridae	4	+	FM, grass
Plains viscacha (*Lagostomus maximus*)	Chinchillidae	4		Grass
Talas tuco tuco (*Ctenomys talarum*)	Ctenomyidae	4		FM
Common fox (*Dusicyon gymnocercus*)	Canidae	4		Varied
Molina's hog-nosed skunk (*Conepatus chinga*)	Mustelidae	4		Varied
Lesser grison (*Galictis cuja*)	Mustelidae	4		Varied
Pampas cat (*Felis colocolo*)	Felidae	3	+	Varied
Geoffroy's cat (*Felis geoffroyi*)	Felidae	4	?	Varied
Marsh deer (*Blastocerus dichotomus*)	Cervidae	4		FM, grass
Pampas deer (*Ozotoceros bezoarticus*)	Cervidae	4		Grass

Note: Taxonomic arrangement of mammals in table as in Nowak (1999) and common and scientific names are based on Wilson and Reeder (1993).
[a] Class: 1 = endemic species; 2 = species with endemic subspecies; 3 = species with population locally restricted to tidal marsh; 4 = generalist.
[b] + = regularly found in salt marsh (salinity ≈ sea water).
[c] Alternate habitat (for generalists) FM = freshwater marsh or other aquatic habitat; grass = grasslands or fields; scrub = scrub habitats; varied = different habitats.
[d] Non-native species.

TABLE 9. Species of tidal-marsh mammals of Europe.

Species	Family	Class[a]	Saltmarsh[b]	Alternate habitat[c]
Eurasian shrew (*Sorex araneus*)	Soricidae	4	+	FM, grass
Eurasian water shrew (*Neomys fodiens*)	Soricidae	4	+	FM, grass
Lesser shrew (*Crocidura sauveolens*)	Soricidae	4	+	Varied
European water vole (*Arvicola terrestris*)	Muridae	4		Varied
Field vole (*Microtus agrestis*)	Muridae	4	+	Varied
Eurasian harvest mouse (*Micromys minutus*)	Muridae	4		FM, grass
Long-tailed field mouse (*Apodemus sylvaticus*)	Muridae	4		Varied
House mouse (*Mus musculus*)	Muridae	4		Varied
Black rat (*Rattus rattus*)	Muridae	4	+	Varied
Brown rat (*Rattus norvegicus*)	Muridae	4	+	Varied
Least weasel (*Mustela nivalis*)	Mustelidae	4	+	Varied
European rabbit (*Oryctolagus cuniculus*)	Leporidae	4	+	Varied
European hare (*Lepus europeus*)	Leporidae	4	+	Varied
Sika deer (*Cervus Nippon*)	Cervidae	4	+	Varied

Note: Taxonomic arrangement of mammals in table as in Nowak (1999) and common and scientific names are based on Wilson and Reeder (1993).
[a] Class: 1 = endemic species; 2 = species with endemic subspecies; 3 = species with population locally restricted to tidal marsh; 4 = generalist.
[b] + = regularly found in salt marsh (salinity ≈ sea water).
[c] Alternate habitat (for generalists) FM = freshwater marsh or other aquatic habitat; grass = grasslands or fields; scrub = scrub habitats; varied = different habitats.

Two subspecies of mink (*Mustela vison lutensis* and *M. v. halilemnetes*), were originally described as being restricted exclusively to the saltmarshes of the Atlantic and Gulf Coasts, respectively (Bangs 1898). *M. v. halilemnetes* was described as being slightly paler than *M. v. lutensis* and somewhat paler than *M. v. vulgivaga* from Louisiana coastal marshes. However, these differences did not seem valid and *M. v. halilemnetes* has been lumped into the more widely distributed *M. v. lutensis* and *M. v. vulgivaga* is now recognized to be more widely distributed throughout Louisiana.

Northern raccoons (*Procyon lotor megalodus*) inhabiting the marshes of southern Louisiana have been described as being more yellowish with a more pronounced mid-dorsal line than adjacent subspecies. Specimens of the swamp rabbit (*Sylvilagus aquaticus littoralis*), from the gulf coasts of Texas, Louisiana, Mississippi, and Alabama, were described as being darker and more reddish brown than the nominate subspecies. The validity of this subspecies has been questioned, however. Specimens of the muskrat (*Ondatra zibethicus*), from the central Gulf Coast, assigned to the subspecies *O. z. rivalicus,* have been described as being darker than specimens from the northcentral and northeastern US (Willner et al.1980).

At least one subspecies of white-tailed deer (*Odocoileus virginianus mcilhennyi*), from the coastal marshes of Louisiana, has been described as being smaller, darker, and larger footed than upland populations (Miller 1928). Because of numerous introductions and relocations since the mid-1950s, current populations of deer in Louisiana do not reflect these patterns of geographic variation.

DISCUSSION

Endemism

Whereas most of the dominant plants of tidal marshes are restricted to tidal marshes or may occur locally in inland saline habitats (Chapman 1977), the terrestrial vertebrate fauna shows relatively little specialization. In fact, given the ecological distinctiveness of both tidal-marsh environment and its flora, the number of described endemic taxa of terrestrial vertebrates is surprisingly small. We know of five species restricted or largely restricted to coastal marshes and only 18 additional species with at least one subspecies so restricted. The 23 species were predominantly birds (11), followed by mammals (nine) and reptiles (three). No endemic taxa of amphibians are known from tidal marshes.

Why North America?

The most striking, and ultimately puzzling, pattern in the occurrence of endemism was its almost complete restriction to North American coastlines (Fig. 1). All but one of the 25 species showing complete or partial endemism were found in North America. The single exception is a subspecies of Australian songbird. Of these, 14 are species (or have subspecies) restricted to the Atlantic or Gulf Coasts and eight are found only along the Pacific Coast.

To be sure, these data are probably biased by the lack of geographic coverage for some areas. More fundamentally because the vast majority of endemic taxa are subspecies, the detection of such forms will vary with the thoroughness with which geographic variation has been assessed in different regions. Certainly, the uncovering of ecological races associated with tidal marshes in South America is still a possibility. However, we believe that differences in the thoroughness of taxonomists to describe subspecies does not account for differences among the faunas of Europe, Australia, and North America. It is likely that the level of descriptive taxonomy for birds and mammals even in South America is sufficiently high that tidal-marsh subspecies would have been described. The significance of saltmarsh subspecies was initially discussed by Grinnell (1913) and Wetmore (1926) was probably aware of these associations because he collected in the South American saltmarshes in 1920–1921.

The concentrations of endemic forms in North America is likely to be real, at least partially, and not just artifacts of collecting or taxonomy. Any explanation needs to account for characteristics of the Atlantic, Gulf Coast, and Pacific Coast tidal marshes, because endemics are associated with all. We have identified four hypotheses that might explain the high levels of endemism in North American tidal marshes:

1. Any post-glacial expansion of tidal marsh habitats in North American coastlines resulted in either higher quality or quantity of tidal-marsh habitat that would in turn be able to support large enough populations of colonizing species to allow rapid diversification to proceed.
2. Tidal-marsh habitats were more stable throughout the Pleistocene resulting in the minimization of the extinction of differentiated forms.
3. Tidal-marsh habitats have shown both greater expansion and stability through time (a combination of the first two hypotheses).

4. The North American fauna contained particular taxa that had characteristics that favored the successful colonization of and diversification in tidal marshes.

The first three hypotheses would result in different temporal patterns of divergence of tidal marsh forms. The first hypothesis would result in most tidal-marsh endemics having diverged over the past 5–10,000 yr as coastal estuarine marshes developed. The greater stability hypothesis 2 should result in a preponderance of taxa that diverged earlier in the Pleistocene. Hypothesis 3, the combined hypothesis, would be supported by a complex pattern of both older and more recently derived taxa.

The results of studies looking at genetic divergences between coastal-marsh vertebrates and upland conspecifics (Chan et al., *this volume*), suggest that the endemic taxa in North American tidal marshes have both ancient and very recent association with tidal marshes, supporting hypothesis 3. Of the 14 species or subspecies for which genetic divergence data are available, eight are Holocene (<8,000 yr BP, i.e., since the last glaciation), four date to the mid-Pleistocene (500,000–1,000,000 yr BP) and three apparently diverged during the Pliocene (1.8–8 million yr BP). This suggests that tidal marshes in North America have both provided stable refugia for few endemics throughout the Pleistocene, but have expanded rapidly enough after the last glacial maximum (LGM) to allow rapid evolution of morphological traits. The tendency, although the sample size is too small for formal analysis, is for the older taxa to have southern distributions (Seaside Sparrow, Savanna Sparrow, diamondback terrapin, and saltmarsh snake), which suggests that southern refugia along the Gulf of Mexico and Mexican coastlines may have played a role in reducing the probability of extinction.

What makes the higher degree of endemism in North America all the more puzzling is that the processes that lead to endemic forms in North America have operated along coastlines that are very different geologically (Malamud-Roam et al., *this volume*). Presently, the Atlantic and Gulf coasts of North America have the greatest aerial extension of coastal tidal marsh. The Pacific Coast is a tectonically younger and more active coastline than the East Coast and supports an overall limited amount of tidal marsh, making opportunities for adaptation and divergence less likely.

The pattern of divergence in North America suggests that in more northern areas, tidal-marsh refugia were not available for most vertebrate taxa and that the Pleistocene was an epoch that saw repeated extinction and recolonization along these coastlines. In contrast, the number of taxa with more southerly distributions suggests that these populations could retreat to tidal marshes through many or all of the Pleistocene glaciations. The discussion in Malamud-Roam et al. (*this volume*) makes it clear that lower sea levels, colder water temperatures, reduced sedimentation from river flow, and ice cover all combine to reduce the extent of tidal marshes during the glacial maxima. Therefore, for potential Pleistocene refugia we should look to areas with shallow protected coastlines, fresh water, sediment input, and moderate ocean temperatures (Chapman 1977). The Mississippi delta-gulf coast, certain areas in the South Atlantic, and the Gulf of California are all likely areas for such refugia (Malamud-Roam, pers. comm.). Conditions in the East were drier and the marsh systems would have received less sediment influx. However, it seems likely that delta and barrier island formations would have persisted. In contrast, southwestern North America saw wetter conditions with greater precipitation during the glacial periods (Thompson et al. 1993) and the Colorado River delta system would have drained the great inland alkali lakes and probably transported more sediment (Sykes 1937); cooler temperatures would have favored more diverse saltmarsh vegetation and shifted the marsh-mangrove transition southward (Chapman 1977). An analysis of these factors may explain the global patterns as well. For example, the major extensions of shallow, protected shorelines in Europe are along the glaciated northern edge of the continent, and most of the major river systems drain northward into these areas. At least from a superficial analysis, conditions appear much more favorable for Pleistocene coastal marsh formation in North America than at least the western Palearctic.

The appropriate taxa hypothesis (hypothesis 4) would predict that endemics would be from a few vertebrate families that possess some identifiable adaptation to tidal-marsh life. For example, the surprising lack of endemic avian taxa in tidal marshes of eastern South America, may in part reflect the paucity of emberizid finches (the main group displaying such endemism) in the Pampean faunal region (J. P. Isacch, pers. comm.). However, in general we believe that the lack of appropriate source taxa is not a convincing argument for why endemic forms are close to absent outside of North America. Although a few vertebrate genera are repeatedly involved in the development of endemic forms, this is to be expected because the occupancy of tidal marshes probably provides a number of ecological filters to possible colonizing species. The important point is that the different genera involved are themselves diverse and unremarkable as far as we know.

SPECIALIZATION IN DIFFERENT CLASSES OF VERTEBRATES

Brackish marshes, located at the higher tidal zones and the upper ends of estuaries, support the largest number of species and presumably require the least specialization to the tidal-marsh environment. We, therefore, consider the proportion of species found in true saltmarshes to be an indication of the degree of specialization on tidal marshes within a higher order taxonomic grouping. We have focused our attention on reptiles, birds and mammals. Comparing these classes of vertebrates, we find mammals comprised the greatest proportion of saltmarsh inhabiting taxa (44%), followed by reptiles (39%), birds (33%), and amphibians (16%).

SPECIALIZATION AND DIFFERENTIATION

We found a clear relationship between the probability that a taxon will show some degree of differentiation and whether it occupies saltmarsh or is found only in brackish marsh. Considering reptiles, birds, and mammals, we find that 97% of the endemic species or endemic subspecies are found in saltmarsh. On the other hand, only 30% of the non-differentiated species occur in saltmarsh. Mammals stand out as having a large number of non-differentiated species that occupy saltmarshes. Removing mammals, only 16% of the non-differentiated vertebrates occur in saltmarshes. These results are non-circular, because it is entirely possible to have coastal marsh taxa that avoid saltmarshes and yet are morphological distinct from non-coastal marsh populations. The Coastal Plain Swamp Sparrow is an example of this. The results do suggest that species that occupy the extreme of the environmental gradient in saltmarshes are the ones most likely to have differentiated.

PATTERNS OF SPECIES RICHNESS

We evaluate regional differences in species richness by comparing the faunas of a single continental coastline. We have data to compare mammals and birds and total tidal-marsh area of both coasts of North America, the coastline of Europe, and that of eastern South America. For the bird analysis, we included only the south coast of Australia, using an approximation of 600 km^2 for total marsh areas. The total number of native tidal-marsh species of birds and mammals was highly correlated across these different coastal areas (Fig. 2a; r^2 = 0.93, N = 4) with the highest species richness in eastern South America and North America and lowest in western North America. Furthermore, the total number of mammal and bird species occupying saltmarsh was also correlated (Fig. 2b; r^2 = 0.73, N = 4). However the number of saltmarsh birds and mammals was poorly correlated with the total number of tidal-marsh species.

The diversity of tidal-marsh mammals and birds appeared to be related to the overall area of tidal marshes along a particular coastline (r^2 = 0.50 for birds; r^2 = 0.40 for mammals). The number of saltmarsh species of mammals (Fig. 3; r^2 = 0.71) and birds (r^2 = 0.55) were more strongly correlated with total tidal marsh area. We have already seen that the tendency for differentiation of taxa was somehow related to the occupancy of true saltmarsh. The total number of endemic species and species with endemic subspecies was related to the number of saltmarsh species (Fig. 4; r^2 = 0.72) and unrelated to the number of tidal marsh species as whole (r^2 = 0.02).

Although general species-area relationships exist that emphasize the overall importance of current marsh area, some caution is necessary. First, these relationships are based on few data points (coastlines) and only birds and mammals. Second, and perhaps more importantly, the

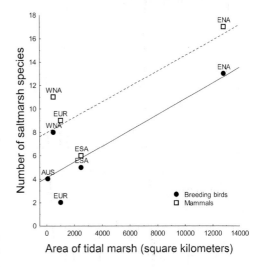

FIGURE 3. The number of species of native mammals and breeding birds plotted against tidal-marsh area for different coastline for saltmarsh species.

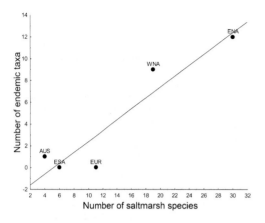

FIGURE 4. For native mammals and breeding birds, the number of endemic species (category 1) or species with at least one endemic subspecies (category 2) plotted against number of total saltmarsh mammal and breeding bird species (only bird species for Australia).

outliers may provide greater insight for future research than the general trends. For overall tidal marsh species, South America appears to have higher species richness than expected just based on marsh area alone. More interesting though is the much-higher-than expected species richness of both saltmarsh species and endemic taxa found in western North America when plotted against marsh area. As we stated in the section on endemic taxa, this cannot be explained by the size and diversity of habitats associated with an undisturbed San Francisco Bay system. A number of the taxa are associated with small marshes in the relatively marsh-free coastline of southern California.

ALTERNATIVE HABITATS

For non-specialist tidal-marsh vertebrates, the dominant alternative habitat is freshwater marsh and related aquatic habitats (49%) with grasslands and agricultural fields contributing another 29%. These figures vary considerably among major classes of vertebrates; for birds, amphibians, and reptiles, freshwater marsh comprises the major alternative habitat for 59–73% of the species. However, in mammals only 30% of the species are found commonly in freshwater marshes and a majority is found in grassland or a variety of other upland habitats. The lack of freshwater marsh species of mammals to colonize tidal marshes perhaps underscores the role of substrate flooding in the presence of small mammals in the most tidal of marshes.

KEY AREAS OF FRESHMARSH-SALTMARSH INTERCHANGE

We have discussed possible refugia for saltmarsh taxa that have persisted through the glaciations. The diversity of organisms that have colonized saltmarshes with or without diversification can be additionally explained by the potential of interaction of coastal marsh and interior marsh faunas. It is unclear whether any of the tidal-marsh regions outside of North America have such areas of extensive potential interchange, but one area to look in South America would be the wetlands associated with the Rio de la Plata connecting to interior wetlands in the La Plata Basin and the flooded pampas. La Plata basin is located in an area where significant freshwater and brackish water interactions may take place; it holds the largest wetlands in the world and has a wide variety flora and fauna associated with grasslands and wetlands. In North America, three areas stand out as having a long history of close contact or interdigitation of major fresh, brackish and saltwater wetlands. First, Odum (1953) and Neill (1958) focused on the Florida peninsula as an area of important interchange between freshwater and marine organisms. The low relief, large amount of freshwater flow, and changing sea levels have contributed to a continuous faunal interchange along the salinity gradient. Present day Florida has a very extensive brackish wetland system and many of the freshwater systems have a high concentration of dissolved chloride, which allows for the inland invasion of typically marine invertebrates. Neill (1958) noted that Florida supported an extraordinarily high number of salt and brackish marsh populations of reptiles and amphibians. Certain birds show freshwater-saltwater distributions where in other places in their range they are restricted to saltmarsh habitats (Seaside Sparrows and Boat-tailed Grackles [*Quizcalus major*]). A second region is the San Francisco Bay-delta-Central Valley wetland systems discussed by Malamud-Roam et al. (*this volume*). Prior to human manipulation of the hydrology of the region and also through the pluvial periods in western US, this region must have formed a large area of interchange between freshwater and brackish water wetlands. Western North America had extensive alkali and saline wetlands which may have provided source taxa for tidal marsh invasions. Finally, the coast of the Gulf of Mexico supports a large portion of the North American oligohaline marshes, particularly in the region of the Mississippi Delta.

ECOLOGICAL DIVERGENCE IN TIDAL-MARSH
POPULATIONS AND SPECIES

Saltmarsh forms often show high degrees of morphological, ecological and physiological differentiation, even though estuarine habitats are localized and not isolated from upland or freshwater habitats. Certain features of tidal-marsh life may provide exceptionally sharp selective gradients which ultimately favor assortative mating with individuals that can survive in saltmarsh habitats. Dunson and Travis (1994) made a convincing argument for the role of physiological adaptation to salinity as a driving force for divergence of saltmarsh animals.

As in the specialized saltmarsh flora, a major set of adaptations that distinguish tidal-marsh specialists is related to living in a highly saline environment. In general, estuarine species do not show as extreme a level of adaptation to saline environments as do more truly marine forms. For example, while sea-going reptiles and pelagic birds have salt glands which concentrate and excrete hypersaline solutions (Heatwole and Taylor 1987), only a few estuarine vertebrate taxa possess specialized salt glands. Clapper Rails have large nasal salt glands (Olson 1997) and the largely freshwater King Rail possesses smaller ones, even when both are raised in similar captive environments (Schmidt-Nielsen and Kim 1964). The fact that even the largely freshwater species possesses a salt gland suggests that the original evolution of salt glands in this rail complex occurred during an ancestral occupation of estuarine habitats prior to the recent colonization by Clapper Rails (Olson 1997). Passerine birds do not possess specialized salt glands, but show adaptation of kidney morphology and renal function, which include enlarged size including an increase in the number of medullary cones and an increase in the proportion of nephrons with loops of Henle. This results in an increased ability to concentrate salt (Goldstein, *this volume*)

By and large, tidal-marsh vertebrates are not as specialized for saline environments as are marine vertebrates. Estuarine snakes (e.g., *Nerodia clarkia*) are not known to have specialized glands, but have a relatively impermeable skin and are behaviorally adapted to obtain all of their moisture from their diet (Pettus 1958, 1963; Conant and Lazell 1973, Dunson and Mazzotti 1989). The behavioral difference is simple—tidal-marsh snakes do not drink salt water, whereas their freshwater relatives have no such inhibition. Nonetheless, the difference is sufficient to cause freshwater water snakes to die in salt water, where the tidal-marsh snakes survive. The same is true for turtles that inhabit tidal marshes—either as a specialist (diamondback terrapin Robinson and Dunson 1975) or generalist (common snapping turtle [*Chelydra serpentine*]; Dunson 1981, 1986)

Research on adaptation to salinity in tidal marsh mammals is more limited. Meadow voles from saltmarshes are apparently physiologically limited to ingesting water with <50% of the salinity of sea water (Getz 1966). They cannot consume large quantities of smooth cordgrass, which has tissue moisture with high salt concentrations. Saltmarsh inhabiting voles may obtain moisture from precipitation, dew, and from the tissue of less salty plants, such as saltmeadow cordgrass. Furthermore, Coulombe (1970) showed that the California vole was better able to consume halophytic plants than its performance on tests with saline drinking water would suggest. Although the physiological mechanism has not been studied, it appears that the salt marsh harvest mouse can survive longer ingesting more saline solutions than does the non-tidal marsh western harvest mouse (Fisler 1963, 1965). In several of the above studies, it was demonstrated that mice from freshwater marshes either showed evidence of weight loss or mortality when presented higher salinity water but see MacMillen (1964). Other adaptations to salinity are behavioral, including the possible use of torpor under conditions of osmotic stress (Coulombe 1970). The presence of physiological and behavioral adaptations in saltmarsh forms suggests that selection against hybrids between inland and salt populations may be intense and forms the basis for ecological separation and perhaps to speciation. This may go a long way towards explaining why differentiated forms tend to be specifically associated with salt rather than brackish tidal marsh.

A sharp discontinuity in predominant substrate color (often related to the concentration of sulphates in sea water) may also provide a strong selective force against hybrids and thus shape morphological divergence, even in the face of ongoing gene flow. As discussed by Grenier and Greenberg (*this volume*), many saltmarsh forms tend to be melanistic, displaying both grayer and blacker coloration. Presumably this is related to local cryptic adaptation to saltmarsh substrates and hence may contribute to enforcing diversification in tidal marsh forms. Since Grinnell's (1913) classic paper on estuarine vertebrates, saltmarsh melanism has been documented in a number of avian and mammalian taxa. We have documented melanistic populations of shrews, voles, and harvest mice in our discussion of endemic mammals. For example, several endemic subspecies of small mammals were originally described as new in southern California partially because of their darker pelage coloration (Von Bloeker 1932). Neill (1958) noted a number of examples of local melanism associated with tidal marsh reptiles and amphibians. More recently, melanistic subspecies of water snakes have been described for the *Spartina* and *Juncus* marshes of coastal North Carolina (Conant and Lazell 1973). Later Gaul (1996), for example, found a strong correlation between degree of melanism and the salinity of the water in a tidal marsh in the common water snakes of coastal North Carolina. The gulf salt marsh snake (*Nerodia clarkia clarkia*) is reported to have more evenly dark dorsal lines with less patterning (Pettus 1963; Myers 1988), which may be an adaptation to the lack of blotchy patterning in tidal muds. It may also, as the author suggested, blend in with the striped patterning of grassy marsh vegetation.

CONCLUSIONS

Although many vertebrates use brackish or saline tidal marshes from time to time, the number of species that are commonly resident in the marshes is generally small and the number that occurs regularly in saltmarshes is even smaller. In mammals and birds the number of species that use tidal marshes along a particular coastline is related to the current amount of tidal marsh. Despite a fairly strong suite of environmental differences, tidal marshes support relatively few endemic species, i.e., species that are wholly or largely restricted to these marshes.

Overall the greatest number of species and the most endemic taxa of tidal marshes are found in eastern North America. Fewer endemic taxa are found along the West Coast of North America, primarily in the San Francisco Bay, but also in the smaller marshes of the southern California and Mexican coastlines. Surprisingly, the only endemic taxon we have been able to locate outside of North America is a subspecies of Australian thornbird. Oscine passerines, and in particular the emberizine sparrows, have the largest number of endemic taxa. Other groups with multiple endemics are the shrews, murid rodents, and colubrid snakes.

The probability that a taxon is endemic or has endemic subspecies in tidal marshes is related to its occupancy of salt as opposed to brackish marshes. This is not surprising considering that the characters that show divergence are behavioral and physiological adaptations to cope with salinity and shifts in coloration to blend in with the dark acid-sulfate soils. These ecosystem characteristics, along with major changes in invertebrate communities and patterns of plant production and seed set co-vary and reinforce differentiation along the fresh to saltmarsh gradient.

Coastlines with no endemics either have little marsh habitat or have been subjected to heavy human disturbance. The lack of tidal-marsh endemics along the temperate Atlantic coastline of South America is probably the most difficult to explain, for despite the overall high species diversity of South American vertebrate faunas and the large number of species reported thus far from tidal marshes, we were unable to locate any endemic taxa associated with tidal marshes of the large estuaries of southeastern coast of the continent. Further research will confirm this pattern or determine that it is an artifact of less thorough collecting and less detailed taxonomic work at the subspecific level.

FUTURE RESEARCH

This paper is clearly a first assessment of diversity and endemism of terrestrial vertebrates in several of the tidal marsh systems throughout the world. For many taxonomic groups and for several regions there is a crucial need for basic inventory of species presence and absence. This would be particularly true for reptiles and amphibians outside of North America, birds and mammals in Asia, and all taxa in South Africa, and parts of the Middle East.

In groups for which alpha-level faunal lists are available or can be pieced together, a need remains for information focusing on more detailed distribution of taxa along a salinity and tidal gradient, together with complementary data on morphological and genetic variation between upland and tidal-marsh habitats.

We were able to piece together some information on the distribution and prevalence of tidal marshes. But for most regions it remains crude, particularly in comparison to the type of information that can be gathered and analyzed with remote-sensing technologies and geographic information system. A comprehensive review of different coastal-marsh types is an essential component to a more sophisticated analysis of marsh biogeography.

Although detailed information of recent and geologic history is available for certain estuarine and marsh systems, this information is scattered and is apparently not available for many coastlines. Much more information on the deep and recent history of coastal-marsh systems is also central to developing evolutionary and biogeographical hypotheses on the formation of tidal marsh assemblages.

Finally, the sharp environmental gradient between upland and freshwater habitats and saltmarsh is actually a composite of several selective gradients working in consort in shaping divergence in tidal marsh populations: salinity, geochemistry of the substrate and resulting coloration, and a shift to a more marine invertebrate prey base. Adaptations to any of these reinforce the selective advantage of assortative mating which leads to ecological- or physiologically based speciation. For these reasons, tidal marshes represent one of the best systems for studying ecological differentiation and speciation.

ACKNOWLEDGMENTS

We obtained information on the distribution of tidal-marsh taxa from a number of workers in the field. We would like to thank J. Marchant and P. Atkinson for providing background information on British saltmarsh birds. L. Joseph, P. Brown, and R. Jaensch provided insight into the saltmarsh birds of Australia. We received valuable information on the birds of east Asian marshes from H. Nijati and N. Moores. Information on South America was provided by Rafael Dias, P. Petracci, S. Claramunt, and R. Fraga. For information provided on eastern North American mammals, we thank N. Moncrief. M. L. Merino, and S. González provided information on South American mammals. M. Delibes, J. Flowerdew, A. Grogan, R. Strachan, R. Trout, D.W. Yalden, provided insights into the saltmarsh mammals of Europe. M. Castellanos helped with mammal range information and figures. D. Lee provided information on reptiles and amphibians of North America. Additionally helpful insights and information were provided by P. Adam, P. Fidelman, W. Lidicker, N. Montgomery, B. Shaffer, R. McDiarmid, M. Bertness, G. Chmura, and O. Iribarne, and A.Shuqing. W. Post provided comments that greatly improved the quality of this manuscript.

EVOLUTION AND CONSERVATION OF TIDAL-MARSH VERTEBRATES: MOLECULAR APPROACHES

YVONNE L. CHAN, CHRISTOPHER E. HILL, JESÚS E. MALDONADO, AND ROBERT C. FLEISCHER

Abstract. The tidal marshes of North America are home to a diverse collection of morphologically differentiated reptiles, birds, and mammals. We reviewed the existing molecular studies on endemic tidal-marsh vertebrates, including turtles, snakes, sparrows, rails, shrews, and rodents. We found both deep and shallow divergences from their nearest upland relatives in all geographic regions. In the Northeast, the Saltmarsh Sharp-tailed Sparrow (*Ammodramus caudacutus*) has probably been isolated from the inland forms of the Nelson's Sharp-tailed Sparrow (*A. nelsoni*), for >600,000 yr, while the salt-marsh form of the Swamp Sparrow (*Melospiza georgiana*) evolved from upland relatives <40,000 yr ago. On the West Coast, saltmarsh forms of the Song Sparrow (*M. melodia*) and the ornate shrew (*Sorex ornatus*) show low levels of genetic differentiation from neighboring upland forms, while the salt marsh harvest mouse (*Reithrodontomys raviventris*) living in the same marshes shows deep genetic divergences from upland forms dated to nearly 4,000,000 yr ago (MYA). On the Gulf Coast, saltmarsh Clapper Rails (*Rallus longirostris*) show either a very recent split from, or high levels of gene flow with freshwater King Rails (*R. elegans*), but Seaside Sparrows (*A. maritimus*) probably diverged from an upland ancestor 1.5–2 MYA and diamondback terrapins (*Malaclemys terrapin*) 7–11 MYA. The timing of those divergences ranges from late Miocene to late Holocene, and suggest a complex history of multiple invasions and differentiations in saltmarshes. Molecular approaches have increased our understanding of the evolutionary origin of these unique forms, revealed the complex patterns of genetic structure within them, and furthered conservation efforts.

Key Words: Adaptation, allozymes, *Ammodramus*, genetic structure, geographic variation, *Malaclemys*, *Melospiza*, microsatellites, mitochondrial DNA, morphology, *Nerodia*, *Rallus*, *Reithrodontomys*, *Sorex*, tidal marsh.

EVOLUCION Y CONSERVACION DE VERTEBRADOS DE MARISMA DE MAREA: ENFOQUES MOLECULARES

Resumen. Las marismas de marea de Norte América son el hogar de una diversa colección de reptiles, aves y mamíferos morfológicamente diferenciados. Revisamos los estudios moleculares existentes de vertebrados endémicos de marismas de marea, incluyendo tortugas, culebras, rascones, gorriones, musarañas y roedores. Encontramos divergencias tanto profundas como poco profundas de sus parientes más cercanos de las tierras más altas en todas las regiones geográficas. En el Noreste, el Gorrión Cola Aguda de marisma salada (*Ammodramus caudacutus*) ha sido probablemente aislado de las formas de las tierras interiores del Gorrión Cola Aguda Nelson (*A. nelsoni*), por mas de 600,000 años, mientras que la forma de marisma salada del Gorrión Pantanero (*Melospiza georgiana*) evolucionó de parientes de tierras más altas hace menos de 40,000 años. En la costa oeste, formas de marismas saladas del Gorrión Cantor (*M. melodia*) y de la musaraña vistosa (*Sorex ornatus*) muestran bajos niveles de diferenciación de formas de vecinos de las tierras mas altas, mientras que el ratón de cultivo de marisma salada (*Reithrodontomys raviventris*) viviendo en las mismas marismas, muestra profundas diferencias genéticas de formas de tierras mas altas, que datan de hace aproximadamente 4,000,000 años (HMA). En la Costa del Golfo, rascones de marisma salada (*Rallus longirostris*) muestran ya sea de una separación muy reciente, o elevados niveles de flujo genético con el Rascón Real (*R. elegans*) de agua fresca, pero los Gorriones Costeros (*A. maritimus*) probablemente divergieron de un ancestro de tierras mas altas 1.5–2 HMA y la tortuga acuática (*Malaclemys terrapin*) divergió hace 7–11 HMA de su forma ancestral. El tiempo en el que transcurrieron esas divergencias de rango del Mioceno tardío al Holoceno tardío, sugieren una historia compleja de invasiones múltiples y diferenciaciones en marismas saladas. Enfoques moleculares han aumentado nuestro entendimiento del origen evolutivo de estas formas únicas, revelando los patrones complejos de la estructura genética entre ellas, y han ayudado al progreso de esfuerzos de conservación.

Saltmarsh bird species, such as the Seaside Sparrow (*Ammodramus maritimus*), the Saltmarsh Sharp-tailed Sparrow (*A. caudacutus*) and the Clapper Rail (*Rallus longirostris*) have been known to science for >200 yr (American Ornithologists' Union 1983). Based on research in the past century, vertebrate zoologists have now described 25 species or morphologically differentiated subspecies of avian, mammalian and reptilian species largely or wholly restricted to tidal marshes (Hay 1908, Grinnell 1909, 1913; Clay 1938; Marshall 1948a, b; Boulenger 1989). Tidal marshes are discrete in their distribution and present a profound environmental disjunction, and therefore may impose intense directional selective pressures on colonizing populations. Much speculation exists over the evolutionary history of these tidal-marsh forms and with the advent of molecular techniques many avenues for exploration have opened. Genetic markers have become important tools for evolutionary and conservation biologists. We can now use genetic markers to determine phylogenetic relationships, clarify the sister taxa of tidal-marsh endemics and to time their divergence from ancestral taxa. This enables the direct testing of evolutionary hypotheses relating to the origin of tidal-marsh endemics and the circumstances under which they diverged. A variety of molecular markers, such as allozymes, restriction fragment length polymorphisms (RFLPs), DNA sequences, and microsatellites have been applied to tidal-marsh taxa, providing estimates of divergence, gene flow, and population differentiation (Table 1).

For conservation biologists, the use of molecular techniques can provide genetic estimates of intraspecific variation important for evaluating the viability and adaptive potential of endangered populations (Smith and Wayne 1996). Molecular markers are also particularly important for determining the taxonomic status of endemic taxa, both for defining conservation units and for advocating appropriate management action. Improper lumping of distinct species can result in underestimates of regional biodiversity and splitting of non-distinct taxa can divert valuable resources and reduce opportunities for proper genetic management (Frankham et al. 2002). Furthermore, molecular techniques can verify hybrids that are suspected based on morphology (Frankham et al. 2002).

In addition to macroevolutionary and conservation questions, tidal-marsh taxa lend themselves to the study of microevolutionary

TABLE 1. A VARIETY OF DIFFERENT MOLECULAR MARKERS, SUCH AS ALLOZYMES, RESTRICTION FRAGMENT LENGTH POLYMORPHISMS (RFLPs), DNA SEQUENCES, AND MICROSATELLITES, HAVE BEEN EXAMINED IN TIDAL MARSH TAXA, PROVIDING ESTIMATES OF AMONG SPECIES AND AMONG POPULATION DIFFERENTIATION.

Molecular marker	Description
Allozymes	Protein allozyme variation is derived from alleles separated based on their charge as they migrate through a gel medium, most often, starch. Individuals are assigned a genotype, and heterozygosity and allele frequencies can be calculated. An important statistic, F_{st}, which is a measure of among population genetic variance, can be calculated from allele frequencies. F_{st} ranges from 0–1, with a value of zero indicating panmixia and lack of genetic differentiation, and a value of 1 indicating fixation of alternate alleles in each population and therefore a lack of gene flow. Estimates of F_{st} based on allozymes for avian taxa are extremely low (average F_{st} = 0.022, SD = 0.011), but are higher for mammalian taxa (average F_{st} = 0.230, SD = 0.183) (Barrowclough 1983).
RFLPs	Restriction enzymes are used to cleave DNA at specific recognition sites and then the DNA fragments are separated by weight on a gel. Often mitochondrial DNA is isolated and is used in order to reduce the number of fragments. Mutations within the recognition sites produce a variable banding pattern and an estimated percent nucleotide sequence divergence can be calculated. For avian taxa, interspecific divergence among congeners have ranged from 0.07%–8.8% (Avise and Lansman 1983).
DNA sequences	With the invention and widespread application of the polymerase chain reaction (PCR), researchers now have the ability to produce sequences affordably and from small amounts of tissue. DNA sequence data provides both phylogenetic information and information on population genetic structure based on allele frequencies. Percent sequence divergence can be used to estimate the time to the most recent common ancestor if the divergence between taxa using DNA sequences is calibrated from the fossil record. The use of mtDNA sequences has become very important for the field of phylogeography (Avise 2000).
Microsatellites	Microsatellites are tandem repeats of simple sequences that occur frequently and at random throughout the genome. These highly polymorphic markers are flanked by unique sequences that serve as ideal sites for the design of primers that can be used for PCR amplification. Since the microsatellite polymorphism is stable and is inherited in a Mendelian fashion these markers can be highly informative. PCR primers can be labeled with one of the four currently available fluorescent dyes. After PCR, the products are separated on acrylamide gels and using a scanning laser and commercially available software, primers labeled with different dyes alleles can be distinguished even when their sizes overlap. Allele sizes are reproducibly and accurately determined. Allele frequencies, heterozygosity, and F_{st} estimates can be calculated.

processes (Avise 2000). First, tidal saltmarsh habitat is discrete and relatively homogeneous simplifying the understanding of the spatial configuration underlying the genetic variation. Second, advances in our understanding of marsh history (Malamud-Roam et al., *this volume*) may allow biologists to estimate the timing of habitat availability for tidal marsh organisms, improving hypotheses for divergence times of upland and tidal marsh forms. Third, tidal-marsh forms are often found in close proximity to conspecific or closely related non-tidal marsh populations, yet face a different set of environmental challenges. The combination of these three factors provides a unique framework within which we can attempt to understand the role of selection on short temporal scales; and the sensitivity of selective forces to gene flow, with the advantage of multiple geographic replicates and with a diverse group of vertebrates that includes reptilian, avian, and mammalian taxa.

In this chapter we present an overview of studies in which molecular markers have been used to study tidal-marsh vertebrate taxa. Much evolutionary work has been devoted to certain groups, such as sparrows in the genera *Ammodramus* and *Melospiza*. Several other studies have focused on determining the taxonomic status and genetic basis for differentiation in endemic saltmarsh populations such as those found in the diamondback terrapins (*Malaclemys terrapin*), water snakes (*Nerodia fasciata* and *N. sipedon*), Clapper Rail (*Rallus longirostris*), Savannah Sparrow (*Passerculus sandwichensis*), shrews (*Sorex ornatus* and *S. vagrans*), and harvest mice (*Reithrodontomys raviventris* and *R. megalotis*). These case studies are followed by a synthesis of patterns that have emerged from the genetic study of tidal-marsh taxa and recommendations for new avenues of research.

A central issue addressed in these case studies is the timing of divergence between tidal-marsh and related non-tidal marsh taxa. Divergence times are based on molecular clock studies that calibrate rates of base pair substitutions using independent fossil or geological evidence. In many cases the studies reviewed here attempted to estimate a divergence date based on mitochondrial DNA (mtDNA) data. In most birds, for example, the cytochrome *b* gene has been found to have a substitution rate of 1–3% per million years (Fleischer and McIntosh 2001), and many authors use 2% per million years as a rough clock for estimating dates of cladogenesis (Klicka and Zink 1997, Avise and Walker 1998). A number of potential problems are associated with applying a molecular clock beyond the taxa for which it was estimated, including rate heterogeneity across lineages, calibration error, and overestimation of divergence time due to ancestral polymorphism (Edwards and Beerli 2000, Arbogast et al. 2002). For example, the overestimation due to ancestral polymorphism for cytochrome *b* in birds is likely to be about 175,000 yr, or on average 12% of avian haplogroup divergence is taken up by ancestral polymorphism (Moore 1995). These problems still need to be addressed in many of these taxa, however, with these caveats in mind, patterns of divergence across tidal marsh taxa may still be comparable.

CASE STUDIES

DIAMONDBACK TERRAPIN

Systematics, distribution, and ecology

The North American diamondback terrapins are medium-sized emydids that exploit and are confined to brackish coastal waters on the eastern coast of the US, from New York state to Texas. Emydid turtles are normally characteristic of freshwater ecosystems in the Americas, Europe, North Africa, and Asia. Although a handful of species have colonized productive estuarine areas, most cannot survive in sea water (Davenport and Wong 1986; Davenport et al. 1992). Unique among emydid turtles, *Malaclemys* is physiologically capable of spending several weeks in sea water without frequent access to fresh water (Gilles-Baillen 1970, Dunson 1985) and has therefore attracted much physiological and ecological study (Robinson and Dunson 1975).

Malaclemys terrapins have a suite of behavioral, physiological and morphological traits that allow them to occupy euryhaline environments ranging in salinity from 11–31 ppt (Dunson 1985). Their wide geographical distribution, perhaps coupled with limited gene flow between brackish-water populations separated by open coast, has led to an unusual degree of recognized subspeciation. The genus is monotypic and seven subspecies have been recognized throughout its range. Pritchard (1979) describes the following subspecies (running from north to south in the species' distribution): *M. t. terrapin* (found from Cape Cod to Cape Hatteras), *M. t. centrata* (a subspecies overlapping with the northern subspecies and stretching to Florida), *M. t. tequesta* (Florida east coast terrapin), *M. t. rhizophora* (an obscure subspecies found in Florida mangroves), *M. t. macrospilota* (the ornate diamondback of the southern part of the Gulf Coast of Florida), *M. t. pileata* (the Mississippi diamondback, distributed to eastern Louisiana from the Florida Panhandle), and finally the Texas diamondback, *M. t. littoralis*, which is distributed from Louisiana to Corpus Christi in southern Texas.

Phylogenetic analysis of morphological and molecular character data indicate map turtles (*Graptemys*) are closely related to *Malaclemys* and are likely a sister taxon (Gaffney and Meylan 1988, Bickham et al. 1996, Lamb and Osentoski 1997). Map turtles are largely riverine and with 12 species, the genus is the largest in the family Emydidae.

Evolutionary history and biogeography

A Pleistocene divergence between *Malaclemys* and *Graptemys* was proposed by Wood (1977) and was investigated by Lamb and Osentoski (1997) using molecular markers. Examination of the cytochrome *b* locus in mtDNA between *Malaclemys* and *Graptemys* revealed a deep divergence (1.54–3.11% sequence divergence). Assuming a cytochrome *b* evolutionary rate of 0.2–0.4% per million years in turtles (calibrated against fossil evidence and biogeographic barriers) Lamb and Osentoski (1997) estimated that *Malaclemys* and *Graptemys* may have diverged from a common ancestor some 7–11 MYA during the late Miocene and not during the Pleistocene. Thus the evolution of this group is the earliest example of divergence of a terrestrial tidal marsh vertebrate from its freshwater ancestor to date.

Genetic structure and within-species processes

Molecular markers have also been used to address genetic differentiation among populations of *Malaclemys* and within species evolutionary processes. Although cytochrome *b* sequence divergence values within *Malaclemys* subspecies range from 0.0–0.38% (Lamb and Ostentoski 1997), terrapins from the Atlantic assemblage north of Cape Canaveral were differentiated by restriction enzyme analysis as well as sequence analysis from a Gulf Coast assemblage from south Florida westward. This phylogeographic split detected within *Malaclemys* by Lamb and Osentoski (1997) supports prior inferences of the distinctness of Atlantic and Gulf Coast diamondback terrapins by Lamb and Avise (1992) who looked at mtDNA restriction site variation (N = 53 from Massachusetts to western Louisiana).

Lamb and Osentoski (1997) proposed that a regional vicariant event resulting from Pleistocene glacial maxima shaped the mtDNA divergence within *Malaclemys*. During Pleistocene glacial maxima, sea levels dropped approximately 150 m in the Gulf of Mexico, exposing extensive portions of the west Florida shelf as well as portions of the Yucatan Peninsula (Poag 1973). This land-mass expansion, coupled with increased aridity in the southeast and hypersaline conditions at the mouth of the gulf, likely isolated the gulf's estuarine ecosystems from those in the Atlantic.

Despite the finding that mtDNA haplotypes differed between Atlantic and Gulf coast populations, microsatellite markers showed low overall genetic differentiation between terrapin populations from New York, North and South Carolina, the Florida Keys, and Texas (S. Hauswaldt, pers. comm.). Analysis of a total sample of 320 individuals at six microsatellite loci provided evidence that East Coast terrapins were more similar to Texas terrapins than either group was to the terrapins from the Florida Keys—a pattern the researchers attributed to the well-documented translocation of Texas terrapins to the Atlantic Coast after the early-20th-century depletion of terrapin stocks by over harvest (S. Hauswaldt, pers. comm.).

In summary, mtDNA variation has demonstrated an ancient divergence for diamondback terrapins from their closest freshwater relative, the map turtles. Furthermore, molecular markers have supported the hypothesis of a Pleistocene vicariant event isolating Atlantic and Gulf coast populations. Finally, anthropogenic translocation of terrapins has left a genetic signature detected using microsatellite loci.

WATER SNAKES

Systematics, distribution, and ecology

Among vertebrates that inhabit tidal marshes, members of the water snake complex (*Nerodia fasciata-sipedon-clarkii*) provide an example of divergence at several levels. As previously recognized, *N. fasciata* of the southeastern US was comprised of six subspecies that could be clearly divided into two groups on the basis of ecology and physiology. The freshwater group (*N. f. fasciata*, *N .f. confluens*, and *N. f. pictiventris*) occupies an extensive area of freshwater habitats along the coastal plain of eastern North America (Conant 1975). The salt-water group (*N. f. clarkii*, *N. f. compressicauda*, and *N. f. taeniata*) inhabits a narrow coastal saltwater zone and are distributed almost continuously from the mid-Atlantic Coast of Florida to southern Texas, including the Florida Keys (Conant 1975). Progressive loss of saltmarsh habitat on the east coast of Florida has resulted in *N. f. taeniata* being listed by USDI Fish and Wildlife Service as a threatened species.

The snakes in the salt-water group are physiologically well adapted to exploit their saline environment (Zug and Dunson 1979; Dunson 1980). The geographic ranges of the saltwater and freshwater groups sometimes overlap in the saltwater-freshwater ecotone, yet, because of their divergent adaptations, they are for the most part microallopatric in these areas (Krakauer 1970). Krakauer (1970) speculated that the existence of hybrid populations of these two ecologically distinct groups would be transient, as any potential advantage resulting from heterosis would be counterbalanced by maladaptation of hybrids to freshwater or saline environments.

The ecological distinctness of these snakes has resulted in controversy regarding their taxonomic status (Clay 1938, Cliburn 1960, Conant 1963), which may be clarified using molecular markers. Analyses of allozymes at 33 protein coding loci by Lawson et al. (1991) showed that except for areas of considerable habitat disturbance, gene flow between the saltmarsh and freshwater groups is very slight or absent and is on the same order as that seen between the freshwater group and *Nerodia sipedon*. Based on these analyses, they recommended elevation of the two groups to species level.

According to Lawson et al. (1991), the evolution of saltwater adaptation in these snakes took place on Floridian Pliocene islands. These islands were probably small and may have been devoid of standing fresh water with a topography and climate that resembled that of the Florida Keys today. They hypothesize expansion and colonization along the shores of the Gulf of Mexico followed the closure of the Suwannee Straits. The freshwater *Nerodia fasciata pictiventris* evolved through a southward expansion into peninsular Florida from an ancestral population that was originally distributed north of the Suwanee Straits (Lawson et al. 1991). If one accepts that the freshwater and saltmarsh groups evolved in allopatry, the transition zones described in Lawson et al. (1991) are the result of secondary contact. The ecological adaptations of saltmarsh snakes and the freshwater group have reached a high degree of specialization and the fusion of these two groups seems unlikely. Rather, with the passage of time, Lawson et al. (1991) expected that the selection gradient between the two habitats would eventually promote development of increasing specialization and further the divergence between them.

Genetic structure and within species processes

In contrast to the large divergence observed at allozyme loci between the freshwater and saltwater group, Lawson et al. (1991) found that the within-group divergence is minimal at the allozyme level. Two morphological characters, head shape and numbers of dorsal scale rows, unite the saltmarsh group *Nerodia fasciata clarkii*, *N. f. compressicauda,* and *N. f. taeniata* and distinguish them from the freshwater subspecies of *N. fasciata*. The molecular evidence supports the close association of these three saltmarsh forms, but the question of whether *N. f. taeniata* arose from ancestral *N. f. clarkii* or ancestral *N. f. compressicauda* could not be resolved by analyses of these data. The genetic distances separating the three taxa are no greater than those found between demes within each.

The Carolina salt marsh snake

The Carolina salt marsh snake (*Nerodia sipedon williamengelsi*), currently listed by the state of North Carolina as a taxon of special concern, is a melanistic water snake endemic to estuarine habitats in coastal North Carolina. It is closely associated with saltmarshes dominated by black needlerush (*Juncus romerianus*), and *Spartina* marsh grasses. Although no formal physiological studies have been conducted to determine if this subspecies has similar adaptations to saline environments as those mentioned above for the saltmarsh group, Conant and Lazell (1973) determined that Carolina salt marsh snake would not drink salt water, a finding similar to that of the *N. f. clarkii*.

In order to clarify the taxonomic status and genetic distinctness of the Carolina salt marsh snake, Gaul (1996) used a combination of molecular and morphological techniques to examine the relationships between this saltmarsh snake and the nominate subspecies *Nerodia sipedon sipedon*, as well as the dynamics of hybridization between the saltmarsh snake and a closely related species, the banded water snake (*Nerodia fasciata*). In a study of restriction endonuclease digests of mtDNA, Gaul (1996) found six unique haplotypes in coastal *N. sipedon*, but no clear distinction was detected between *N. s. williamengelsi* and *N. s. sipedon*. However, analysis of morphological characters revealed statistically significant differences between the two subpecies in numbers of ventral scales, subcaudal scales, and lateral bars. Two morphological characters, ventral scales and lateral bars, showed evidence of clinal variation and appear to correspond closely to estuarine salinity gradients. Evidence for hybridization between *Nerodia sipedon williamengelsi* and *N. fasciata* was observed in five specimens; mtDNA variation observed in these hybrids suggests that hybridization events between these two species are bi-directional. In his review of the *N. sipedon-fasciata* complex, Conant (1963) speculated that interbreeding between the two forms represented introgressive hybridization, resulting from habitat alteration rather than evidence of conspecificity. However, further studies need to be done to establish the phylogenetic relationships of this subspecies. This is an example in which no evidence of genetic differentiation from a freshwater ancestor has occurred, despite significant morphological, ecological, and physiological adaptations.

CLAPPER RAIL AND KING RAIL

Systematics, distribution, and ecology

Clapper Rails are characteristic of tidal-salt and brackish-marsh habitat along the coasts of North America—and locally in freshwater marshes along the Colorado River and in the Imperial Valley, extending into mangrove swamps as far south as Peru and Brazil (Eddleman and Conway 1998). The closest inland relative to the Clapper Rail is the King Rail, which inhabits freshwater and brackish marshes, swamps, and rice fields (Ehrlich et al. 1988). Clapper Rails form a superspecies with the King Rail and another close relative, the Plain-flanked Rail (*Rallus wetmorei*), which occupies saltmarsh and mangrove habitats syntopically with Clapper Rails in Venezuela. Based on mitochondrial ATPase8 sequences from a museum specimen, *R. wetmorei* falls within a clade containing both King and Clapper rail sequences (B. Slikas, pers. comm.), and in fact is identical in sequence to King Rail and many Clapper Rails. No other *Rallus* species appear to be closely related to this group of rails.

Clapper Rails hybridize readily with King Rails in habitats of intermediate salinity on the East and Gulf coasts (Meanly and Weatherbee 1962) and the California subspecies of Clapper Rails were at one time considered subspecies of King Rails. Some authors, such as Ripley (1977) favor conspecific status, but the American Ornithologists' Union (1983) favored recognition as distinct species. Olson (1997) believed that the hybridization is limited to areas of intermediate salinity and does not result in enough introgression to justify merging of the taxa into one species. He showed that the width of the interorbital bridge, the region of the skull in which the salt glands occur, was greater in King than in Clapper rails. Furthermore, Olson provided evidence that this trait was stable and genetically based by comparing interorbital widths between Clapper Rails reared in captivity in freshwater with wild ones from salt-marshes. He found the width was intermediate in one hybrid specimen.

Evolutionary history and biogeography

King and Clapper rails from the Gulf Coast showed very low levels of divergence in both allozymes (37 loci) and mtDNA (15 endonucleases; Avise and Zink 1988; King Rail, N = 10; Clapper Rail, N = 7). Allozymes showed no fixed differences and only one significant difference in allele frequency between the two species. The low allozyme divergence is expected given what has been found in birds generally (Barrowclough 1983), but the lack of divergence at mtDNA was surprising. Using 15 endonucleases Avise and Zink (1988) found an estimated sequence divergence of 0.6% in the mtDNA between King and Clapper rails. Interestingly, they also found evidence of intraspecific size polymorphism and more than one type per cell (heteroplasmy) in their mtDNA.

The low sequence divergence in King and Clapper rails has been confirmed by further research (R. C. Fleischer, unpubl. data). King and Clapper rails differed by 0.8% in cytochrome *b* sequences (while both differed from Virginia Rails [*Rallus limicola*] by 7-8%) and a tree rooted by the Virginia Rail showed the King Rail falling within the Clapper Rail clade. King and Clapper rails differed by an average of 0.3% for the central domain of the control region (275 base pairs), but King Rails (N = 4) had sequences identical to those of Gulf Coast (N = 4) and East Coast (N = 4) Clapper Rails. Thus the differences between the two species are minor and generally at the level of variation often found within populations of a single species. These results suggest that the morphological and ecological differences noted by Olson (1997) and others may be very recently evolved, or are maintained despite a large level of hybridization.

Conservation genetics

Twenty-one subspecies of Clapper Rails are described, with six occurring in the United States. The 21 subspecies have been divided into three groups based on plumage and geography: *crepitans* on the Atlantic and Gulf coasts of North America and the Caribbean; *obsoletus* in California and Northern Mexico; and *longirostris* in South America. Although most East Coast populations are abundant, West Coast populations are limited by substantial recent losses of habitat and three subspecies in the western US are listed as endangered (California Clapper Rail [*Rallus longirostris obsoletus*],

Light-footed Clapper Rail [*R. l. levipes*], and Yuma Clapper Rail [*R. l. yumanensis*]; Eddleman and Conway 1998).

In order to clarify the taxonomic status of the endangered subspecies of Clapper Rails, R. C. Fleischer (unpubl. data) compared mtDNA sequence variation between *Rallus longirostris levipes* (N = 7) and *R. l. yumanensis* (N = 4) of southern California and among samples from the subspecies *R. l. obsoletus* (N = 3) of northern California, *R. l. crepitans* of the eastern US (N = 4), and *R. l. saturatus* of the Gulf of Mexico (N = 4). No differences in central domain control region or cytochrome *b* mtDNA sequences were found among *R. l. yumanensis*, *R. l. obsoletus*, and *R. l. levipes*. No differences were found between *R. l. crepitans* and *R. l. saturatus*. Only a single base difference (in the control region sequence) separated the eastern and western samples indicating a divergence of <0.2% across the continent for the two gene regions. Thus mtDNA data provide little support for any of the subspecies occurring in North America.

Fleischer et al. (1995) and Nusser et al. (1996) also studied the genetics of two of these subspecies, *Rallus longirostris yumanensis* and *R. l. levipes* using minisatellite and randomly amplified polymorphic DNA (RAPD) markers to examine the extent of genetic variation within populations. They assessed variation among four disjunct marsh populations ranging from San Diego to Point Mugu along the southern California coast.

Fleischer et al. (1995) found extremely low minisatellite variation within the four isolated coastal marsh *Rallus longirostris levipes* populations (estimated heterozygosities of 25–42%), while the single sample of *R. l. yumanensis* yielded a more typical heterozygosity of 72%. Interestingly, the band-sharing among populations indicated a very low level of divergence (high gene flow) among the four coastal populations, but a substantially lower level of gene flow between the two subspecies (Fleischer et al. 1995). The RAPD analysis of the same birds (Nusser et al. 1996) revealed extremely low levels of variation in both subspecies (16 polymorphic bands out of 1,338 scored), and almost no divergence between them taking all loci into account (0.23% divergence). Patterns based on the 16 polymorphic bands revealed greater similarity among the four *R. l. levipes* populations than between these populations and *R. l. yumanensis*.

In summary, members of the *Rallus longirostris-elegans* complex are found across North America and exhibit limited divergence between species and among subspecies of Clapper Rails. A comprehensive study that looks further at relationships within and between both King and Clapper rails is needed to investigate the hypothesis of the invasion of Clapper Rails into the tidal-marsh habitat and morphological and genetic differentiation from populations of King Rails. An investigation of the degree of hybridization and back-crossing between the two species is also important. Additional genetic markers, such as microsatellites or AFLPs (amplified fragment length polymorphism) need to be developed to allow us to determine the patterns of colonization of North America. Furthermore, genetic studies indicate that isolated coastal populations of *R. l. levipes* exhibit low levels of heterozygosity and may therefore be in danger of reduced viability and inbreeding depression.

SALTMARSH SHARP-TAILED SPARROW AND SEASIDE SPARROW

Systematics, distribution, and ecology

The genus *Ammodramus* contains the only two species of bird that are essentially endemic to tidal marshes (Greenberg and Maldonado, *this volume*): the Saltmarsh Sharp-tailed Sparrow and the Seaside Sparrow. We use the term essentially, because the latter species has some subspecies, including the extinct Dusky Seaside Sparrow (*Ammodramus maritimus nigrescens*) in east Florida and the endangered Cape Sable Seaside Sparrow (*A. m. mirabilis*) in south Florida that occurred or occur locally in freshwater marsh and flooded prairie as well as brackish and saltmarsh. Nelson's Sharp-tailed Sparrow (*A. nelsoni*), which until recently (American Ornithologists' Union 1983) was considered conspecific with the Saltmarsh Sharp-tailed Sparrow, also has two subspecies that are associated with coastal marshes.

Taxa within the seaside-sharp-tail group of sparrows have received various levels of taxonomic recognition. Different subspecies or groups of subspecies have been elevated to or demoted from species status. As many as five and as few as two species have been recognized. In this account, we capitalize the names of the three currently recognized species, and sharp-tailed sparrows refers to Nelson's and Saltmarsh Sharp-tailed sparrows together. Figure 1 shows species names and relationships.

The Seaside Sparrow breeds in the Gulf and Atlantic coast saltmarshes from New Hampshire south to Florida and west to Texas. Nine subspecies of Seaside Sparrow are currently recognized, of which two were formerly accorded species status (American Ornithologists' Union 1957, 1973), but it is likely that the number

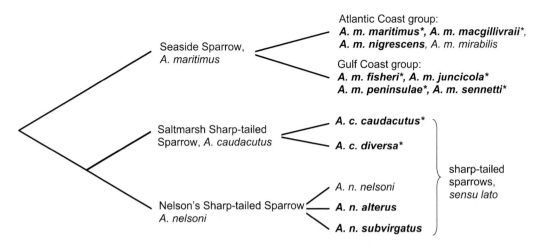

FIGURE 1. Relationships among species and subspecies of marsh-nesting *Ammodramus* sparrows. Taxa in bold inhabit tidal marshes; tidal-marsh obligates are indicated with asterisks. *A. nelsoni nelsoni* inhabits mid-continent freshwater marshes and *A. melodia mirabilis* inhabits freshwater prairies in Florida. Major mtDNA lineages in sharp-tailed sparrows (1.2% different; Rising and Avise 1993) are currently recognized as full species, while major mtDNA lineages in Seaside Sparrows (1.0%; Avise and Nelson 1989) are not, and are labeled groups above. Although this diagram is consistent with existing molecular data (J. Klicka, unpubl. data), it is provided primarily to aid the reader in following the more detailed discussion of named taxa.

of recognized subspecies should be reduced (Post and Greenlaw 1994). Populations from New Jersey north (*Ammodramus maritimus maritimus*) are largely migratory and winter in the southeastern US. South of the Chesapeake Bay, Seaside Sparrow populations are year-round residents and include the subspecies *A. m. macgillivraii* from the mid-Atlantic states to Georgia, *A. m. nigrescens* and *A. m. mirabilis* in east and south Florida respectively, and *A. m. peninsulae*, *A. m. juncicola*, *A. m. fisheri* and *A. m. sennetti* along the Gulf Coast from Florida to Texas. With the exception of the subspecies on the Florida Peninsula, Seaside Sparrows are saltmarsh obligates, breeding, wintering, and even migrating via saltmarsh habitat.

The Saltmarsh Sharp-tailed Sparrow breeds in saltmarshes from southern Maine to Virginia. Two weakly differentiated subspecies are recognized: *Ammodramus caudacutus caudacutus* from Maine to New Jersey and *A. c. diversus* from New Jersey to the Delmarva Peninsula. Most individuals migrate to the southern US to winter along the Atlantic Coast and, rarely, the Gulf Coast (Post 1998). Saltmarsh Sharp-tailed Sparrows are also saltmarsh obligates, and their range is as linear and patchy as that of the Seaside Sparrow.

Nelson's Sharp-tailed Sparrow encompasses three subspecies, one of which, *Ammodramus nelsoni nelsoni*, breeds in interior freshwater marshes. A second, *A. n. alterus*, breeds in coastal marshes along James Bay and Hudson Bay. The third, *A. n. subvirgatus*, breeds from the Gulf of St. Lawrence to southwestern Maine. *A. n. subvirgatus* breeds mostly in saltmarshes, but perhaps partly due to differences in habitat availability in the northern part of its range, can be found in brackish and fresh estuarine marshes as well. All three subspecies are migratory, wintering in the coastal marshes of the southeastern US, with the interior subspecies wintering largely along the Gulf Coast (Post 1998; J. S. Greenlaw, pers. comm.).

Evolutionary history, and biogeography

The systematics, biogeography, and degree of population isolation of the *Ammodramus* sparrows has been addressed repeatedly in the past century using morphology, behavior, and molecular genetic tools such as allozymes, mtDNA RFLPs, microsatellite DNA polymorphism and mtDNA sequencing. This offers the possibility of understanding timing and evolutionary dynamics involved in what may have been multiple transitions by upland forms to tidal marsh endemism.

Zink and Avise (1990) examined the phylogenetic relationship of eight *Ammodramus* species based on trees derived from mtDNA RFLPs (16 restriction enzymes) and allozymes (30 enzyme loci, 24 variable), and using the Savannah Sparrow as an outgroup. LeConte's (*A. lecontei*), Seaside and sharp-tailed sparrows consistently formed a wetlands clade in this analysis (Rising 1996). Although the Savannah Sparrow may not be an appropriate outgroup to *Ammodramus* as a whole (Avise et al. 1980, Zink and Avise 1990, Carson and Spicer 2003) (*Ammodramus* as currently constituted is probably not monophyletic), Zink and Avise's (1990) work provides the best framework so far for understanding the relationships within the genus. MtDNA data and allozyme data differed on the exact relationships within the wet grassland clade so it remains unclear if sharp-tails share a more recent common ancestor with LeConte's or Seaside sparrows.

Prior to the Zink and Avise phylogenies, two scenarios were proposed for the evolution of the saltmarsh *Ammodramus*. Beecher (1955) invoked late Pleistocene and Holocene isolation by postglacial marine embayments that formed over depressed river valleys as the mechanism leading to diversification of the Seaside from sharp-tailed sparrows, which in turn were derived from Savannah Sparrows. Funderburg and Quay (1983) suggested that Seaside Sparrows evolved from Savannah Sparrows in the late Pliocene and early Pleistocene.

In a scenario more consistent with the molecular phylogenies Greenlaw (1993) proposed two cycles of vicariance and differentiation in sharp-tailed sparrows in the Pleistocene. Depending on whether sharp-tails were derived from Seaside or LeConte's sparrows, Greenlaw presents two alternative scenarios. In the first, Seaside Sparrows invade tidal habitats on the East Coast and later give rise to the southern coastal form of sharp-tail (now *Ammodramus caudacutus*), which then colonizes inland, freshwater habitats, evolving into *A. nelsoni*. The inland form then reinvades coastal habitat at a later time, becoming the present-day subspecies *A. n. subvirgatus*. This scenario involves two invasions of tidal habitat—an early invasion by Seaside Sparrows, and a very recent reinvasion of saltmarshes by *A. n. subvirgatus*.

Greenlaw's alternative scenario posits LeConte's Sparrow as the sister taxon to sharp-tails. In this scenario, LeConte's Sparrow gives rise to an inland, freshwater form of sharp-tail (now *Ammodramus nelsoni*). This form invades coastal habitats in mid-Pleistocene and gives rise to *A. caudacutus*. Later, the inland form invades tidal habitats a second time, giving rise to *A. n. subvirgatus*. This second scenario involves three separate invasions of tidal habitat by *Ammodramus* sparrows: (1) the inland form of the sharp-tail gives rise to *A. caudacutus*, (2) the inland form later gives rise to *A. n. subvirgatus*, and (3) Seaside Sparrows invade tidal marshes independently of sharp-tails. In both cases, the current zone of sympatry between *A. n. subvirgatus* and *A. c. caudacutus* in southwestern Maine is hypothesized to represent secondary contact.

The uncertain phylogeny of the *Ammodramus* genus as a whole complicates speculation about the invasion of tidal marsh habitat by Seaside and/or sharp-tailed sparrows. The ancestral condition of emberizines is terrestrial—at least one invasion of tidal marshes must have occurred. If contact between *A. n. subvirgatus* and *A. c. caudacutus* is secondary, and if *A. n. subvirgatus* is descended from inland forms, as put forth by Greenlaw (1993) at least two invasions of tidal marshes have occurred. If sharp-tailed sparrows share a more recent common ancestor with LeConte's Sparrows than with Seaside Sparrows, then three separate invasions may have occurred. At this point the genetic evidence appears to rebut the scenarios that hypothesize a close link between Seaside and Savannah sparrows. At the very least, Seaside Sparrows are more closely related to both LeConte's and sharp-tailed sparrows than they are to Savannah Sparrows. However, the exact order of branching within the LeConte's-Seaside-sharp-tailed clade is not yet clear, nor is the closest relative outside that group. To some extent we are still awaiting a more comprehensive phylogeny of the Emberizidae that includes the marsh-nesting *Ammodramus* species in the context of close and distant relatives before we can distinguish between some of these scenarios, and put speculation about the original shift(s) by the ancestor(s) of these species to tidal marsh endemism on more solid ground.

No source disputes that within the marsh-nesting sparrows the Nelson's Sharp-tailed Sparrow and the Saltmarsh Sharp-tailed Sparrow are each other's closest relatives. In fact, until recently the two forms were considered conspecific (American Ornithologists' Union 1983). Greenlaw (1993) documented both morphological (plumage coloration, bill size) and behavioral (song and display) differences between a northern group of sharp-tailed sparrows (the subspecies *nelsoni*, *alterus*, and *subvirgatus*, at that time still part of *Ammodramus caudacutus*) and a southern group (subspecies *caudacutus* and *diversus*). Rising and Avise (1993) used mtDNA RFLPs and skeletal morphology to study differences among the subspecies and populations of sharp-tailed sparrows (N = 220 individuals, 12 sites throughout the range, 20 restriction enzymes) and discovered a deep division (average of 1.2% sequence divergence) between northern and southern forms, and much less divergence (0.2%) within each region. Assuming a rate of mtDNA RFLP sequence divergence of 2%/1,000,000 yr (Shields and

Wilson 1987), the gap between northern and southern forms corresponds to a separation of approximately 600,000 yr duration. The mtDNA groups correspond to Greenlaw's behavioral division between northern and southern forms, and skeletal characters also distinguished northern from southern sharp-tails (Rising and Avise 1993). Also, observations from marshes in southern Maine, where the northern mtDNA forms (represented by *A. n. subvirgatus*) and the southern forms (*A. c. caudacutus*) come in contact, suggested that assortative mating is the rule where the forms are sympatric (Greenlaw 1993).

Following these studies, the American Ornithologists' Union (1995) recognized each form as a full species, the northern form as Nelson's Sharp-tailed Sparrow and the southern as Saltmarsh Sharp-tailed Sparrow. The presumed date of the split based on mtDNA differences between these two species (600,000 yr before present, mid-Pleistocene) is inconsistent with the hypothesis put forward by Beecher (1955), who had proposed the early Holocene (about 8,000 yr before the present) as the date of sharp-tailed sparrow diversification into present forms. Likewise, the division of Seaside Sparrows into two similarly ancient groups argues against the timing suggested by Beecher (1955) and Funderburg and Quay (1983) for late Pleistocene-Holocene diversification of Seaside Sparrows.

Genetic structure and within-species processes

In the same way that sharp-tailed sparrows proved to have a deep genetic divide between two groups, Seaside Sparrows also show a major genetic split. However, unlike in sharp-tails, the genetic divide in Seaside Sparrows does not appear to be consonant with any morphological or behavioral divide. In a broad-ranging survey, Avise and Nelson (1989) assayed 40 individuals from 10 locations throughout the Seaside Seaside Sparrow's range, digesting mtDNA with 18 informative restriction enzymes. Nelson et al. (2000) assayed mtDNA of four individuals of the threatened Cape Sable Seaside Sparrow (*Ammodramus m. mirabilis*) with the same enzymes. These studies found a group of Atlantic Coast birds that was quite divergent in mtDNA genome from a second group along the Gulf Coast. Mitochondrial sequence divergence was estimated at 1% between the two groups, corresponding to separate evolutionary trajectories for these two groups over the last 500,000 yr (Avise and Nelson 1989). Sequence divergence among individuals on each coast averaged about 0.2%. Both of the isolated and distinctively marked Florida forms, the endangered *A. m. mirabilis* and the extinct *A. m. nigrescens*, belonged to the Atlantic Coast group (Avise and Nelson 1989, Nelson et al. 2000), and in fact, nothing from mtDNA typing suggested that either subspecies was particularly divergent from other Atlantic Coast forms.

Because of the concordance between phenotypic and molecular characters, the American Ornithologists' Union (1995) now recognizes two species of sharp-tailed sparrows. With only the mtDNA divergence data showing an ancient divergence, the Gulf and Atlantic coast populations of Seaside Sparrows have not been accorded species status, and in fact the divergence between Atlantic and Gulf coast Seaside Sparrows receives no special taxonomic recognition at all. MtDNA is well suited to tracing gene genealogies, but multiple nuclear loci are particularly useful in monitoring gene diversity and gene flow between populations. The patchy and isolated nature of tidal-marsh habitats and the small population sizes within patches may make saltmarsh sparrows particularly prone to loss of genetic diversity due to bottlenecks. Seutin and Simon (1988) observed one example of this effect when perhaps as few as three individuals of *subvirgatus* type Nelson's Sharp-tailed Sparrows established a breeding colony on fresh water in the St. Lawrence River, Quebec, in 1980, 200 km from the nearest known breeding site. This colony increased to perhaps 50 individuals over the next 6 yr. Three males collected in 1986 displayed complete uniformity at 46 enzyme loci, and also showed no variation in protein banding patterns from isoelectric focusing (Seutin and Simon 1988).

C. E. Hill (unpubl. data) examined population differentiation in three populations of Seaside Sparrows (N = 15–61 individuals/population), using four microsatellite loci. One of those populations was from St. Vincent National Wildlife Refuge in Apalachicola, Florida (representing the Gulf Coast mtDNA group), and two were from South Carolina, from marshes near Georgetown and near Charleston (about 90 km from each other, and roughly 600 km from the Florida population). As would be predicted by the mtDNA studies or by an isolation-by-distance model, the two South Carolina populations were more similar to each other than either was to the Florida population. However, the two South Carolina populations also showed significant differences from each other in allelic frequencies at microsatellite loci, even though separated by less than 100 km, suggesting low levels of gene flow between Seaside Sparrow populations even over short distances.

Conservation genetics

Understanding the genetic structure of *Ammodramus* sparrow populations can guide the captive breeding of highly endangered forms, reveal which populations are affected by hybridization, and also help in setting conservation priorities for less threatened forms. Knowledge of the genetic divide between Gulf and Atlantic coast Seaside Sparrows would have better informed the selection of a breeding female to mate with the only surviving Dusky Seaside Sparrow. We now know that Dusky Seaside Sparrows had Atlantic, not Gulf, mtDNA and that, had the male Dusky Seaside Sparrows survived long enough to complete the breeding program, all the offspring from the captive breeding program would have had Gulf Coast mtDNA (Avise and Nelson 1989), although having the wrong mtDNA might not have hurt the viability of those offspring (Zink and Kale 1995).

In another genetic study with conservation implications, Shriver (2002) used five polymorphic microsatellite loci to estimate gene flow across the zone of sympatry of Nelson's (*Ammodramus n. subvirgatus*) and Saltmarsh (*A. c. caudacutus*) Sharp-tailed Sparrows in southern Maine. He sampled sparrows from five breeding locales, ranging from Lubec, Maine (allopatric *A. n. subvirgatus*), to three marshes in southwestern Maine, where both species occur, to Prudence Island, Rhode Island (allopatric *A. c. caudacutus*). In the area of sympatry, he classified 19 of 89 birds by plumage as hybrids. Of the apparently pure parental types in the zone of sympatry, 29 birds appeared to be Nelson's Sharp-tailed Sparrows, and 41 appeared to be Saltmarsh Sharp-tailed Sparrows. Genetic analysis confirmed that the hybrids were intermediate in allelic frequencies between the allopatric parental forms. In addition, the putative parental types within the zone of sympatry carried, on average, 25% of their microsatellite genotype from the other species. Few genetic barriers apparently exist between the two newly recognized species of sharp-tailed sparrows. Introgression may be slowed by the narrow nature of the hybrid contact, but *A. n. subvirgatus* genes have still been found in *A. c. caudacutus* populations as far south as Parker River, Massachusetts, which Shriver argues should perhaps increase the conservation priority for Saltmarsh Sharp-tailed Sparrows in the area unaffected by introgression (Shriver 2002).

The genetic exchange across the hybrid zone between the two species of sharp-tailed sparrows also brings back unanswered questions about the biogeography of sharp-tailed and Seaside Sparrows in the Pleistocene glaciations. If the two mtDNA defined clades of sharp-tails interbreed so readily today, then how have the two clades maintained their distinctiveness through 600,000 yr and several Pleistocene glacial advances? Were the two groups confined to distinct refugia in each glacial advance and have only come into contact in the Holocene? Perhaps the ancestors of Nelson's Sharp-tailed Sparrows were in freshwater refugia and the ancestors of Seaside Sharp-tailed Sparrows on the coast (Greenlaw and Rising 1994). The same question could be asked about the Gulf Coast versus Atlantic Coast seaside sparrow mtDNA groups: how did they remain separate through 500,000 yr of Pleistocene glaciation and of sea-level rise and fall, when the distribution of salt-marsh habitat was radically different from the present distribution?

In summary, the Seaside Sparrow, the Saltmarsh Sharp-tailed Sparrow, and the *subvirgatus* subspecies of the Nelson's Sharp-tailed Sparrow have all invaded the tidal-marsh habitat and are fully or largely endemic to it. Since their invasion of saltmarshes, the forces that have determined their evolutionary trajectories may have included the linear and patchy nature of their habitat and a corresponding tendency to isolation in small habitat islands, as well as apparent strong selection, especially on plumage color. The role of glacial advances and the ways in which the sparrows were assorted into refugia during times of glacial advance are not entirely clear. There are divisions apparent from mtDNA analyses dating back 500,000–600,000 yr in both Seaside and sharp-tailed sparrows. Within each mtDNA group, subspecies are distinctive at least in plumage (Seasides), and sometimes in plumage, ecology, and morphology (sharp-tails). The two species of sharp-tails retain their ability to interbreed, and do interbreed in New England; the two mtDNA groups of Seaside Sparrows are currently allopatric. One subspecies of Seaside Sparrow is extinct, another is endangered, and the restricted and fragmented range of all these birds means local populations or entire subspecies may be quite vulnerable to extirpation, as happened with the Dusky Seaside Sparrow.

SAVANNAH SPARROW

Systematics, distribution, and ecology

Savannah Sparrows range across North America from Alaska to central Mexico and are characteristic of open habitats including grasslands, meadows, and agricultural fields. Seventeen subspecies are recognized. Although most populations of Savannah Sparrows are migratory, five or six of the subspecies are resident or partially migratory in saltmarshes in California or Mexico and two are resident in coastal Sonora and Sinaloa (Wheelwright and Rising 1993, Rising 2001). These tidal-marsh subspecies can be divided into two groups, the large-billed group (*Passerculus sandwichensis rostratus*) and the Belding's group (*P. s. beldingi*). The large-billed Savannah Sparrows are distinctive enough morphologically to have once been considered a species (American Ornithologists' Union 1931). They are characterized by a large decurved culmen and are pale with diffuse streaking (Taber 1968b, Pyle 1997). Tidal marsh subspecies belonging to the large-billed group consist of *P. s. rostratus* (breeds on the Gulf Coast of northeast Baja California and northwest Sonora) and *P. s. atratus* (resident on the coast of central Sonora south to central Sinaloa). The Belding's group is characterized by long, straight, slender bills and is darkly streaked (Taber 1968a, Pyle 1997). The Belding's group consists of *P. s. beldingi* (resident on the Pacific coast from Morro Bay, California, south to El Rosario, Baja California), *P. s. annulus* (resident around the shores of Bahia Sebastian Vizcaino, Baja California) and *P. s. guttatus* (resident around Laguna San Ignacio) and *P. s. magdalenae* (resident around Bahia Magdalena; Wheelwright and Rising 1993). Breeding Savannah Sparrows increase clinally in size along the Pacific Coast from south to north (Rising 2001).

Genetic differentiation of tidal-marsh endemic

In a preliminary molecular study, Zink et al. (1991) compared three specimens of *Passerculus sandwichensis rostratus* using 20 restriction endonucleases and found a large amount of genetic differentiation between *P. s. rostratus* and non-tidal marsh subspecies in California (N = 3) and Louisiana (N = 5). They estimated 1.7% sequence divergence between non-marsh and *P. s. rostratus* (Zink et al. 1991). In a more comprehensive study Zink et al. (in press) sequenced two mitochondrial genes, ND2 and ND3 from five sites in Baja California, and coastal Sonora, and compared them with eight continental populations (total N = 112). They found the saltmarsh populations to be genetically distinct from continental populations (average nucleotide sequence divergence was 2%, F_{st} = 0.063, P < 0.001). They recommended based on genetic, morphological, and behavioral differences that the saltmarsh and typical Savannah Sparrows be considered separate species (Zink et al., in press).

Zink et al. (in press) further suggested that at one time three isolated populations of Savannah Sparrows existed, belonging to three clades. Since then, two of the clades which are found in continental Savannah Sparrows have become admixed, however the third clade belonging to the saltmarsh Savannah Sparrows have remained isolated either by habitat barriers or geographical distance. Using a calibration of 2% sequence divergence per million years (Shields and Wilson 1987, Tarr and Fleischer 1993), *P. s. rostratus* has been isolated for at least 750,000 yr from typical populations of Savannah Sparrows (Zink et al. 1991). The lack of geographic structure within saltmarsh Savannah Sparrows suggests that there is either widespread gene flow throughout the range of saltmarsh Savannah Sparrows, or that the colonization of different saltmarsh habitats are too recent for effective sorting of lineages (Zink et al. in press).

Song Sparrow and Swamp Sparrow

Sparrows of the genus *Melospiza* are widespread in shrubby habitats and wetlands throughout North America. Of the three species in the genus, the Song Sparrow (*Melospiza melodia*) and the Swamp Sparrow (*M. georgiana*) have colonized tidal-marsh habitats and have recognized tidal marsh subspecies along the shores of San Francisco Bay and the mid-Atlantic estuaries, respectively. The Swamp Sparrow probably split (along with the boreal-breeding Lincoln's Sparrow [*Melospiza lincolnii*]) from a common ancestor shared with Song Sparrows in the early to mid-Pleistocene (Zink and Blackwell 1996).

Systematics, distribution, and ecology

Song Sparrows are widespread inhabitants of moist habitats across North America. Extremely variable across their range, various authorities have recognized from 24–30 diagnosable subspecies (American Ornithologists' Union 1957, Arcese et al. 2002). Three of those subspecies are resident in tidal-saltmarsh habitat in the San Francisco Bay region, each occupying one of three sub-bays of the greater San Francisco Bay. The Samuel's Song Sparrow (*Melospiza melodia samuelis*) is resident in San Pablo Bay, the Suisun Song Sparrow (*M. m. maxillaris*) in Suisun Bay, and the Alameda Song Sparrow (*M. m. pusillula*) in south San Francisco Bay (Fig. 2). These Song Sparrow subspecies differ markedly from each other and related non-tidal marsh subspecies in plumage and size. The Samuel's Song Sparrow is small in size and blackish olive in dorsal coloration (Marshall 1948b). The Alameda Song Sparrow is slightly smaller than the Samuel's

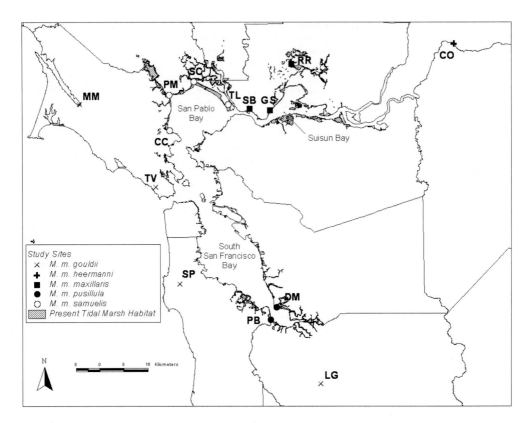

FIGURE 2. Map of San Francisco Bay region showing sampling sites from Chan and Arcese (2003) and the tidal-marsh habitat of three tidal-salt-marsh subspecies of Song Sparrow (*M. melodia samuelis*, *M. m. maxillaris*, and *M. m. pusillula*). The range of *M. m. samuelis* is San Pablo Bay (sites—China Camp State Park, Marin Co. [CC], Petaluma River Mouth, Sonoma Co. [PM], Sonoma Creek, Sonoma Co. [SC], and Triangle Levy, Sonoma Co. [TL]), *M. m. maxillaris* is Suisun Bay (sites—Benicia State Recreation Area, Solano Co. [SB], Goodyear Slough, Solano Co. [GS], and Rush Ranch Open Space, Solano Co. [RR]), and *M. m. pusillula* is San Francisco Bay (sites—Palo Alto Baylands, Santa Clara Co. [PB], Dumbarton Marsh, Alameda Co. [DM]). *M. m. gouldii* occurs in upland habitats surrounding the bay (sites—Mark's Marsh, Marin Co. [MM], Tennessee Valley, Marin Co. [TV], San Pedro Valley County Park, San Mateo Co. [SP], and Los Gatos Creek County Park [LG]); *M. m. heermanni* occurs to the east of Suisun Bay (site—Cosumnes River Preserve, Sacramento Co. [CO]) (San Francisco Estuary Institute. 2000).

Song Sparrow, has a yellowish-grey dorsal color and is the only Song Sparrow subspecies with a yellowish wash to the belly (Ridgeway 1899, Marshall 1948b). Suisun Song Sparrow is the largest of the tidal-marsh subspecies with a laterally flared bill at the nostrils (Marshall 1948b). The Suisun Bay birds have the largest relative bill size of any North American Song Sparrow. The closest upland species are the Marin Song Sparrow (*M. m. gouldii*) found along the California coast (Grinnell and Miller 1944) and the Heerman's Song Sparrow (*M. m. heermani*) resident in central and southwest California (previously *M. m. mailliardi*; Arcese et al. 2002) whose range meets the Suisun Song Sparrow at the eastern edge of Suisun Bay.

Swamp Sparrows are typical inhabitants of freshwater and brackish marshes across the eastern US and Canada (Mowbray 1997). In contrast to the extremely polytypic Song Sparrow, the Swamp Sparrow consists of only three subspecies, one of which, the Coastal Plain Swamp Sparrow (*Melospiza georgiana nigrescens*), is endemic to tidally influenced brackish marshes and freshwater marshes along the mid-Atlantic Coast (Wetherbee 1968). The Coastal Plain Swamp Sparrow is distinguished by a larger bill, grayer plumage, and increased black coloration on the head and nape (Greenberg and Droege 1990).

Evolutionary history and biogeography

Song Sparrows show no geographic structure in mitochondrial haplotypes corresponding to the geographic variation in plumage and other aspects of morphology captured in the currently recognized subspecies (Zink and Dittmann 1993, Fry and Zink 1998). The lack of geographic structure in mitochondrial haplotypes may result from the expansion of the Song Sparrow from multiple refugia after the glaciers receded (Zink and Dittmann 1993, Fry and Zink 1998). To further explore the probable recency of the evolution of geographic variation in Song Sparrows in general, Chan and Arcese (2002) examined genetic structure within and among tidal-marsh populations. This analysis used nine microsatellite loci to examine differentiation among six tidal marsh populations (two per tidal marsh subspecies) and three nearby upland populations (total N = 215, average of 22 birds per population; Fig. 2). Overall genetic differentiation was extremely low (F_{st} = 0.0217, P < 0.001) indicating a large amount of gene flow (7.78 immigrants per generation; Chan and Arcese 2002). Despite the low amount of differentiation, they found the Alameda Song Sparrow was genetically distinct from both upland subspecies and other tidal-marsh subspecies. However, they did not find statistically significant differences in allele frequencies between the Samuel's Song Sparrow, Suisun Song Sparrow, and the Heerman's Song Sparrow (Chan and Arcese 2002). The lack of differentiation among those subspecies, coupled with the pattern of variation at microsatellite loci found by Chan and Arcese (2002) suggest that San Pablo Bay and Suisun Bay tidal-marsh sparrows were the result of a recent invasion from the Central Valley of California.

This hypothesis is concordant with the geological history of the tidal marshes in the San Francisco Bay region, which provide some insight into the evolutionary history of the tidal-marsh Song Sparrow subspecies. Evidence from core samples indicates at least four episodes of emergence and submergence from river valley to estuary on the continental shelf that is now San Francisco Bay (Atwater 1979). The most recent submergence occurred approximately 10,000 yr ago, when what is now San Francisco Bay was a riparian valley through which the Sacramento and San Joaquin rivers flowed (Atwater 1979). The current water level was reached about 5,000 yr ago (Atwater 1979), but the tidal marshes surrounding San Francisco Bay are probably 4,000–6,000 yr old with south San Francisco Bay marshes being a bit younger, approximately 2,000 yr old (Atwater et al. 1979).

The origin of the Alameda Song Sparrow is not clear because the sister group was not identified in their study; however, the pattern of relatedness between populations suggests a different colonization into the tidal marsh habitat for the Alameda Song Sparrow. Therefore, it appears that two separate invasions to the tidal-marsh habitat in San Francisco Bay occurred, not subspecific differentiation within a single colonization.

Despite the morphological divergence apparent in all three tidal-marsh Song Sparrow subspecies (Marshall 1948b; Chan and Arcese 2002, 2003), only one subspecies was differentiable at a neutral, rapidly evolving marker. Chan and Arcese (2002) noted four possible explanations for this discrepancy: (1) recent divergence of tidal marsh subspecies and lack of differentiation due to inadequate lineage sorting—possibly due to large effective population size, (2) high current gene flow and strong selection on morphological or plumage characters resulting in a decoupling of neutral loci and quantitative loci, (3) high current gene flow resulting in introgression between previously differentiated subspecies, or (4) high gene flow with a large environmental component to morphological and plumage development. Further research involving common garden experiments would aid greatly in differentiating between these scenarios.

As with Song Sparrow subspecies, molecular studies on the Coastal Plain Swamp Sparrow have failed to find marked differentiation at nuclear (Balaban 1988) or mitochondrial loci (Greenberg et al. 1998). Greenberg et al. (1998) sequenced a total of 641 base pairs of mtDNA, including the hypervariable mtDNA control region, COII/t-lys/ATPase8, and ND2. They found extremely low levels of genetic variation (mean sequence divergence = 0.21%) and population differentiation (F_{st} = 0.057, P = 0.208) from 29 Swamp Sparrows, including individuals from two populations of the Coastal Plain Swamp Sparrow (near Delaware Bay, N = 7; near Chesapeake Bay, N = 4), two populations of the two inland subspecies (Garrett County, Maryland N = 5; Clay County, Minnesota N = 2), and several migrant populations (total N = 11). More recent analyses of five microsatellite loci (R. C. Fleischer et al., unpubl. data) provide additional support for a very low level of genetic divergence between the Coastal Plain Swamp Sparrows and the inland subspecies.

The low level of variation indicates a recent coalescence of mtDNA haplotypes in Swamp Sparrows, which is estimated based on ATPase8 at 40,000 yr (Greenberg et al. 1998). Given that most of the current range of Swamp Sparrows was covered by glaciers in the past 10,000–15,000 yr, the Coastal Plain Swamp Sparrow appears to have differentiated morphologically in a very short amount of time (Greenberg et al. 1998). The question of whether the morphological differentiation is mainly genetic or environmental remains problematic, but nestling Coastal Plain Swamp Sparrows that were hand reared in an aviary with only fresh water developed stereotypical plumage patterns and morphology (R. Greenberg, pers. comm.). Recent analyses of the melanocortin-1 receptor gene (L. Gibbs et al., unpubl. data), known to result in darker plumage patterns in a wide range of avian taxa (Theron et al. 2001, Mundy et al. 2004), have not revealed a relationship for this species.

In summary, morphological divergence among endemic tidal-marsh subspecies of *Melospiza* sparrows and their freshwater relatives appear to be recently evolved or strongly environmentally influenced. Little support was found for genetic differentiation of tidal marsh subspecies in this genus.

ORNATE SHREW AND WANDERING SHREW

Systematics, distribution, and ecology

Populations of both the ornate shrew (*Sorex ornatus*) and the wandering shrew (*Sorex vagrans*) are found in coastal marshes of the West Coast of North America. The ornate shrew is a rare species restricted to coastal marshes and riparian communities of California, from 39° N latitude southward discontinuously to the tip of Baja California, Mexico. Subspecies of the ornate shrew often were described in the past by body size and pelage coloration which may be the result of environmental induction rather than genetically based differences, and sometimes based on only one or two specimens (Owen and Hoffmann 1983). However, the validity of the nine named subspecies of ornate shrews has recently been confirmed using univariate and multivariate statistical analyses of cranial measurements (Maldonado et al. 2004). Wandering shrews occur in northwestern North America down the Coast Range and Sierra Nevada of northern California. At least one subspecies of wandering shrew is known from the marshes of San Francisco Bay. Because of their short life span (Rudd 1953, Newman 1976), semi-fossorial habit, habitat specialization, high metabolism (McNab 1991), and small size, dispersal between patches of mesic habitat is limited and the high degree of local geographic morphological and genetic variation in shrews is expected. Furthermore, faced with a high abundance of invertebrate food, shrew populations can achieve high local densities in tidal marshes. Therefore, the evolution of multiple saltmarsh subspecies in this genus is also not surprising.

Although the existence of nine morphologically distinct subspecies of ornate shrew is well founded, a molecular genetic analysis of this species using mtDNA and allozymes, separates the species into southern, central, and northern clades (Fig. 3). Hence the patterns of morphological and genetic variation in this species are not concordant. Furthermore, genetic analysis puts into question the species status of certain morphologically ascribed subspecies, in the San Francisco Bay marshes. Therefore, we will develop our analysis of the evolutionary relationship of these species based on a recent molecular analysis (Maldonado et al. 2001; Fig. 3).

Patterns of genetic differentiation in saltmarsh habitats

In the genetic analysis of the cytochrome *b* region of the mtDNA of ornate shrews, Maldonado et al (2001) found 24 different haplotypes in 20 populations. Except for three population groupings, all populations had unique haplotypes. The occurrence of unique haplotypes in most localities suggests that genetic subdivision is a common characteristic of ornate shrews throughout most of their range. Fourteen of the 20 populations of ornate shrews were fixed for a single unique haplotype (392 base pairs). The remaining six populations had two to four different haplotypes. Interestingly, two of the subspecies endemic to tidal marshes (Grizzly Island [*Sorex ornatus sinuosus*—population 2 in Fig. 2] with Rush Ranch [*S. o. californicus*—population 5], and Los Banos [*S. o. californicus*—population 8]) with Salinas [*S. o. salarius*—population 10]; Fig. 3) were not significantly differentiated from the more widely distributed subspecies and have all haplotypes in common, implying that they are part of the same interbreeding population. In addition, even though pairwise computations of F_{st} using an analysis of molecular variance (AMOVA) indicate that most ornate shrew populations are significantly differentiated relative to a random collection of genotypes, once again these two populations, restricted to tidal marshes, are included among those that are not genetically differentiated (Grizzly Island versus Rush Ranch, Salinas versus Los Banos, and Salinas versus Sierra Nevada). These populations were also not significantly differentiated in the allozyme analysis. Several populations from different groupings were geographically distant but showed small genetic distances. The clearest example occurs in the central populations where the Salinas population is approximately 300 km away from the Los Banos populations, but the average Nei's allozyme genetic distance

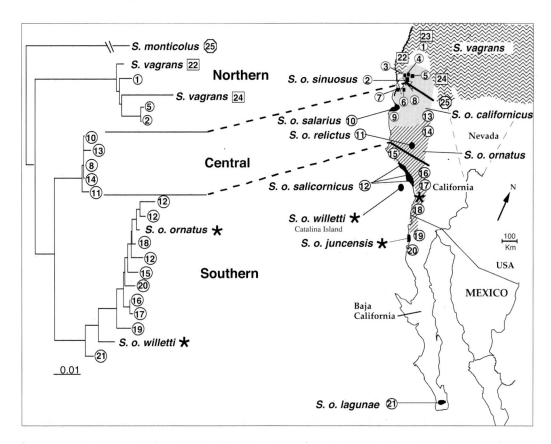

FIGURE 3. Map of the southwestern US and northwestern Mexico showing locations of populations sampled in Maldonado et al. (2004). Distribution of nine subspecies of ornate shrew (*Sorex ornatus*) is indicated. Thick lines indicate subdivisions based on genetic analyses (neighbor-joining tree based on average sequence divergence between populations, from Maldonado et al. 2001). Asterisks mark populations not sampled for morphometric study. Boxes indicate populations of wandering shrews (*S. vagrans*) and circles indicate populations of ornate shrews. The montane shrew population used as an outgroup is indicated with an octagon.

between them is only 0.009, and they share the same cytochrome *b* haplotypes. In contrast, the distance between Salinas and Grizzly Island and Rush Ranch is 150 km smaller, but their average allozyme distance and sequence divergence values are more than ten times greater (0.059 and 0.054, respectively). Considering genetic differentiation between marsh populations and their closest sampled inland relatives, there is a relatively large sequence divergence (1.2–1.3%) from Rush Ranch-Grizzly Island (*Sorex ornatus*) to Bodega Bay (the closest sampled population being wandering shrews.

Evolutionary history and biogeography

As suggested by low rates of gene flow, shrews are poor dispersers, and the imprint of past events may be long retained in present day populations. Clades have a high genetic divergence (4.2–4.9% cytochrome *b* sequence divergence), suggesting a relatively long evolutionary independence from one another. Based on molecular data, populations in the northern clade diverged from the central and southern populations >1 MYA, and genetically are more similar to neighboring populations of wandering shrews. Northern clade ornate shrew haplotypes from Grizzly Island, Rush Ranch, and Dye Creek localities are grouped with those attributable to wandering shrews from Bodega Bay (population 22), Mt. Shasta (population 23), and the Sweetwater Mountains (population 24). Since ornate shrews could not be genetically differentiated from the wandering shrew in the northern region, Maldonado et al (2001) hypothesized that northern populations of the ornate shrew may be unique lowland and coastal forms of the wandering shrew that have converged independently on the morphology of southern and central California ornate shrews. However, by analyzing skull morphology, Maldonado et al (2004) showed that ornate and wandering shrews, as well as the closely related montane shrew (*Sorex monticolus*) are well differentiated. Shrews from the northern region have morphology similar to ornate, and not wandering or montane shrews. Within the ornate shrews, populations across the range differ in morphology. However, morphological differentiation is not concordant with the deep tripartite pattern of genetic differentiation. A neighbor-joining tree of all populations based on a between groups F-matrix derived from a discriminate function analysis (Fig. 4) did not show a clustering of the three regions suggested in the genetic analysis. Similarly, populations pertaining to neighboring populations of shrews from tidal and upland areas were not located in the same or neighboring branches.

Previous evolutionary hypotheses concerning the radiation of shrews have drawn on the conventional wisdom that Pleistocene climatic cycles precipitated a large portion of speciation events between extant sister taxa (Findley 1955). The tripartite division of ornate shrew clades dates to the early Pleistocene and does not reflect isolation in recent ice-age refugia. In contrast, past patterns of genetic divergence within clades appear to be erased by population contraction during inter-glacials and re-established during glacial period expansions and suggest that ice-age effects may have more pronounced impact on regional within-clade diversity than on speciation (Maldonado et al. 2001). Furthermore, the patterns of morphological differentiation observed among tidal marsh populations of ornate shrews may be the result of local adaptation with low levels of genetic differentiation.

SALT MARSH HARVEST MOUSE AND WESTERN HARVEST MOUSE

Systematics, distribution, and ecology

Two species of harvest mice show at least some evidence of differentiation in tidal marshes along the Pacific Coast. The salt marsh harvest mouse (*Reithrodontomys raviventris*) was first described by Dixon (1908) and is the only mammalian species endemic to tidal marshes (Greenberg and Maldonado, *this volume*). Two subspecies are restricted to the salt and brackish marshes of the San Francisco Bay region: *R. r. raviventris* (South Bay up to and including Corte Madera and Richmond) and *R. r. halicoetes* (North Bay and Suisun Bay; Hall 1981). Its highly restricted geographic range led biologists to believe that it became isolated during the formation of saltmarshes in the San Francisco Bay region (Fisler 1965).

The western harvest mouse (*Reithrodontomy megalotis*) is distributed widely in central and western North America. Populations occupy

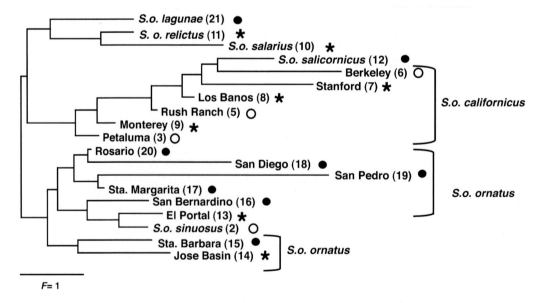

FIGURE 4. Neighbor-joining tree based on the between groups F-matrix (df = 17, 358) derived from a discriminant function analysis of 19 populations of ornate shrews (*Sorex ornatus*) (Modified from Maldonado et al. 2004). Symbols denote geographic assignment of the populations based on genetic data as follows: southern, central*, and northern regions. Locality codes are in () and correspond to Fig. 3.

coastal saltmarshes along San Francisco Bay, southern California, and Monterey Bay. These latter two populations have been, respectively, described as *R. m. limicola* (Von Bloeker 1932) and *R. m. distichlis* (Von Bloeker 1937), based on darker pelage coloration than the more widespread grassland form (*R. m. longicaudus*) that occurs in the adjacent upland habitats. Fisler (1965) found a similar tendency to dark pelage coloration in western harvest mice inhabiting the San Francisco Bay area; however, Pearson (1951) could not find differences between *R. m. distichlis* and *R. m. longicaudus*. Furthermore, Collins and George (1990) found no significant morphological or allozymic evidence to support the continued recognition of *R. m. limicola* in southern California. The results of their genetic and phenetic analysis suggest that within populations of *R. megalotis* in southern California no historical geographic units exist. Rather, broad phenetic overlap occurs among the samples suggesting that only gradual, small-scale phenetic change occurs among *R. megalotis* populations in mainland southern California.

Evolutionary history, and biogeography

Originally, the specific status of salt marsh harvest mouse was based on its sympatry with its then putative sister taxa, the western harvest mouse (Hooper 1952, Fisler 1965, Shellhammer 1967). However, analyses of karyotypic and allozymic data presented by Hood et al. (1984) and Nelson et al. (1984) suggested that salt marsh harvest mouse was both most closely related to the allopatric plains harvest mouse (*Reithrodontomys montanus*) from which it is distinct at the species level. This was supported by cytochrome *b* sequence data (Bell et al. 2001), where genetic distances separating salt marsh harvest mouse from plains harvest mouse and western harvest mouse ranged from 13.50% and 14.75%, respectively. These values are greater than those for other currently recognized biological species such as Sumichrast's harvest mouse (*Reithrodontomys sumichrasti*) and the western harvest mouse (9.79%) and Sumichrast's harvest mouse and Zacatecas harvest mouse (*Reithrodontomys zacatecae*) (8.55%). Using a sequence divergence value of 3.5% per million years, Bell et al. (2001) estimated that the salt marsh harvest mouse and plains harvest mouse diverged from a common ancestor 3.9 ± 0.7 MYA. This places the divergence of the salt marsh harvest mouse as far more ancient than the formation of saltmarshes around San Francisco Bay (Malamud-Roam, *this volume*).

Perhaps the salt marsh harvest mouse diverged from the plains harvest mouse when the Monterey Bay was a drainage basin. Approximately 5 MYA large areas in central California were covered by sea (Wahrhaftig and Birman 1965). Throughout much of the period that followed, the San Joaquin Valley was a wide seaway rather than the present day continental river valley and suitable mesic habitats for harvest mice were restricted to the margin of this seaway. One of the most profound barriers developed at the point were the present day central valley drains into the Pacific Ocean near Monterey Bay and where the largest marine canyon of the Pacific coast of North America is found (Yanev 1980). Throughout most of the Pliocene-Pleistocene, this area was a vast embayment that likely prevented dispersal for small vertebrates (Peabody and Savage 1958). A number of studies (Barrowclough et al. 1999, Rodriguez-Robles et al. 1999, Conroy and Cook 2000, Bronikowski and Arnold 2001, Maldonado et al. 2001, Rodriguez-Robles et al. 2001) have identified a deep phylogeographic break in northern California, often at or north of San Francisco Bay. In the case of the salt marsh harvest mouse its distribution is exclusively in the San Francisco Bay area and occurs in sympatry with the western harvest mouse also endemic to the tidal marshes of the San Francisco Bay area. Perhaps when the barriers to dispersal disappeared, the less specialized western harvest mouse expanded its range into this area and could have excluded the expansion of the upland adapted plains harvest mouse. The salt marsh harvest mouse became restricted to tidal marsh habitat where it outcompetes the western harvest mouse. Studies by Fisler (1965) suggest that the salt marsh harvest mouse is better adapted to tidal marshes than the western harvest mouse. For example, the salt marsh harvest mouse can drink and tolerate salt water, has a water-repellant pelage, and is a better swimmer. Possibly due to convergent evolution from occupying similar habitats, the western harvest mouse and the salt marsh harvest mouse are more similar morphologically than the salt marsh harvest mouse is to the plains harvest mouse.

Intraspecific differentiation in the salt marsh harvest mouse appears to be very low, although data exist for only two individuals: one sample from Tolay Creek representing the northern subspecies (*R. r. halicoetes*), and one sample from Newark representing the southern subspecies (*R. r. raviventris*; Bell et al. 2001). The Tamura-Nei genetic distance values were low (0.0018) suggesting at lack of differentiation at the intraspecific level.

SYNTHESIS

MOLECULES VERSUS MORPHOLOGY

Historically, the study of geographic variation in phenotype and morphology has been central to understanding evolutionary processes, under the assumption that morphology reflected an underlying genetic structure (Mayr 1942, Mayr 1963). In some cases, e.g., Swamp Sparrows, Clapper Rails, ornate shrews, and water snakes, morphological differentiation and differentiation at neutral loci are not concordant. In other cases, morphology and genetics only coincide some of the time. Within the Seaside Sparrows and sharp-tailed sparrows, the most prominent genetic finding is that each is divided into two widely divergent groups based on mtDNA (Avise and Nelson 1989, Rising and Avise 1993). In sharp-tailed sparrows, behavioral, and to some extent morphological, evidence corroborated this split (Greenlaw 1993), but in Seaside Sparrows, the two groups were not recognized before the genetic work was done, and no known morphological or behavioral differences separate the groups. Cases exist, however, such as in the Savannah Sparrows, sharp-tailed sparrows, and salt marsh harvest mice, where morphologically endemic species or subspecies were identified and their dissimilarities confirmed with surprisingly large amounts of genetic divergence at neutral loci.

Several explanations can account for the lack of concordance between molecules and morphology. As has been noted in ornate shrews (Maldonado et al. 2001, 2004) and Song Sparrows (Smith 1993), phenotypic differences in integument coloration, body size, and other morphological characters may be primarily developmental rather than genetic in origin. Alternately, evolutionary patterns at neutral loci and loci under selection may differ. With selection on quantitative trait loci, phenotype could diverge rapidly, while neutral loci diverge at a rate proportional to the effective population size. This lineage sorting may be incomplete, even in well-established species (see Funk and Omland 2003 for a review).

This is an area where few studies have explored and quantified those differences. We are still trying to understand with common garden and reciprocal-transplant experiments whether the differences between upland and tidal-marsh endemics are genetically based. Indications in Clapper Rails and Swamp Sparrows are that differences in phenotypic traits such as interorbital bridge widths and plumage are primarily genetically based (Olson 1997; R. Greenberg, unpubl. data), and in Song Sparrows, that morphology may be somewhat plastic (Smith 1993). Further studies with quantitative trait loci and on functional genes are promising avenues for future research in determining the strength of selection on tidal-marsh endemics.

ORIGIN OF TIDAL-MARSH TAXA

Genetic data combined with molecular-clock calibrations provide an estimate of the timing of divergence of taxa that occupy tidal marshes. Because these data only pertain to the accumulation of mutations in neutral markers in populations, they provide an estimate of the amount of time these taxa have experienced a unique evolutionary trajectory. These estimates do not provide insight into how long populations may have occupied tidal marshes, particularly if these populations have gone through ancient or recent periods of genetic connection with non-tidal marsh populations. Nor do they provide an estimate of how long divergent taxa may have been restricted to tidal marshes. But as a starting point, they provide a picture of the timing of the initiation of genetic divergence.

With the exception of terrapins and salt marsh harvest mice, the estimates of past population coalescence fall within the time period defined by the late Pleistocene to possibly as late as the Holocene. Within this time frame, a majority of taxa analyzed show divergence in the late Pleistocene with a smaller number dating back to the late Pliocene or early Pleistocene. Presumably, a pattern of Pleistocene colonization and differentiation characterizes other temperate zone habitats as well. However, we have been unable to find a similar habitat-based cross-taxa analysis to this one.

The ages of these splits vary by taxon within every region. For example, in northeastern North America, the Saltmarsh Sharp-tailed Sparrow is estimated to have been independent from the inland forms of the Nelson's Sharp-tailed Sparrow for 600,000 yr (Rising and Avise 1993), while the saltmarsh form of the Swamp Sparrow evolved from upland relatives <40,000 yr ago (Greenberg et al. 1998). On the West Coast, saltmarsh forms of the Song Sparrow and ornate shrew are barely distinguishable at genetic loci from neighboring upland forms (Maldonado et al. 2001, Chan and Arcese 2002), while the salt marsh harvest mouse living in the same marshes split from its closest upland relative nearly 4 MYA (Bell et al. 2001). On the Gulf Coast, Clapper Rails of saltmarshes widely share mtDNA haplotypes with King Rails of freshwater marshes (Avise and Zink

1988; R. C. Fleischer, unpubl. data), but Seaside Sparrows probably diverged from an upland ancestor 1.5–2 MYA (Rising and Avise 1993), and diamondback terrapins probably did so at 7–11 MYA (Lamb and Osentoski 1997). The timing of those divergences ranges from Holocene to early Pliocene to late Miocene, and suggest a complex history of multiple invasions and differentiations in saltmarshes.

One factor that may have aided differentiation of tidal-marsh taxa is the restricted dispersal indicated by behavioral and ecological studies in many species. In many cases ecological studies have noted higher philopatry and reduced dispersal of tidal-marsh endemics compared with closely related upland species. In ornate shrews, limits to physiological tolerance of inhospitable habitats has been noted (McNab 1991, Maldonado et al. 2001). In Savannah Sparrows, most of the tidal-marsh forms are non-migratory, compared with migratory upland forms. In Song Sparrows, ecological studies suggested that drift may have played a large role in their differentiation, with tidal-marsh forms showing some of the shortest dispersal distances recorded for a Song Sparrow as well as shortened wings (Marshall 1948a).

Low dispersal, combined with the resource-rich, homogenous landscape of the tidal-marsh habitat, may reduce gene flow to other habitats and increase local adaptation. In Red-winged Blackbirds (*Agelaius phoeniceus*), for example, genetic distance as measured from allozymes between populations in Sacramento and San Francisco bays (214 km apart) was 10 times as great as the genetic distance between Florida and Oregon (Gavin et al. 1991). Gavin et al. (1991) hypothesized that philopatry and non-migratory behavior may have caused this differentiation and discussed the possibility that the brackish environment in which they live may enforce a selective regime that could reduce immigration or emigration to other habitats.

The geological history of the tidal marshes may provide some insight into the evolutionary origin of tidal-marsh taxa. For many of the taxa, the last Pleistocene glacial maximum was hypothesized to play a role in their diversification, and in some cases, such as the Swamp Sparrow, the molecular evidence supports this hypothesis (Greenberg et al. 1998). In other examples such as the salt marsh harvest mouse and diamondback terrapin, diversification predates the Pleistocene. Although the last glacial maxima may not have played a large role in the speciation of tidal-marsh endemics, in the Gulf and Atlantic coasts, genes have recorded the geological history as revealed by comparative phylogeography of Seaside Sparrows and diamondback terrapins (Rising and Avise 1993, Lamb and Osentoski 1997).

Furthermore, perhaps differences in the stability of the habitat over time may provide clues as to how quickly phenotypic differentiation can occur. For the most part, though, a concordance is lacking between geology and differentiation with respect to the origination of tidal-marsh taxa. For example, in the case of the salt marsh harvest mouse, the amount of divergence from its presumed sister taxon greatly exceeds the age of its current habitat. The entire range of the salt marsh harvest mouse is now restricted to tidal-marsh habitats surrounding San Francisco Bay; however, those habitats did not exist as such 10,000 yr ago, and have gone through periodic inundations as sea levels rose and receded during glaciations. It remains a challenge to match our new understanding of the evolutionary dynamics of tidal-marsh vertebrates with geologic reconstructions of coastal-marsh history.

CONSERVATION GENETICS

The uniqueness of tidal marsh forms, in combination with the rapid destruction and development of tidal marsh habitat in North America, has led to efforts for their protection and proper management. The preponderance of taxa that show significant divergence in morphology and life history with difference in molecular makers demands that we assess the definition of important conservation taxonomic units when approaching tidal-marsh conservation.

The U.S. Endangered Species Act of 1973 together with the Distinct Population Segment (DPS) amendment in 1978 protects distinct species, subspecies, and populations. Since designation provides large financial resources as well as immediate protection from hunting, habitat exploitation, and other anthropogenic impacts that threaten population viability, much debate surrounds the criteria used to designate units of management and conservation (O'Brien and Mayr 1991). See also Fraser and Bernatchez (2001) for a discussion of conservation units.

Several definitions for conservation units (O'Brien and Mayr 1991, Moritz 1994) attempt to incorporate meaningful criteria to identify groups of populations with distinct evolutionary potential. Most of these are based on phylogenetic distinctness (subspecies—Avise and Ball 1990, O'Brien and Mayr 1991, Ball and Avise 1992; evolutionarily significant units (ESUs)—Ryder 1986, Moritz 1994, Moritz et al. 1995). This is because phylogenetic partitioning results from accumulation of differences due to the lack of gene flow (O'Brien and Mayr 1991).

However, the emphasis on genetic criteria has been criticized, and an approach that incorporates adaptive differences based on ecological and genetic exchangeability is advocated by Crandall et al. (2000).

Although results from genetic studies led directly to the designation of species status for the Nelson's and Saltmarsh sharp-tailed sparrows, in Savannah and Seaside sparrows, large genetic divergence has not yet been accompanied by specific status, although few would question their recognition as ESUs. In Clapper Rails, genetic studies alone provided little basis for the recognition of a taxon as a distinct species, or even as ESUs; however, their taxonomic status has not been modified because their distinctness remains defendable based on morphological and habitat differences. The same is true for the Coastal Plain Swamp Sparrow, which is morphologically and ecologically well-differentiated from freshwater subspecies, but shows no reciprocal monophyly of mtDNA sequences nor significant allele frequency differences. Since the criteria used to define taxonomic and conservation units are still being debated and because many tidal marsh taxa are threatened, their status will likely remain contentious as well.

Besides contributing to defining conservation units, molecular studies can improve genetic management of endemic tidal marsh taxa. The naturally fragmented and linear nature of the tidal-marsh habitat has the possibility of increasing intraspecific genetic structure, complicating proper management and restoration programs. For example in Seaside Sparrows, being aware of the phylogeographic structure underlying the species would have better informed attempts to save the Dusky Seaside Sparrow. Populations with low heterozygosity, as indicated by molecular studies, such as those of the Light-footed Clapper Rail, should be studied carefully for indications of inbreeding depression (Keller and Waller 2002), and if in danger of extinction may benefit from management efforts that increase gene flow and genetic variability, and prevent genetic erosion.

However, increasing gene flow and variability may also be harmful to the integrity of tidal-marsh endemics. Anthropogenic impacts, such as habitat disturbance and climatic change have the potential to change selective forces in the marshes over short-term ecological timescales and may increase introgression (Takekawa et al. chapter 11, *this volume*). Conant (1963) speculated that interbreeding between the water snakes (*Nerodia sipedon williamengelsi* and *N. fasciata*) represented introgressive hybridization resulting from habitat alteration. In San Francisco Bay, the one genetically distinct subspecies of Song Sparrow (*Melospiza melodia pusillula*), may be in danger of introgression from the upland subspecies nearby if salinity changes in the tidal saltmarshes due to urbanization.

Furthermore, natural hybridization may reduce the range of the strict saltmarsh endemics, increasing priority for non-introgressed populations. Although the zone of hybridization appears to be geographically stable in Sharp-tailed Sparrows in southern Maine (Montagna 1940, 1942; Rising and Avise 1993, Shriver 2002), and the front where hybridization can occur is narrow due to the linear nature of tidal marsh habitat, there is recognizable introgression of *Ammodramus nelsoni subvirgatus* genes into *A. caudacutus caudacutus* populations as far south as Parker River, Maine (Shriver 2002). Shriver (2002) points out that this introgression reduces the known range of pure *A. c. caudacutus*, and perhaps should increase conservation priority in the restricted range of this species.

As habitat destruction and global climatic change continue to reduce and fragment tidal marsh habitat and their containing populations, the use of molecular markers will be even more essential to implement for conservation.

Future Studies

The use of molecular approaches to study tidal-marsh taxa has just begun. Many of the DNA sequence and RFLP studies described in this review are from the mitochondrial genome, which has advantages as well as disadvantages. MtDNA has a high mutation rate, maternal mode of transmission, and lack of recombination (Avise 2000). A high mutation rate is particularly valuable for detecting variation among closely related taxa and a lack of recombination facilitates phylogenetic reconstruction. However, the entire mitochondrial genome is a single locus resulting in one part of the mitochondrial genome influencing all others through linkage, preventing any site from being independent. Furthermore, the maternal inheritance of mtDNA can produce incongruence between population history and gene history when sex differences occur in dispersal or fitness (Ballard and Whitlock 2004). In order to better resolve the population history of endemic tidal-marsh taxa many of these studies need to be extended to nuclear loci. Furthermore there are also more powerful analytical tools for reconstructing population history, such as coalescent approaches that have not been widely applied to tidal-marsh endemics.

Coalescent approaches are particularly appropriate techniques for understanding the evolutionary history of tidal-marsh taxa. The coalescent is a stochastic process that can be used to model past population demographic history and provides a statistical framework for data analysis (see review in Rosenberg and Nordborg 2002). Analysis of genetic polymorphism data utilizing coalescent approaches enables us to address competing demographic and biogeographic hypotheses; it can be used within a hypothesis testing framework to test alternate scenarios for colonization and subsequent differentiation. Hypotheses based on geological reconstruction can be tested with genetic data providing a better understanding of the process of differentiation in endemic taxa. Issues that can be more readily addressed with coalescent approaches are: The number of colonization events, the need for a bottleneck for differentiation, and the lack of subsequent gene flow. Furthermore, combined with likelihood methods, coalescent approaches allow not just hypothesis testing, but parameter estimation; the effective population size, extent of gene flow, and timing of divergence.

The tidal-marsh habitat provides a unique field laboratory to study intraspecific spatial genetic structure and metapopulation dynamics. Because tidal saltmarshes are discrete and often patchily distributed, they are ideal places to study the extent and amount of gene flow between patches. The relative homogeneity of the habitat lends itself to straightforward quantification of population sizes for comparison to genetic estimates of effective population size, providing many of the necessary parameters for population genetic models. In addition, their discrete and patchy distribution may act to reduce the ability of genetically based divergence to evolve in local tidal-marsh populations. This distribution results in a large edge effect that may reduce the isolation of tidal marsh and other habitats. Furthermore, as documented by Malamud Roam et al. (*this volume*), tidal marshes have been highly unstable in their location and extension throughout recent geological history. The ice sheets covered high-latitude tidal marshes and many of the largest areas of estuarine tidal marsh located at mid-latitudes are a result of flooding from the sea-level rise associated with the melting of Pleistocene glaciers. However, lower sea levels during the glacial periods may have exposed more coastal plain and increased tidal marshes in some areas. It will also be interesting to document genetic changes in tidal-marsh organisms in response to potential changes in saltmarsh distributions that may occur with changes in sea level due to global warming.

Furthermore, several endemic taxa have yet to be examined using molecular approaches, such as the California Black Rail (*Laterallus jamaicensis coturniculus*), and subspecies of Marsh Wren (*Cistothorus palustris*), Common Yellowthroat (*Geothlypis trichas*), California vole (*Microtus californicus*), meadow vole (*Microtus pennsylvanicus*), and masked shrew (*Sorex cinereus*).

The possibility of studies of selection on functional genes is particularly exciting. For example, it may prove important to test the hypothesis set forth by Gavin et al (1991) that the large amounts of genetic differentiation of Red-winged Blackbirds from the San Francisco Bay and the Salton Sea may be the result of physiological adaptations to saltmarsh environments that prevent or reduce interchange with populations living in upland or freshwater environments. Furthermore, this hypothesis could be tested by examining physiological tolerance for salt water and genetic evidence for gene flow among populations of various vertebrates found in salt and freshwater environments.

The study of tidal-marsh taxa is a special opportunity as part of the larger paradigm of geographic variation. Here we have the opportunity to synthesize how different vertebrate groups, respond to the same environmental pressures, such as salinity (Goldstein, *this volume*), periodic flooding (Reinert, *this volume*), substrate color, and available food (Grenier and Greenberg, *this volume*) with a variety of different ages and extent of gene flow. Molecular approaches, combined with morphological, ecological and behavioral studies, will continue to improve our understanding and aid conservation of tidal marsh vertebrates in the future.

ACKNOWLEDGMENTS

We thank R. Greenberg, S. Droege, and M. V. McDonald for organizing and for inviting us to participate in this symposium. R. Greenberg, M. V. McDonald and A. Hill provided helpful suggestions and comments on earlier drafts that greatly improved the quality of this manuscript. R. Gaul provided useful information on his M.S. thesis work on the Carolina salt marsh snake. S. Hauswaldt provided unpublished information on terrapin microsatellite variation, B. Slikas provided unpublished information on rail systematics and J. Klicka provided unpublished information on *Ammodramus* systematics. We also thank three anonymous reviewers who provided helpful comments and suggestions that improved the quality of this manuscript.

ADAPTATION TO TIDAL MARSHES

Salt marsh harvest mouse (*Reithrodontomys raviventris*)
Drawing by Julie Zickefoose

AVIAN NESTING RESPONSE TO TIDAL-MARSH FLOODING: LITERATURE REVIEW AND A CASE FOR ADAPTATION IN THE RED-WINGED BLACKBIRD

STEVEN E. REINERT

Abstract. Throughout the coastal US, nests of birds breeding in saltmarshes are subject to periodic flooding by spring and storm tides. I found documentation of nest loss to tidal flooding for nine species of terrestrial saltmarsh birds and nine species of waterbirds. A review of adaptations to periodic tidal inundations in these species revealed four general categories of responses: (1) placement of nests such that they exceed the elevation of tides or float on the surface of rising flood-waters, (2) nest-repair and egg-retrieval behaviors that keep eggs in nests during and after floods, (3) rapid post-flood renesting which enables the nesting cycle to be completed just prior to the encroachment of the next flooding spring tide, and (4) timing of the breeding season to avoid periods of peak seasonal tidal amplitude. Adaptations were more advanced in species that had longer exposure to marine environments; they were most developed in colonially nesting gulls and terns for which substantive evidence indicates that environmental cues (peak high tides) are used to time and place nests such as to avoid tidal flooding. Nest repair behaviors during and following flood tides were most advanced in the Laughing Gull (*Larus atricilla*) and Clapper Rail (*Rallus longirostris*); rails generally exhibited more sophisticated post-flood responses than sparrows. To determine how a population of Red-winged Blackbirds (*Agelaius phoeniceus*) nesting in a saltmarsh in Rhode Island responded to monthly spring-tide flooding events, I studied the population during 1982–1985. Spring tides destroyed 34% of active nests, and overall nest success values were among the lowest reported for this widely studied species. Forty-four percent of successful nests were renesting attempts by females that lost early nests to tidal flooding and initiated a replacement nest within 48 hr. Females usually laid the first egg in the new nest 5 d following flooding, thus enabling young to fledge just prior to the encroachment of the next lunar flood tide 27–29 d later. My data demonstrate that salt-marsh nesting Red-winged Blackbirds employ responses to flooding similar to those exhibited by obligate *Ammodramus* sparrows, and suggest that the study population is representative of a larger population of saltmarsh inhabiting Red-winged Blackbirds that is ecologically isolated from regional populations nesting in non-tidal habitats.

Key Words: adaptations, *Agelaius phoeniceus*, *Ammodramus*, *Larus*, *Melospiza melodia*, nest survival, *Rallus longirostris,* reproductive success, saltmarsh, *Sterna*, tidal flooding.

RESPUESTAS DE ANIDACIÓN DE AVES A LA INUNDACIÓN DE MARISMAS DE MAREA: REVISIÓN BIBLIOGRÁFICA Y UN CASO PARA ADAPTACIÓN DEL MIRLO DE ALAS ROJAS

Resumen. Por toda la costa de los EU nidos de aves reproductoras en marismas saladas están sujetas a inundaciones periódicas por manantiales y mareas de tormenta. Encontré documentación de pérdida de nidos por inundaciones de marea para nueve especies de aves terrestres de marisma salada y nueve especies de aves acuáticas. Una revisión de las adaptaciones a las inundaciones periódicas de marea en estas especies, revelaron cuatro categorías generales de respuestas: (1) la colocación de nidos en la cual ellos rebasaron la elevación de las mareas o flotaron en la superficie de torrentes de agua emergiendo, (2) reparación de nido y comportamientos de recuperación de huevos que mantienen los huevos en los nidos durante y después de las inundaciones, (3) veloz reanidación de post-inundación, la cual permite que el ciclo de anidación sea completado justo antes de la ocupación de la siguiente inundación por la marea, y (4) el tiempo en que transcurre la época de reproducción para evitar periodos de estación de mareas de amplitud más alta. Las adaptaciones fueron más avanzadas en las especies que estaban mas expuestas a ambientes marinos; ellas eran mas desarrolladas en anidaciones coloniales de gaviotas y charranes, para las cuales evidencia sustancial indica que señales del medio ambiente (el punto mas alto de mareas altas) son utilizadas en tiempo para colocar nidos, como para evitar inundaciones de marea. Comportamientos de reparación de nido durante y seguido de las inundaciones de marea, fueron mas avanzadas en la Gaviota Reidora (*Larus atricilla*) y el Rascón (*Rallus longirostris*); los rascones generalmente mostraron respuestas pos-inundación más sofisticadas que los gorriones. Para determinar como una población de Mirlos de Alas Rojas (*Agelaius phoeniceus*) anidando en un marisma salada en la Isla de Rhode respondió a eventos mensuales de inundaciones de marea, estudié la población durante 1982–1985. Las mareas de muelle destruyeron el 34% de los nidos activos, y sobre todos los valores de nidos exitosos se encontró el más bajo reportado por esta especie ampliamente estudiada. Cuarenta y cuatro por ciento de nidos exitosos fueron intentos de reanidación por hembras que perdieron nidos antes de las inundaciones por marea e iniciaron un reemplazo de nido dentro de 48 horas. Las hembras usualmente ponen su primer huevo en el nuevo nido 5 d en seguida de la inundación, permitiendo así que los juveniles emplumen justo antes de la invasión de la siguiente inundación de marea lunar 27–29 d después. Mis datos demuestran que los Mirlos de Alas Rojas de marisma salada en anidación, emplean respuestas similares a las inundaciones a aquellas exhibidas por *Ammodramus* gorriones, y sugieren que la población de estudio es representativa de una población mayor de habitantes Mirlos de Alas Rojas de marisma salada, la cual esta ecológicamente aislada de poblaciones regionales anidando en habitats de no marea.

Despite their intrinsic productivity and proximity to productive estuarine marine systems, tidal marshes provide nesting habitat for relatively few species of birds (Greenberg and Maldonado, *this volume*). Flooding, both lunar-driven tidal inundations (spring tides), and less predictable storm-driven events, may constitute a primary reason for the lack of diversity of tidal marsh breeders. Flooding events regularly inundate and destroy nests of all breeding species on a given saltmarsh (Kale 1965, Burger 1979, DeRagon 1988, Marshall and Reinert 1990, Shriver 2002). Nest loss to tidal flooding has been documented for at least nine species of terrestrial saltmarsh birds and nine species of shorebirds, waterfowl, and colonial waterbirds (Table 1). The constant threat of inundation combined with pressures from nest-robbing aerial predators such as crows (*Corvus* spp.) and grackles (*Quiscalus* spp.) presents a paradox for terrestrial breeding species—how to place their nests as high as possible to minimize flooding risks, while providing adequate vegetative cover over nests in herbaceous plant communities that are short (<1 m) in stature (Johnston 1956a, Post et al. 1983). Further, the risks of flooding for all coastal nesting birds may be heightened in future decades due to increasing rates of surface flooding from rising sea levels (Shriver 2002; Erwin et al., *this volume*).

For the few terrestrial vertebrate species that have evolved as saltmarsh specialists, and that therefore must regularly negotiate flooded habitats, the payoff is an environment featuring minimal interspecific competition (Post et al. 1983, Powell and Collier 1998) and an abundance of animal and plant food resources. Indeed, Post and Greenlaw (1982) determined that a single female Saltmarsh Sharp-tailed Sparrow (*Ammodramus caudacutus*), raising her young with no assistance from males, matched the reproductive output of a pair of Seaside Sparrows (*Ammodramus maritimus*) with no apparent cost to her survivorship. Post and Greenlaw (1982) concluded that food was not limiting for either species, and, that in saltmarshes, events such as floods could be the principal factors in checking population levels. Post et al. (1983) also concluded, based on abundance of seeds and invertebrate animals, and lack of competition from other terrestrial bird species, that food was not limiting to Seaside Sparrows on their breeding marshes on Long Island, New York, and in a Florida Gulf Coast site.

A further advantage of nesting on the saltmarsh may be a relatively low incidence of nest parasitism by cowbirds (*Molothrus* spp.). Parasitism of nests was rare for a saltmarsh nesting race of Song Sparrow (*Melospiza melodia samueli*) in San Francisco Bay, and Johnston (1956a) suggested that this could relate to cowbirds being unfamiliar with saltmarsh habitats. Nesting studies of Seaside Sparrows (Marshall and Reinert 1990), Saltmarsh Sharp-tailed Sparrows (DeRagon 1988), and Red-winged Blackbirds (*Agelaius phoeniceus*; this paper) in New England all found no incidents of cowbird parasitism (Greenberg et al., *this volume*).

Bird species that nest in the saltmarsh display an array of adaptive responses to tidal inundations that enable them to survive in the tidal environment and thus partake of the high food availability and potentially low nest parasitism rates. The responses fall into four categories: (1) placement of nests such that they exceed the elevation of tides or float on the surface of rising flood-waters, (2) nest-repair and egg-retrieval behaviors that keep eggs in nests during and after floods, combined with resumed incubation of eggs post immersion, (3) rapid post-flood renesting which enables the nesting cycle to be completed just prior to the encroachment of the next flooding spring tide, and (4) timing of the breeding season to avoid periods of peak seasonal tidal amplitude.

The first section of this paper reviews these responses in both terrestrial bird species, for which the marsh habitats constitute both the nesting and principle foraging grounds, and in colonial waterbirds which nest on the marshes but regularly forage in adjacent habitats. I then explore how adaptations to local tidal conditions can evolve in tidal-marsh populations of a non-specialized species, the Red-winged Blackbird.

The Red-winged Blackbird is an abundant species nesting in freshwater wetland and upland habitats throughout North America (American Ornithologists' Union 1998). In the northeastern US, the Red-winged Blackbird commonly nests in the high-marsh zone of the saltmarsh in habitats dominated by smooth cordgrass (*Spartina alterniflora*). In conjunction with studies of Seaside and Saltmarsh Sharp-tailed sparrows (DeRagon 1988), I conducted a

TABLE 1. SALT MARSH BREEDING BIRDS THAT HAVE LOST NESTS TO TIDAL FLOODING.

Species	State(s)	Nesting habitat flooded	References
White Ibis (*Eudocimus albus*)	SC	Black needlerush (*Juncus roemarianus*)	Frederick 1987.
Mallard (*Anas platyrhynchos*)	NJ, RI	Smooth cordgrass (*Spartina alterniflora*), salt meadow cordgrass (*S. patens*)	Burger 1979, observation of author.
American Black Duck (*Anas rubripes*)	MA, RI	Smooth cordgrass	Observation of author.
Clapper Rail (*Rallus longirostris*)	CA, GA, MS, NJ, VG	Hairy gumweed (*Grindelia humilis*), smooth cordgrass, California cordgrass (*S. foliosa*), high marsh habitats	Kozicky and Schmidt 1949, Stewart 1951, Zucca 1954, Mangold 1974, Jackson 1983, Massey et al. 1984, Schwarzbach et al. 2006.
Virginia Rail (*Rallus limicola*)	NY	Smooth cordgrass	Post and Enders 1970.
Willet (*Catoptrophorus semipalmatus*)	not avail.	High-marsh habitats	Lowther et al. 2001.
Laughing Gull (*Larus atricilla*)	NJ	Smooth cordgrass, salt meadow cordgrass, often on mats of debris	Bongiorno 1970; Montevecchi 1975, 1978; Burger 1979, Burger and Shisler 1980.
Herring Gull (*Larus argentatus*)	NJ	Smooth cordgrass, high-marsh habitats; some on mats of debris	Burger 1977, 1979.
Common Tern (*Sterna hirundo*)	NJ	Smooth cordgrass, high-marsh habitats, often on mats of debris	Burger and Lesser 1978, Storey 1978, Burger 1979, Buckley and Buckley 1982.
Forster's Tern (*Sterna forsteri*)	MD, VA	High-marsh habitats, often on mats of debris	Storey 1978.
Black Skimmer (*Rynchops niger*)	NJ	*Spartina* spp., often on mats of debris	Burger 1982.
Marsh Wren (*Cistotorus palustris*)	GA, RI	Smooth cordgrass	Kale 1965, observation of author.
Savannah Sparrow (*Passerculus sandwichensis*)	CA	Virginia glasswort (*Salicornia virginica*), bushy pickleweed (*S. subterminalis*)	A. Powell, pers. comm.
Nelson's Sharp-tailed Sparrow (*Ammodramus nelsoni*)	ME	High-marsh habitats	Shriver 2002.
Saltmarsh Sharp-tailed Sparrow (*Ammodramus caudacutus*)	ME, NJ, RI	High-marsh habitats	Woolfenden 1956, DeRagon 1988, DiQuinzio et al. 2002, Shriver 2002.
Seaside Sparrow (*Ammodramus maritimus*)	FL, MA, NJ, NY, RI,	Smooth cordgrass, high-marsh habitats	Woolfenden 1956, Greenlaw 1983, Post 1974, Post et al. 1983, Marshall and Reinert 1990.
Song Sparrow (*Melospiza melodia*)	CA	Woody saltwort (*Salicornia ambigua*)	Marshall 1948a, Johnston 1956a.
Red-winged Blackbird (*Agelaius phoeniceus*)	RI	Smooth cordgrass	Tilton 1987, this study.

4-yr study of the nesting ecology of Red-winged Blackbirds occupying a smooth cordgrass saltmarsh in Rhode Island to determine how this facultative nesting species responded to monthly spring tide flooding events. This work revealed life-history characteristics similar to those used by obligate Seaside and Saltmarsh Sharp-tailed sparrows in responding to monthly tidal inundations. Those traits and others relating to a compressed nesting cycle represent substantial deviations from Red-winged Blackbird behaviors in non-tidal habitats throughout the mid-latitudes of North America. The second section of this paper presents findings from my field research wherein I compare Red-winged Blackbird responses to flooding to those of other saltmarsh species, and suggest that Red-winged Blackbirds in coastal New England comprise an ecologically isolated population adapted to the saltmarsh environment.

ADAPTIVE RESPONSES TO TIDAL FLOODING IN SALTMARSH NESTING BIRDS

Adaptations to Tidal Flooding in Colonial Waterbirds

As aquatic specialists that have resided in coastal habitats since pre-Pleistocene times, colonially nesting larids of the Atlantic Coast would be expected to exhibit relatively sophisticated adaptations to the environmental extremes characteristic of marine habitats (Burger 1979, Frederick 1987). These ground-nesting species are extremely vulnerable to mammalian predation, and thus coastal saltmarsh islands, devoid of such predators, form their principal nesting habitat. The cost of this reduced predation risk is the frequent occurrence of wind- and storm-driven tidal-flooding events that have catastrophic effects on all nesting species. A single washout event can destroy all or most nests of all nesting species, including White Ibis (*Eudocimus albus*), Laughing Gull (*Larus atricilla*), Herring Gull (*Larus argentatus*), Common Tern (*Sterna hirundo*), Forster's Tern (*Sterna forsteri*), and Black Skimmer (*Rynchops niger*) (Storey 1978; Burger 1979, 1982; Frederick 1987).

A high degree of plasticity in reacting to the unstable environmental conditions of marine environments is the overriding factor relating to nest success (Bongiorno 1970, Storey 1978, Buckley and Buckley 1982). Observations of Laughing Gulls (Bongiorno 1970; Montevecchi 1975, 1978; Burger and Shisler 1980) and Common Terns (Storey 1978) indicate that those species assess peak tidal heights during the pre-laying period and use those cues to place nests at high elevations. Evidence presented by Bongiorno (1970) and Burger and Shisler (1980) suggest that Laughing Gulls use proximate cues of marshgrass height and structure in selecting relatively high areas of islands for nest placement. Larids also achieve high-elevation nest sites by building tall nest structures, and by placing nests on elevated mats of marsh debris, usually windrows of dead eel grass (*Zostera*) and smooth cordgrass left on the marsh surface by winter storms. Placement on mats not only elevates the nests of gulls and terns, but when exceptionally high tides exceed their elevation, the mats float on the flood-waters leaving nests intact and dry. If the mat is not washed off the island, the nests atop them remain safe (Bongiorno 1970; Montevecchi 1975, 1978; Burger and Lesser 1978, Burger 1979, Storey 1978, Buckley and Buckley 1982). Common Tern nests built individually on relatively large platforms of dead vegetative material also floated safely, and adults remained in attendance regardless of their altered, post-flood location (Buckley and Buckley 1982). Montevecchi (1975) suggested that the ability to assess the timing of peak tides allows Laughing Gulls to synchronize their nesting with the onset of the lunar cycle.

Despite the array of flood avoidance strategies employed by colonial-nesting larids, storm and/or wind driven waters are often high and turbulent enough to damage nest structures, wash out eggs, float nest-supporting mats off island, and drown pipping eggs and recently hatched young. Eggs are especially vulnerable to floating off during washouts in later developmental stages, as their buoyancy increases during the course of incubation (Nol and Blokpoel 1983). The eggs of many species of saltmarsh birds—White Ibis, Clapper Rail (*Rallus longirostris*), Laughing and Herring gulls, Common Tern, Song Sparrow, and Red-winged Blackbird—can tolerate immersions of limited duration in salt water (Johnston 1956b, Burger 1979, Ward and Burger 1980, Frederick 1987). In controlled experiments, >60% of Laughing and Herring gull embryos, at varying stages of development survived to pipping after immersions of up to 120 min in salt water (Burger 1979, Ward and Burger 1980). Thus, post-inundation, larids employ several tactics to enable the continued development of remaining eggs: (1) repair of nests during and after floods to keep eggs above water and in the nest, (2) construction of nests around washed-out eggs, (3) incubation of eggs outside of the nest, and (4) retrieval of eggs back to the nest (Burger 1977, 1979; Burger and Lesser 1978, Buckley and Buckley 1982).

Renesting after nest loss is common among colonially nesting larids, and is sometimes accompanied by colony relocation (Burger and Lesser 1978, Montevecchi 1978, Storey 1978, Buckley and Buckley 1982). Storey (1978) determined that the renesting response was especially well developed in Forster's Tern which (1) initiated nesting early enabling more seasonal nesting attempts, (2) initiated new clutches more quickly than Common Terns, thereby reducing risks from future flood events, and (3) produced relatively large second clutches. Storey considered those responses adaptations to the tidal environment.

REPRODUCTIVE SUCCESS AND FLOODING RESPONSES OF TERRESTRIAL SALTMARSH BIRDS

Ammodramus Sparrows of the Eastern United States

The Seaside Sparrow and Saltmarsh Sharp-tailed Sparrow are the only passerine species in the eastern US for which nesting activities are largely restricted to saltmarshes. Nests of Seaside Sparrows are typically placed in smooth cordgrass habitats (Woolfenden 1956, Greenlaw 1983, Post et al. 1983, Marshall and Reinert 1990), whereas Saltmarsh Sharp-tailed Sparrows (hereafter, saltmarsh sparrow) commonly place nests in stands or mixed communties of smooth cordgrass, saltmeadow cordgrass (*Spartina patens*), salt grass (*Distichlis spicata*), black needlerush (*Juncus gerardi*), and marsh elder (*Iva frutescens*) (Woolfenden 1956, DeRagon 1988, DiQuinzio et al. 2002, Shriver 2002). In Rhode Island, DeRagon (1988) demonstrated that saltmarsh sparrow nests elevated in the stems of smooth cordgrass plants achieved the same elevation as nests that were placed in higher areas of the marsh. Nests of saltmarsh sparrows in Rhode Island were most often covered above by tufts of saltmeadow cordgrass, salt grass, or black needlerush, while Seaside Sparrows wove canopies of smooth cordgrass leaves over their nests.

Flooding was a significant cause of nest mortality in all populations of these species studied in the northeast (Table 2). In a smooth cordgrass dominated saltmarsh on the Gulf Coast of Florida, some Seaside Sparrow nests were destroyed by flooding associated with storms, but most unsuccessful nests there were victims of rice rat (*Oryzomys palustris*) predation (Post et al. 1983). Flooding events that destroy sparrow nests in the marshes of southern Long Island occur when high tides are associated with storms or onshore winds; spring tides alone did not typically reach the elevation of nests (Post 1974, Post et al. 1983, Post and Greenlaw 1994).

TABLE 2. Reproductive success and nest mortality factors for Seaside and Saltmarsh Sharp-tailed sparrows in the northeastern United States.

Site	State	Years	N	Nest success[a]	Fledging probability[b]	Percent nest mortality			Reference
						Flood	Predation	Other	
Seaside Sparrow (*Ammodramus maritimus*)									
Long Island unaltered marshes	NY	1970–1971	>100	47[c]		36	10		Post 1974.
Long Island ditched marshes	NY	1970–1971	>15	66[c]		26	7		Post 1974.
Long Island	NY	1977–1978	144		0.35				Post et al. 1983.
Gulf hammock	FL	1979–1980	77		0.03	<50	>50		Post et al. 1983.
Allens Pond	MA	1984–1985	60	38	0.32	86	11	3	Marshall and Reinert 1990.
Saltmarsh Sharp-tailed Sparrow (*Ammodramus caudacutus*)									
100 Acre Cove	RI	1981–1982	172	59		63	11	7	DeRagon 1988.
Long Island	NY	1977–1978	238	47	0.12–0.37	28, 44[d]	38	16	Post and Greenlaw 1982.
Prudence Island	RI	1998	23	22		78	11		DiQuinzio 1999.

[a] Percent of nests fledging at least one young.
[b] Probability that an egg will produce a fledgling; calculated using method of Mayfield (1975).
[c] Percent of eggs that produced fledglings (nest success not reported).
[d] Percent failed nests lost to flooding for years 1977 and 1978, respectively.

In Rhode Island, spring-tide flooding accounted for less than half of saltmarsh sparrow nest mortality events in marshes with restricted tidal flow (DiQunizio et al. 2002). Nest success was relatively low for Seaside and saltmarsh sparrows in marshes with unrestricted, or relatively unrestricted tidal flows in southern New England where new-moon spring tides regularly exceeded the elevation of sparrow nests (Seaside Sparrow at Allens Pond, Marshall and Reinert 1990; saltmarsh sparrows at HAC, DeRagon 1988; saltmarsh sparrows at Prudence Island and Galilee, post-restoration, DiQuinzio 1999; Table 2). The proportion of failed nests attributable to spring-tide flooding events at those sites was 86%, 63%, 78%, and 91%, respectively.

In Rhode Island and Maine, respectively, DeRagon (1988) and Shriver (2002) found that flooding during new moon spring tides destroyed the majority of early saltmarsh sparrow nests. Females responded to nest destruction by immediately initiating a replacement nest. These events served to synchronize nesting activities, and most female saltmarsh sparrows were successful in completing their nesting cycle prior to the encroachment of flooding tides associated with the following new-moon period.

While saltmarsh sparrows in Rhode Island did not situate replacement nests at higher elevations following nest loss to flooding, they did place nests in the highest elevations of the high-marsh community. DeRagon (1988) and Shriver (2002) both found that the mean substrate elevation at nest sites was significantly higher (5 cm in Rhode Island) than at random points in the same plant communities. DeRagon determined that females avoided the lower 40% of that community's vertical range when selecting nest sites, and while such fine-scale selection would rarely impact nest survival during the flooding tides of new moons, such small differences in nest elevation could enable a nest to avoid flooding by lesser amplitude full-moon spring tides. The data of DiQunizio et al. (2002) suggested that in response to increased tidal flow resulting from marsh restoration efforts in a Rhode Island marsh (Galilee pre-, post-restoration; Table 2), sparrows modified their nest placement by nesting in taller vegetation and building nests higher above the substrate.

In a Massachusetts saltmarsh (Allens Pond; Table 2) Marshall and Reinert (1990) documented a nesting-cycle response to spring-tide flooding events by Seaside Sparrows that is very similar to that described above for saltmarsh sparrows. The simultaneous loss of many early season nests due to new-moon spring tide inundations acted to synchronize the subsequent nesting attempts of the unsuccessful pairs. Seaside Sparrows had first eggs in replacement nests in a mean of 6.25 ± 1.7 d from the date of nest destruction, and the young of renesting pairs fledged, nearly simultaneously, just prior to or at the time of the next spring tide. Seaside Sparrow nest cycles extended as little as 1 d due to large clutches, or protracted incubation or nestling periods, were subject to destruction if replacement nests were not high enough to avoid the tides. Marshall and Reinert (1990) witnessed nestling Seaside Sparrows climb from their nests to avoid being drowned by spring-tide inundations, and in one nest they observed a nestling drown two siblings while elevating its own body above the flood waters. At Allens Pond, Seaside Sparrow replacement nests were higher than earlier nests as a result of the seasonal growth of the smooth cordgrass plants that supported them, and the probability of nests fledging young was greatest for such late season nests that were synchronized to the tidal cycle.

The renesting response of *Ammodramus* sparrows is not necessarily an adaptive reaction to life in the tidal marsh. Some passerines nesting in upland habitats have similarly short renesting periods (Song Sparrow, 5 d, Nice 1937; Gray Catbird [*Dumetella carolinensis*], 5.05 d, Northern Cardinal [*Cardinalis cardinalis*], 5.5 d, Scott et al. 1987) suggesting that for many passerines a minimized renesting interval may maximize reproductive output regardless of habitat type. However, Shriver (2002) found that because saltmarsh sparrow females in a Maine saltmarsh renested <3 d after a nest destroying new-moon tide, they had a nest-success rate that was 41% greater than Nelson's Sharp-tailed Sparrow (*Ammodramus nelsoni*), which initiated nesting >10 d after nest loss in the same study area. Because throughout its range Nelson's Sharp-tailed Sparrow nests primarily in freshwater habitats, Shriver concluded that saltmarsh sparrows were better adapted to the tidal environment.

SALTMARSH SONG SPARROWS OF SAN FRANCISCO BAY

Two races of the Song Sparrow (*Melospiza melodia pusillula* and *M. m. samuelis*; Marshall 1948a) are endemic to saltmarsh habitats of the San Francisco Bay area of California. The *samuelis* race occupies emergent saltmarsh habitats dominated by California cordgrass (*Spartina foliosa*), woody saltwort (*Salicornia ambigua*),California gum plant (*Grindelia cuneifolia*), and salt grass. Its nests are placed in all those plant types, but most commonly in saltwort and gum plant, both <1 m in height (Johnston 1956a). Nests are placed as high as possible in the vegetation (25–30 cm) such as to still provide cover above, and Johnston (1956a) noted that this strategy enabled some nests to avoid inundation. Flooding spring tides occurred throughout the breeding season (March–July) and often destroyed nests—tidal inundations accounted for the destruction of 12% of eggs laid in the average year, and egg mortality was as high as 24% in some years (Johnston 1956b). N. Nur (pers. comm.) documented a renesting response in San Francisco Bay Song Sparrows that mirrors that of their East Coast *Ammodramus* counterparts—destructive flooding tides synchronized the nesting activities of females which promptly renested. This response enabled completion of the nest cycle prior to the onset of the nest flooding spring tide which followed 27–29 d later.

Johnston (1956a) determined that the saltmarsh Song Sparrow races in San Francisco Bay bred earlier by 15 d than did their upland nesting counterparts at identical latitudes. Because tidal amplitude increases over the course of the breeding season in San Francisco Bay, Johnston concluded that the early nesting represented an adaptation in the bay area birds to avoid the season's highest tides. In one year >60% of the eggs laid by a *samuelis* population in the whole season had fledged young prior to the first serious tidal event. Thus, for San Francisco Bay Song Sparrows, early nesting to avoid the season's highest tides was a critical factor in determining the reproductive success of *samuelis* (Johnston 1956a). As witnessed for the *Ammodramus* sparrows, nestling Song Sparrows were seen climbing from nests to avoid being drowned by rising tide waters, and vegetative growth throughout the breeding season enabled placement of late season renests higher over the substrate. This latter factor enabled nest elevations to keep pace with the progressively higher tides of the breeding season. Johnston (1956b) noted that some eggs in nests that were inundated for limited periods of time survived to hatching.

THE CLAPPER RAIL

The Clapper Rail nests in low- (California and smooth cordgrass), and high-marsh habitats of the saltmarsh throughout the coastal US (Eddleman and Conway 1998), and because of its propensity to nest at low elevations flooding is the most significant nest mortality factor (Stewart 1951, Burger 1979, Andrews 1980, Massey et al. 1984, Eddleman and Conway 1998). The platform nests of the Clapper Rail are bound to their cordgrass support stems, and thus do not typically float on rising tide waters as do those of saltmarsh nesting larids (Mangold 1974, Burger 1979, Andrews 1980). In southern California, however, Massey et al. (1984:71) reported that for the Light-footed Clapper Rail (*Rallus longirostris levipes*), the tall stems of California cordgrass "not only provided cover but allowed the nest to float upwards in place during a high tide." Clapper Rails compensate for their nest placement in low-elevation habitats by building high nest structures (Burger 1979). Andrews (1980) determined that rails in New Jersey placed their nests in relatively tall vegetation that allowed them to maximize nest elevation. He described rail nests as tall columns of nest material that elevated eggs 15–64 cm above the ground, and concluded that such construction was an adaptation to the demands of the saltmarsh habitat. Kozicky and Schmidt (1949) noted that a difference in nest height of only 5–8 cm could be critical to rail nesting success, and especially during the hatching period. Further, rails were observed hurriedly building their nests higher in the midst of flood-tide events (Andrews 1980, Jackson 1983). In field experiments, Burger (1979) demonstrated that Clapper Rails were able to perceive damage to their nests and immediately and rapidly rebuild them, and that their ability to do so was more advanced than in most larid and waterfowl species. These nest-building responses are important to reproductive success, since if a flood tide does inundate a rail nest, and eggs are not washed out, females will continue to incubate them and at least some eggs will survive to hatching (Kozicky and Schmidt 1949, Mangold 1974). Further, Clapper Rails are known to retrieve eggs that are washed out to some distance from the nest (Burger 1979).

Following nest loss to flooding or predation, Clapper Rails persistently renest (Stewart 1951, Mangold 1974, Andrews 1980). Rails may nest five, or even more times in one season, and in one population in San Francisco Bay, one-half of the nesting population renested at least once during the breeding season (Eddleman and Conway 1998). Eddleman and Conway (1998) speculated that their ability to repeatedly renest allowed rail populations in good habitat to recover rapidly after catastrophic flooding events. Andrews (1980) further suggested that, as for marsh nesting sparrows, late season nests are placed higher due to the seasonal growth of the supporting vegetation, thus reducing flooding risks.

AGE OF SALTMARSH TAXA AND ADAPTATION TO TIDAL FLOODING

Only colonially nesting larids exhibit substantive evidence that environmental cues (peak high tides) are used to time and place nests to avoid tidal flooding. The nesting activities of terrestrial species, including *Ammodramus* sparrows, Song Sparrow, and Clapper Rail, also become synchronized to the spring-tide cycle, but only after initial nests are destroyed by an early season flooding event. Among larids, the ability to repair nests after damage inflicted by flooding tides, and of eggs to survive immersions in salt water, was most highly developed in the Laughing Gull, a saltmarsh specialist, and Ward and Burger (1980) attributed this to a longer evolutionary history of nesting in the saltmarsh environment. Similarly, the Saltmarsh Sharp-tailed Sparrow has a much longer evolutionary history as a saltmarsh specialist than does the Nelson's Sharp-tailed Sparrow (Rising and Avise 1993), which probably explains the more rapid nesting response of the former after flooding tides (Shriver 2002). This explanation may apply also to the highly developed responses to flooding—including nest building and repair, and egg retrieval—seen in the Clapper Rail relative to saltmarsh passerines. However, the history of Clapper Rails in saltmarsh habitats remains unclear (Chan et al., *this volume*).

NESTING ADAPTATIONS OF RED-WINGED BLACKBIRDS TO TIDAL-MARSH HABITAT: A CASE STUDY

In the northeastern US coastal nesting populations of the Red-winged Blackbird are sometimes found breeding in saltmarshes alongside Seaside and Saltmarsh Sharp-tailed sparrows (Reinert et al. 1981, Post et al. 1983, Reinert and Mello 1995). Yet, despite the voluminous literature on the species, little is known about the breeding ecology of saltmarsh inhabiting Red-winged Blackbirds, and in particular how they are able to nest successfully despite spring tides which inundate marshes in New England every 27–29 d. Saltmarsh Red-winged Blackbirds are not known to be morphologically distinct, and Red-winged Blackbirds show little geographic structure in neutral genetic markers, such as MtDNA (Ball et al. 1998). However, research on various taxa (Chan et al., *this volume*) shows that adaptations can occur in tidal-marsh populations that have recently differentiated or face ongoing gene flow from upland populations.

Because of the high predictability of tidal flooding events in New England saltmarshes, an evolutionary response is feasible, and selection for a contracted nesting cycle period should be strong considering the drastic consequences of a delayed response. I tested the hypothesis that such local adaptations exist using data collected from my 4-yr field study of Red-winged Blackbird reproductive success in a Rhode Island saltmarsh. My nesting data were supplemented with parental food-provisioning data collected on the same marsh by Tilton (1987). I predicted that female Red-winged Blackbirds nesting in saltmarshes in New England have adapted to tidal cycles with a reduced renesting interval, and nestling maturation is accelerated by higher parental feeding rates, including the male's more frequent participation in provisioning.

METHODS

STUDY SITE

The study site abuts Hundred Acre Cove (HAC), a 40-ha embayment of the Barrington River estuary in Barrington, Rhode Island. The 32-ha marsh is bordered by the open water of the estuary to the south, and by deciduous forest and/or stands of common reed (*Phragmites australis*) to the north, east, and west. The Red-winged Blackbird study population inhabits a 30–60 m wide zone of smooth cordgrass bordering a 2.7 ha permanent pool centered within the study marsh. Smooth cordgrass stands, 40–80 cm in height, are poorly drained throughout, and standing water was present at the base of most stands. Small patches of saltmeadow habitat were interspersed among the smooth cordgrass habitat, and north and west of the 2.7 ha central pool, networks of smaller (<100 m^2) pools were interspersed among the smooth cordgrass stands forming additional open water-smooth cordgrass ecotones. A more detailed description of the study site is available in DeRagon (1988).

FIELD METHODS

Beginning with annual onset of nest building in early May, the study site was visited on an almost daily basis during the breeding seasons of 1982–1985. Nests were checked regularly the last 3 yr of the study, at intervals of every other day between spring-tide periods, and then daily for the several days before and after maximum spring-tide inundations. Nest contents and productivity were scored during these visits; a nest was considered successful if at least one young fledged from it.

The following nest measurements were taken most years (although not every measurement was taken for every nest every year): nest height above substrate, surrounding vegetation height, distance to nearest open water, nest-bowl depth, and nest-rim elevation relative to mean sea level. The latter was determined in two ways: in 1982 and 1983: by rod and transit conducted shortly after the breeding season; and in 1985 by my chalked stake technique. In the latter, I determined nest elevation by (1) placing a chalked stake at each nest and at a reference marker of known elevation, (2) measuring the distance from each nest's rim and the marker to the chalk line left after a flooding spring tide, and (3) using the measurements from the reference marker to calculate the distance from the marker elevation to the rim elevation of each nest, thereby establishing the rim elevation relative to mean sea level (MSL).

In 1984, seven adult female Red-winged Blackbirds were captured in mist nets and marked with USDI Fish and Wildlife Service aluminum bands on one leg and two colored, celluloid bands on the opposite leg. In 1983, I banded 11 nestlings (nine of which fledged) from four nests, and in 1984, I banded 29 nestlings (25 of which fledged) from 10 nests.

Despite not having banded every female, the small size of my study population combined with conservative deductions allowed me to assign nearly all of the nests to specific females and thereby ascertain nesting chronology. The basis for my deduced assignment was that if an egg was laid in a newly constructed nest within 7 d of the loss of the nest of an unmarked female in the same territory, I assumed that the owner of the lost nest and the new nest was the same. While this supposition did not guarantee correct identity, my determinations were facilitated by low female/male ratios (Table 3), and were corroborated by data on marked females.

Hypothesizing that water inundation does not necessarily result in immediate embryo death, I wished to determine the number of days after laying that an egg would remain non-buoyant, and thus resistant to flooding mortality. For this determination, I used three nests found during the nest-building period in 1985 and marked each egg with its laying sequence. I then checked the nests daily (beginning 2 d after the last egg was deposited), and at each visit placed eggs individually in a beaker of water collected at the nest site and scored their floatability. The daily check was continued until all eggs in the clutch floated to the surface.

I established the elevation of marsh-flooding tides relative to mean sea level by placing a chalked oak stake, covered with a perforated PVC pipe, in the interior of a large pool. I then subtracted the distance from the chalk mark left by the peak elevation of flood water to the top of the stake, from the known elevation of the top of the stake to establish the tidal height relative to MSL. This method yielded the peak elevation of flood tides for 33 d during the 4-yr study. To estimate the elevation of tides for nights when the tide was of insufficient height to leave a chalk mark, or when the gauge was not checked, I developed a regression model of my 33 measurements on a single predictor variable: the measured tidal elevation at a NOAA sampling station located in the Providence River of the same Narragansett Bay estuarine system, and approximately 7 km from the study site. The resulting regression equation ($r^2 = 0.90$, $P < 0.0001$) predicted tidal elevation at my study site.

TABLE 3. POPULATION DEMOGRAPHICS AND NEST-SUCCESS DATA FOR RED-WINGED BLACKBIRDS OVER FOUR YEARS.

Variable	1982	1983	1984	1985	4-yr mean
N males	14	11	9	9	10.8
N females	16	13	13	14	14.0
N 2-, 3-, and 0- female males (totals)	4, 0, 1	2, 0, 0	2, 1, 0	3, 1, 0	11, 2, 1
Mean females/male	1.1	1.3	1.5	1.6	1.3
Mean nests/female (range)	2.0 (1–5)	2.5 (1–5)	1.7 (1–3)	2.0 (1–4)	2.0
Mean nests/territory	2.3	2.9	2.4	3.1	2.7
Total active nests[a]	32	32	22	28	28.5
Date first egg	5/17	5/19	5/16	5/12	-
Date first young	6/9	6/6	6/1	5/25	-
Date first fledgling	7/13	6/25	6/17	6/26	-
N (%) successful nests	7 (21.9)	6 (18.8)	12 (54.5)	7 (25.0)	8 (28.1)
N nests depredated (%, c, m, u)[b]	13 (40.1, 0, 6, 7)	8 (25.0, 0, 1, 7)	8 (36.4, 0, 4, 4)	8 (28.6, 1, 1, 6)	9 (32.4, 1, 12, 24)
N (%) nests lost to flooding	10 (31.2)	15 (46.9)	2 (9.1)	12 (42.8)	10 (34.2)
N (%) nests abandoned	1 (3.1)	2 (6.2)	0 (0)	1 (3.6)	1 (3.5)
N (%) nests lost to unknown cause	1 (3.1)	1 (3.1)	0 (0)	0 (0)	0.5 (1.8)
Total fledglings	18	12	33	11	18.5
Mean fledglings/nest (range)	0.56 (0–3)	0.38 (0–3)	1.50 (0–4)	0.39 (0–3)	0.65
Mean fledglings/female (range)	1.12 (0–3)	0.92 (0–3)	2.54 (0–4)	0.79 (0–3)	1.32
Mean fledglings/male (range)	1.29 (0–5)	1.09 (0–3)	3.67 (1–8)	1.22 (0–6)	1.71
N renests	18	19	9	14	15

[a] N nests reaching at least egg-laying stage.
[b] In parentheses: % nests lost to predation, number lost to crows (c), mammalian predators (m), unknown predators (u).

FOOD DELIVERIES TO NESTLINGS

To compare rates of food deliveries to nestlings by parent Red-winged Blackbirds on my study site in Rhode Island to non-tidal populations, I used data collected for birds at my study site by Tilton in 1984 and 1985 (see Tilton 1987 for methodology). Data from my study site were then compared, using the paired t-test (pairing on nestling day), to data on nestling provisioning from freshwater-wetland/upland populations in Indiana and Wisconsin—in Indiana (Patterson 1991; 1974 and 1975) and Wisconsin (Yasukawa et al. 1990; 1984–1987). Each of those authors presented mean food-delivery rates, by nestling day. For the t-tests, I used the mean of the 1984 and 1985 data from Rhode Island (Tilton 1987) as the food-delivery rates were not significantly different between years. Rates were different between the 2 yr at the Indiana site (Patterson 1991), and thus I made separate comparisons between the combined data for Rhode Island and each of the 2 yr for Indiana. I used data from male-assisted females only for the Indiana site because those values were not significantly different than, and very similar to, those for unassisted females at any nestling age in either year (Patterson 1991:3–4). The data from Wisconsin (Yasukawa et al. 1990) were presented cumulatively for 1984–1987. I also computed, by nestling day, the percent of total food deliveries made by males, and likewise compared those values between my saltmarsh study site and the Indiana and Wisconsin populations.

To compare the percent of successful nests at my study area to the cumulative values reported for marsh and upland habitats by Beletsky (1996), I used the two-sample test of proportions. Mean values are presented with standard deviations throughout. All statistics were performed with Stata v. 7 (Stata Corp., College Station, TX).

RESULTS

The breeding population of Red-winged Blackbirds varied from nine to 14 males and 13–16 females annually over the 4-yr period. The maximum number of polygynous males in any year was four. No male paired with more than three females, and males paired with an average of 1.3 females overall. One unpaired male defended a territory in 1982. Over the 4-yr study, the first egg, hatching, and fledgling dates were 12 May, 25 May, and 17 June, respectively (Table 3).

Four of seven females banded in 1984 returned to the study area to breed in the following year. Of the 34 banded nestlings that fledged in 1983 and 1984, only one, a first-year male, was subsequently encountered. This male did not establish a territory, however, and was not sighted again.

Red-winged Blackbirds nested exclusively within the cover of irregularly flooded smooth cordgrass stands. Early season nests were placed among brown, persistent stems from the prior year's growing season; later nests were placed in new-growth smooth cordgrass plants. Fifty-one (81%) of 63 nests measured were placed within 1.5 m of the open water edge of a pool, over standing water, where the tallest stands of smooth cordgrass occurred. The mean height above the substrate of 74 nests measured was 39.6 cm (± 19.2); the mean height of the vegetation surrounding 50 nests measured was 58.3 cm (± 15.8).

Red-winged Blackbird females built nests in 2–4 d. They laid one egg per day, and two (3.3%), three (38.0%), or four eggs (58.7%) per clutch ($\bar{x}$ = 3.60 ± 0.49 for 92 nests for which I suspected that no eggs were lost to flooding or predation). The incubation period, measured as the interval from the laying of the penultimate egg to the hatching of the last young, was 13 d for 20 (77%) of 26 nests with complete data ($\bar{x}$ = 12.9 ± 0.7). The mean incubation period of five nests inundated during the egg laying and early incubation period (12.8 d) was not significantly different than for the 21 nests which were not flooded (12.9 d; t = 0.17, P = 0.87). The nestling period, measured as the interval from the hatching of the first young to the fledging of the last young, was 11 d for 12 (48%) of 25 nests with complete data ($\bar{x}$ = 10.8 ± 0.9, range = 9–12). The hatching period, calculated as the number of days from the first to last egg hatched, ranged from <1 d (for clutches of two–four eggs) to 3 d (for two clutches of four eggs) ($\bar{x}$ = 1.6 ± 0.7). The modal nesting cycle period (days from first egg laid to last young fledged) calculated for 17 nests with complete data was 23 d ($\bar{x}$ = 24.1 ± 1.4, range = 22–27).

For 65 nestlings in 28 successful nests with adequate chronological data, I determined the number of days in the nest under the assumption that the first-hatched young were the first to fledge. One young (1.5%) fledged after 8 d in the nest, 13 (20.0%) after 9 d, 29 (44.6%) after 10 d, 17 (26.2%) after 11 d, and 5 (7.7%) after 12 d. The mean number of days in the nest for the 65 nestlings was 10.2 (± 0.9). Of the 28 nests, 8-d-old young fledged from one (3.6%), 9-d-old from 11 (39.3%), 10-d-old from 17 (60.7%), 11-d-old from 13 (46.5%), and 12-d-old from five nests (17.9%).

The mean depth of the nest bowl (rim to bowl bottom) for 66 nests measured was 60 mm (± 6). The mean elevation of the rims of 80 nests measured was 105 cm above MSL (± 5); the mean elevation of the bowl bottom of 63 nests for which both nest elevation and bowl depth were measured was 99.6 cm (± 4.2).

Three eggs, the last deposited in each of the three experimental nests, were immersed 2 d after their lay date, after 2 d of incubation (incubation began with the laying of the penultimate egg) and all three sank. When immersed 1 d later, one sank, one sank slowly, and one floated. All three eggs floated on day three, after 4 d of incubation.

Three eggs, the penultimate of each clutch, were immersed 3 d after their lay date, after 3 d of incubation all three sank. When they were immersed 1 d later, all three eggs floated. Three eggs, the second laid of two four-egg clutches, and the first laid of one three-egg clutch, were immersed 4 d after their lay date, and after 3 d of incubation all three floated. Two eggs, the first laid of the two, four-egg clutches, were immersed 5 d after their lay date, and after 3 d of incubation both floated. Thus, Red-winged Blackbird eggs with developing embryos became buoyant during the fourth or fifth day of incubation, with earlier laid eggs floating prior to the penultimate and ultimate eggs of the clutch.

Of the 114 active Red-winged Blackbird nests found during this 4-yr study, 59 (52%) were renesting attempts; three (5%) represented the fifth nesting attempt of individual females during a single breeding season, five (8%) represented the fourth nesting attempt, 14 (24%) represented the third nesting attempt, and 37 (63%) the second nesting attempt. Female Red-winged Blackbirds deposited at least one egg in a mean of 1.81 (± 0.99) nests per year.

Of the 37 nests for which I determined a renest interval during the 1983–1985 breeding seasons, the first egg in 23 nests (62%) was laid on the fifth day after nest loss, in five nests (14%) on the sixth day, in two nests (5%) on the seventh day, and in seven nests (19%) on the eighth or later day after nest loss. Of eight renestings among seven color-banded females in 1984 and 1985, the new nest of each was built in the same territory as the previous.

Female Red-winged Blackbirds delivered food to nestlings (Tilton 1987) at a significantly greater rate than did females at an Indiana freshwater wetland site (Patterson 1991) in both 1974 and 1975, and a Wisconsin freshwater wetland-upland site (Yasukawa et al. 1990) for the cumulative years 1984–1987 (Table 4; Fig. 1). Male Red-winged Blackbirds delivered food to nestlings at a significantly greater rate than at the Indiana site in 1974, but not in 1975, and at a significantly greater rate than in the Wisconsin population (Table 4; Fig. 1). The mean percent of food trips made by males was significantly greater at the Indiana (Patterson 1991) site than at the Rhode Island saltmarsh site (Tilton 1987) in 1975 (mean in Rhode Island = 25.4 ± 14.4; mean Indiana = 47.1 ± 21.4; t = -5.13, df = 9, P < 0.001), but not in 1974 (mean in Rhode Island = 23.1 ± 15.7; mean in Indiana = 25.1 ± 18.7; t = -0.47, df = 10, P = 0.65). The mean percent of food trips made by males was significantly greater at the Rhode Island site than the Wisconsin site (Yasukawa et al. 1990) (mean in Rhode Island = 23.1 ± 15.7; mean in Wisconsin = 15.4 ± 12.07; t = 2.63, df = 10, P = 0.03).

REPRODUCTIVE SUCCESS AND MORTALITY FACTORS

Female Red-winged Blackbirds laid at least one egg in 114 nests over the 4-yr study period.

TABLE 4. RATES OF FOOD DELIVERIES TO NESTLINGS BY PARENT RED-WINGED BLACKBIRDS.

Statistical comparisons[a]	Study area	Time period	Mean trips/ nestling/hour
Females			
1	Rhode Island[b]	1984–1985	5.3 ± 2.9
2	Indiana[c]	1974	2.1 ± 0.8
3	Indiana[c]	1975	2.9 ± 0.5
4	Wisconsin[d]	1984–1987	2.7 ± 0.9
Males			
5	Rhode Island[b]	1984–1985	1.5 ± 1.2
6	Indiana[c]	1974	0.6 ± 0.4
7	Indiana[c]	1975	1.4 ± 0.7
8	Wisconsin[d]	1984–1987	0.5 ± 0.4

[a] Paired t-test results, 1:2, t = 4.67, df = 10, P < 0.001; 1:3, t = 3.32, df = 9, P < 0.01; 1:4, t = 3.82, df = 10, P < 0.01; 5:6, t = 3.15, df = 10, P = 0.01; 5:7, t = 0.77, df = 9, P = 0.46; 5:8, t = 3.56, df = 10, P < 0.01.
[b] Tilton (1987).
[c] Patterson (1991).
[d] Yasukawa et al. (1990).

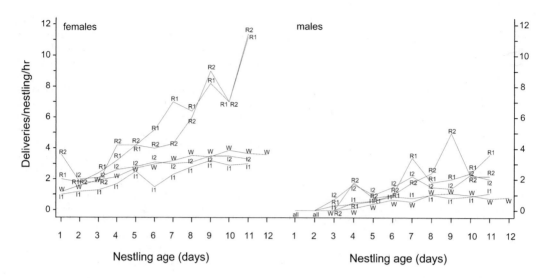

FIGURE 1. Food-delivery rates to nestlings by male and female Red-winged Blackbirds at Hundred Acre Cove salt marsh (Tilton 1987) in 1984 (R1) and 1985 (R2), at a freshwater wetland population in Indiana (Patterson 1991) in 1974 (I1) and 1975 (I2), and at a freshwater wetland-upland population in Wisconsin (Yasukawa et al. 1990) for the combined period 1984–1987 (W).

Seventy-four young fledged from 32 successful nests (1–4 fledglings/nest). The mean numbers of young fledged per nest, per female, and per male were 0.6, 1.3, and 1.7, respectively. Eighty-two nests were not successful due to tidal flooding (39 nests, 48%), predation (37 nests, 45%), nest abandonment (excluding post-flood abandonment, four nests, 5%), and unknown causes (two nests, 2%; Table 3).

Nest loss to flooding occurred during marsh inundations accompanying spring tides, which occur at new moon phases every 27–29 d in coastal New England. Such tides destroy nests by drowning nestlings, and by dispersing eggs that float from nests when the water level rises over the rim. Of the ten spring tides that occurred during the four breeding seasons of the study period, eight exceeded the mean elevation of Red-winged Blackbird nest rims (105 ± 5 cm), and all exceeded the mean elevation of nest bowl bottoms (100 ± 4 cm). The mean elevation of the ten spring tides (112 ± 8 cm) exceeded the elevation of 73 (91%) of the 80 Red-winged Blackbird nest rims for which elevation data were available, and the highest spring tides were well above the elevation of the highest Red-winged Blackbird nests (Fig. 2). At least 56 (49%) of the 114 active nests monitored in this study were completely submerged beneath spring tide waters while they were active.

Spring tides occurring during June, the peak of the breeding season, destroyed 69% (nine of 13) of active nests in 1982 (peak-tide elevation = 114 cm), 67% (8 of 12) of active nests in 1983 (peak-tide elevation = 109 cm), 20% (two of 10) of active nests in 1984 (peak-tide elevation = 106 cm), and 85% (11 of 13) of active nests in 1985 (peak-tide elevation = 114 cm; Table 3). Of the 39 nests lost to flooding, one was at the egg laying stage, 15 were at the incubation stage, five were at the incubation-nestling (hatching) stage, and 18 were at the nestling stage. Sixty-two eggs and 60 nestlings were lost to flooding tides.

Nest predation was witnessed twice when American Crows (*Corvus brachyrhynchos*) were seen removing 4–5 d old and 8–9 d old nestlings from nests (one nestling survived). We assumed that 12 nests torn from their support stems were the victims of mammalian predation, probably northern raccoon (*Procyon lotor*). Twenty-five additional depredated nests found intact and empty were likely the victims of American Crows, Common Grackles (*Quiscalus quiscula*), and mink (*Mustela vison*). A breakdown of predator types by year is presented in Table 3. Six nests were depredated at the egg laying stage, 27 at the incubation stage, and three at the nestling stage; 102 eggs and 12 nestlings were lost to predation. I rarely saw Brown-headed Cowbirds (*Molothrus ater*) on the marsh, and no cowbird eggs were found in Red-winged Blackbird nests.

CHARACTERISTICS OF SUCCESSFUL NESTS

Successful nests were not depredated, and were: (1) started during or immediately after a spring tide such that young fledged prior to the high waters of the spring tide to follow 1 mo later, (2) inundated by spring tide waters during the egg laying or early in the incubation period, and thus the eggs did not float out of the nests, or (3) higher than the peak level of a spring tide (Table 5). Fourteen (44%) of the 32 successful nests resulted from female Red-winged Blackbirds immediately renesting after losing a nest to a May or June spring tide flooding event (Table 5). For each of those 14 nests, the first egg was laid in the new nest on the fifth day after nest loss. This uniform renesting response served to synchronize the nesting activities of all Red-winged Blackbirds (as well as Seaside and Saltmarsh Sharp-tailed sparrows) which lost nests to a spring tide. By initiating the new clutch on the fifth day after nest loss, the young of the subsequent clutch (if the nest was not depredated) were able to fledge (or at least climb onto stems above the nest) just prior to the onset of the floodwaters associated with the next spring tide which occurs 27–29 d after the last (Fig. 3). Indeed, for many of those successful nests renesting intervals of only 1 or 2 d longer than the modal 5-d period would have resulted in the drowning of some or all of the nestlings that fledged. Five other successful nests were started during or right after a spring tide: three were first nesting attempts and two were post predation attempts (Table 5).

Of 56 active nests that were inundated for 1–3 successive days by a spring tide, 17 (30%) — 11 of 13 in the egg-laying stage, three of eight in the incubation stage, and three of 17 in the nestling stage — survived the flood event(s). Eggs of nests in the egg-laying stage

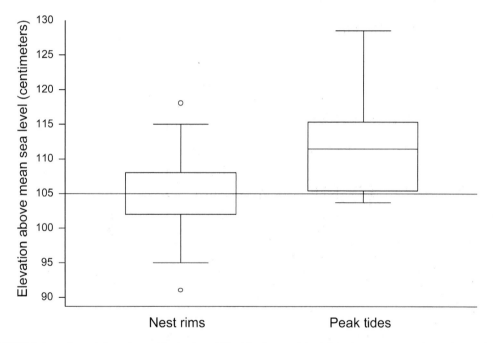

FIGURE 2. Box plots of elevations of Red-winged Blackbird nests (N = 80) and new-moon spring tides (N = 10) at Hundred Acre Cove saltmarsh. The horizontal line marks the median (and mean) elevation of nests.

TABLE 5. CATEGORIES OF CONDITIONS RESULTING IN SUCCESSFUL NESTS.

	Nest attempt				Total nests (%)
	1	2	3	5	
Post-flood renesting		10	3	1	14 (44)
Flooded, eggs did not float	4	2	1		7 (22)
Timing: nest built at spring tide	1	2			3 (9)
Timing: predation at spring tide			2		2 (6)
Spring tide too low to flood nests	4	2			6 (19)
Total nests (%)	9 (28)	16 (50)	6 (19)	1 (3)	32 (100)

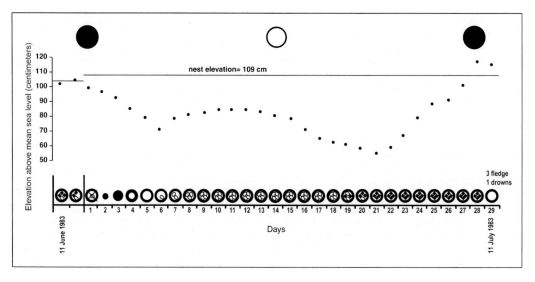

FIGURE 3. Synchronization of the lunar-tidal cycle and Red-winged Blackbird nesting cycle at Hundred Acre Cove salt marsh. Plots to the right of the scale show the elevation of peak daily tides relative to nest elevations. Below the plots are the contents, by day, of two successive nests (diamonds represent nestlings) of one female and how they were impacted by flooding tides. The flood waters associated with an early June new-moon spring tide—which held three young and one infertile egg prior to flood encroachment—destroyed this female's first nest, and those of most other breeding songbirds on the marsh. Subsequent nesting activities of affected females of all species were synchronized to the tidal cycle. Although this female (1) started building a replacement nest within 1 d of destruction, (2) deposited a first egg in the replacement nest only 5 d after losing the first, and (3) completed her nest cycle (first egg to last young fledged) in the modal (for this study site) 23-d period, only three of her four nestlings avoided drowning when the next new-moon flood tide encroached 28 d after the first.

did not float and had the greatest probability (90%) of surviving an inundation, whereas eggs and young in inundated nests in the hatching period did not survive. Of 10 nests inundated during the egg-laying stage, but that did not lose all or any eggs, only one nest (containing two eggs) was abandoned. Of five nests with two–four eggs inundated early in the incubation stage, but that did not lose all or any eggs, two (with two and three eggs, respectively) were abandoned. Three nests that reached the nestling phase were inundated leaving one infertile egg in the nest; all of those nests were abandoned. Overall, seven (22%) of the 32 successful nests survived after having their eggs immersed for one–three successive nights. Thus, the propensity of female Red-winged Blackbirds to continue to incubate eggs after an inundation event was a key factor in nest survival at the study site.

Six nests survived a spring tide because their rim elevation exceeded that of the peak floodwaters. Overall, among nests for which a rim elevation was measured, the mean elevation of 17 that survived a flooding tide (107 ± 4 cm) was significantly higher than of 33 that did not survive an inundation event (104 ± 4 cm; $t = -2.45$, $P = 0.02$).

DISCUSSION

REPRODUCTIVE SUCCESS OF SALTMARSH-NESTING RED-WINGED BLACKBIRDS

Though few references in the peer-reviewed literature describe Red-winged Blackbirds nesting in emergent-saltmarsh habitats, this is a common occurrence in the coastal Northeast, and especially in stands of smooth cordgrass (Reinert et al. 1981, Post et al. 1983, Reinert and Mello 1995). Avian nests occupying emergent-saltmarsh habitats are subject to flooding by high waters associated with coastal storms, or new-moon spring tides, which occur every 27–29 d in the Northeast. Because the majority of new-moon tides exceed the height of passerine nests, most active nests during flood periods are destroyed when eggs float from nests or young drown. In my study population, tidal-flooding was the principal cause of nest loss overall, and in 3 of 4 yr (Table 3). Predation was the second most important mortality factor, overall, and the principle cause of nest failure in 1984 when only one spring tide reached the elevation of Red-winged Blackbird nests. Additionally, a few nests were lost to abandonment/starvation and unknown causes (Table 3).

During this study, nest success in 3 of 4 yr (21.9%, 18.8%, and 25.0%, respectively for the years 1982, 1983, and 1985; Table 3) was lower than any such value reported in a review of 27 Red-winged Blackbird studies (Beletsky 1996). The 4-yr mean nest success at HAC (28.1%) is among the lowest of reported values and was significantly lower than the overall mean of 40.2% for marsh habitats ($z = 2.63$, $P < 0.01$), and 46.4% for upland habitats ($z = 3.73$, $P < 0.001$) reported by Beletsky (1996).

The mean number of young fledged per nest at HAC was also relatively low. The 4-yr mean of 0.65 ranks close to the lowest reported in the reviews of Searcy and Yasukawa (1995) and Beletsky (1996), and was substantially lower than the overall means calculated for marsh (1.23) and upland (1.15) habitats in a meta-analysis by Beletsky (1996; Table 9.1). The mean number of fledglings per nest of 0.38 and 0.39 at HAC for the years 1983 and 1985, respectively, were lower than the values reported in any of the 18 studies reviewed by Searcy and Yasukawa (1995; Table 4.4), or 27 studies reviewed by Beletsky (1996).

The exceptionally low reproductive success I documented at HAC is not surprising considering the combined effects of predation and tidal flooding in this population. Beletsky (1996) summarized the causes of nest failure for 15 Red-winged Blackbird studies conducted in marsh habitats, and three studies conducted in upland habitats. Predation was the principal cause of nest failure for all of the marsh populations, and for two of the three upland populations. Starvation was the principal cause of failure in the third upland population. In only one population (Blakley 1976) did factors relating to natural catastrophes, in this case a wind storm, account for >9% of nest losses (Blakley 1976). As in other populations, Red-winged Blackbirds at HAC experienced substantial nest losses to predation; from 25–40% of nests were lost annually to predators over the 4-yr study, with an overall mean of 32.5% (Table 3). The percent of nests lost to predators averaged over the three upland populations summarized by Beletsky (1996, Table 9.2) was 32.6%, and over the 14 marsh populations, 45.2%. At HAC, however, Red-winged Blackbirds were further subjected to monthly tidal inundations resulting in the exceptionally low reproductive success exhibited.

ADAPTATIONS AND RESPONSES TO TIDAL FLOODING

Nest placement

Although the nesting cycle response to flooding of Red-winged Blackbirds at HAC was similar to that documented for Seaside and saltmarsh sparrows, the overall nesting success of Red-winged Blackbirds was substantially lower than that of saltmarsh sparrows nesting at HAC (DeRagon 1988) and of Seaside Sparrows nesting in a nearby Massachusetts saltmarsh (Marshall and Reinert 1990) (Red-winged Blackbirds 28%, Seaside Sparrow 38%, saltmarsh sparrow 59%; Table 2). This is attributable to the low rates of nest predation for Seaside Sparrows (11% of failed nests) and saltmarsh (11%) sparrows at those respective marshes compared to Red-winged Blackbirds (45%). Nests of Seaside Sparrows and saltmarsh sparrows are nearly always built beneath a tuft or canopy of marsh-grass vegetation, and thus are not visible to avian predators flying over. In contrast, Red-winged Blackbird nests are open above and thus were easily detected by avian predators. At HAC, Red-winged Blackbirds and Seaside Sparrows nested in close proximity to one another in shared stands of smooth cordgrass. Nests of Red-winged Blackbirds were on average 4 cm higher ($t = -3.3$, $P = 0.001$) than those of Seaside Sparrows (DeRagon 1988, this study), probably because sparrows had to situate their nests lower in the vegetation to enable the weaving of a canopy above. Thus, although Red-winged Blackbirds may reduce flooding risks by nesting higher in the plants, this did not offset the increased vulnerability of their nests to predation resulting from a lack of cover above. The survival advantage exhibited by the *Ammodramus* sparrows may exist because, having persisted for longer periods in saltmarsh habitats, they have adapted superior predator avoidance mechanisms.

Post-immersion egg survival

My experiment with three clutches revealed that during an inundation event at HAC Red-winged Blackbird eggs will not float from nests until the third or fourth day of incubation. Female Red-winged Blackbirds in the egg laying or early incubation period did not abandon nests after an inundation event if two or more eggs remained in the nest. Eggs remained viable after being inundated, and indeed the chance timing of early season nests such that egg laying-early/incubation periods and flooding new moon tides coincided, resulted in 22% of the total successful nests over the 4-yr study (Table 5). The viability of eggs after salt-water inundations has been demonstrated for several other species of birds nesting in saltmarshes, including White Ibis (Frederick 1987), Clapper Rail (Kozicky and Schmidt 1949, Mangold 1974), Herring Gull (Ward and Burger 1980),

Laughing Gull (Burger 1979), Common Tern (Burger 1979), and *samuelis* Song Sparrows of the San Francisco Bay area (Johnston 1956b). Burger (1979) and Ward and Burger (1980) conducted salt-water immersion experiments with eggs of Laughing and Herring gulls and determined that in general, egg survival of Laughing Gulls, which have long nested in tidal-marsh habitats, was greater than that of Herring Gulls which have only recently colonized saltmarsh islands. Those experiments indicate that Laughing Gull eggs are better adapted to marine environments, and suggest similar controlled comparisons among populations of Red-winged Blackbirds and Song Sparrows occupying saltmarsh, freshwater wetland, and upland habitats, and between obligate saltmarsh-nesting species such as Seaside Sparrow and Clapper Rail, and closely related species that do not use, or rarely use, saltmarsh habitats, such as Le Conte's Sparrow (*Ammodramus leconteii*) and King Rail (*Rallus elegans*).

Contracted nesting cycle

I predicted, based on the potentially disastrous results of an extended nesting period, that relative to other habitats, selection in saltmarsh populations would favor attributes consistent with a shortened nesting cycle and renesting interval. In the following paragraphs I apply data collected at HAC by Tilton (1987) and me to test my predictions.

At HAC, Red-winged Blackbirds completed their nest cycle (first egg laid to last young fledged) in a mean of 24.1 d ($\pm$ 1.4 d, N = 17). I compared the nest cycle period of the HAC population with similarly derived periods for a freshwater wetland-upland population in Wisconsin provided by K. Yasukawa (pers. comm.). Though periods from three-egg clutches were slightly shorter in Wisconsin (23.6 $\pm$ 1.4, N = 31, Wisconsin; 24.1 $\pm$ 1.6, N = 9, HAC; P = 0.37), the nest-cycle period from four-egg clutches was nearly 1 d shorter at HAC vs. Wisconsin, and this difference approached statistical significance (24.9 $\pm$ 1.3, N = 49, Wisconsin; 24.0 $\pm$ 1.3, N = 8, HAC; P = 0.08) Thus, although these results are not conclusive, they do suggest, consistent with my prediction, a shorter nest cycle period for nests of the modal clutch size.

To further explore the hypothesis that the nesting cycle of the saltmarsh population is compressed relative to populations in other habitats, I employed a period-by-period approach. I excluded the nest-building period in this analysis because it is highly variable within and among populations (1–8 d), and nests can be built in 1 d when necessary and thus egg production, and not nest construction, will limit contraction of the nesting cycle (Case and Hewitt 1963, Yasukawa and Searcy 1995, Beletsky 1996).

Red-winged Blackbirds could shorten their nesting cycle by laying fewer eggs, which would potentially contract the laying, incubation, and nestling periods. Indeed, saltmarsh populations of Song Sparrows (Johnston 1956a) and Swamp Sparrows (*Melospiza georgiana*; Greenberg and Droege 1990) have significantly smaller clutch sizes than populations occupying non-tidal habitats. At HAC, Red-winged Blackbirds laid one egg per day and clutches ranged from two–four eggs with a mean of 3.55. In a meta-analysis of 20 studies conducted by Dyer et al. (1977), mean clutch sizes ranged from 2.43–3.70 with an overall mean of 3.28. The meta-analysis of Martin (1995) yielded a mean clutch size of 3.49. Thus, though this analysis does not control for latitudinal variation in clutch size, the evidence available is not consistent with my hypothesis that clutch sizes at the saltmarsh site were smaller than the norm for the species.

The mean incubation period for 26 nests with complete data at HAC was 12.85 d; one half of females completed incubation in 13 d. Martin's (1995) meta-analysis yielded a mean from the literature of 12.6 d. In his review, Beletsky (1996; Table 6.2) reports incubation periods from four studies conducted in upland and freshwater wetland habitats: 10–12 d (New York), 11–14 d (California), 11–13 d (Washington), and 10–12 d (Illinois). Nero (1984) reported an incubation period of 11–12 d for his freshwater marsh population in Wisconsin. Thus, my data are not consistent with a contracted incubation period.

Martin (1995) demonstrated that among parulids and emberizids, nestling periods were shorter in species that nested in habitats with the greatest predation pressure. It follows that habitat-specific nest mortality factors could shape variability in nestling periods. A review of the available data (Table 6) reveals that Red-winged Blackbird nestlings at my site, on average, leave the nest earlier than at non-tidal sites. Other studies have demonstrated that Red-winged Blackbird nestlings will leave the nest as early as the ninth day of their lives only when disturbed (Allen 1914, Beer and Tibbitts 1950, Case and Hewitt 1963; Table 6). At HAC, >20% of nestlings distributed among 43% of successful nests fledged at 8 or 9 d of age. At approximately 20% of successful nests, premature fledging of one or more young resulted from rising tidal waters which forced young to climb into vegetation surrounding the nest to avoid drowning. Predators forced one young

TABLE 6. REPORTED NESTLING PERIODS OF RED-WINGED BLACKBIRDS.

Duration in days	Region; habitat	Notes	Reference
9–11	New York; freshwater wetland	"On the ninth [day]…The young can fly short distances, however, and can not be kept in the nest if once frightened or removed. If the nest has become polluted, as frequently occurs when it has become greatly compressed by the growing vegetation, they may leave of their own accord on this day. On the tenth the stronger of the young leave and climb to near-by supports… If the nest is approached, all leave, but otherwise the weaker remain until the eleventh day…when all scatter to the vegetation in the immediate vicinity."	Allen (1914: 100–101).
9–13	Wisconsin; freshwater marsh	"On the 9th and subsequent days a disturbance is apt to cause them to leave the nest. Normally the young leave on the 11th or 12th day but in case of cold weather may remain in the nest until the 13th day."	Beer and Tibbitts (1950:73).
9–14	New York; cattail (freshwater) marshes and upland habitats	"If disturbed, nestlings left the nest on the 9th day after hatching. If undisturbed, the stronger nestlings left the nest on the 10th day, and weaker and smaller nestlings left on the 11th or 12th day. In inclement weather, many nestlings remained in the nest up to 14 days."	Case and Hewitt (1963:14).
9.2 females 9.7 males	Michigan; freshwater marsh and upland habitats	"The mean duration of nestling life is shorter for females than males (9.2 vs. 9.7 days)." Note, the nestlings measured were handled on every day in the nest, thus their fledging dates were probably premature.	Holcomb and Twiest (1970:301).
11	Michigan; freshwater marsh/bog	"…for the 11 d of nestling life…"	Fiala and Congdon (1983:644).
11	Illinois; freshwater marsh and bog	"Birds fledge approximately eleven days after hatching."	Strehl and White (1986:179).
11	Ontario; cattail (freshwater) marsh	"…to fledging (usually day 11)…"	Muldal et al. (1986: 108).
11–12	Washington; freshwater marshes	"Fledging occurred usually at 11 to 12 days of age."	Beletsky and Orians (1990:607).
10.8	Wisconsin; freshwater-wetland/upland habitats	Nestling period period (days from first young hatched to last young fledged) for 31 3-egg clutches = 10.6 d (± 1.0); for 49 4-egg clutches = 10.8 (± 1.0).	K. Yasukawa (pers. comm.)
8–12	Rhode Island; saltmarsh	Mean nestling period (days from first young hatched to last young fledged) = 10.8 d (± 0.9, range = 9–12 d, mode = 11 d [48% of 25 nests]). Mean days in nest per nestling = 10.2 (± 0.9, range = 8–12 d, mode = 10 d [45% of 65 nestlings]); see Results section).	This study.

each from two other nests to fledge prematurely. However, the remainder of nests from which young fledged at 8 or 9 d showed no evidence of disturbance.

Whittingham and Robertson (1994) found that when Red-winged Blackbird nestlings received food at a heightened rate due to male participation in provisioning, the mass of the young at 8 d of age was significantly greater than that of nestlings fed by the female alone. Such a head start could equate to the difference between life and death for nestling songbirds in saltmarshes. Tilton's (1987) data on parental food provisioning at HAC suggests that selection for early development of young has favored male participation in the feeding of nestlings, and accelerated food delivery rates to young by both male and female parents.

At HAC, Tilton (1987) found that 11 out of 12 males fed young, which greatly exceeds the highest reported proportion from inland populations (Whittingham and Robertson 1994, Searcy and Yasukawa 1995, Beletsky 1996). Further, because the mean harem size at HAC (1.3) is among the lowest reported (Searcy and Yasukawa 1995, Beletsky 1996), and most males that provision restrict their feeding to one nest only, or no more than one nest simultaneously (Beletsky and Orians 1990, Yasukawa et al. 1990, 1993; Patterson 1991), the percent of total nests provisioned by males at HAC is also relatively high.

At HAC, not only did most males assist females in feeding young, but the rates of food deliveries by females were significantly faster than females occupying upland and freshwater wetland habitats in Indiana (Patterson 1991) and Wisconsin (Yasukawa et al. 1990) (Table 4; Fig. 1). This finding is consistent with several Red-winged Blackbird studies that demonstrated that females at male-assisted nests do not reduce their provisioning rates as a result of receiving assistance from the male (Muldal et al. 1986, Whittingham 1989, Beletsky and Orians 1990, Yasukawa et al. 1990, Patterson 1991, Whittingham and Robertson 1994). Additionally, male provisioning rates at HAC were significantly greater than at the Indiana site in 1 of 2 yr (no difference in the other year), and significantly greater than at the Wisconsin site (Table 4; Fig. 1). While the suggestion is that the survival value of accelerated nestling development has selected for an accelerated provisioning rate, the high availability of food in saltmarsh habitats (Post and Greenlaw 1982, Post et al. 1983) may also play a role. Indeed, at HAC 70% of food trips were completed in the smooth cordgrass dominated habitats within territory boundaries (Tilton 1987). More likely, the two factors (selection and food abundance) are intertwined.

Thus, nestling provisioning behaviors at HAC (Tilton 1987) are consistent with the hypothesis that saltmarsh populations of the Red-winged Blackbird are adapted to produce young capable of fledging early—often by climbing up smooth cordgrass stems surrounding the nest—and that this ability is achieved by an enhanced nutritional status of nestlings. This is enabled by a high percentage of nests at which males assist females in the feeding of the nestlings, and a relatively high rate of food deliveries by both male and female parents. That more frequent feeding results in better conditioned young is clear, as several studies have shown that the starvation rate is lower, and the fledging success higher, for pair-fed Red-winged Blackbird nestlings vs. young fed by their mothers alone (Beletsky and Orians 1990, Yasukawa et al. 1990, Patterson 1991, Whittingham and Robertson 1994).

Renesting interval

In 62% (23 of 37) of documented cases, female Red-winged Blackbirds at HAC deposited the first egg in a new nest on the fifth day following the loss of an earlier active nest, and in 81% of cases the first egg of the replacement nest was deposited within 7 d of the earlier nest destruction event. Renest intervals of Red-winged Blackbirds as short as 4 d have been documented by Dolbeer (1976) and Beletsky and Orians (1996). Because Red-winged Blackbirds in my study area, and others (Picman 1981, Beletsky and Orians 1991), commonly produced four or five clutches within a breeding season, the energetic demands of renesting do not appear to be limiting.

Despite the ability to rapidly renest, and its apparent low cost, the available data on renest intervals reveals that not all female Red-winged Blackbirds respond to nest destruction by immediately initiating a new nest. In an Ohio old field, Dolbeer (1976) determined a mean renest interval of 9.7 d (range = 4–30 d) for 17 renests by a minimum of 16 females in 1973, and a mean of 12.1 d (range = 4–29 d) for 10 renests by a minimum of nine females in 1974. Yasukawa (pers. comm.) determined a mean renest interval of 6.1 d for 55 females renesting on the same territory, and 5.9 d for nine females that moved between territories following a nest loss. Of 877 intervals (following unsuccessful first nests only) documented in Washington by Beletsky and Orians (1996), 16% were of 5 d or fewer, 41% were of 6–10 d, and the remainder were >11 d. Thus, while timely renest responses are the norm for the populations for which data are available, females from most populations studied do not immediately initiate the replacement nest. Clearly, selection for near minimum nest intervals at HAC would be strong, as nests of females that delay initiating a new nest—even for only 1–3 d—after the loss of a nest to tidal flooding will likely be destroyed by the next spring tide should it survive to that stage. Indeed, most females at HAC initiated replacement nests immediately after suffering the loss of a nest. This contrast in renesting periods between saltmarsh and non-tidal populations of Red-winged Blackbirds parallels similar comparisons made in a Maine saltmarsh by Shriver (2002) between the Saltmarsh and Nelson's Sharp-tailed sparrows.

Renest intervals of the saltmarsh sparrow, an obligate saltmarsh species, were significantly shorter than those of Nelson's, a species for which freshwater marshes comprise the dominant breeding habitat. Shriver (2002) concluded that saltmarsh sparrows had likely evolved for a longer time with the predictable flooding effects of tides and the lunar cycle, and thus had developed, as demonstrated here for the Red-winged Blackbird, a relatively rapid post-flood renesting response.

CONCLUSIONS

Flooding is an important selective factor for shaping both the behavior and life history of birds that breed in tidal marshes. Taxa with long evolutionary histories in tidal-marsh systems show a variety of adaptations from nest construction and placement to the timing of breeding and the ability to rapidly renest. Even more interesting is the incipient behavioral adaptations of birds in tidal-marsh-breeding populations that otherwise show no local morphological adaptations to tidal marshes and are not known to be genetically distinct. The Red-winged Blackbirds of New England marshes clearly show an ability to nest rapidly in response to flooding events, which is facilitated by changes in nestling care and feeding. Such local adaptation to tidal-marsh conditions in an otherwise undifferentiated population is not surprising given the amount of differentiation that has been documented in a variety of taxa in the absence of underlying genetic divergence (Chan et al., *this volume*). Future research should focus on the degree to which the behavioral responses to flooding in tidal marsh passerines are facultative or genetically based.

ACKNOWLEDGMENTS

I am indebted to F. C. Golet and W. R. DeRagon for their help with banding efforts, various aspects of data collection, and in strategizing field methods. Help in the field was also provided by M. Janowkski, G. Shaughnessy, J. Sullivan, M. A. Tilton, and C. Wilson. L. Beletsky, A. Powell and N. Nur shared insights on Red-winged Blackbird, Savannah Sparrow, and Song Sparrow behaviors, respectively. I thank R. Greenberg, M. V. McDonald, and P. Paton for their reviews of an earlier draft of this paper. Special thanks go to K. Yasukawa for his thoughtful review, and for sharing unpublished data from his studies of Red-winged Blackbirds in Wisconsin. This research was supported in part by the NOAA Office of Sea Grant, U. S. Department of Commerce, Grant No. NA79AA-D-00096.

FLOODING AND PREDATION: TRADE-OFFS IN THE NESTING ECOLOGY OF TIDAL-MARSH SPARROWS

Russell Greenberg, Christopher Elphick, J. Cully Nordby, Carina Gjerdrum, Hildie Spautz, Gregory Shriver, Barbara Schmeling, Brian Olsen, Peter Marra, Nadav Nur, and Maiken Winter

Abstract. Tidal-marsh vertebrates experience two distinct challenges to successful reproduction: inundation of the soil with water, which is variable and often unpredictable, and the simple vegetative structure, which offers few safe havens from predation. We review both published and unpublished studies of tidal-marsh birds and their relatives to determine if overall nest success is lower and if predation and flooding are higher than in non-tidal-marsh relatives. In addition, we examine information on clutch size, breeding season, and nest location for differences between tidal and non-tidal taxa. Overall, we find little support for the idea that the additive effects of flood- and predation-loss leave tidal-marsh sparrows with a net high nest loss rate compared to ecologically comparable sparrows. In part, the two sources of mortality are negatively correlated and hence, at least partly compensatory. Flooding is an important cause of mortality in some populations, notably those along the north Atlantic Coast and south San Francisco Bay. However, in general, predation is the most important source of mortality in tidal-marsh populations. The importance of predation may be masked in populations that also suffer high rates of flood-related nest loss. Within tidal-marsh sparrows, clutch size is lower at sites with higher predation rates. The effect of other sources of mortality and latitude disappear when the variables are entered in a step-wise regression. If predation does effect variation in clutch size in tidal-marsh species, it probably is a result of the effect lower brood size has on nest conspicuousness rather than a bet-hedging strategy against high nest loss. Overall, clutch size is relatively low in tidal-marsh forms, although these comparisons are often confounded by other variables, such as latitude, altitude, or continentality of climate. Nesting seasons tend to be longer in tidal-marsh birds. However, few studies have quantified annual nest success and, hence, partitioned the role of clutch size versus season length in determining between population variation in overall reproductive success.

Key Words: *Ammodramus,* avian life history, *Melospiza,* nest predation, nest success, saltmarsh.

INUNDACIÓN Y DEPREDACIÓN: INTERCAMBIOS EN LA ECOLOGÍA DE ANIDACIÓN DE GORRIONES DE MARISMA DE MAREA

Resumen. Los vertebrados de marisma de marea experimentan dos retos distintos para reproducirse exitosamente: inundación del suelo con agua, el cual es variable y a menudo impredecible, y la estructura vegetativa simple, la cual ofrece pocos refugios seguros para la depredación. Revisamos estudios tanto publicados como no publicados de aves de marisma de marea y sus parientes, con el fin de determinar si la totalidad del éxito de nidos es menor si la depredación y la inundación son mas altas que en los parientes que no son de marisma de marea. Además, examinamos información del tamaño de la nidada, época de apareamiento, y localización de nidos para las diferencias entre taxa de marea y de no marea. Sobre todo, encontramos poco respaldo para la idea de que efectos aditivos de pérdida de inundación- y depredación- dejan a los gorriones de marisma de marea con un grado de pérdida de nido neto alto, comparado a gorriones ecológicamente comparables. En parte, las dos fuentes de mortandad están negativamente correlacionadas y por ello, al menos parcialmente compensatoria. Las inundaciones son una causa de mortandad importante en algunas poblaciones, notablemente en aquellas a lo largo de la costa noratlántica y en al sur la Bahía de San Francisco. Sin embargo, en general, la depredación es la fuente más importante de mortandad en poblaciones de marisma de marea. La importancia de la depredación quizás se encuentre enmascarada en poblaciones que también sufren altos grados de pérdida de los nidos relacionada a inundaciones. Dentro de los gorriones de marisma de marea, el tamaño de la nidada es mas baja en sitios con mayores grados de depredación. El efecto de otras fuentes de mortandad y latitud desaparecen cuando las variables son ingresadas en regresión de paso acertado. Si es que la depredación afecta la variación en el tamaño de la nidada en especies de marisma de marea, probablemente sea el resultado del efecto que el tamaño menor de cría tiene sobre lo evidente que es el nido, en vez de una estrategia de apuesta-protectiva en contraste a una alta pérdida de nido. Sobre todo, el tamaño de la nidada es relativamente baja en formas de marisma de marea, a pesar que estas comparaciones son confundidas a menudo por otras variables, tales como latitud, altitud, o continentalidad del clima. Las épocas de anidación tienden a ser mas largas en aves de marisma de marea. Sin embargo, pocos estudios han cuantificado éxitos anuales de nidos, y por ello, han dividido el rol del tamaño de la nidada contra la longitud de la época por determinar entre la variación de la población en el éxito reproductivo total.

Tidal marshes are characterized by simple vegetative structure and both regular and irregular inundation with surface water, which can at least occasionally cover most or all vegetation. In this habitat, any vertebrate that relies upon nests for successful reproduction is faced with two sources of nest failure, thereby forcing adaptive compromises in nest structure, placement, and the timing of reproduction. First, placing the nest on or close to the substrate will increase the probability of flooding. Second, placing the nest higher in the vegetation will reduce the cover to hide the nest from potential predators. Flooding is also a problem for birds nesting in non-tidal wetlands. However, with the exception of storm-caused flooding, in most cases water levels remain relatively stable over the time scales required for completing a successful nest attempt. Tidal-marsh species face a generally more variable and often less predictable maximum water level (Reinert, *this volume*). This lack of predictability may force birds to place nests higher than would be optimal in the short-term to minimize the chance of episodic catastrophic flooding.

The clearest indication that adjustments are made in nesting behavior in response to tidal-marsh conditions is the nearly universal tendency for the nests to be higher in the vegetation than those of their non-tidal-marsh relatives, which are either facultative or obligatory ground nesters. According to Johnston (1956a), all tidal-marsh Song Sparrow (*Melospiza melodia*) nests are elevated off the ground and attached to marsh vegetation. Arcese et al. (2002), on the other hand, reports that most Song Sparrow nests are on the ground. Nice (1937) found a tendency in Ohio for early season Song Sparrow nests to be placed on the ground, in fact, over two-thirds of first and second attempts were ground nests. Coastal Plain Swamp Sparrow (*Melospiza georgiana nigrescens*) nests average 30 cm above the ground (SD = 6 cm; Nest height measurements were from ground to top of nest cup in all studies listed.) and no ground nests have been reported in >400 nests located (B. Olsen, unpubl. data). In contrast, non-tidal-marsh Swamp Sparrow (*M. g. georgiana*) nests are frequently placed on the ground; 30% of nests were so located in Rhode Island (Ellis 1980) and western Maryland (B. Olsen, unpubl. data). The third *Melospiza* species, Lincoln's Sparrow (*M. lincolnii*), characteristically nests on the ground in boggy vegetation (Ammon 1995).

The tendency to place nests at a more elevated site also characterizes tidal-marsh *Ammodramus* and related genera. Seaside Sparrow (*Ammodramus maritima*) nests are elevated, on the average, between 14–28 cm above the substrate (varying between population studied; Post and Greenlaw 1994); Salt-marsh Sharp-tailed Sparrow (*Ammodramus caudicatus*) nests are also elevated, ranging from 1–25 cm above the substrate (Greenlaw and Rising 1994; C. Elphick, unpubl. data) with a mean of 12 cm estimated from a population in Connecticut (C. Elphick, unpubl. data). In contrast to the Seaside and Saltmarsh Sharp-tailed sparrows, both of which are found only in tidal marshes, grassland breeding *Ammodramus*, (including the LeConte's Sparrow [*A. lecontei*], probable sister taxa to the tidal-marsh species) nest almost entirely on the ground (M. Winter, unpubl. data; Vickery 1996, Lowther 1996, Green et al. 2002). Furthermore, Savannah Sparrows (*Passerculus sandwichensis*) characteristically nest on the ground (Wheelwright and Rising 1993), but Belding's Savannah Sparrow (*P. s. beldingi*), a subspecies restricted to saltmarshes, usually builds nests that are elevated a few centimeters (Davis et al. 1984; A. Powell, pers. comm.).

The consistent elevation of the nest to avoid flooding compared to non-tidal-marsh relatives probably increases vulnerability of nests to predation if tidal marshes follow the general pattern of relatively high success for nests located on the ground compared with other strata (Martin 1993). The potential impact of two major sources of nest mortality with the seemingly mutually exclusive counter strategies of raising and lowering nest heights would seem a good area to examine the role of adaptive compromise in tidal-marsh sparrow life history. Furthermore, understanding how these forces shape nesting strategies is essential to predicting the effects of changes in predation pressure, hydrology, and sea level on the population of endemic birds. However, although the effects of flooding and predation on reproductive success and have been addressed in studies of individual species, no overview of the nesting biology of tidal-marsh birds has been published. In this paper we provide such an overview of information from published and unpublished studies to examine the following hypotheses: (1) nest success is generally lower in tidal-marsh taxa than in comparable upland or freshwater marsh taxa, (2) predation, particularly from aerial predators is high because the potential of flooding, forces birds to build nests in higher strata where the nests are more vulnerable to predation (Martin 1993), and (3) the frequency of flooding events causing nest mortality will be higher than for sparrow populations in grassland, or even freshwater marsh. In addition, we will examine differences in reproductive parameters, such as breeding season and clutch size that also might be shaped by the tidal-marsh environment.

METHODS

STUDIES USED

This paper is based on analyses of data from 16 studies of individual populations of salt-marsh sparrows. The sample sizes of nests for nest-success calculations range from 18–1,616 per study and total 5,154 nests. Sample sizes for clutch-size estimates range from 18–1,086 and total 3,713 clutches. All studies were multi-year, averaging 3 yr in duration. Some of the studies are summarized in published articles or unpublished dissertations, but we also incorporated data from several unpublished studies by authors of this paper. Published studies include Johnston (1956b) on Song Sparrows; Greenberg and Droege (1990) on Coastal Plain Swamp Sparrow; Post and Greenlaw (1982), DeRagon (1988), and Shriver (2002) on Saltmarsh Sharp-tailed Sparrows; and Post et al. (1983) and Post and Greenlaw (1982) on Seaside Sparrow. DiQuinzio et al. (2001) presented multi-year data on nest success in Saltmarsh Sharp-tailed Sparrows, but because this project was explicitly focused on the effects of a change in marsh management (opening the marsh to a more natural tidal regime), we did not use those data. Comparative studies of non-tidal-marsh relatives include Reinert (1979) and Ellis (1980) on Swamp Sparrows in Rhode Island peat bogs. Data were also obtained from less comprehensive studies cited in *Birds of North America* (Poole 2006).

Unpublished data (six–eight seasons) for Song Sparrows were provided by Point Reyes Bird Observatory (PRBO) from three sites in San Pablo and two in Suisun Bay, and an additional three sites for south San Francisco Bay by J. C. Nordby and A. N. Cohen (two seasons). In addition, R. F. Johnston provided his original data sheets for three seasons of his Song Sparrow study (Johnston 1956b) which allowed us to calculate and use Mayfield exposure estimates (Mayfield 1961, 1975) instead of relying upon originally published crude nest-survival values. For non-tidal-marsh Song Sparrow populations, unpublished data on nest success and clutch size for a 23-yr study at Palomarin Ranch, Marin County, California, were made available by PRBO, as were 2 yr of data on clutch size and nest success from the Consumnes River in the Central Valley south of Sacramento, California. Information about nesting Belding's Savannah Sparrows was provided by J. Williams and A. Powell. Data on Swamp Sparrows in Woodland Beach, Delaware and vicinity, and (for an interior population) Garrett County in extreme western Maryland by B. Olsen. Unpublished data on Seaside and Saltmarsh Sharp-tailed sparrows were provided by B. Schmeling and P. Marra for the Blackwater Wildlife Refuge in Maryland, and by C. Gjerdrum and C. Elphick for Coastal Connecticut. Unpublished data on the LeConte's Sparrow, the putative sister taxa to Seaside and Sharp-tailed sparrows, were provided by M. Winter.

PARAMETERS

We assembled 16 population studies of tidal-marsh sparrows that present data on several key attributes of reproduction: percentage total nest failure, percentage of failure due to predation, flooding, abandonment and other causes, brood parasitism, and length of the nesting season. We then used linear regression and ANCOVA (Statsoft, Inc. 2003) to explore the relationship between clutch size, nest success, failure cause, and location variables (tidal amplitude and latitude). In almost all cases, overall nest success is determined by a modified Mayfield method to correct for differences in the period of nest exposure after detection by researchers. For one study (B. Schmeling and P. Marra, unpubl. data) program MARK (White and Burnham 1999) was used, which provides daily nest-survival estimate similar to the Mayfield method. For the two studies for which only uncorrected nest success was available (DeRagon 1988, Marshall and Reinert 1990), we used the predicted value of corrected nest success based on the overall regression between Mayfield values and uncorrected values for the remaining 14 studies (by the formula: Mayfield loss = 28.5 + 0.69 × uncorrected loss; $r^2 = 0.60$, $P = 0.0006$). Overall nest failure due to flooding, predation, and abandonment was determined by calculating the proportions of total losses for each of these causes and multiplying these by the total Mayfield loss. Clutch size and standard error of clutch size were either previously reported or calculated for each study population. Breeding-season length is measured as the number of days between initiation of the first and last clutch in a population without any attempt to weight by seasonal distribution of efforts (Ricklefs and Bloom 1977). Nest parasitism rates are presented as the proportion of nests with at least one Brown-headed Cowbird (*Molothrus ater*) egg or nestling. For each site, we determined its latitude and, based on 2003 tide tables (National Oceanic and Atmospheric Administration 2004a, b), the mean and maximum tidal range. The latter two parameters were highly correlated, so we conducted all analyses using mean tidal range.

RESULTS

MELOSPIZA SPARROWS

Nest success

Tidal-marsh populations of Song Sparrows had higher nest failure rates than interior population (see Table 1 for values for individual studies). The mean nest failure rates were 90, 81, and 85% for the Suisun, San Pablo, and south San Francisco Bay subspecies, respectively, compared to 73% at Palomarin and 75% at Consumnes River. However, we estimated a failure rate of only 48% for the population of the San Pablo Song Sparrow (*Melospiza melodia samuelis*) studied by Johnston (1956b) in the San Pablo Marsh in Richmond, California 1952–1955. His study site is not one sampled in the more recent studies, but the considerably higher success (mostly due to much lower predation rates) provides the tantalizing possibility that nest success has declined in the 50 yr between the studies.

Coastal Plain Swamp Sparrows also had a relatively high nest failures rate compared to non-tidal populations of Swamp Sparrows. The coastal populations averaged 81% over 3 yr and five sites (R. Greenberg and B. Olsen, unpubl. data) compared to 63% for the Rhode Island (Ellis 1980) and 53% in western Maryland (B. Olsen, unpubl. data) for the interior populations.

Causes of failure

Nest predation was almost the exclusive reason for nest failure in Song Sparrows at Palomarin accounting for 97% of nest failures. This translates to an absolute predation rate of 71%, which is comparable to the predation rates for the tidal-marsh populations, averaging 75.6, 56.6, and 35.7% for the Suisun, San Pablo, and south San Francisco Bay sites, respectively. Failure due to flooding effected 1.7, 9.2, and 29.4% of the nests in the three embayments. Johnston (1956b) found a comparable rate of nest flooding (10.5%) for San Pablo Bay, but much lower predation rates (19.5%) than any of the recently monitored populations.

Between-year variation in predation rates was relatively much lower than that for flood loss. The mean coefficient of variation (CV) for six populations with four or more years of data was 0.21 for predation (0.19–0.36) and 1.27 for flooding (0.53–3.0) and the CVs for predation and flooding are significantly different (Wilcoxon signed rank test; $Z = 2.2$, $P = 0.03$). Analyzing each population and season as a separate observation, failure rates were strongly negatively related to predation rates ($r^2 = 0.71$) and unrelated to flood loss ($r^2 = 0.05$). In this analysis, failure due to predation and flooding are negatively related ($r^2 = 0.30$; Fig. 1a).

Nest loss in both the inland and tidal-marsh Swamp Sparrow populations studied was mostly due to predation (100 and 83%, respectively) with 13% of the Coastal Plain Swamp Sparrow nests lost to flooding. Flood losses were a result of heavy rains and winds rather than tidal inundations acting alone (B. Olsen, pers. obs.). Western Maryland populations lost no nests to flooding during our study. However, other published reports indicate flooding can be an important, episodic source of nest loss in Swamp Sparrows (Mowbray 1997), particularly early in the breeding season. Ellis (1980), for example, reported an

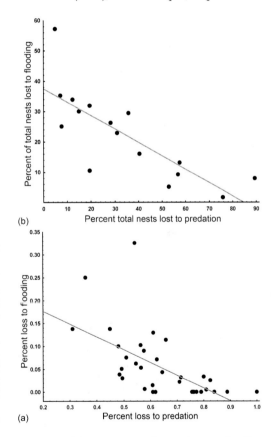

FIGURE 1. (a) Percentage of nests lost to flooding versus predation in San Francisco Bay Song Sparrows. Each point represents a different year within a particular study site. (b) Percentage of nests lost to flooding versus predation across all 16 studies of tidal-marsh sparrows. Each point represents the average value for a study. In the case of San Francisco Bay, for Figs. 3–6, single sites have been pooled for each of the embayments (San Pablo, San Francisco, and Suisun bays).

TABLE 1. NESTING PARAMETERS FOR STUDIES OF TIDAL-MARSH SPARROWS.

State	Species[a]	Site	Reference	Years of study	N of nests	N of nests for clutch size	Latitude of study	Tidal range (meters)	Total percent loss	Percent predation	Percent flood loss	Mean clutch size	SD clutch size	Cowbird percent nests	Breeding season length (days)
CA	SOSP	Suisun	PRBO	6	1,027	667	38.2	1.1	90	76	2	2.95	0.6	1.9	99
CA	SOSP	San Pablo	PRBO	7	1,626	1,086	38.2	1.4	81	57	9	3.07	0.6	1.3	97
CA	SOSP	San Pablo	Johnston (1956b)	4	157	157	37.8	1.3	49	20	11	3.18	0.5	1	114
CA	SOSP	South Bay	C. Nordby (unpubl. data)	2	364	149	37.5	1.8	85	36	29	3.14	0.59	13.1	140
FL	SESP	Gulf	Post and Greenlaw (1982)	2	108	108	28.2	0.8	95	89	8	3.02	0.31	0	158
NY	SESP	Oak Beach	Post and Greenlaw (1982)	2	298	298	41.5	0.2	65	15	30	3.7	0.58	0	77
CT	SESP	Various	C. Gjerdrum and C. Elphick (unpubl. data)	2	19	19	41.3	1.1	36	8	25	4	0.8	0	70
MD	SESP	Blackwater	P. Marra and B. Shmeling (unpubl. data)	2	245	151	38.3	0.8	63	58	5	3.35	0.85	0	73
MA	SESP	Buzzards Bay	Marshall and Reinert (1990)	2	60	55	41.5	1.2	71	5	57	3.9		0	52
ME	SMSTS	Scarborough Marsh	Shriver (2002)	3	69	69	43.1	2.5	54	12	34	3.58	0.95	0	58
MD	SMSTS	Blackwater	P. Marra and B. Shmeling (unpubl. data)	2	18	18	38.3	0.8	64	40	16	3.34	0.77	0	71
RI	SMSTS	Various	DeRagon 1988	2	172			1.2	56	7	35			0	72
CT	SMSTS	Various	Gjerdrum and Elphick (unpubl. data)	2	122	122	41.3	1.1	77	31	23	3.7	0.7	0	65
NY	SMSTS	Oak Beach	Post and Greenlaw (1982)	4	240	176	37.5	0.2	73	28	26	3.87	0.63	0	80
ME	NESTS	Scarborough Marsh	Shriver (2002)	3	53	53	43.1	2.5	73	19	32	3.49	0.93	0	55
CA	BeSS	Pt. Magu	J. Williams (unpubl. data)	4	255	356	33					3.2	0.1		
DE	CPSS	Woodland Beach	B. Olsen and R. Greenberg (unpubl. data)	4	400	400	39.3	1.4	71	58	13	3.2	0.6	0.03	110

[a] Species abbreviations are SOSP (Song Sparrow), SESP (Seaside Sparrow), SMSTS (Saltmarsh Sharp tailed Sparrow), NESTS (Nelson's Sharp tailed Sparrow), BeSS (Belding's Savannah Sparrow), CPSS (Coastal Plain Swamp Sparrow).

average of 8.5% of nests loss due to flooding in a Rhode Island peat bog.

Brood parasitism appears to be rare and patchy in tidal-marsh *Melospiza* sparrows and lower than found in non-tidal-marsh populations. Song Sparrow populations had an average parasitism rates of approximately 1.6% (this is the average of the yearly average for the five San Pablo and Suisun Bay populations), with the exception of the southern San Francisco Bay populations of *M. m. pusillula*, where 13.5% of the nests were parasitized. This latter value is by far the highest rate reported for any tidal-marsh sparrow population (Table 1). The parasitism rates for the other subspecies of saltmarsh Song Sparrows may have been underestimated because in the PRBO studies only nestling cowbirds (and not eggs) were identified. However, the *pusillula* populations were on very small marsh patches, which may have created a much larger edge effect than that found in other studies. Arcese et al (2002) reported that Song Sparrows are a preferred host of cowbirds and parasitism rates in the Pacific region ranged from 5.5–9.9%. In contrast, the average for the Palomarin site was only 2.4%, demonstrating that variation can be substantial even among upland populations.

Only 0.4% (two of 436) of all Coastal Plain Swamp Sparrow nests located were parasitized (R. Greenberg, unpubl. data). Brood parasitism is often episodic and local — the two instances of brood parasitism in the Coastal Plain Swamp Sparrow were for two nests within 50 m of each other in one, 2-wk period. Mowbray (1997) reported that for four studies of interior Swamp Sparrows in eastern North America (883 nests) that parasitism rates averaged 14.9% ($SD = 12.3$).

Clutch size

Johnston (1954) was the first to note that Song Sparrows nesting in tidal marshes had small clutches compared to their upland counterparts in a paper on latitudinal variation in clutch size in West Coast populations of the species. His analysis was based on data from 545 completed clutches taken from öological collections and the field notes of various field biologists. He found that clutch size generally increased with latitude from 3.05 in Baja California to 4.17 in Alaska. He compared sample of 143 clutches from non-tidal-marsh birds in north-central California (37.5–39° N) with 86 clutches from *Melspiza melodia pusillula* (south San Francisco Bay) and 48 from *M. m. samuelis* (San Pablo Bay). Although only mean values are presented, the clutches from the saltmarsh populations averaged 3.31 and 3.28, respectively which is considerably smaller than 3.53 for the non-salt-marsh birds of northern California and 3.71 for non-saltmarsh birds of south-central California. The difference in clutch size persisted even when only first clutches were compared (to eliminate an effect of seasonal change in clutch size). He later published mean clutch size of 3.2 ($SD = 0.6$) based on 147 nests from the population of *M. m samuelis* he studied for 4 yr

Recent studies provide access to much larger sample size for expanding this analysis. Studies in Suisun and San Pablo bays demonstrate that these populations have a slightly, but significantly smaller clutch size than a population monitored in coastal scrub habitat in the Palomarin population. The latter population had an average clutch size of 3.20 ($SD = 0.61$, $N = 597$) over an 8-yr study, whereas populations in Suisun Bay averaged 3.07 and San Pablo Bay 2.95. Three populations monitored for 2 yr in south San Francisco Bay had a clutch size of 3.14 ($SD = 3.59$, $N = 149$) which was not significantly different than the Palomarin site. However, an interior population in central California (Consumnes River) had a clutch size of 3.65, which was significantly higher than all of the tidal-marsh and coastal-upland populations.

Greenberg and Droege (1990) compared clutch size in Coastal Plain Swamp Sparrows breeding in Black Marsh, Baltimore County, Maryland, to data from nest cards for the nominate subspecies in Pennsylvania and New York. They found the tidal-marsh clutch size was significantly smaller (3.25 versus 4.1). Other field studies have found clutch sizes for the interior subspecies averaging 3.9 (Mowbry 1997). Further work in Delaware showed also showed a mean clutch size of 3.28 ($SD = 0.6$, 255 nests) for tidal-marsh nesting Swamp Sparrows compared 3.59 ($SD = 0.6$, 65 nests) for the closest inland population in western Maryland (Garrett County).

Breeding season

Johnston (1954) found that nesting began and ended progressively later in Song Sparrows as one moved north or higher in elevation. Set against this was a much earlier breeding peak in the saltmarsh populations than in comparable non-saltmarsh populations. For example, both *M. m. pusilulla* and *M. m. samuelis* initiated 50–60% of their clutches by early March, whereas this value was <10% for other California populations (including those in southern California). As a result, the saltmarsh forms have a long breeding season, lasting 95 and 120 d for *M. m. pusillula* and *M. m. samuelis*, respectively. Non-saltmarsh populations in north-central California, in contrast, had a breeding season of approximately 91 d. These estimates are based on pooled data from many different years, which may provide a high estimate of nesting season, where the duration of the season is relatively constant, but the initiation date is highly variable. Johnston's study on the San Pablo Bay population showed an average breeding season length of 114 d, which is comparable to his earlier pooled estimates.

More recently collected data support the idea of both an earlier and longer breeding season in the saltmarsh populations. PRBO data showed first clutches occurring during the second or third week of March for the Suisun Bay and San Pablo Song sparrows compared to the first week in April at Palomarin. The estimated breeding season was 99 and 97 d for the Suisun and San Pablo Bay populations, respectively and only 88 d for Palomarin. Data from *M. m. pusillula* (J. C. Nordby and A. N. Cohen, unpubl. data) indicate a nesting season beginning in late February and averaging 140 d. It is unclear what accounts for variation between studies. Coastal Plain Swamp Sparrows also have a substantially longer breeding season than nearby interior populations (110 versus 85 d). The longer breeding season is mainly a result of an extension of nesting activity at the end of the season (well into August), where the interior populations cease reproductive activities in mid-July.

AMMODRAMUS SPARROWS

Nest success

Average nest loss for the 11 study populations of tidal-marsh *Ammodramus* was 66% (SE = 8.3%). Nest loss was similar between Seaside ($\bar{x}$ = 66, SE = 9.4), Saltmarsh Sharp-tailed ($\bar{x}$ = 65, SE = 4.5), and Nelson's Sharp-tailed sparrow ($\bar{x}$ = 73, N= 1). LeConte's Sparrows, the closest non-tidal-marsh relative of the Seaside and Sharp-tailed sparrows, had a nest loss rate of 47% (N = 50 nests) over a set of northern prairie study sites (M. Winter, unpubl. data). Other studies of grassland *Ammodramus* report nest loss values between 50–80% (Vickery 1996, Green et al. 2002, Herkert et al. 2002).

Causes of nest failure

The mean loss to predation for all 11 tidal-marsh *Ammodramus* populations was 27.9% (SE = 7.6) which is similar to the loss due to flooding 26.5% (SE = 14.3). Sharp-tailed and Seaside sparrows showed no significant interspecific difference in the amount of loss to either factor and both factors comprised over 75% of all nest loss. In contrast, predation comprised 94% known causes for nest loss in LeConte's Sparrows and overall nest loss to predation was 59.0%. This latter value is unlikely to have come from the same distribution as for the tidal-marsh *Ammodramus* which has a 99% confidence limit of 52.1.

Clutch size

Mean clutch size for Saltmarsh Sharp-tailed Sparrow and Seaside sparrows are similar (3.62 and 3.59, respectively) as is the single value for coastally breeding Nelson's Sharp-tailed Sparrow (3.49). Clutch size in tidal-marsh *Ammodramus* declines with increasing nest loss to predation (r^2 = 0.69, P = 0.003, N = 11; Fig. 3). Clutch size in these species also increases with latitude (r^2 = 0.42, P = 0.01; Fig. 4.). Because, predation rate decreases with latitude (r^2 = 0.89), it is impossible to tease apart the relative importance of latitude and predation in explaining variation in clutch size.

Clutch size in grassland *Ammodramus* is larger than tidal-marsh congeners even at equivalent latitudes. For example, clutch size in LeConte's Sparrows from the northern prairie region has been reported to be 4.51 (SE = 0.10; M. Winters, unpubl. data) and 4.53 (Lowther 1996). Although the northern prairie study area is farther north than the northernmost sites for which clutch size has been determined for tidal-marsh *Ammodramus*, these values are similar to those found for other grassland *Ammodramus* species found at in more southerly areas. McNair (1987) found an average of 4.4 eggs per clutch for Grasshopper Sparrows (*Ammodramus savannarum*) based on egg-slip data from a large portion of the species' range. A study site in West Virginia (which would be at the latitude of the Maryland coastal studies) reported an annual mean varying from 4.1–4.5 eggs/clutch (Wray et al. 1982). Finally, mean clutch sizes from three studies of the Henslow's Sparrow (*Ammodramus henslowii*) ranged from 3.8–4.2 eggs.

Season length

Nest season length varies from 52–97 d and averages 67 (SD = 9.5) d across all tidal-marsh *Ammodramus* populations. Season length is strongly and negatively correlated with latitude (r^2 = 0.79, P = 0.0006, N = 11). The nesting season of LeConte's Sparrows in wet prairies was 54 d — comparable to the northern most populations of Sharp-tailed Sparrows in coastal marshes (52 d).

Brood parasitism

None of the 11 studies of the tidal-marsh *Ammodramus* reported cowbird parasitism

(N = 1,404 nests); apparently, no substantiated reports exist for either Seaside Sparrow or saltmarsh populations of sharp-tailed sparrow (Post and Greenlaw 1994, Greenlaw and Rising 1994). These zero values can be compared to other *Ammodramus* and grassland sparrows in general based on a recent review of parasitism values in grassland birds (Shaffer et al. 2003). The average parasitism rate for 28 studies of four *Ammodramus* species (LeConte's Sparrow, Henslow Sparrow, Grasshopper Sparrow, and Baird's Sparrow [*A. bairdii*]) was 16.5 (SD = 15.2%, N = 1,162 nests)) with 93% of the studies reporting values >0. The mean value for 59 studies of four additional grassland sparrow species (Lark Sparrow [*Chondestes grammacus*], Vesper Sparrow [*Poocetes gramineus*], Savannah Sparrow, and Chesnut-collared Longspur [*Calcarius ornatus*]) was 14.3% (SD = 14.6%, with 85% of the values >0). A nested ANOVA (*Ammodramus* vs. non-*Ammodramus*, irrespective of habitat with species nested within the two taxonomic groupings) showed no significant difference between the two groups or between species within a group ($F_{1,80}$ = 0.085; $F_{5,80}$ = 1.1). However, a nested ANOVA with tidal-marsh *Ammodramus* included as a third group showed a significant difference among groups ($F_{2,89}$ = 4.11, P = 0.02) with parasitism rates for tidal-marsh *Ammodramus* significantly lower for grassland *Ammodramus* (P = 0.003) and other grassland sparrows (P = 0.006) based on Bonferroni's post hoc test. We conclude that tidal-marsh *Ammodramus* have much lower nest parasitism rates than is typical for grassland sparrows and the grassland *Ammodramus* have rates consistent with grassland sparrows as a whole.

OVERALL PATTERNS

Nest success

The mean overall nest failure for 16 tidal-marsh sparrow populations was 66% (SD = 19%), with approximately 80% of failure caused by predation (36.4% of total nests) and flooding (22%). As we saw with Song Sparrow populations, predation rates are negatively related to flood loss rates (r^2 = -0.42, P = 0.009; Fig. 2). Nest failure tended to be higher in populations of *Melospiza* than *Ammodramus*, but not significantly so (75.5 versus 66.2%; t = 0.8, df = 14, P = 0.29). Nest success was weakly and positively related to latitude (r^2 = 0.26, P < 0.05) and unrelated to mean tidal range (r^2 = 0.03). Almost half of the between population variance in nest loss can be explained by predation (Fig. 2; r^2 = 0.45, P = 0.008), whereas none is related to flood loss (r^2 = 0.06).

Losses due to flooding

Loss to flooding is highly variable between populations of tidal-marsh sparrows, averaging 22.3 (SD = 14.4%) and ranging from 1.7–57.3% of all nests. Overall, nest loss to flooding is unrelated to mean tidal range (r^2 = 0.09) and is weakly related to latitude (r^2 = 0.29, P = 0.04). Loss due to flooding is not well documented for interior *Ammodramus*, but may occur in early season nests of LeConte's Sparrow (M. Winters, unpubl.data; Lowther 1996).

Losses due to predation

Average loss to predation averages almost twice as high as loss to flooding (χ^2 = 34.5, SD = 5.5%) with considerable between population variation (values range from 4.8–89.3%). Predation rate is significantly related to latitude (r^2 = 0.56, P = 0.001). Flood loss and predation are strongly, negatively related (r^2 = 0.63; Fig. 1b).

Clutch size

Average clutch size for all tidal-marsh sparrow populations considered was 3.38 eggs—Song and Swamp sparrows both averaged 3.25, Saltmarsh Sharp-tailed Sparrows averaged 3.64 eggs, Seaside Sparrows averaged 3.58 eggs, and Belding's Savannah Sparrows averaged 3.20 eggs. Although the sample size is limited, the three *Ammodramus* have considerably higher clutch sizes (3.6) than the two *Melospiza* (3.2). Because of this taxonomically related variation, subsequent analysis will use genus as a categorical variable in ANCOVA.

Clutch size shows a striking relationship to the amount of predation (Fig. 3; r^2 = 0.63, P < 0.0004). An ANCOVA for heterogeneity in slopes based on clutch size as the dependent, predation rate as the independent, and genera as the grouping variables showed a significant genera vs. latitude and predation rate interaction ($F_{2,13}$ = 14.2, P = 0.0009). Within *Ammodramus*, the relationship between clutch size and predation rate is strong and significant (r^2 = 0.73, P = 0.0004), with clutch size declining as predation rate increases. In contrast, clutch size is not significantly correlated with predation rate in the five *Melospiza* populations (r^2 = 0.51, P = 0.17). In contrast to predation, clutch size shows no relationship to loss due to flooding (r^2 = 0.07) for tidal-marsh sparrows as a whole.

Clutch size also varies with latitude, in this case positively, within saltmarsh sparrows, but the relationship is much weaker (Fig. 4; r^2 = 0.33, P < 0.02). Once again, heterogeneity in this relationship can be found between populations

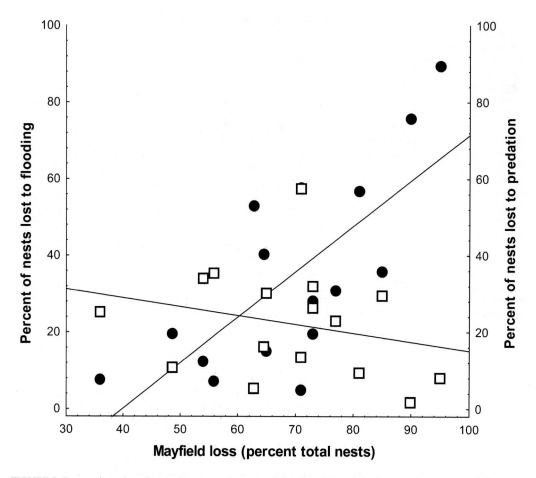

FIGURE 2. Proportion of total nests lost to predation and flooding plotted against crude success rate for tidal-marsh sparrow populations. Each point represents the average for a population (Table 1). Crude nest success is highly correlated with predation rates ($r^2 = 0.44$) but not flood loss ($r^2 = 0.0$).

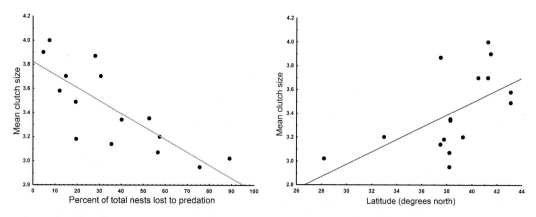

FIGURE 3. Mean clutch size plotted against percentage loss to predation across populations of tidal-marsh sparrows.

FIGURE 4. Mean clutch size plotted against latitude across populations of tidal-marsh sparrows.

of *Ammodramus* and others (*Melospiza* and *Passerculus*); the ANCOVA for heterogeneity in slopes shows a significant interaction exists between genus and latitude ($F_{3,11} = 3.8$, $P = 0.05$). When predation rate, flooding rate, and latitude are entered into a multiple regression (Statsoft 2004) with clutch size as the dependent variable, only predation rate is included as a significant variable ($r^2 = 0.75$, $P < 0.001$). The variables show some intercorrelation; however, collinearity is not a problem because the correlation coefficients are moderate (0.48–0.79) and the tolerances are between 0.3 and 0.62 (Belsley 1980). These results suggest that although clutch size varies with latitude, that predation pressure is a more important factor in explaining differences in clutch size among populations.

Brood parasitism

Brood parasitism is generally low, averaging 1.1% of nests across the 16 populations. In fact, brood parasitism by cowbirds has not been found for any tidal-marsh nesting *Ammodramus* and is reported to average 1.6% or less for all populations of *Melospiza* except those in south San Francisco Bay.

Nesting season

The period between the first and last nest initiations is a relatively crude index of the nesting season length, being very sensitive to the extreme tails of the seasonal distribution of nesting efforts. Despite the short-comings of this indicator, the general patterns associated with nesting season are quite clear. Nesting season is, expectedly, related to latitude. With all data included, the relationship is significant, but weak ($r^2 = 0.26$, $P = 0.04$). The Florida population of Seaside Sparrow is a clear outlier to the pattern found in more northerly populations. With this population removed from the analysis, the relationship is much stronger (Fig. 5; $r^2 = 0.53$, $P = 0.003$). As stated in the analysis of the individual genera, the relationship between breeding season and latitude is particularly striking within the *Ammodramus* and is not significant in *Melozpiza*. However, the latter genus lacks widely distributed populations along a single coastline.

DISCUSSION

OVERALL NEST SUCCESS

Overall nest success is generally low in tidal-marsh birds, averaging about 31% but ranging from 5–64% between populations. But this average is similar to that found for grassland or

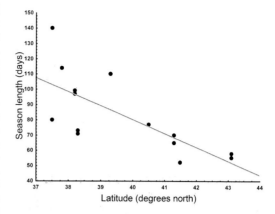

FIGURE 5. Mean annual breeding season (initiation of first and last clutch) plotted against latitude across populations of tidal-marsh sparrows.

shrub-nesting sparrows, which generally range from 20–50%. Therefore, the hypothesis that tidal-marsh sparrows facing the duel threat of predation and flooding suffer an inordinately high level of nest loss is not supported.

Few data are available to compare inland and tidal-marsh *Melospiza*, but they support the hypothesis that overall nest loss is higher in the tidal-marsh populations. The values for recent studies of *Melospiza*, particularly Song Sparrows in San Francisco Bay are quite high (82–95%) for a temperate-zone, open-nesting passerine (Martin 1993). This is intriguing because the *Melospiza* are recent colonists of saltmarshes (Chan et al, *this volume*) and, at least in the case of the Song Sparrow, may have made a more profound ecological shift to occupy saltmarshes from upland scrub and riparian habitats than did the *Ammodramus*. Hence one might predict that they would be less well adapted to marsh environments. Also, the relatively sparse cover provided by *Salicornia* marshes, where Pacific Coast Song Sparrows live, may make hiding nests more difficult than is the case for *Ammodramus* sparrows nesting in *Spartina*.

However, the data showing low nest success in Song Sparrows have all been collected in the last decade. The older data set from Johnston (1956b) suggests that nest success may have been much higher in the past. These data also suggest that predation, in particular, has increased dramatically, which is consistent with the invasion of new potential predators and the general increase in population of feral animals in and around bayside marshes, as well as the increasing fragmentation of the marshes themselves (Takekawa et al., *chapter 11, this volume*). Whether the low nest success of tidal-marsh Song Sparrows, or even that of other tidal-marsh species, is a result of very recent environmental changes or is a more long-term characteristic of these taxa is a fundamental question to answer. If low nest success due to predation is a recent phenomenon related to recent human activity, then it will be less appropriate to invoke this as a factor that has shaped tidal-marsh sparrow life history.

As noted by Reinert (*this volume*) for tidal-marsh Red-winged Blackbirds, cowbird parasitism rates are low for tidal-marsh sparrows, particularly for East Coast populations of *Ammodramus*. The low parasitism rates could result from the lack of perches (Post and Greenlaw 1994), the isolation from habitats where cowbirds can feed, or, perhaps more interestingly, from the lack of tolerance to saline conditions in nestling or fledgling cowbirds and, hence, selection pressure adults against searching in tidal-marshes for nests. With regard to the first hypothesis, grasslands are often reported to have relatively low parasitism rates. As we reported above, the studies of grassland *Ammodramus* suggest that while often low, parasitism rates are quite variable—which is distinctly different from the invariably low values for tidal-marsh sparrows. Many tidal marshes are adjacent to agricultural or suburban areas and have edge vegetation with trees and other elevated posts. The idea that selection has shaped an aversion of cowbirds from entering tidal marshes to search for nests is an attractive and testable hypothesis.

The Importance of Predation and Flooding

Despite the challenge of locating a nest safe from tidal and storm-driven inundation, predation accounts for more nest loss than does flooding in most tidal-marsh sparrows. Furthermore, both within-and across-species analyses show that variation in nest success between localities is correlated with nest predation rates, but not loss of nests to flooding or other factors. Flooding, however, can be a critical factor within certain populations, especially Saltmarsh Sharp-tailed Sparrows. With the exception of tidal marshes along the northeastern coast of North America (Shriver 2002; Reinert, *this volume*), loss to flooding is a less predictable and more variable cause of nest loss than predation for tidal-marsh sparrows, even within sites (as exemplified by the San Francisco Bay Song Sparrow data). High predation rates coupled with a substantial and variable source of density-independent mortality (flooding) may make the life history of tidal-marsh sparrows unusual for a temperate song bird.

Further investigation into the how these qualitatively different sources of mortality shape the life history of tidal-marsh birds could make a profound contribution to avian life history studies. Based on a cross-population regression, we found a negative relationship between predation loss and flood loss both between populations of Song Sparrows and across all tidal-marsh populations. Along these lines, DiQuinzio et al. (2002) studied a single population of Saltmarsh Sharp-tailed Sparrows in a Rhode Island marsh where natural tidal flow was reinstated. They found a shift in the major source of nest failure from predation to flooding over the subsequent 5-yr period. In this case, overall nest failure increased dramatically with a rise in the number of flooding events.

Two possible explanations for the negative relationship between nest loss to predation and flooding are: (1) a trade-off exists between nest placement that reduces predation versus one that lowers the probability of flood loss, or (2) the sources of mortality are compensatory so that nests that are flooded have reduced exposure to possible predation. Thus, the true potential impact of predation is underestimated in populations that experience flooding and the difference in predation between tidal and the non-tidal population is underestimated in these data.

Clutch Size in Tidal-marsh Populations

Clutch size showed a very strong relationship to nest predation loss. Within a regression analysis, predation loss was a much stronger predictor of clutch size than was latitude or other sources of nest loss. This provides evidence to suggest that predation selects for smaller clutch size among tidal-marsh sparrow populations. The strong role of predation would be consistent with the hypothesis of some workers that food is not an important limiting factor shaping reproductive strategies in tidal-marsh sparrows (Post and Greenlaw 1982). However, we need more direct evidence on the possible role of food in shaping such life-history parameters as clutch size in tidal-marsh sparrows. Although latitude, which might correlate with both increasing day length (Lack 1947) and a greater seasonal peak in food resources (Ricklefs 1980) does not show as strong a relationship with clutch size as does predation loss, it is possible that a more direct index of food availability might be a better predictor.

Accepting the importance of predation in selecting for smaller clutch sizes within tidal-marsh birds, we can evaluate two hypotheses originally proposed for small clutch size in tropical birds for why this might be the case. First, smaller clutches (and particularly, smaller resulting broods) attract less attention from visual predators, such as mammals and birds (Skutch 1949). Second, in the face of a high probability of nest loss and a long breeding season, the best strategy might be to reduce the investment in individual clutches and save reserves to maximize a female's ability to lay and care for multiple clutches and broods, in the chance that one will survive (Foster 1974). The fact that predation loss alone shows a much stronger relationship with clutch size than does overall nest loss, and that flood loss is actually positively related to clutch size, would suggest that the predation-reduction hypothesis would be the most likely to explain the relationship between predation levels and clutch size in tidal-marsh sparrows.

As a group, tidal-marsh sparrows have relatively small clutches, averaging 3.5 across all populations, compared to inland populations of closely related taxa. This result accords with previously published suggestions that tidal-marsh sparrows have unusually small clutches (Johnston 1954, Greenberg and Droege 1990). Such a difference is clutch size is also found between the freshwater-marsh King Rail (*Rallus elegans*) and the saltmarsh-breeding Clapper Rail (*R. longirostris*; Meanly 1992, Eddleman and Conway 1998). Unfortunately few comparative data are available from related and ecologically similar species, where latitude, elevation, and other environmental differences not associated with habitat differences are not confounding the comparison. For example, the Belding's Savannah Sparrow has the smallest clutch size of North American Savannah Sparrows, but it is also located at the lowest latitude (Wheelwright and Rising 1993). We have too few data to make perhaps the most appropriate comparisons between the subspecies of Nelson's Sharp-tailed Sparrow, Saltmarsh Sharp-tailed, and Seaside sparrows have clutches between 3.5–4 eggs even toward the northern end of their breeding distribution, which is considerably smaller than the 4.6 clutch size reported for the related LeConte's Sparrow. However, the latter data come from 5° further north than the coastal *Ammodramus* and from a regional with otherwise highly continental climate. The higher clutch size is also found in populations of less closely related *Ammodramus* at latitudes of 38–39° N, which is comparable to the study sites for saltmarsh species. These populations are still at more interior sites with more continental climates. It has been hypothesized that populations facing environments with a smaller difference between summer maximum and winter minimum resources will have smaller clutch sizes. For example, Cody (1968) suggested that birds in areas with more equable climates have smaller clutch size because with large non-breeding carrying capacity and high adult survivorship, fewer high-quality young will be better able to fight for vacancies in the population structure.

The strongest test of the difference in clutch size between tidal-marsh and grassland sparrows comes from the classic analysis of clutch-size trends in San Francisco Bay Song Sparrows completed by Johnston (1954). His discovery that the tidal-marsh populations have smaller clutches than comparable (same latitude) populations from west-central California is largely supported by our analyses of larger samples observed in marshes in the last decade. The birds in San Pablo Bay, for example, show a significantly smaller (0.2 eggs) clutch size than populations 20 km west in coastal scrub of Marin County. Clutch sizes from interior central California are larger by 0.6 eggs. Only the data from the south San Francisco Bay population (3.18 eggs) are similar to the west Marin County data and they are still lower than the interior population. Another appropriate comparison comes from the inland and tidal-marsh populations of Coastal Plains Swamp Sparrows in Maryland, where clutches average 0.3 eggs greater in the interior populations. Seaside Sparrows in Florida provide yet another possible comparison where latitude is controlled. Unfortunately the data are contradictory. Post and Greenlaw (1994) report a substantially higher clutch size in the non-tidal-marsh populations of *A. m. nigrescens* and *A. m. mirabilis* (3.5 vs. 3.1). However the much larger data set for *A. m. mirabilis* published by Lockwood et al. (1997) includes a mean clutch size of 3.1—which is identical to the tidal-marsh populations.

OTHER ASPECTS OF TIDAL-MARSH SPARROW LIFE HISTORY

Sparrow demography will only be complete when we have studies of life history focusing less on a single parameter, such as clutch size and nest survivorship, and more on integrating these into a broader understanding of life history as a whole (Young 1996, Martin et al. 2000). The length of the breeding season, changes in survival probability within a breeding season, probability of surviving to another breeding season, dispersal strategies, and opportunities for successfully fledged young to enter the breeding population are all factors that shape life history over and above what we have discussed here. In the long run, our understanding of saltmarsh sparrow life history will require long-term data on annual productivity, survivorship, and juvenile dispersal success for multiple populations. However, to date, few studies have measured or estimated these components of tidal-marsh sparrow fitness for entire breeding seasons.

Annual reproductive success (as measured by the number of young fledged/pair/season) has been found to vary from the extremely low value of 0.6 for a Florida population of Seaside Sparrows (Post and Greenlaw 1994), to the moderately low values of 2.1 young/pair in the Belding's Savannah Sparrows, to 2.3 in the Coastal Plain Swamp Sparrow, and finally to higher values of 4.7 for the San Pablo Song Sparrow (Johnston 1956b) and 4.3 and 4.7 for Saltmarsh Sharp-tailed and Seaside sparrows, respectively, in New York (Post and Greenlaw 1982). Modal values of two–five young per pair appear to be typical of temperate zone songbirds (Wray et al. 1982) and we will need more data to see how components of reproductive strategy contribute to between population variation in productivity (Ricklefs and Bloom 1977).

It could also be argued that the less seasonal climate and productivity of tidal marshes might allow for higher adult survival in their endemic sparrows than is found in upland relatives. As has been argued for tropical birds (Young 1996, Martin et al. 2000), this might select for lower reproductive effort within a breeding season and hence smaller clutch sizes. However, too few data exist to even begin to estimate survivorship patterns in saltmarsh passerines. Those estimates that are available are equivocal on this point. Average (2 yr) return rate of adult Saltmarsh Sharp-tailed Sparrows was approximately 57% (Post and Greenlaw 1982). This value is similar to those estimated for Seaside Sparrows (6 yr of data) from the same area, which ranged from 40–60% (Post and Greenlaw 1994). De Quinzio et al. (2001) found an annual adult survivorship of 60% the first year of a 4-yr study of Saltmarsh Sharp-tailed Sparrows in Rhode Island. The survivorship dropped to approximately 35% for the following seasons, but this was probably affected by large changes in marsh hydrology. Johnston reported annual return rates averaging 53% for the San Pablo Song Sparrows. Two studies report survivorship for saltmarsh sparrows that are substantially higher than is typical for temperate songbirds: Post et al. (1982) reported 85.7% for a resident population along the Gulf Coast of Florida and Grenier (pers. comm.) found an overall annual survival of 80.2% for a population of San Pablo Song Sparrows. The latter study and that of DiQuinzio et al. (2001) were the only ones to use mark-recapture models in their estimation procedures.

Two studies have suggested that natal dispersal may be more localized in saltmarsh sparrows. DiQuinzio et al. (2001) estimated that approximately 35% of the Saltmarsh Sharp-tailed Sparrows settled locally in their natal marsh. Johnston(1956a) plotted the settlement pattern of young San Pablo Song Sparrows and found that young appeared to disperse a shorter distance than was found for a Song Sparrow population in Ohio (Nice 1937). Both studies emphasized that the patchy distribution of marshes and their stability (in the short run) may favor local dispersal. This in turn may reduce selection on producing many fledglings, but increase selection on producing fewer high-quality young.

SMALL MAMMALS: THE NEST ECOLOGY FRONTIER

It should be noted that the challenges of avoiding flooding and predation in tidal marshes face other saltmarsh birds (Reinert, *this volume*) and other terrestrial vertebrates. For example, small mammals in tidal marshes are largely terrestrial and depend upon nests for breeding and resting. The natural-history literature, however, strongly suggests that nest structure and placement differ in tidal-marsh populations and species when compared to upland relatives. Fisler (1965) noted that the subspecies of saltmarsh harvest mouse (*Reithrodontomys raviventris halicoetes*) that lived in less tidal, brackish marshes made nests similar to the upland species (*R. megalotis*). However, the subspecies associated with highly tidal saltmarshes (*R. r. raviventris*) probably did not make nests, but used abandoned bird nests (which would be much smaller and elevated off the substrate). Similarly, that in the marsh rice rat (*Oryzomys palustris*) of eastern marshes, which normally constructs a nest of woven grass and sedges on the ground at the base of shrubs, nests may attach nests to marsh vegetation in areas that are flooded at high tide and use the elevated nests of Marsh Wrens (*Cistothorus palustris*; Wolfe 1982). Johnston and Rudd (1957) described the breeding nests as being fairly substantial (8–24 cm × 6–12 cm × 4–6 cm) and placed under or in a cavity of an object on the substrate. Resting nests were found to be invariably elevated in the *Salicornia*. Interestingly, the authors note that a substantial portion of the young monitored in their study were lost due to tidal flooding. However, the sample size was small. Other small mammals that occupy tidal marshes that regularly build nests include voles (*Microtus*) and cotton rats (*Sigmodon*), but we are unable to ascertain any special properties of tidal-marsh nests in these genera. Adaptation for reproduction in tidal-marsh mammals certainly is an area that could use further research.

CONCLUSIONS AND FUTURE STUDY

1. Tidal-marsh sparrows suffer from relatively low nest success, particularly populations in the genus *Melospiza*.
2. Overall, predation is the greatest source of nest loss, but flooding can be locally important. The two sources of mortality show a negative relationship in frequency of occurrence. This may simply be a result of flooding removing nests from the pool of potentially depredated nests, or it could indicate a trade-off between strategies to reduce one cause of failure or another. Predation rates are the best predictor of between population variation in nest-success rates. Nest loss due to flooding is variable and often unpredictable within and between seasons.
3. Clutch size is strongly related to predation rate, with smaller clutches occurring where nest predation rates are high. Overall nest success is much more weakly correlated with clutch size, which increases with flood loss. This suggests that the effect of predation on clutch size is specific to reducing predation and that clutch reduction as a bet-hedging strategy against frequent nest loss is a less likely explanation.
4. Clutch size is also related to latitude, but much more weakly than it is to predation. In a multiple regression analysis, only predation is included as a significant independent variable.
5. Brood parasitism is very low. It is unknown from *Ammodramus* species and generally around 1% for tidal-marsh *Melospiza*. Traditional explanations focus on the isolation of marsh from cowbird habitat and the lack of elevated perches. However, these explanations are not completely convincing and the possibility that saltmarsh searching in cowbirds has been selected against because cowbird young would survive poorly needs further experimental research.
6. Evolutionary explanations for patterns in clutch size and nest success need to be treated with some caution given the much higher nest survival and lower nest predation rates found in a saltmarsh Song Sparrow population in San Francisco Bay 50 yr prior to other studies cited here.
7. Data on other aspects of tidal-marsh sparrow demography, such as total fledging success/pair/season, survivorship, dispersal, and life-time reproductive success are absent or spotty. Studies suggesting shorter dispersal are tantalizing. More local dispersal may be part of a reproductive strategy where young compete to enter already dense local populations. The tidal-marsh sparrows would be an excellent system for long-term, comparative demographic studies focusing on how life-history parameters vary with biotic (predation) and abiotic (flooding) sources of mortality and how these relationships might change with sea-level rise.
8. The constraints of flood avoidance and predation need more detailed analysis through correlational studies with seasonal changes in nest placement. This topic would be particularly amenable to experimental manipulations. Sparrows would provide ideal study species for such research, but it must be remembered that other bird species and small mammals must solve the problem of when and where to place nest structures in tidal marshes to minimize mortality.

ACKNOWLEDGMENTS

We would like to thank Sam Droege and Scott Sillett for comments on a draft of this paper. R. F. Johnston provided original field data from his classic study of Song Sparrows.

OSMOREGULATORY BIOLOGY OF SALTMARSH PASSERINES

DAVID L. GOLDSTEIN

Abstract. In North America, several taxa (species or subspecies) of sparrows in the family Emberizidae are characteristic of saltmarshes. The fact that recognizable avian taxa are associated with, and perhaps restricted to, saltmarshes suggests that these habitats impose significant selective pressures. A likely candidate for this selective agent is the demand placed on homeostasis by a limited supply of fresh water and a possibly high intake of salt. A number of studies in the laboratory document that saltmarsh sparrows differ from their upland conspecific relatives. Saltmarsh residents tend to drink more, to tolerate saltier water, and to diminish drinking rates at high salt concentrations, in contrast to non-saltmarsh birds. The kidneys of sparrows from saltmarshes are large, enhanced particularly in medullary mass associated with an increased number of nephrons with loops of Henle. They may also have enhanced urine concentrating ability. These features are consistent with expectations for birds that drink salty water. Nevertheless, little evidence exists that such intake actually occurs in the field; just a single study has indicated an increased urine flow in birds freshly captured in the field, and direct measures of water or sodium intake that could corroborate this hypothesis are lacking. Studies of physiological function in the field, and of the relative roles of inheritance versus environment in determining osmoregulatory capabilities in birds, would help to resolve the question of how important osmoregulation is in restricting saltmarsh sparrows to that habitat.

Key Words: Avian kidneys, Emberizidae, *Passerculus*, salinity tolerance, saltmarsh birds.

BIOLOGÍA OSMOREGULATORIA DE COLORINES DE MARISMA SALADO

Resumen. En Norte América, varias taxa (especies o subespecies) de gorrión en la familia Emberizidae son características de marismas saladas. El hecho de que taxa avícola reconocible se encuentre asociada con, y quizás restringida a marismas saladas, sugiere que estos habitats imponen presiones selectivas significativas. Un candidato parecido para este agente selectivo es la demanda localizada en homeostasis por un limitado suministro de agua fresca y una posible toma alta de sal. Un número de estudios en el laboratorio documentan que los gorriones de marisma salada difieren de sus parientes conespecíficos de tierras más altas. Los residentes de marismas saladas tienden a beber más, para tolerar aguas más saladas, y tienden a disminuir las proporciones de beber a unas concentraciones altas de sal, en contraste a las aves que no son de marismas saladas. Los riñones de los gorriones de marismas saladas son más largos, amplificados particularmente en la masa medular, asociada con un número incrementado de nefrones con lazos de Henle. Quizás también hayan aumentado su habilidad de concentración de urina. Estas características son consistentes con las expectativas de aves que beben agua salada. No obstante, existe poca evidencia de que dicha entrada de hecho suceda en el campo; solo un estudio ha indicado un flujo incrementado de urina en aves recientemente capturadas en el campo, y mediciones directas de entradas de agua y sodio, lo cual comprueba de lo que carece esta hipótesis. Estudios de función fisiológica de campo, y lo relacionado a los roles relativos de herencia contra ambiente para determinar las capacidades osmoregulatorias de las aves, ayudaría a resolver la pregunta de qué tan importante es la osmoregulación en restringir gorriones de marismas salados al habitat.

Because most tidal marshes are inundated with sea water, one of the fundamental adaptive challenges for successful colonizing organisms is the ability to tolerate salty fluids or to find alternative sources of water. Among vertebrates, the ability to survive these conditions may derive from three classes of traits, which roughly reflect the degree of specialization to marine life (Dunson and Travis 1994). First, many organisms rely on behaviors that minimize their exposure to or intake of salt water. For example, species and subspecies of water snakes avoid drinking salt water, relying instead on fluids obtained from osmoregulating prey items (Pettus 1958, Dunson 1980). Likewise, the herbivorous meadow vole (*Microtus pennsylvanicus*), which apparently can not tolerate ingesting salty water, satisfies its water needs by consuming dew and precipitation and selectively eating grasses with low salt content (Getz 1966). Second, physiological adaptations may be based on existing organs and structures. Examples of this strategy include integumentary adaptations in some estuarine snakes and turtles that reduce the fluxes of sodium and water between animal and environment (Dunson 1980). Likewise, adaptations in kidney structure and function may allow certain saltmarsh forms of mice (Fisler 1962, 1963; MacMillen 1964) and sparrows to better tolerate ingesting saline water. Finally, the most specialized marine forms have evolved novel features—in particular, salt glands—that help to produce and excrete a concentrated salt solution. Salt glands are normally associated with completely marine forms, but are also found in some estuarine and saltmarsh residents like rails (Olson 1997), crocodiles (Dunson and Mazzotti 1989), and diamondback terrapins (*Malaclemys terrapin*; Robinson and Dunson 1976, Hart and Lee, *this volume*).

Although saltmarshes tend to be relatively low in biodiversity, some groups of organisms show particularly high levels of evolutionary success in occupying these ecosystems. Passerines, particularly New World sparrows in the family Emberizidae, are one such group that has repeatedly invaded coastal marshes throughout the Pleistocene (Chan et al., *this volume*). To help understand how they have achieved this success requires exploration of the details of salinity tolerance or avoidance. In this paper I examine the possible adaptations that might underlie the success of saltmarsh sparrows.

THE CHALLENGE POSED TO BIRDS BY SALTMARSHES

Birds and mammals are the only vertebrates capable of producing urine that is hyperosmotic to plasma. In both groups this capability derives from the presence of loops of Henle in the kidneys. These structures are part of a counter-current multiplication system that generates a renal medullary osmotic gradient, and this gradient is used to extract water from the urine.

Despite sharing this physiological and anatomical basis of function, birds and mammals differ in their ability to concentrate urine. In mammals, maximum urine concentration is typically >1,000 mosM and may reach seven times this value in small desert rodents, more than 25 times the concentration of blood plasma (Beuchat 1990). In contrast, birds typically can concentrate urine only to 600–1,000 mosM, two–three times plasma osmolality (Goldstein and Braun 1989). Moreover, this maximum avian urine concentration is no more concentrated than seawater, with a concentration of about 1,000 mosM (Fig. 1).

For birds living in saltmarshes, the available water is saline, sometimes even more concentrated than seawater (Fig. 1). The question arises: can birds tolerate drinking these saline waters? Birds obligatorily lose body water through respiratory and cutaneous evaporation. Thus, it would seem that if a bird drank seawater, sole reliance on urinary excretion of ingested salts would impose a net water loss. This is the dilemma facing passerine birds inhabiting saltmarshes.

Despite this, a number of passerine taxa are characteristic inhabitants of North American saltmarshes. The Seaside Sparrow (*Ammodramus maritimus*) and Saltmarsh Sharp-tailed Sparrow (*A. caudacutus*) are the species most associated with saltmarshes. In several other species, recognizable subspecies of otherwise freshwater or upland species are saltmarsh inhabitants. These include representatives of the Savannah Sparrow (*Passerculus sandwichensis*), Nelson's Sharp-tailed Sparrow (*Ammodramus nelsoni*), Song Sparrow (*Melospiza melodia*), and Swamp Sparrow (*Melospiza georgiana*), all members of the Emberizidae. In South America, species of *Cinclodes* in the family Furnariidae are found along shorelines, where they include osmoconforming marine mollusks in their diet and likely incur substantial seawater loads. The physiological features of these species, many of which may be shared with emberizids, are just beginning to receive study (Sabat 2000, Sabat and Martínez Del Río 2002). Here, I restrict discussion to the emberizids of North American saltmarshes.

The fact that recognizable avian taxa are found associated with, and perhaps restricted to saltmarshes, suggests that these habitats impose significant selective pressures. This supposition is further supported by the observation that some behavioral, morphological, and physiological traits are shared by a number of saltmarsh taxa (Bartholomew and Cade 1963, Greenberg and Droege 1990). Thus, some suite of factors apparently induces taxonomic differentiation, largely restricts saltmarsh sparrows to saltmarshes, and constrains other forms of these same species from colonizing saltmarshes. One candidate for this selective agent is the demand placed on homeostasis by a limited supply of fresh water and a possibly high intake of salt. For example, Greenberg and Droege (1990) note that even tidal-marsh Swamp Sparrows (*Melospiza georgiana*) do not breed in areas with waters >50% seawater concentration.

SALT GLANDS: A POTENTIAL (BUT MISSING) SOLUTION

Cephalic salt-excreting glands that compensate for the limited urinary concentrating ability have evolved in many birds that live in marine environments. These glands are capable of secreting solutions of nearly pure NaCl at concentrations that may exceed those of seawater. Included among the avian orders with salt glands are the truly marine groups, like Procelariiformes and Sphenisciformes, and those that are more sporadically or facultatively marine, such as Charadriiformes, ducks, and herons. Functional salt glands also occur in

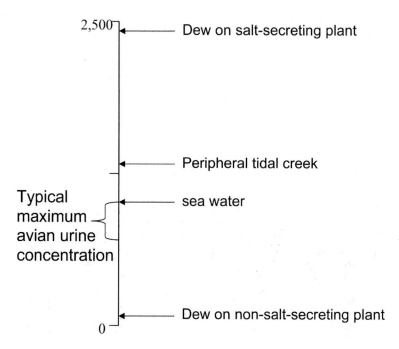

FIGURE 1. A comparison of the concentrations of water sources in a tidal marsh (measured at Bahia San Quintin, Baja California, Mexico; Goldstein et al. 1990) with the concentrating ability of the avian kidney.

some non-aquatic species, like young roadrunners (*Geococcyx*; Ohmart 1972) and several Falconiformes (Cade and Greenwald 1966). However, passerines lack functional salt glands; even those inhabiting saltmarshes or marine shores must rely on the kidneys and intestinal tract for eliminating dietary salt.

OTHER MORPHOLOGICAL FEATURES

At present, morphological attributes are used to distinguish subspecies of saltmarsh sparrows. These features are shared by several taxa (Greenberg and Droege 1990). Saltmarsh sparrows tend to have large, narrow beaks and have less rusty coloration in their plumage. It is not clear—indeed, it is unlikely—that any of these features have a role in osmoregulation. In some birds that feed on marine foods, beak morphology is specialized to function as a filter, limiting the intake of salt water (Mahoney and Jehl 1985, Janes 1997). No evidence, however, supports this function for the large beaks of saltmarsh sparrows.

DRINKING

WATER CONSUMPTION IN THE LABORATORY

The consumption of salty water has been evaluated in several species of saltmarsh sparrows. Patterns appear to be similar whether drinking diluted seawater or solutions of NaCl (Basham and Mewaldt 1987). Bartholomew and Cade (1963) delineated general patterns of drinking rate in response to increasing salinity of the water. Like their upland relatives, saltmarsh sparrows given a choice of fresh or saline water typically prefer fresh water and consume relatively small quantities of saline (Bartholomew and Cade 1963, Poulson 1969). However, when given just a single drinking solution, whether fresh water or saline, saltmarsh residents tend to drink more water than non-saltmarsh sparrows (Bartholomew and Cade 1963; Figs. 2 and 3). Moreover, the saltmarsh birds generally decrease consumption as salinity increases. (See Fig. 3 for an exception to these generalizations). Saltmarsh sparrows can maintain body mass when drinking more highly concentrated salt solutions than is tolerated by non-saltmarsh forms.

Saltmarsh and non-saltmarsh sparrows differ in the relation between drinking rate and tolerable saline concentration (Fig. 4). Non-saltmarsh sparrows drink maximally at saline concentrations above the maximum they can tolerate; that is, they increase consumption even while incurring a net loss of body water (and hence body mass) apparently in an effort to excrete the ingested salt. In contrast, saltmarsh sparrows drink maximally at concentrations

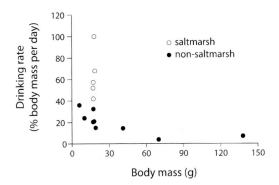

FIGURE 2. Freshwater drinking rates in saltmarsh emberizids (open circles) compared with other birds (filled circles). The five highest drinking rates are saltmarsh emberizids. Data from Bartholomew and Cade (1963).

below the highest they can tolerate, and at least some can maintain body mass even while drinking NaCl solutions with osmolality equivalent to full-strength seawater.

It is also possible that saltmarsh sparrows actually require more salt in their diet. Large-billed Savannah Sparrows (*Passerculus sandwichensis rostratus*) were able to resist desiccation, as evidenced by maintenance of body mass, much better after drinking seawater for several days than after drinking distilled water (Cade and Bartholomew 1959). This effect could derive from an elevated obligatory salt loss, perhaps resulting from an enhanced abundance of salt-wasting unlooped nephrons in the kidneys, and this might contribute to the restriction of saltmarsh taxa to that habitat. It is not known whether the differences in drinking patterns between saltmarsh and non-saltmarsh sparrows are genetically encoded or perhaps induced by environmental factors such as salt intake during growth.

WATER CONSUMPTION UNDER NATURAL CONDITIONS

In the laboratory, saltmarsh sparrows drink relatively large volumes of salt water. Does this occur also in the field? Poulson (1969) argued that the correlation between urine concentrating ability and salinity of waters available in the field does suggest that free-living saltmarsh passerines are likely to acquire much of their water intake from saline waters. Yet evidence to support this view is lacking. Indeed, Williams and Dwinnel (1990) reported that they had not observed Belding's Savannah Sparrows (*Passerculus sandwichensis beldingi*) drink seawater in the field.

Measures of energy and water flux have been used to estimate drinking rates in the field in one subspecies of saltmarsh passerine,

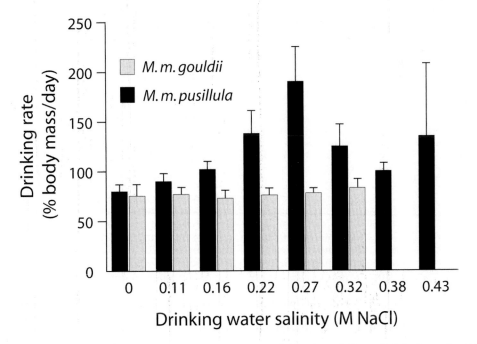

FIGURE 3. Drinking rates of two subspecies of Song Sparrow (*Melospiza melodia pusillula* [saltmarsh resident; solid bars] and *M. m. gouldii* [non-saltmarsh resident; open bars]) given solutions of varying NaCl concentration in the laboratory. Data from Basham and Mewaldt (1987).

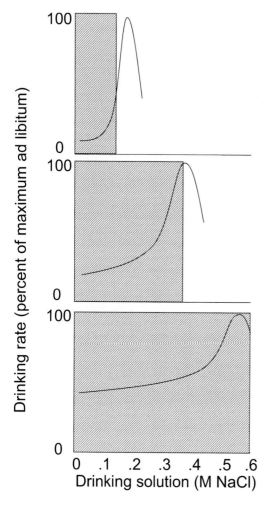

FIGURE 4. Generalized patterns of drinking in response to varying molarity of NaCl in the drinking water. Patterns are depicted for birds ranging from less salt-tolerant (top panel) to more salt-tolerant (lower panel). The solid line represents the pattern of drinking; the hatched regions indicate the range of salinities over which the birds can maintain body mass (Poulson 1969).

Belding's Savannah Sparrows inhabiting the saltmarshes of Baja California (Williams and Dwinnel 1990). In these birds, total water fluxes in the field, including water intake deriving from drinking, feeding, and oxidative metabolism, have been measured from the turnover of tritiated water. These water turnover rates are similar to those predicted from the allometry of water turnover rates in a variety of non-saltmarsh species (Williams et al. 1993). Energy intake in these birds was also measured using doubly labeled water (Williams and Nagy 1984, 1985; Speakman 1997). This measure, combined with an analysis of the water and energy content of their diet, was used to estimate water intake from sources other than drinking. The result suggested that diet and metabolic water could account for all but about 3.5 ml of water intake. The authors conjectured that this additional water may have derived from sparrows drinking dew (Fig. 1), which was available each morning.

Other methods that might yield further insight into drinking patterns in the field include the simultaneous measure of water and sodium fluxes (Goldstein and Bradshaw 1998), analysis of osmotic concentrations of gut contents (Sabat and Martínez del Rio 2002), and diet analysis using stable isotopes (Sabat and Martínez del Rio 2002). Studies at times of year other than the breeding season might also prove instructive. None of these approaches has been applied to saltmarsh sparrows. Still, field-caught birds have osmoregulatory organs that appear to be suited for a high intake of salt and water. Together with the data on drinking in the laboratory, it remains a reasonable conjecture, though undocumented, that saltmarsh sparrows in the field consume salty water.

FOOD SELECTION

Foods available to sparrows in saltmarshes are likely to vary substantially in salt content. Foods taken directly from the vegetation, including seeds and insects, are probably no more salty than those found in other, non-marine, habitats. On the other hand, some saltmarsh plants do actively excrete salt onto their surfaces, and invertebrates like crustaceans, mollusks, or annelids that live in the briny waters are likely to osmoconform with those waters and so have elevated salt content (Withers 1992). Thus, choice of diet may well dictate the salt load ingested.

Data on diet choice in saltmarsh sparrows suggest significant variability both within and among species. For example, during the breeding season the Saltmarsh Sharp-tailed Sparrow consumes mostly animal foods, including large proportions of insects but also amphipods (23.7%) and mollusks (3.6%) that could have more marine-like body fluids. In contrast, this same species switches to 30% plant parts during non-breeding months (Greenlaw and Rising 1994). Invertebrates can vary substantially in salt content. For example, terrestrial mollusks (e.g., terrestrial snails) may be water rich and low in salt, whereas marine gastropods have body fluids as concentrated as seawater; proper identification of such prey is critical to evaluating the osmoregulatory implications of diet choice.

MIGRATION AND SEDENTARINESS

The migratory habits of saltmarsh sparrows vary. Some, like Saltmarsh Sharp-tailed Sparrows and northern populations of Seaside Sparrows, are migratory. Notably, though, even when they leave their northern breeding grounds for regions further south they remain in tidal marsh habitat. Tidal-marsh populations of Swamp Sparrows also leave their breeding grounds in the autumn, but remain in tidal marshes along the Carolina coast (Greenberg et al., in press). For several other species, though, the saltmarsh sparrows are sedentary. This is true for the coastal populations and subspecies of Song Sparrows and Savannah Sparrows, and it is also true for southern populations of Seaside Sparrows. Indeed, for Savannah Sparrows, the saltmarsh variants are among the only non-migratory subspecies, although some Large-billed Savannah Sparrows are known to move north and out of saltmarshes in the winter. The sedentary habits of many saltmarsh sparrows may have evolved in association with the contraction of saltmarsh habitats, along with the physiological specialization of the birds. The strict association with saltmarshes also implies an inability to invade other habitats. The extent to which this results from osmoregulatory constraints is not resolved.

EXCRETORY ORGANS

KIDNEYS

Kidney structure

Birds have unique kidneys, with structures intermediate between reptilian and mammalian. As noted above, their ability to produce urine that is hyperosmotic to plasma derives from the presence of a renal medulla containing loops of Henle. However, only some avian nephrons, 25% or fewer, possess these loops, and the great majority are unlooped and therefore unable to concentrate urine (Goldstein and Braun 1986, 1989). The proportions and numbers of these nephron types vary among species (Goldstein and Braun 1989). What might one hypothesize for saltmarsh birds?

Kidneys of saltmarsh birds should have high populations of loopless nephrons. As noted above, these relatively short and simple nephrons do not contribute to the urine concentrating ability and might be thought of as water and salt wasting. Thus, to cite an extreme example, some hummingbirds, which ingest highly dilute nectar and must excrete water loads while conserving scarce electrolytes, have few or no loops of Henle (Beuchat et al. 1999). In contrast, arid-adapted species, which must conserve water, appear to have a reduced proportion of loopless nephrons (Thomas and Robin 1977). One might expect that in this regard saltmarsh species would more closely resemble hummingbirds—they need to excrete substantial volume loads, judging from their high drinking rates in the laboratory, and thus their kidneys should feature a large number of loopless nephrons (though see Sabat and Martínez del Rio [2002] for a possible contrary example).

Saltmarsh birds also should have substantial numbers of nephrons with well developed loops of Henle. In contrast to hummingbirds, the fluids ingested by saltmarsh sparrows may contain abundant electrolytes. Ingestion of hyperosmotic fluids would require the concentrating ability conferred by looped nephrons. Thus, if saltmarsh birds rely on salty drinking water they would require the features of both loopless and looped nephrons, and an abundant representation of both nephrons types should result. Together, this implies that saltmarsh passerines should have relatively large kidneys.

The morphology of individual nephrons also is variable. In particular, the principle of a countercurrent multiplier predicts that longer loops of Henle should confer a greater ability to generate a medullary osmotic gradient, and thereby a greater ability to extract water and concentrate the urine. Birds that ingest saline waters and lack salt glands therefore might be expected to have long loops of Henle, permitting solutes to be excreted in a minimal water volume. Overall, then, we predict large kidneys with long loops of Henle in these species.

Aspects of kidney structure have been evaluated in several populations of saltmarsh Savannah Sparrow (Table 1; Poulson 1965, Johnson and Ohmart 1973, Johnson 1974), and preliminary reports for *Cinclodes* suggest parallel findings. Kidney mass is relatively large in saltmarsh passerines, at least in part from an enlargement of the renal medulla (Casotti and Braun 2000). This translates into high values of relative medullary thickness (medullary length relative to kidney mass), an index used to compare medullary development across species (Johnson 1974). Interestingly, this enlargement appears to entail an increase in the number of medullary cones, but not in their length (Poulson 1965). It is less clear whether the large kidney size also reflect a large mass of cortical tubule elements. In a comparison across species, Belding's Savannah Sparrows had similar cortical mass (including elevated mass of proximal tubules and reduced distal tubules) compared with House Sparrows

TABLE 1. ASPECTS OF KIDNEY MORPHOLOGY IN SALTMARSH AND NON-SALTMARSH SPARROWS.

	Upland sparrow[a]	Saltmarsh Savannah Sparrow (*Passerculus sandwichensis beldingi*)[a]
Kidney mass (g)	0.21	0.34
Relative number of medullary cones[b]	10	20
Medulla volume (mm^3)[c]	6.8	15.2
Relative length of Henle's loop[d]	2.6	2.5

Note: Data from Poulson (1965), Johnson and Mugaas (1972), Goldstein et al. (1990), and Casotti and Braun (2000).
[a] Upland sparrows include Song Sparrows (for medullary volume) and, for the other three variables, Savannah Sparrows (*Passerculus sandwichensis brooksi* and an unidentified subspecies). The Savannah Sparrow varieties did not differ in body mass.
[b] Medullary cone abundance expressed as the number of units of medullary cone seen per histological section of kidney, as described by Poulson (1965).
[c] Song Sparrows and Savannah Sparrows had nearly identical total kidney volumes, 129.4 and 129.8 mm^3, respectively (Casotti and Braun, 2000).
[d] Relative length calculated as (mean length of the medullary cones × 10) divided by the cube root of kidney volume. See Johnson (1974).

(*Passer domesticus*) and Song Sparrows (Casotti and Braun 2000). Intraspecific analyses of these variables, e.g., comparing saltmarsh and non-saltmarsh Savannah Sparrows, are not available. Thus, saltmarsh passerines have kidney structures only partly consistent with predictions. Nevertheless, the large kidneys, incorporating well-developed cortex and medulla, are most consistent with expectations for handling large amounts of ingested salt water.

Kidney function in the laboratory

Only a few measures of excretory function have been made in saltmarsh sparrows. These entail the collection of fluid voided by birds drinking water of different salinities. Because the ureters empty into the posterior intestine in birds, where urine composition can be modified even at high urine flow rates (Laverty and Wideman 1989), voided fluid probably does not represent the output of the kidneys.

The most notable finding in these studies is a report that Belding's Savannah Sparrows drinking 0.6 M NaCl excreted a fluid with a Cl$^-$ concentration of 960 meq/L and a total osmolality of ~2000 mosmol/kg, more than five times the mean plasma osmotic concentration (Table 2; Poulson and Bartholomew 1962). This is the highest osmolality reported for avian urine and is often quoted in the literature. Nevertheless, I am cautious about this datum—birds in this experiment had variable plasma osmolalities, including values up to 610 mosmol kg^{-1}, far above normal. They were also drinking copious volumes of saline (about 20 ml d^{-1}) and it is possible that fluid could have passed through the gut without full absorption, so that excreted fluid represented a mix of urine and this gut fluid. No other aspects of renal function have been evaluated in saltmarsh sparrows.

Kidney function in the field

A single study has explored renal output from saltmarsh sparrows in the field (Table 3; Goldstein et al. 1990). Belding's Savannah Sparrows in the tidal marshes of Baja California produced relatively copious urine flow, about five times that produced by an upland subspecies of Savannah Sparrow captured at the same time in the scrub surrounding the marsh. The osmotic concentration of this urine was also about 20% higher in the saltmarsh birds, but in no case near the maximal values reported by Poulson and Bartholomew (1962).

LOWER INTESTINE

In birds, urine empties from the ureters into the cloaca, from where it may move by reverse peristalsis into the colon (Goldstein and Skadhauge 1999). The colon has the ability to modify the urine in a variety of ways, including uptake of organic molecules, transport of electrolytes between blood and lumen, and reabsorption or secretion of water. Moreover, the capacities, characteristics, and structural

TABLE 2. URINE CONCENTRATING ABILITY IN SELECTED PASSERINES.[a]

Species	Maximum urine/plasma osmotic ratio
Saltmarsh Savannah Sparrow (*Passerculus sandwichensis beldingi*)	5.8 (see text)
Non-saltmarsh Savannah Sparrow (*Passerculus sandwichensis brooksi*)	3.2
House Finch (*Carpodacus mexicanus*)	2.4
Zebra Finch (*Taeniopygia guttata*)	2.8

[a] Data from Poulson and Bartholomew (1962) and Goldstein and Braun (1989).

TABLE 3. Urine production by saltmarsh and upland Savannah Sparrows.[a]

	Saltmarsh	Upland
Urine flow (µl/h)	500	100
Urine osmolality	575	485
Plasma osmolality	350	340

[a] Data from Goldstein et al. (1990).

bases of these transport functions are altered in response to dietary salt content. In the chicken (*Gallus gallus*), for example, birds on low-salt diets have enhanced capacity for lower intestinal salt absorption, the absorption becomes insensitive to the presence of organic substrates like amino acids and glucose, and the gut surface area is magnified by the development of an extensive apical brush border (Table 4; Elbrønd et al. 1993).

The responsiveness of the lower intestine of saltmarsh sparrows to salt intake under controlled conditions remains little studied. However, Savannah Sparrows of Baja California provide an illustration of these patterns under field conditions (Goldstein et al. 1990). Belding's Savannah Sparrow, the saltmarsh resident, had a colonic epithelium with relatively smooth mucosal surface, providing a small surface area indicative of a low transport capacity as would be expected if electrolytes were abundantly available. In contrast, birds of a presumably migratory, non-saltmarsh subspecies found at the same time in nearby upland habitat had colonic epithelia with extensive micro-villous folding on the mucosal surface, providing substantial re-absorptive surface area as needed for salt conservation (Fig. 5). Again, these findings provide indirect evidence that saltmarsh sparrows have elevated salt intake.

OSMOREGULATION IN SALTMARSH MAMMALS

As noted in the introduction, passerine birds are not the only terrestrial vertebrates that might be challenged by the osmoregulatory demands of saltmarshes. Because mammals are the only non-avian vertebrates capable of producing urine that is hyperosmotic to plasma, it may be instructive to examine the role of salt tolerance in the distributions of mammalian saltmarsh races and species. Several taxa of small rodents—species or subspecies—inhabit and appear to be restricted to these environments (Shellhammer et al. 1982, Woods et al. 1982, Bias and Morrison 1999). One might hypothesize that a salty environment would be less of a challenge to homeostasis in small mammals than in birds. Small rodents typically can concentrate their urine to several times the osmotic concentration of seawater (Beuchat 1990), and a substantial component of mammalian urinary solutes is NaCl. Thus, even if the rodents drank saline water they should be able to eliminate the salts in a lesser volume, resulting in a net gain of pure water. Indeed, work by MacMillen (1964) on western harvest mice (*Reithrodontomys megalotis*) and Fisler (1963) on salt marsh harvest mice (*R. raviventris*) showed that rodents from saltmarshes can survive drinking sea water. Upland subspecies of *Reithrodontomys* may (Fisler 1963) or may not (MacMillen 1964) be less tolerant of highly saline solutions than their saltmarsh counterparts. Nevertheless, several studies suggest that salinity does limit the distribution of small mammals in saltmarsh environments. For example, meadow voles cannot tolerate water with salt concentrations >50% sea water, and Getz (1966) suggested that selective herbivory of plants with lower salt concentrations, as well as the use of dew, allows voles to occupy saltmarshes.

Physiological features other than tolerance of simple NaCl solutions may well contribute to these patterns. California voles (*Microtus californicus*) in saltmarshes were better able to ingest succulent halophytes like *Salicornia* than seemingly more salt tolerant cricitid rodents, perhaps because of special features of the digestive processes (Coulombe 1970). Moreover, in contrast to the results of laboratory drinking experiments, harvest mice were averse to eating plants with high salt content, perhaps because they contained cathartic ions (Coulombe 1970). Thus, food selection and use of dew as a water source may be important mechanisms for tolerating the saltmarsh environment; torpor may allow mammals to survive periods of osmoregulatory stress.

TABLE 4. The effect of high- vs. low-salt diet on luminal morphology of the hen lower intestine (coprodeum).[a]

	High NaCl diet	Low NaCl diet
Na transport	0–1 µmol/cm2h	7–12 µmol/cm2h
Apical surface area	86 cm^2	202 cm^2
Number of microvilli	35 × 10^9	71 × 10^9
Length of microvilli	7,289 m	19,738 m

[a] Data from Elbrønd et al. (1993).

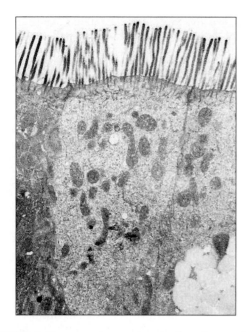 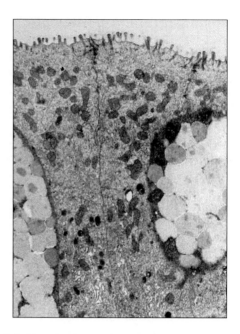

FIGURE 5. Electron micrographs of colons from upland (left) and saltmarsh Savannah Sparrows. Note in particular the enhanced elaboration of the microvilli in upland birds, associated with their greater need for intestinal salt uptake.

CONCLUSIONS

Species, subspecies, and populations of several emberizid sparrows are resident in North American saltmarshes. The fact that these taxa are distinct from conspecifics residing outside of saltmarshes suggests that one or more selective agents have induced differentiation in these habitats. Moreover, the sedentary habits of many of these taxa suggest that they may be constrained from leaving the saltmarshes, and perhaps that non-saltmarsh taxa are unable to invade. In this review, I have examined the evidence that osmoregulatory demands, associated with a dearth of fresh water, an abundance of salt, or both, may be the selective agent acting on these birds.

Overall, evidence from laboratory studies indicates that the osmoregulatory biology of saltmarsh sparrows is specialized. Saltmarsh residents drink and tolerate more salty drinking solutions. Moreover, several lines of evidence from the field, including urine flow rates and morphological features of osmoregulatory organs, provide circumstantial evidence for intake of salt water in the field. Direct evidence for that intake is lacking.

The question remains as to whether the osmoregulatory capacities of saltmarsh birds restrict them to those habitats. At least one study implies that saltmarsh sparrows may actually require more salt in their diet, at least if they are to tolerate times of water restriction. A few studies of saltmarsh mammals indicate that other physiological challenges, such as effects of ions other than Na^+ and Cl^-, may be important. It remains unclear whether osmoregulatory physiology can explain the exclusion of non-saltmarsh taxa from the marshes. Other explanations, from food availability to social interactions, are also possible.

We also do not know the extent to which osmoregulatory features of saltmarsh sparrows are genetically determined. No one has yet attempted common garden experiments with and without salty diets, in which saltmarsh and non-saltmarsh birds are reared under common conditions.

Emberizid sparrows may be among the best indicator species for evaluating the health of saltmarshes. An understanding of the physiological and behavioral traits responsible for constraining the birds to that habitat may prove a valuable tool in understanding the processes that create and define saltmarshes.

SOCIAL BEHAVIOR OF NORTH AMERICAN TIDAL-MARSH VERTEBRATES

M. Victoria McDonald and Russell Greenberg

Abstract. We examine the relationship between the social behavior of terrestrial vertebrates and the unique biophysical characteristics of tidal marshes with emphasis on birds, particularly those species and subspecies restricted to tidal marshes. However, where relevant, examples from mammals are also included. Tidal marshes are structurally and floristically simple habitats that are highly productive and spatially variable in quality, a variability that is magnified by the local effects of tidal inundation. Such conditions are thought to contribute to the evolution of polygynous mating systems. Over half (five species) of the saltmarsh-breeding species are commonly polygynous. This distribution of mating systems is not appreciably different from closely related species or subspecies of freshwater marsh or grassland habitats. The distribution of breeding territories in some tidal-marsh birds show a strong tendency to aggregate around certain habitat features, leaving regularly flooded tidal flats as shared feeding areas. In addition, some mammals show a tendency to form aggregations and an increase in social tolerance, partly in response to the forced crowding the tidal flooding can impose. The unusual non-territorial spacing behavior and related scramble, polygynous mating system of the sharp-tailed sparrows (*Ammodramus* spp.). may be an endpoint in the tendency for sparrows to occur in local high-density semi-colonial conditions, where males are no longer to defend territories. The sharp-tailed sparrow social system may have evolved in response to its subordinate relationship with syntopic Seaside Sparrows (*Ammodramus maritimus*). In this structurally simple, but zonal habitat, interspecific territoriality and avoidance seem to be well developed among passerine and possibly small mammal taxa.

Key Words: *Ammodramus*, bird, breeding patterns, mammal, mating systems, salt marsh, spatial distributions.

COMPORTAMIENTO SOCIAL DE VERTEBRADOS DE MARISMA DE MAREA NORTE AMERICANOS

Resumen. Examinamos la relación entre el comportamiento social de vertebrados terrestres y las características biofísicas únicas de marismas de marea, haciendo énfasis en aves, particularmente aquellas especies y subespecies restringidas a marismas de marea. Sin embargo, donde era relevante, también fueron incluidos ejemplos de mamíferos. Los marismas de marea son habitats simples estructural y florísticamente, los cuales son altamente productivos y espacialmente variables en cualidad, una variabilidad que es magnificada por los efectos locales de inundaciones de marea. Dichas condiciones, se piensa que contribuyen a la evolución de sistemas de apareamiento poliginuos. Por encima (cinco especies) las especies reproductoras de marisma salada son poliginuas. Esta distribución de sistemas de apareamiento no es apreciablemente diferente de especies o subespecies cercanamente relacionadas de marisma de agua fresca o habitats de pastizal. La distribución de territorios de reproducción en algunas aves de marisma de marea, muestran una fuerte tendencia a acumular ciertas características alrededor de su hábitat, dejando por lo regular las planicies inundadas por marea como áreas de alimentación compartidas. Además, algunos mamíferos muestran una tendencia a formar agrupaciones y un incremento en la tolerancia social, en parte en respuesta al amontonamiento forzado que impone la inundación de marea. El comportamiento espacial no territorial inusual y la lucha relacionada, sistema poligineo de apareamiento de los gorriones cola aguda quizás sean un punto final en la tendencia en los gorriones que acontece en condiciones de densidad local-alta y semi-colonial, en donde los machos ya no defienden el territorio. El sistema social del gorrión cola aguda quizás haya evolucionado en respuesta a su relación subordinada con gorriones costeros sinotópicos (*Ammodramus maritimus*). En este estructuralmente simple, pero hábitat de zona, territorialidad ínterespecifica y evitación parecen estar bien desarrolladas entre los colorines y posiblemente en taxa mamífera pequeña.

Studies of the social behavior of marsh birds over the past 40 yr have formed the basis of theories for the evolution of territorial polygyny (Verner and Willson 1966, Orians 1969; Searcy and Yasukawa 1989, 1995), coloniality (Orians and Christman 1968) and interspecific territoriality (Orians and Willson 1964; Murray 1971, 1981). However, most of this research has focused on bird populations in freshwater marshes, and the biophysical environment of tidal marshes is quite distinct from interior marshes in ways that might further influence the evolution of social behavior. Social systems have been investigated in a few tidal-marsh species, most notably species in the genus *Ammodramus* (Post and Greenlaw 1982, Greenlaw and Post 1985). To date, no comprehensive overview of social systems of tidal-marsh birds has been published. This paper will provide such an overview from a behavioral ecology and evolutionary point of view, wherein we pose two questions: What is the nature and extent of variation of the social behavior observed in tidal salt marsh animals? What saltmarsh environmental features may lead to favoring or disfavoring certain social strategies?

METHODS

Because most detailed behavioral studies of tidal-marsh vertebrates have been directed at birds, particularly passerines, this paper will have a strong avian focus. However, some important observations have been made on other vertebrates, particularly small mammals, and these will be discussed as well.

The analyses in this paper are based on the descriptions and classifications of social behavior for North American saltmarsh birds based on the literature. Bird species are classified by migratory status, breeding territorial system, patchiness of distribution, and mating system in Tables 1 and 2; they are divided between salt-marsh species (Table 1) and saltmarsh relatives (Table 2). We believe the more scant information on mammalian social systems would not support such a classification scheme, but we have incorporated examples from small-mammal studies in the body of the paper.

KEY CHARACTERISTICS OF TIDAL MARSHES

Saltmarshes, as contrasted to inland habitats, have four key attributes which might shape the evolution and expression of social behavior: (1) higher food abundance, particularly for species dependent upon invertebrates, (2) lower seasonality in resource abundance, (3) lower structural and floristic diversity, and (4) habitat quality variability dependent on patterns of tidal flooding, which in turn influences vegetation cover.

HIGH FOOD ABUNDANCE

Tidal marshes are known for their high primary productivity (Adam 1990), which should ultimately lead to high food abundance for herbivores, detritivores, and their predators. Post and Greenlaw (1982) hypothesized that food is rarely, if ever, a limiting factor shaping the population dynamics or social systems of tidal-marsh birds, at least during the breeding season. By this hypothesis, individuals should not be expected to adopt a strategy that provides greater access to food because increasing

TABLE 1. BREEDING SOCIAL SYSTEM ATTRIBUTES OF NORTH AMERICAN TIDAL-MARSH BIRDS.

Species	Migratory status[a]	Territorial system[b]	Patchiness[c]	Mating system[d]	References
Clapper Rail (*Rallus longirostrus*)	R-M	2	P	M	Eddleman and Conway (1998).
Willett (*Catoptrophus semipalmatus*)	M	2		M	Howe (1982).
Marsh Wren (*Cistothorus palustris*)	R-M	1	P	P	Kale (1965).
Common Yellowthroat (*Geothlypis trichas*)	R-M	1		M	Foster (1977).
Song Sparrow (*Melospiza melodia*)	R	2	P	M	Johnston (1956a, b).
Swamp Sparrow (*Melospiza georgiana*)	M	1	P	M	Greenberg and Droege (1990).
Seaside Sparrow (*Ammodramus maritimus*)	R-M	2	P	M	Post and Greenlaw (1994).
Nelson's Sharp-tailed Sparrow (*Ammodramus nelsoni*)	M	3	P	P	Shriver (2002).
Salt Marsh Sharp-tailed Sparrow (*Ammodramus caudacutus*)	M	3	P	P	Greenlaw (1993).
Savannah Sparrow (*Passerculus sandwichensis*)	R	1?		M	Powell and Collier (1998), J. Williams (pers. comm.).
Red-winged Blackbird (*Agelaius phoeniceus*)	R-M	2	P	P	Reinert (*this volume*).
Boat-tailed Grackle (*Quiscalus major*)	R-M	3	P	P	Post et al. (1996).

[a] Migratory status (M = migrant, R = resident, and R-M = partial migrant).
[b] Breeding territorial system (1 = all-purpose territory, 2 = nesting territory with discontinuous supplementary feeding areas, and 3 = non-territorial).
[c] Patchy distribution (P = high clumped).
[d] Mating system (M = monogamous, P = polygynous).

TABLE 2. BREEDING SOCIAL SYSTEMS OF NON-TIDAL MARSH-RELATIVES OF NORTH AMERICAN TIDAL-MARSH SPECIES.

Species	Migratory status[a]	Territorial system[b]	Patchiness[c]	Mating system[d]	References
King Rail (*Rallus elegans*)	R–M	1	?	M	Meanley (1992).
Willet (*Catoptrophus semipalmatus*)	M	1		M	Lowther et al. (2001).
Marsh Wren (*Cistothorus palustris*)	R–M	1	P	P	Kroodsma and Verner (1997).
Common Yellowthroat (*Geothlypis trichas*)	R–M	1		M	Guzy and Ritchison (1999).
Song Sparrow (*Melospiza melodia*)	R–M	1		M	Arcese et al. (2002).
Swamp Sparrow (*Melospiza georgiana*)	M	1	P	M	McDonald (pers. obs.).
LeConte's Sparrow (*Ammodramus leconteii*)	M	1	P	M	Murray (1969), Lowther (1996).
Nelson's Sharp-tailed Sparrow (*Ammodramus nelsoni*)	M	3	P	P	Murray (1969).
Savannah Sparrow (*Passerculus sandwichensis*)	R–M	1		M/P	Wheelright and Rising (1993).
Red-winged Blackbird (*Agelaius phoeniculus*)	M	2	P	P	Yasukawa and Searcy (1995).
Boat-tailed Grackle (*Quiscalus major*)	R–M	3	P	P	Post et al. (1996).

[a] Migratory status (M = migrant, R = resident, and R–M = partial migrant).
[b] Breeding territorial system (1 = all-purpose territory, 2 = nesting territory with discontinuous supplementary feeding areas, and 3 = non-territorial).
[c] Patchy distribution (P = high clumped).
[d] Mating system (M = monogamous, P = polygynous).

access to food, as opposed to other resources, such as safe nesting sites, will have a negligible effect on reproductive output. Food availability has rarely been quantified, and as far we know food abundance has not been experimentally manipulated to test its importance as a factor in strategies of tidal-marsh birds. Furthermore, assessment of food abundance has to factor in prey quality related to such things as size, distribution, digestibility and salt content of prey items. Post et al. (1982) present some circumstantial evidence that supports the food non-limitation hypothesis. The evidence is three-fold: (1) based on diet analysis and arthropod sampling, prey items used by sparrows are in ample supply, (2) for their focal species (Seaside Sparrow [*Ammodramus maritimus*] and sharp-tailed sparrows), male provisioning is either non-existent (sharp-tailed sparrows) or not necessary when experimentally eliminated (Seaside Sparrow), (3) the focal species have similar diets which appear to be broad and change opportunistically with little selection for diet specialization. Although each of these lines of evidence has merit, the assertion that food is not a critical resource for tidal marsh birds has not been critically tested at a general level.

LOW SEASONALITY

Because of their coastal locality, tidal marshes tend to have a less seasonal climate than interior marshes, grasslands, and other similar habitats. Furthermore, much of the productivity is based on inputs from marine systems, which also show reduced seasonality in productivity compared to temperate terrestrial habitats. Although the phenology of food resources per se for birds and mammals have not been monitored on an annual basis, it would be reasonable to hypothesize that seasonality of resources is reduced which might have a number of indirect effects on social systems, e.g., longer breeding seasons and increased residency.

STRUCTURAL AND FLORISTIC SIMPLICITY

Being flat, wet, and open grassland areas, tidal marshes share features known to influence social behavior in similar ecosystems such as fresh-water marshes, grassland, and tundra. However, the appearance of homogeneity and low variation in vegetation form may be biased by a human perspective; to a tidal-marsh bird or mammal, great variation may exist for what humans may perceive as only subtle nuances in microhabitat. For example, clumps of grass, such as a rush (*Juncus*) tussock, slightly elevated and/or separated from the other vegetation serve as song posts advantageous in defining defending critical territorial boundaries (McDonald 1986).

SALINITY AS A PHYSIOLOGICAL BARRIER

The effect of salinity on social behavior is probably manifested primarily through the effect that it has on habitat structure and floristic diversity. Salinity may favor certain plants, such as pickleweed (*Salicornia virginia*; Padgett-Flohr and Isakson 2003) and affect the stature of such plants (see Geissel et al. 1988 for *Salicornia*). Saltmarshes tend to be simple in structure and plant species composition even compared to other wetlands habitats (Mitsch and Gooselink 2000). Because salinity may provide a physiological barrier to potential colonizing species, it is likely that it contributes to the low-species diversity of vertebrates in tidal marshes. Low diversity combined with abundant food resource often leads to high densities of a few dominant species, which is a dominant force shaping the social environment for tidal marsh species.

PATCHY HABITAT QUALITY

Tidal marshes are generally zonal (Mitsch and Gosselink 2000) in their vegetation patterns. Because tidal marshes are dominated by one or a few species, zonal shifts in these dominants may create a pervasive change in habitat quality, such as a shift from pickerelweed to cordgrass (*Spartina*) cover. Within these zonal patterns, small topographic variation may have a large impact on the availability of food, cover, and the propensity of areas to be flooded by regular tides and stochastic flooding. Stochastic flooding of tidal marshes, caused by the combination of wind, rain, and tidal influx, is frequent and sometimes destructive enough to expect that some behavioral adaptations have evolved. One adaptation would be the ability to deal with temporarily crowding with minimal stress and energy expenditure. At a longer time scale, individuals may select areas to nest that have a lower probability of inundation for the reproductive period.

Tidal sloughs, serving as conduits for water and its payload of nutrients and flora, are typically part of the natural tidal-marsh landscape. In some marshes, artificial channels or ditches are present in addition to or instead of natural water channels. Daily or twice-daily tides bring marine waters, which can be advantageous (e.g., replenishing water-associated food sources for prey species, such as crabs [*Uca* spp.], and providing escape from predators for swimming and diving small mammals) or detrimental, e.g., bringing aquatic predators closer to nests. Tidal sloughs may also act as landmarks for territorial and home-range boundaries.

RESULTS

MATE SELECTION AND MATING PATTERNS

Less than half (five of 11) of the saltmarsh species (Table 1), and a similar portion of their non-saltmarsh relatives (four–five of 11) are polygynous (Table 2). These values are based on a behavioral assessment of mating systems, rather than one based on genetic paternity. We know of a single published study of the frequency of extra-pair paternity in a salt-marsh passerine (Seaside Sparrow; Hill and Post 2005).

The Red-winged Blackbird (*Agelaius phoeniceus*), Marsh Wren (*Cistothorus palustris*) Boat-tailed Grackle (*Quisculus major*), and two species of sharp-tailed sparrows (Saltmarsh Sharp-tailed Sparrow [*Ammodramus caudacutus*] and Nelson's Sharp-tailed Sparrow [*A. nelsoni*]) have mating systems that differ from monogamy. The first two species commonly display territorial polygyny with two or more females nesting on the territory of some males. The percentage of male Marsh Wrens attracting greater than one female, however, was relatively low in the one tidal-marsh population studied. Kale (1965) found only 5% of males had more than one mate compared to 12–50% in non-tidal marshes (Kroodsma and Verner 1997). Reinert (*this volume*) found that tidal-marsh Red-winged Blackbirds had a small average harem size (1.3) compared to most inland populations studied. He attributes the reduced harem size in saltmarsh populations on the need for males to assist in feeding young to shorten the nesting cycle in the face of tidal flooding.

Sharp-tailed sparrows of both species have truly unusual mating systems for temperate-zone passerines. The Saltmarsh Sharp-tailed Sparrow's system is aptly described as a form of scramble competition polygyny (Post and Greenlaw 1982). In this system, males survey successive areas from exposed perches as they actively roam their home ranges where females are likely to be found and attempt to intercept and copulate with them, usually through forced mating. One to several males may converge on a single female at the same time, or several solitary males may successively interact with a female during a short period. The mating system of the Nelson's Sharp-tailed Sparrow differs from that of the Saltmarsh Sharp-tailed Sparrow in the greater importance of male-male dominance interactions in determining mating success. A small proportion of males may perform a disproportionate share of copulations (Gilbert 1981, Greenlaw and Rising 1994). A dominant male may fight with and chase away other males when a female is present, and then follow her (Greenlaw and Rising 1994).

Due to the importance of direct male-male competition in sharp-tailed sparrows, it might be hypothesized that size dimorphism is greater than in related monogamous species, a divergence driven by the importance of dominance in male-male interactions. However, based on measurements presented in the literature (Post and Greenlaw 1994, Greenlaw and Rising 1994), this does not seem to be the case for *Ammodramus* sparrows. The potentially heightened importance of sperm competition between males in this species (Greenlaw and Rising 1994) compared to monogamous species might affect patterns of sperm production. In this case, cloacal protuberances in males of some polygynous or promiscuous avian species, such as the sharp-tailed sparrows are unusually large relative to their overall body size. The monogamous Seaside Sparrow, which is substantially larger than sharp-tailed sparrows overall, has a smaller protuberance (Greenlaw and Rising 1994). This intriguing difference between the species needs further exploration with more data on cloacal protuberance but does suggest an important line of comparative research.

Boat-tailed Grackles also displays an unusual mating system for a temperate zone songbird (Post 1992, Post et al. 1996), showing similarities to the sharp-tailed sparrow systems. Males establish dominance hierarchies in the non-breeding season, usually away from colony sites. Females nest in dense colonies in marsh islands or isolated trees, which then attract numerous males. The male's mating success is determined by a strong dominance hierarchy where an alpha male (the identity of whom is often very stable from season to season) garners a plurality ($\approx$ 25%) of the successful mating attempts. As in the sharp-tailed sparrows, males do not defend a territory but compete directly for females.

These few exceptions aside, social monogamy prevails among tidal-marsh species and subspecies. For some bird species monogamy seems to be obligatory, because both sexes are needed to complete incubation (e.g., Clapper Rails [*Rallus longirostris*]; Oney 1954, Eddleman and Conway 1998). In other species, some underlying plasticity is suggested by a small percentage of males have more than one female nesting on their territory. A low frequency of polygyny (<2%) has been reported for tidal-marsh Swamp Sparrow (*Melospiza georgiana*) and Song Sparrow (*M. melodia*; L. J. Grenier, J C. Nordby, and H. Spautz, pers. comm.), a value which is typical for many temperate zone songbirds. Even in the Seaside Sparrow in which polygyny has never been recorded under natural circumstances, males have been induced experimentally (by removing males during the eggs stage or by using testosterone implants) to accept more than one mate (Greenlaw and Post 1985, McDonald 1986, respectively).

Hypotheses on evolution of monogamy in Seaside Sparrow were tested by Greenlaw and Post (1985), integrating male-removal experiments, measurements of territory quality, nesting data, and behavioral observations. Experimentally induced bigamy (using hormone implants) in New York (Greenlaw and Post 1985) and Florida (McDonald 1986) indicate that male Seaside Sparrow will accept second mates. Post and Greenlaw concluded that male help is advantageous but not necessary for female reproductive success, and that female-female aggression in Seaside Sparrows is probably not important in maintaining monogamy. Although territory quality was highly variable in the New York Seaside Sparrow populations studied, a polygyny threshold evidently was not exceeded. The authors suggest this was because either resources (food and nest sites) were not limiting, or because females could compensate for the effects of resource food inequality among territories by feeding at distant sites with impunity outside their mates' territories. It is interesting that Hill and Post (2005) report that extra-pair paternity in a locally dense population of Seaside Sparrows was quite low (11% of nestlings) compared to other emberizids. They argue that this is mediated by a high degree of female aggression toward territorial intrusions of non-mate males, and females apparently do not accept extra-pair copulations in shared feeding areas. This suggests that females may need the help of committed mates, after all. A moderately low value of extra-pair paternity (18%) was found in a dense population of San Pablo Song Sparrows (*Melospiza melodia samuelis*; L. Grenier, pers. comm.).

Ratios of mated to unmated birds vary with habitat quality even in the monogamous Seaside Sparrow. For example, in dense breeding populations of New York living in unaltered and hence presumably higher quality marshes, a significantly higher proportion of males were mated, as compared to an artificially ditched marsh (poorer quality) with a low sparrow density. Similarly, at a well-studied Florida Gulf Coast site, the proportion of unmated male Seaside Sparrows varied in frequency, comprising about 10–25% of the territorial males in different parts of study area in the years 1980–1987 (Greenlaw and Post 1985; McDonald 1986, pers. obs.). Again, variation in incidence of unmated males between populations was attributed to habitat suitability (Post et al. 1983, Greenlaw and Post 1985, Post and Greenlaw 1994).

Maintenance and duration of pair bonds

The rate of selecting the same mate in subsequent years can only be approximated for most species, and little is known of winter social structure of tidal-marsh sparrows. West Coast subspecies are non-migratory, as are Clapper Rail and Seaside Sparrows in the Southeast. In Seaside Sparrows, females of migratory populations returning in spring may or may not mate with their previous mate, perhaps depending on nest-site quality of male territories which varies annually (Post 1974, Post and Greenlaw 1994). In non-migratory populations of Seaside Sparrows, the pair bond appears to be maintained through the year, because both former pair members stay in or around male's former breeding territory (McDonald, pers. obs.; Post and Greenlaw 1994). Similarly, Johnston (1956b) found that resident adult Song Sparrows maintained a non-breeding home range in the vicinity or their breeding territory and these birds were joined in the winter by one or two immatures.

PARENTAL CARE

Animals in resource-rich environments such as tidal saltmarshes should have lessened energetic demands on parents allowing more flexibility in social systems and more instances of single-parent responsibility for the nesting, brooding, and caring for fledged young. With food plentiful, other resource competitions shift to higher prominence. Nest sites and mates are more in demand, and male participation in feeding young may not be as vital in tidal-marsh birds as in non-marsh counterparts. However, complete male emancipation from care of eggs and young has only been found in the sharp-tailed sparrows and Boat-tailed Grackle. Bi-parental incubation occurs in tidal-marsh rails, and biparental feeding of young is found in all species of passerines except the sharp-tailed sparrows, Boat-tailed Grackles, and some Red-winged Blackbirds.

SPATIAL ASPECTS OF TIDAL-MARSH POPULATIONS

Territory clumping and social aggregations

With the exception of male sharp-tailed sparrows, all tidal-saltmarsh birds are territorial during the nesting season. Seven of 11 species or populations diverge from classic all-purpose breeding territories in ways we will discuss below (Table 1). Only three of 11 of the non-saltmarsh populations show such divergence (Table 2).

As with other wetland systems, territory size and density vary considerably, even within a single population. In general, tidal-marsh birds are known for achieving some remarkably high densities with commensurately small territories. For example, Marsh Wren, aggressively defend territories from 60 to >10,000 m^2 (Kroodsma and Verner 1997). Territory size for the saltmarsh population in Georgia studied by Kale (1965) was on the small end of this range, averaging approximately 60–100 m^2 depending upon year and site.

A high proportion of saltmarsh-breeding species (eight of 11) are reported to show highly patchy distribution during the breeding season; this proportion is six of 11 for the non-saltmarsh relatives. A high abundance of food combined with patchily distributed areas safe from predation and flooding might lead to aggregations of high-density nesting territories and sometimes a separation of nesting territories from feeding areas. For example, Johnston (1956b) reports that if the entire tidal marsh was considered potential habitat, then the average density of Song Sparrows would be approximately 2.5 pairs/ha^{-1}. This value however, does not take into account that the actual defended area by breeding Song Sparrows was restricted to the taller vegetation along tidal sloughs. Johnston estimated territory density at approximately 20–25 pairs ha^{-1}. Seaside Sparrows are also noted for their local dense clustering of territories (Post 1974). In northern populations territory size can range from approximately 20 to >10,000 m^2 in a single marsh.

Territory size varies consistently according to location within the range of the species. Territory size in northern populations of Seaside Sparrows are small compared to southern populations (Post et al. 1983) and, conversely, in the sharp-tailed sparrow, home-range size (because this species is not strictly territorial) of northern males is much larger than that of birds in southern populations (Gilbert 1981). Within a region, habitat quality is important in determining the amount of clumping in territories. Post (1974) found the tendency to form high-density territory clusters was more pronounced in unaltered than in ditched marshes. Olsen (unpubl. data) compared territory size in Swamp Sparrows (*Melospiza georgiana*) in marshes with different levels of topographic relief created by muskrat (*Ondatra zibethicus*) activity. The more hummocky marsh supported clustered territories as small as 60 m^2, where the less topographically diverse swamp had a much more evenly distributed set of larger ($\approx$ 1 ha) territories.

Clumping of territories is not restricted to passerines, because Clapper Rails are reported to have highly variable territory sizes with some as small as 0.1 ha, and the distribution has been described as colonial suggesting territory clustering (Eddleman and Conway 1998). The tendency to aggregate is well developed in the non-territorial sharp-tailed sparrows (Greenlaw and Rising 1994). Nesting females (and hence competing males as well) tend to aggregate in what has been described as a colonial system.

Separation of nesting and feeding areas

At least eight of 11 saltmarsh species show a tendency to have separate feeding areas and nesting territories (Table 1). Of the non-saltmarsh relatives, only one of 11 species, the Red-winged Blackbird (Table 2), has been shown to have such a pattern of breeding season space use. In the case of Seaside (Post 1974) and Song sparrows (Grenier and Nordby, unpubl. data), as well as Clapper Rails, shared feeding areas are in more open, less vegetated parts of the marsh itself (Eddleman and Conway 1998). The eastern Willet (*Catoptrophus semipalmatus*), however, has nesting territories of about 0.5–1 ha within saltmarshes, but often travels to nearby intertidal mudflats and beaches to forage (Howe 1982).

Non-territorial aggregations

Refuge from tidal flooding can greatly restrict available habitat for short periods of time. Some tidal vertebrates may occur in local aggregations during high waters. Sibley (1955), for example, reported finding a flock of >100 emberizid sparrows, predominantly Song Sparrows, along a levee in a south San Francisco Bay tidal marsh experiencing a flood tide. The formations of such aggregations would require immigration of sparrows from a fairly large area. Johnston (1955) subsequently noted that Song Sparrows in an undiked tidal marsh move within their normal winter home range to areas above the flood-tide level. Johnston suggested that the flocking is a facultative response to human-altered marsh hydrology. West Coast populations of Song Sparrows are generally not highly social in the winter, but do form flocks, particularly when snowfall restricts available habitat in the winter (Greenberg, pers. obs.). Therefore, Song Sparrows may be sufficiently plastic in their social behavior to respond to unpredictable events, such as tidal flooding in a human restricted habitat.

Fisler (1965), based on his experience with captive animals, found that the salt marsh harvest mouse (*Reithrodontomys raviventris*) displayed a generally less aggressive and more socially tolerant disposition than did the western harvest mouse (*R. megalotus*), which may be a behavioral adaptation to the frequent, but short-lived crowding that is imposed on the species. Johnston (1957) reported that small mammals in San Francisco Bay marshes, including harvest mice, are able to escape the direct effect of flooding by climbing on emergent vegetation, and that crowding above the flood line is rare. Padgett-Flohr and Isakson (2003) showed that the tendency to clump in salt marsh harvest mouse was a seasonal phenomenon, occurring during the breeding and immediate post-breeding period, but that the mice tend to occur in aggregations associated with mid-saline conditions thus avoiding both extremes in salinity. This more persistent patchy distribution may also explain the high social tolerance found in this harvest mouse species. Finally, Harris (1953) found that both marsh rice rats (*Oryzomys palustris*) and meadow voles (*Microtus pennsylvanicus*) found refuge at muskrat houses and feeding platforms during extreme high tides in brackish *Spartina* marshes on the Chesapeake Bay.

The tendency to aggregate in high-marsh zones or near the marsh ecotone was demonstrated in the Suisan shrews (*Sorex ornatus sinuosus*; Hays and Lidicker 2000). During the winter, they live in distinct social groups consisting of a single adult male, several adult females, and sub-adults. Even when the adult male died, these units persisted. With the onset of the breeding season groups were integrated by outsider adult males, and the result was an almost complete change-over in group membership (Hays and Lidicker 2000). The same study found shrew population densities highest where the marsh bordered the adjacent grassland, and that wintering sub-adult males mostly occupied areas of the marsh below high-tide level.

The importance of refuges from flooding

Availability, exploitability, and perhaps in extreme crowding, the defensibility of temporary refuges in times of high waters may be one of the most significant resources for most marsh animals and probably more so for mammals than birds. The role of marsh-upland ecotone in providing temporary or even seasonal refuge needs to be examined as land development in many areas moves closer to the actual marsh edge. The necessity of refugia also varies with major stochastic events in addition to flooding. Although generally rare, saltmarsh fires (Gabrey and Afton 2000) and prolonged freezing with resulting ice floes both drastically change the vegetation profile for one to two seasons following the event (Post 1974). Several mammal studies suggest refuges are more important in unmodified marshes than in muted (tidal extremes mitigated due to dikes) marshes (Padgett-Flohr and Isakson 2003, Kruchek 2004).

Data for mammals indicate that apart from serving merely as temporary locations for high-water escape, these ecotonal areas apparently can function both as population sources and sinks. For example, juvenile marsh rice rats tend to be excluded from saltmarsh and are forced into the upland old-field vegetation at the marsh border (Kruchek 2004). Other studies have found that age and sex ratios in areas of peripheral to saltmarshes vary seasonally (Hays and Lidicker 2000), in response to flooding (Hays and Lidicker 2000), weather (Kruckek 2004), and sometimes in response to density (Geissel et al. 1988). Finding from these individual studies invite further research on social interactions and habitat use in tidal marsh-upland systems.

Interspecific territoriality and avoidance

Interspecific territoriality or interspecific avoidance is often associated with structurally simple environments, such as grasslands and marshes (Murray 1969). High food abundance of tidal marshes may initially attract individuals from a variety of species to feed in or colonize tidal marshes, but low-structural diversity provides few ways for generalized insectivores to diverge in their fundamental foraging niche. Furthermore, because selection in the form of nesting mortality is so high and because the source of the mortality is nearly identical regardless of species or even class, we would expect a higher level of intra-specific nest-site competition in tidal marshes as compared to other habitats.

The interactions between *Ammodramus* species provide the best example of interspecific dominance and avoidance. Seaside Sparrows are dominant to sharp-tailed sparrows, which they regularly chase and supplant from nesting areas. The average distance between nests of the two species is greater than within species even where abundance is in approximate parity (Post 1974). Seaside Sparrows also supplant sharp-tailed sparrows in shared feeding areas. Although sharp-tailed sparrows are not territorial, the behavioral evidence suggests that aggression from Seaside Sparrows may cause sharp-tailed sparrows to avoid certain nesting areas.

The aggressively mediated spatial segregation between these species may be unique among tidal-marsh sparrows, where generally only one species is found in a particular marsh. However, where Swamp and Seaside sparrows co-occur, territories show almost no overlap. This avoidance appears to be mediated by distinct yet subtle differences in vegetation preference rather than by behavioral interactions (R. Greenberg, pers. obs.).

Spatial exclusion and avoidance has been documented among several small mammals of saltmarshes. Based on inferences from capture patterns during a population crash of California voles (*Microtus californicus*), Geissel et al. (1988) proposed that the salt marsh harvest mouse is a fugitive species that avoids spatial overlap with the dominant vole populations. The pattern fits an included niche model where the more salt-tolerant harvest mice can always take refuge in the lower more saline marshes, but expand into higher, grassier marshes in the absence of voles. If this pattern of physiological tolerance and competitive interaction is correct, it is similar to the relationship between lower marsh and upper marsh and upland plants, where the competitively superior forms are less able to colonize more saline-marsh zones. Although the gradient underlying competition in this case appears to be marsh salinity, a similar pattern of avoidance has been described between upland populations of western harvest mouse and California vole, where habitat disturbance is the driving habitat feature (Blaustein 1980). In the latter species pair, evidence from trap avoidance suggests that harvest mice avoid the odor produced by voles. The avoidance behavior between voles and salt marsh harvest mouse reported by Geissel et al (1988) needs further investigation both based on field distributions and behavioral interactions in the laboratory. Other researchers have not found such segregation between the species (Padgett-Flohr and Isakson 2003).

Avoidance also occurs between more distantly related taxa. Seaside Sparrows tend to avoid areas dominated by rushes (*Juncus*) in marshes along the southern Atlantic and Gulf Coasts and avoid spatial overlap with marsh rice rats. The relationship between the two species is complex, however, because rice rats are also major predators on sparrow nests (Post 1981). Guttenspergen and Nordby (*this volume*) discuss a similar possible interaction between Marsh Wrens and other passerines (notably Song Sparrows). The egg-puncture behavior of Marsh Wrens discourages other birds from nesting in their vicinity. This phenomenon has been well documented in freshwater marshes (Picman 1984).

DISCUSSION

SOCIAL ADAPTATIONS IN TIDAL-MARSH VERTEBRATES

To support a hypothesis of social adaptation to tidal marshes, the basic question is to what extent are their social behaviors ancestral, i.e., behaviors carried over from their immediate non-marsh dwelling ancestors (Searcy et al.

1999); and to what extent are the behaviors we now observe in tidal-marsh-dwelling species derived? If new social patterns are consistently detected in tidal-marsh birds, then we need to determine if there are underlying, genetically based differences specific behavioral traits of if the differences reflect facultative shifts within the behavioral repertoire of the non-tidal-marsh population. Although a thorough phylogenetic analysis of social system patterns probably cannot be accomplished at this point, we can make some specific comparisons between tidal-marsh taxa and their sister taxa.

It is unclear if any consistent difference trend can be found in the comparison of mating systems between tidal-marsh forms and their relatives. Coastal populations of both species of sharp-tailed sparrows apparently share their non-territorial and polygynous mating system with the inland subspecies of the Nelson's Sharp-tailed Sparrows (Murray 1969), although our understanding of the social system of the latter remains sketchy. Further evidence that the Sharp-tailed Sparrow mating system is not a specialized adaptation to tidal marsh is the observation that the Aquatic Warbler (*Acrocephalus paludicola*), which breeds in European freshwater sedge-fern bogs, is the song bird with the social system most similar to the Sharp-tailed Sparrow (Schulze-Hagen et al. 1999). Similarly, Boat-tailed Grackles are not restricted to salt marsh for breeding and non-salt marsh populations show a similar social system to those in saltmarshes (Post et al. 1996). The fact that these unusual mating systems are shared between salt- and freshwater-marsh populations does not exclude the possibility that this social system evolved in tidal marshes and characterizes the inland population as a result of a very recent colonization event, but it does show that the factors that maintain it are probably shared between interior and coastal marshes. Finally, in Savannah Sparrow (*Passerculus sandwichensis*), polygyny has been reported for some non-saltmarsh populations (Rising 1989, 2001, Wheelwright and Rising 1993), but not for the saltmarsh subspecies (J. B. Williams, pers. comm.).

Polygyny is frequent, but does not prevail in tidal-marsh birds. Two studies of polygynous Marsh Wrens and Red-winged Blackbirds suggest that, harem sizes are smaller than for interior populations of the same species. Furthermore, it is now unclear how common polygynous systems are in marsh birds in general. Surprisingly few attempts have been made to synthesize data relating breeding systems to marsh habitats incorporating the entire New World fauna. The most complete survey was published by Greenlaw (1989), which includes both North and South American species. Based on this geographically broader view, Greenlaw suggests that polygyny is not disproportionately represented in marshland passerines. Although polygyny prevails in marshland passerines in North America, monogamy is dominant in South America, and polygyny there is rare. Greenlaw also points out that in Europe the correlation between polygyny in bird species in general, and marshes in particular, is not evident (Von Haartman 1969; but see a more comprehensive recent analysis of one taxonomic group by Leisler et al. 2002). Greenlaw (1989) argues that the expectation that polygyny should be common in marshland passerine birds is based on the assumption that a single mechanism, i.e., female choice in relation to polygyny thresholds, determines the mating systems in marsh habitats. Greenlaw (1989) points out that much of the pattern found in North American birds is found in species of a single family, Icteridae. He further argues that multiple routes exist in the evolution of avian polygyny. Outside of North America, one cannot presume that an observed correlation between polygyny and marshes has any significance concerning the importance of particular mechanisms (e.g., female choice in relation to polygyny threshold) that can account for a given mating system).

A polygyny-threshold model for the development of polygynous systems in marsh birds relies upon their being high productivity (so that males can be emancipated from provisioning and other forms of parental care) and highly variable habitat quality or a distinct advantage for nest protection conferring to females for nesting in close proximity to conspecifics (Searcy and Yasukawa 1989). High food abundance and patchy distribution of females characterize monogamous tidal-marsh-dwelling Song, Swamp and Seaside sparrows. In tidal-marsh sparrows and grackles, the increase in local density of territorial males may prevent individual males from being able to defend territories of sufficient quality and size to attract multiple mates. Therefore, it appears that attempts to be polygynous would be swamped.

As stated above, territorial clumping and semi-coloniality does appear to be a consistent feature of the spacing behavior of saltmarsh passerines. However, such behavior has often been noted in marsh birds in general, and it is unclear if tidal marshes display an unusually high tendency toward this behavior. For example, the LeConte's Sparrow (*Ammodramus lecontii*; Lowther 1996), the closest living relative of the Seaside Sparrow and sharp-tailed sparrows is noted for often having very small territories that can be patchily distributed in prairie marshes. Murray (1969), however, found based on a direct comparison on his northern-prairie study site, that Nelson's Sharp-tailed Sparrow was much more prone to display a semi-colonial distribution than the LeConte's Sparrow. The locally high densities found in coastal plain Swamp Sparrows are, if anything, more pronounced in inland populations of this species (Greenberg, pers. obs.).

Separation of nesting and feeding territories is well developed among tidal-marsh birds and probably reflects the ephemeral availability of productive of some inter-tidal microhabitats. It terms of the expression of this pattern in non-tidal marsh relatives, the LeConte's Sparrow apparently does not have separate feeding areas (Murray 1969, Lowther 1996); however, this species remains poorly studied. Separate feeding areas have also not been reported in other inland species of *Ammodramus* (Vickery 1996, Green et al. 2002, Herkert et al. 2002). Non-tidal populations of Song Sparrows also do not show a tendency to have shared feeding areas (Nice 1937, Arcese et al. 2002). Shared feeding areas are not known for the interior marsh-nesting populations of the Willett (Lowther et al. 2001). King Rails are not known to have shared feeding areas during the breeding season (Meanley 1992), but space use of this species is also poorly known.

At this point, the partial separation of undefended feeding areas and defended nesting territories does seem to be a consistent characteristic of tidal-marsh populations or species when compared to related non-tidal-marsh forms. This pattern of space use probably reflects the highly variable quality of marsh vegetation and location for providing nesting areas safe from flooding and predation and the fact that the best areas for reproductive activities are decoupled from areas that have the highest abundance of accessible food. In addition, it might reflect the fact that microhabitat selection for feeding is much more flexible (not being constrained by nest location) and can respond to the rapidly changing face of the tidal marsh.

Non-territorial scramble polygyny clearly distinguishes the social system of the sharp-tailed sparrows as the most derived and unusual of the tidal-marsh-breeding species. As Greenlaw (1989) suggests, the underpinning of the system is that neither potential nesting areas for females or females are economically defensible. As a working hypothesis this system may be viewed as an endpoint in the tendency of tidal-marsh females to form nesting aggregations, to the point where the density of territorial males and non-territorial intruders prevent males from defending territories. With no unique territorial resources, males must compete directly for access to receptive females. However, other saltmarsh sparrows occur in high densities with males being able to maintain nesting territories. The additional factor in the sharp-tailed sparrow system may be the species social subordination to congeners. This may force congregations in certain habitat types and the presence of clusters of non-territorial males may minimize the ability of other species to dominate individual sharp-tailed sparrows (Murray 1971, 1981). In this sense, the sharp-tailed sparrow system seems to be an endpoint for various behavioral features expressed to varying degrees in other tidal marsh species. Boat-tailed Grackles represent, perhaps, another extreme where colonial females are economically defensible because dominant males can easily repel competitors and predators from the colony site.

DIRECTIONS AND OPPORTUNITIES FOR FURTHER RESEARCH ON THE EVOLUTION OF SOCIAL SYSTEMS OF TIDAL-MARSH ANIMALS

As this paper shows, with the exception of a few well-studied species and populations, we are still in the descriptive stage in understanding social systems and behaviors of tidal-marsh vertebrates. At this time, the most detailed comparative information comes from birds, with less known about mammals and virtually nothing known about reptiles. Outside of more basic descriptive research, focusing on comparisons between tidal-marsh and related taxa, more conceptual areas of research exist for which tidal-marsh vertebrates should prove an interesting and tractable system.

The most salient feature of tidal-marsh social systems is the tendency to form aggregations of individuals or territories. The underlying fitness trade-offs associated with these patterns has only been occasionally explored. In particular, what is the nature of the trade-off between risk of flooding and lack of nesting cover on one hand, and the competition and density dependent increase in predation that might ensue from settling in crowded but safer habitats? How do uncoupled gradients in food availability, cover, and presence of high and unflooded substrates shape decisions on where to settle? Since a gradient of dispersion patterns is present in many of the taxa and the underlying habitat structure and floristics is relatively simple, the tidal-marsh system would provide excellent opportunities to examine the forces driving these patterns. This could be done both within and between populations and species to develop general hypotheses.

The consequences of these clumped spatial patterns in terms of other aspects of social behavior, particularly communication and mating systems, would be a fascinating area of inquiry. For example, how does the tendency to occur in very high local densities affect patterns of both short-distance and long distance acoustic signals in birds? It has been hypothesized that females give a nest-departure call in marsh or grassland environments where populations achieve high densities (McDonald and Greenberg 1991). We proposed that females need to communicate to their mates when they leave the nest to minimize aggressive harassment. Another question relating clumping behavior to communication is how the tendency to form dense aggregations shapes the relationship between males and females and the dynamics of extra-pair mating. Surprisingly, we know almost nothing about the actual genetic contribution of different males and females to their putative offspring for any of the terrestrial tidal-marsh vertebrates.

In addition to the social environment, tidal marshes offer an opportunity to study how the physical environment might shape communication signals (Morton 1975, Wiley 1991). The presence of locally differentiated populations allows comparative work on how the structure of the environment and the local microclimate might shape acoustic signals in birds. This could be accomplished through detailed analysis of the signal and reciprocal playback experiments using recorded vocalizations from tidal-marsh and inland populations. Furthermore, the mode of signal presentation, e.g., frequency of vocalization, use of perches, and use of flight songs should be a fruitful area of inquiry.

Reliance on scent communication should be modified in tidal-marsh dwellers due to the daily inundation of water that could dilute or dissolve olfactory cues. Three sets of predictions, not necessarily mutually exclusive, can be generated and are open for investigation due to no known data on the subject: (1) scent trails and markers are chemically adapted so as to be less prone to water wash-out, and also perhaps resistant to chemical alternation by saline solutes, (2) scent use is relied on less, in general, in tidal-marsh animals as compared to their non-marsh counterparts, and (3) scent application and function use in navigation and communication are temporarily adapted to tidal cycles, e.g., preferentially placed on stems above high-tide levels, or relied on as reproductive readiness cues during the middle of the lunar cycles.

Given the simple habitat structure and often high abundance achieved by a few dominant species, tidal marshes clearly provide a good system for investigating interspecific behavioral partitioning of space. In particular, the avoidance behavior hypothesized between some species of small mammals would be an opportune focus for integrated ecological work on patterns of distributions and behavioral studies of interspecific dominance and communication.

Finally, most research on social behavior has focused on breeding-season events. Only the broadest picture is available for the social interactions of vertebrates during the non-breeding season. The pattern of resource availability may be most distinct between tidal-marsh and upland habitats at this time of year. Future integrations of our understanding of non-breeding and breeding social systems in migratory and non-migratory populations will contribute to this rapidly growing area of ornithological research.

ACKNOWLEDGMENTS

MVM is indebted to W. Post and J. S. Greenlaw for providing ideas, instruction, use of their study sites, and sharing their data early in her career as a student of avian behavior on tidal saltmarshes. J. W. Hardy and T. Webber also encouraged and guided her during her graduate years at the University of Florida. E. S. Morton has always been a source of ideas on the evolutionary behavioral ecology of vertebrates, especially birds. The Smithsonian Conservation and Research Center and the Department of Biology, University of Central Arkansas provided resources during the writing of this paper.

TROPHIC ADAPTATIONS IN SPARROWS AND OTHER VERTEBRATES OF TIDAL MARSHES

J. LETITIA GRENIER AND RUSSELL GREENBERG

Abstract. Tidal marshes present trophic challenges to terrestrial vertebrates in terms of both the abiotic (tidal flow and salinity) and biotic (vegetative structure and food resources) environments. Although primary productivity is high in tidal marshes, supporting an abundance of terrestrial and intertidal invertebrates, seeds, and fruit are far less abundant than in comparable interior habitats. In response to these food resources, terrestrial vertebrates in tidal marshes tend to be either herbivores or predators on invertebrates, including marine taxa. We examine the trophic adaptations of sparrows in the subfamily Emberizinae, the vertebrate group that shows the greatest amount of divergence associated with colonizing tidal marshes. Across several different evolutionary clades, tidal-marsh sparrows tend to be heavier and have significantly longer and narrower bills than their closest non-tidal-marsh relatives. The morphological divergence is greatest in taxa with longer associations with tidal marshes. We hypothesize that longer, narrower bills are an adaptation to greater year-round feeding on invertebrates, particularly benthic marine invertebrates, and a reduced dependence upon seeds. Tidal-marsh sparrows and a number of other terrestrial vertebrates share an additional indirect adaptation to foraging in tidal marshes. Most specialized taxa exhibit grayer and blacker dorsal coloration (particularly in brackish upper estuaries) than their closest non-tidal-marsh relatives. We propose that this consistent shift in dorsal coloration provides camouflage to terrestrial vertebrates foraging on exposed tidal sediments, which tend to be grayish to blackish in color due to the prevalence of iron sulfides, rather than the iron oxides common to more aerobic sediments in interior habitats.

Key Words: bill morphology, biogeographic rule, ecological speciation, tidal marsh, trophic adaptation.

ADAPTACIONES TROFICAS EN GORRIONES Y EN OTROS VERTEBRADOS DE MARISMAS DE MAREA

Resumen. Los marismas de marea presentan retos tróficos para vertebrados terrestres en términos ambientales, tanto abióticos (flujo de la marea y salinidad) y bióticos (estructura vegetativa y recursos alimenticios). A pesar de que la productividad primaria es alta en marismas de marea, soportar una abundancia de invertebrados terrestres e intermareales, semillas, y frutas son mucho menos abundantes que en habitats interiores comparables. En respuesta a estos recursos de alimento, vertebrados terrestres en marismas de marea tienden a ser ya sea herbívoros, o depredadores de invertebrados, incluida taxa marina. Examinamos las adaptaciones tróficas de gorriones en la subfamilia Emberizinae, el grupo vertebrado que muestra la mayor cantidad de divergencia asociada con la colonización de marismas de marea. A través de varias clades diferentes evolucionadas, los gorriones de marisma de marea tienden a ser mas fuertes y tienen significativamente picos mas largos y mas delgados, que sus parientes mas cercanos de marismas que no son de marea. La diferencia morfológica es mayor en taxa con asociaciones más amplias con marismas de marea. Hacemos una hipótesis de que picos más largos y más delgados son una adaptación a una alimentación de invertebrados durante todo el año, particularmente invertebrados bénticos marinos, y una dependencia reducida a semillas. Los gorriones de marisma de marea y un número más de otros vertebrados terrestres comparten una adaptación adicional indirecta al forrajeo de marismas de marea. La mayoría de la taxa especializada, exhibe una coloración dorsal más grisácea y negra (principalmente en estuarios salobres más altos) que sus parientes más cercanos que no son de marismas de marea. Proponemos que este consistente cambio en la coloración en el dorso, provee camuflaje a los vertebrados terrestres que forrajean en sedimentos expuestos por la marea, los cuales tienden a ser grisáceos a negros de color, debido a la prevalecencia de sulfuros de hierro, en lugar de óxidos de hierro, comunes en sedimentos mas aeróbicos en habitats del interior.

Tidal marshes present profound adaptive challenges to terrestrial vertebrates that attempt to colonize them. The physical influence of tidal cycles and the chemical influence of salinity combine to create a wetland ecosystem where the benthic environment has strong marine characteristics, yet the vegetative layers resemble freshwater marsh habitats (Chabreck 1988). The frequency of tidal inundation varies across marsh habitats by elevation, creating a gradient of floristically distinct zones (Eleuterius 1990, Faber 1996). At the lowest elevation, unvegetated tidal sloughs with periodically exposed mud offer another habitat absent from other wetlands.

Although tidal marshes can be quite productive, providing ample trophic opportunities to support dense vertebrate populations, these marshes are generally quite restricted in the area they cover relative to interior habitats (Chapman 1977). Therefore, the population size of a terrestrial vertebrate inhabiting a tidal marsh is likely to be small compared to population size in upland habitats. Finally, most tidal marshes formed well after the receding of the ice during the last glaciation and, thus, were colonized by upland species over a relatively short span of time from an evolutionary perspective (Malamud-Roam et al., *this volume*). The influence of glaciation probably means that tidal-marsh taxa suffer from repeated episodes of local extinction or significant population bottlenecks.

The abundant food resources and sharp environmental gradient between marsh and upland favor local adaptation of tidal-marsh taxa, while the ephemeral nature of tidal-marsh habitats and their geographically restricted distribution inhibit differentiation. Thus, the empirical question remains: How much adaptive differentiation characterizes tidal-marsh vertebrates, and how do these adaptations develop?

PROPOSED TAXON CYCLE FOR TIDAL-MARSH VERTEBRATES

Trophic adaptation to tidal-marsh resources probably occurs in several stages, paralleling taxon cycles proposed for the evolution of biota in other systems (Ricklefs and Bermingham 2002). First, as estuaries form and tidal-marsh vegetation develops, animals that can withstand the physical challenges colonize the emerging marsh to take advantage of abundant food and few competitors. Second, the lack of competitors in tidal marshes allows for niche expansion and ensuing increased variation in trophic-related characters (e.g., diet, bill, legs, and feet). Third, if gene flow is reduced between tidal-marsh and non-tidal-marsh conspecifics, selection would drive adaptation to the novel tidal-marsh conditions. Initial diversification (from step 2) would provide the genetic variance for selection to act upon. Gene flow is likely to be reduced in situations where tidal-marsh populations are geographically isolated. Where populations are parapatric, assortative mating by habitat, which is easy to invoke for territorial passerines, may allow speciation to occur even in the face of relatively high levels of gene flow (Rice and Hostert 1993). In this case, diversifying selection for efficient exploitation of tidal marshes and upland foods may create resource polymorphisms in trophic-related characters (Skulason and Smith 1995). These polymorphisms could then evolve into genetically differentiated populations if assortative mating occurs (Smith et al. 1997). Where upland and tidal-marsh populations were allopatric at the time of marsh colonization, directional selection may cause rapid divergence of the marsh populations. Possible examples of taxa that underwent speciation following differentiation resulting from adaptation to the marsh environment are the Clapper Rail (*Rallus longirostrus*), Seaside Sparrow (*Ammodramus maritimus*), the Saltmarsh Sharp-tailed Sparrow (*Ammodramus caudacutus*), the salt marsh harvest mouse (*Reithrodontomys raviventris*), and the salt marsh snake (*Nerodia clarkii*).

After a certain point, specialization for tidal marshes might hinder competitive ability in the habitat of origin, thus further reducing the possibility of gene flow and favoring the local adaptation to tidal-marsh conditions. Eventually, extreme specialization for saltmarshes might reduce competitive ability even in brackish marshes, allowing a second wave of invasions in brackish marshes by upland colonists.

THE COMPARATIVE APPROACH TO THE STUDY OF TIDAL-MARSH ADAPTATIONS

A time-honored approach to studying adaptation is examining patterns of phenotypic variation among unrelated taxa across a similar environmental gradient. The approach of relating geographic variation to causal explanations of adaptation in vertebrates has led to the development of a number of biogeographic rules that are both useful and controversial (Zink and Remsen 1986).

Differentiation of tidal-marsh taxa is prevalent in vertebrates along the Pacific and Atlantic coasts of North America, particularly among sparrows of the subfamily Emberizinae. Ten sparrow species or well-marked subspecies have been described as endemic to tidal marshes. Because emberizids have colonized tidal marshes at various times and along different coastlines, we focus mainly on the trophic adaptations of this particular group of vertebrates. However, we discuss other terrestrial-vertebrate taxa when comparisons are appropriate.

The mere correlation between geographic variation and occupancy of a particular habitat does not prove that tidal-marsh populations are responding adaptively to a selective gradient associated with that habitat. In order to use this comparative approach to develop an adaptive hypothesis for a particular character, we look for the following lines of evidence: (1) A similar pattern of phenotypic differentiation across unrelated taxa (convergence), (2) a testable explanation of how the convergence confers a fitness advantage, (3) a genetic basis for the phenotypic differences, and (4) a correlation between the magnitude of divergence in a character and the extent of time a particular taxa has inhabited a particular ecosystem.

CONVERGENCE IN BILL DIFFERENTIATION IN TIDAL-MARSH PASSERINES

Bill morphology is an evolutionarily labile feature in birds that is sensitive to selection due to changes in foraging substrate or diet. Geographic variation in bill size and shape can be an adaptive response to differences in food resources between habitats (Bardwell et al. 2001). Bill shape relates to handling time and preferences for certain foods (Hrabar et al. 2002). Bowman (1961) used functional analysis and correlations to show that bill size and shape are intimately tied to the proportion and size of seeds and invertebrates in the diet. Boag and Grant (1981) quantified the selection intensity and corresponding change in bill depth when a drought caused a shift in the size of seeds available to the Galápagos Medium Ground-Finch (*Geospiza fortis*). While many of these studies focused on adaptations in bill depth for seed-eating, bill length has been related to speed of closure and, hence, the mobility of animal prey handled (Beecher 1962, Ashmole 1968, Greenberg 1981). Bill size and shape are also associated with the structures from which prey is captured (Bowman 1961). Thus, bill morphology is a likely trait to undergo adaptation to the trophic environment in tidal-marsh sparrows. Natural selection may also act upon leg and foot proportions, perhaps to improve foraging efficiency (Schluter and Smith 1986).

Murray (1969) found convergence in bill length of tidal-marsh *Ammodramus* taxa, and Greenberg and Droege (1990) showed that the pattern of longer and overall larger bills in tidal marshes compared to non-tidal-marsh relatives extends to other sparrow genera. The latter authors found that the overall volume of Coastal Plain Swamp Sparrow (*Melospiza georgiana nigrescens*) bills in tidal marshes was significantly greater than the bill volume of the interior subspecies *M. g. georgiana* (Fig. 1). They provided a short review of the literature for Seaside Sparrows, sharp-tailed sparrows (*Ammodramus caudacutus* and *A. nelsoni*), Song Sparrows (*Melospiza melodia*), and Savannah Sparrows (*Passerculus sandwichensis*), pointing out that in each species the taxa found in tidal marshes have larger bills. Grenier and Greenberg (2005) extended this line of study by measuring museum specimens from 10 sparrow taxa endemic to tidal marshes, their closest upland relatives, and 20 other taxa in the family Emberizidae to determine the relationship between bill length, bill depth, and body mass in these groups. Tidal-marsh birds tended to be heavier and had significantly longer and deeper bills than their nearest kin outside of this habitat. Even when differences in body size between the two groups were factored out, tidal-marsh sparrows had longer bills (Fig. 2). Tidal-marsh sparrows had relatively longer bills with respect to bill depth as well. The authors also compared tarsus length between tidal-marsh birds and sister taxa, but found no significant differences.

These studies provide support for a new biogeographic rule (albeit one of limited taxonomic scope)—songbirds resident in tidal marshes have longer, narrower bills than interior relatives. The basis of this rule is a trophic adaptation, which stands in contrast to other established biogeographic rules (e.g., Allen's and Bergman's) with physiological premises.

HYPOTHESIZED FUNCTION OF DIFFERENTIATED BILLS

Based on our understanding of tidal-marsh food resources and the functional morphology of passerine bills, we hypothesize that the differentiation in bill size and shape is an adaptation for increased consumption of animal foods, particularly marine invertebrates, and

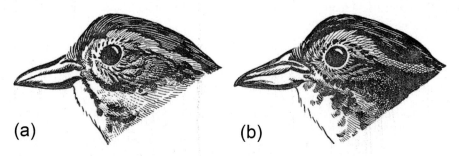

FIGURE 1. From Greenberg and Droege (1990). Line drawings of typical adult male *Melospiza georgiana georgiana* (a) and *M. g. nigrescens* (b). Note the longer bill and darker crown plumage of *nigrescens*.

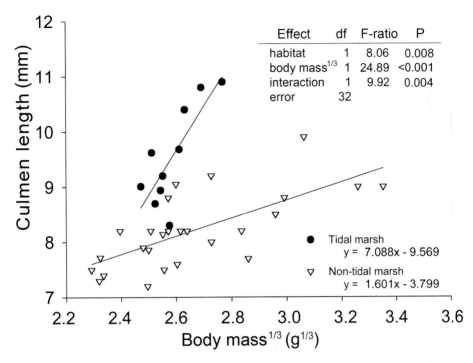

FIGURE 2. From Grenier and Greenberg (2005). Relationship between culmen length and body mass for 36 taxa in Emberizidae. Tidal-marsh birds have longer bills relative to body mass, and bill length increases more quickly with body mass in tidal-marsh taxa.

a concomitant decrease in eating seeds. Short, deep bills are used for cracking large, hard seeds, while bird species with purely animal diets have long, thin bills (Beecher 1951).

These bills designed for carnivory are often used as forceps to probe for and capture invertebrates in crevices under bark (Bowman 1961). Various researchers have proposed that tidal-marsh sparrows use their long bills to probe for invertebrates in the sediment (Post and Greenlaw 1994, Allison 1995). In marshes fringing northern San Francisco Bay, amphipods (*Traskorchestia traskiana*) likely form a large part of the diet for the San Pablo Song Sparrow (*Melospiza melodia samuelis*; Grenier 2004). S. Obrebski and G. H. Irwin (unpubl. data) found that these amphipods burrow into the substrate among the roots of the pickleweed (*Salicornia virginica*) or in cracks in the substrate during neap-tide periods, probably to avoid desiccation. Long bills may be valuable for extracting invertebrates such as amphipods from cracks that form when the high marsh has not been inundated for some time. A study of forest passerines (Brandl et al. 1994), concluded that species with longer, less-deep bills consumed a broader phylogenetic spectrum of prey taxa, because they were able to extract new prey types from under bark. Similarly, tidal-marsh sparrows consume a broader diversity of prey than their upland kin by eating marine invertebrates, possibly by the same mechanism of being able to extract them from crevices.

As well as being useful for probing, long, thin bills may be an adaptation to handle more mobile prey. Functional mechanics analysis indicates that the tips of longer bills can be moved more quickly (Beecher 1962), and ecologists have surmised, therefore, that longer bills may close more rapidly on prey (Ashmole 1968), making them more efficient for capturing active prey (Greenberg 1981). Invertebrates are more active than seeds, so the modified bills of tidal-marsh sparrows may be adapted for efficiently capturing animal prey throughout the year.

TROPHIC OPPORTUNITIES AND CHALLENGES IN TIDAL MARSHES

Relatively few species of terrestrial vertebrates reside in tidal marshes, despite the rich food resources they offer (Chabreck 1988), suggesting that adaptation is required before tidal marshes can be exploited successfully. The greatest challenge for terrestrial vertebrates invading tidal marshes is probably the high salt content of the food and water in these areas (Sabat 2000). Inability to adapt to hyper-saline conditions may exclude most songbird families from tidal-marsh residence. Salt tolerance, which mainly relates to adaptations for processing the food after it is eaten, is well documented for certain saltmarsh sparrows and the salt marsh harvest mouse (Goldstein, *this volume*; Fisler 1965).

Tidal marshes support an abundant invertebrate community that includes marine taxa, such as crustaceans, polychaetes, and mollusks, as well as terrestrial arthropods, mainly spiders and insects (Daiber 1982). In a study of the relationship between invertebrate abundance and avian foraging, Clarke et al. (1984) excluded birds from pools and the area around them in a Massachusetts saltmarshes in July and August, with little effect on the abundance of invertebrates in the exclosures. These data suggest that invertebrate resources are extremely abundant at certain times of year relative to the needs of avian predators, implying bottom-up rather than top-down control of food resources.

The low plant-species diversity and the high degree to which perennial plants rely upon vegetative propagation results in a limited diversity and abundance of seeds (Leck 1989) and no animal-dispersed fruits (Mitsch and Gosselink 2000) in tidal marshes. Based on studies on both North American coasts, Leck (1989) reported that few species were present in saltmarsh-soil seed banks (13-17 species) and seed abundance was low (700-900 seeds/m^2) compared to freshwater marshes and grasslands (30-40 species, 1,000-3,000 seeds/m^2). Furthermore, regular tidal scouring of the substrate limits seed accumulation (Adam 1990). However, saltmarsh plants are highly productive (Mitsch and Gosselink 2000), so food resources are abundant for herbivorous animals.

PATTERNS IN THE DIET OF TIDAL-MARSH SPARROWS

A comparison of diet by Greenlaw and Rising (1994) between the closely related Nelson's Sharp-tailed Sparrow (*Ammodramus nelsoni*), which breeds in the interior of the continent as well as in tidal marshes, and the Saltmarsh Sharp-tailed Sparrow, which breeds exclusively in tidal marshes, exemplifies how the feeding niche differs in the tidal marsh habitat. Stomach contents from Nelson's Sharp-tailed Sparrows breeding in wet areas of New Brunswick and North Dakota, and migrating along the East Coast of the US showed that these birds eat mainly insects (78% of relative total volume, N = 15) and spiders (14%) from May–August, with seeds becoming very important post-breeding (73%, N = 11), and some mollusks consumed during fall migration in coastal marshes (19%, N = 7). Nelson's Sharp-tailed Sparrow takes a wide variety of insects, suggesting little specialization on particular prey. The winter diet is not known. Contents of Saltmarsh Sharp-tailed Sparrow stomachs indicated a different trend in seasonal diet. Although insects were important, the proportion was lower in the breeding season (44%, N = 20) and stayed about the same in the fall (36%, N =12). However, plant matter, which was absent from the breeding diet, made up only 30% of fall foods. A major addition to the diet (amphipods, 24% in both seasons) replaced summer insects and fall seeds for these tidal-marsh sparrows (Greenlaw and Rising 1994).

The Seaside Sparrow, an obligate tidal-marsh species, consumes a diet similar to that of its close relative, the Saltmarsh Sharp-tailed Sparrow (Post and Greenlaw 1982). Also a generalist, the Seaside Sparrow eats insects, spiders, amphipods, mollusks, crabs, and marine worms during the breeding season and adds seeds as a major food source during winter (Martin et al. 1961). Martin et al. (1961) took a small sample (N = 6) of stomachs from New Jersey late in the breeding season and found a heavy reliance on marine invertebrates—36% crab and 24% snails by volume.

Fewer data on Swamp Sparrow diet have been published than for the other tidal-marsh songbirds. Like the upland Sharp-tailed Sparrows, inland Swamp Sparrows (*Melospiza georgiana georgiana* and *M. g. ericrypta*) consume mainly insects during breeding and seeds during fall and winter (Mowbray 1997). However, the proportion of animal matter in the inland Swamp Sparrow diet is relatively higher year-round (Mowbray 1997). The diet of the tidal-marsh subspecies has not been studied, but Greenberg and Droege (1990) suggest that these populations may consume marine invertebrates as a major food source.

The detailed summary of Savannah Sparrow diets by Wheelwright and Rising (1993) shows that this species follows the same pattern. The subspecies outside of tidal marshes consume insects, spiders, mollusks, seeds, and fruit, with insects predominating during breeding and seeds and fruit more important during the off-season. A survey of 1,098 stomachs of breeding Savannah Sparrows from 31 localities across North America showed that crustaceans, which were present in 2% of the stomachs from upland locales, were found in 30% of the tidal marsh stomachs (predominantly fiddler crabs [*Uca*]). Mollusks were also more frequent in the stomachs of saltmarsh birds, and insects were less common than in the upland samples (Wheelwright and Rising 1993).

The breeding season diet of non-tidal-marsh Song Sparrows is dominated by a wide variety of insects with the addition of a few other invertebrates and seeds, and seeds become the main food source in fall and winter (Arcese et al. 2002). Aldrich (1984) found that tidal-marsh Song Sparrows eat more animal matter than upland conspecifics, and this result was confirmed by a stable-isotope study that showed that San Pablo Song Sparrows from a tidal marsh in northern San Francisco Bay assimilated little, if any, plant matter during the breeding season (Grenier 2004). Behavioral data from the same study indicated that San Pablo Song Sparrows fed heavily in areas where the abundant invertebrate biomass was marine snails and amphipods, and sparrows frequently carried amphipods to nestlings. Comparing the stomach contents of 233 Song Sparrows from a variety of habitats around San Francisco Bay, Marshall (1948a) found that the fall diet of saltmarsh birds continued to be mainly animals with the addition of some *Grindelia* seeds and *Spartina* flowers, while brackish marsh and terrestrial populations consume mainly seeds supplemented by insects.

In summary, tidal-marsh sparrows and their close relatives are generalists that consume both animal matter and seeds. However, tidal-marsh birds tend to eat a greater proportion of invertebrates and a correspondingly lesser proportion of seeds than their inland relatives. This difference may be most pronounced in autumn and winter. Tidal-marsh sparrows also consume more marine invertebrates and fewer insects and spiders than upland sparrows.

TROPHIC PATTERNS IN OTHER TIDAL-MARSH VERTEBRATES

Beyond sparrows, few studies have focused on trophic specialization in tidal-marsh vertebrates. As discussed in detail for sparrows, the available data suggest that other birds in tidal marshes reduce consumption of seeds in favor of increased use of invertebrates, including intertidal taxa. Clapper Rails, which dwell in saltmarshes, and King Rails (*Rallus elegans*), which prefer brackish to freshwater marshes, both feed predominantly on animal matter, particularly crustaceans. However, King Rails (as well as freshwater populations of the Yuma Clapper Rail [*R. l. yumanensis*]) eat much larger quantities of insects and seeds than do Clapper Rails, particularly in the autumn and winter (Meanley 1992). Meanwhile, saltmarsh Clapper Rails eat primarily crabs and snails (Eddleman and Conway 1998). A study of the fall diet of Soras (*Porzana carolina*) in Connecticut showed that in freshwater marshes, birds consumed large quantities of seeds, whereas birds in saltmarshes fed almost entirely on invertebrates (Webster 1964). Similarly, the stomach contents of diamondback terrapins (*Malaclemys terrapin*), the only turtle specialized on saltmarshes, consist almost exclusively of marine mollusks (Ernst and Barbour 1989).

Given that tidal marshes show strong similarity in vegetation to inland marshes and other habitats, we would expect species that are restricted to the vegetation layer to show less pronounced diet shifts than those that use the marine substrate. Few data are available to test this prediction, but a small amount of information seems to support the idea. Marsh Wrens (*Cistothorus palustris*) provide a good example of a species that forages almost entirely from reed, grass, and shrub layers of the tidal marsh. The detailed study of diet by Kale (1965) for the subspecies in *Spartina* marshes in Georgia shows that Marsh Wrens consume primarily insects and spiders, but a small portion of stomach samples contained invertebrates, such as crabs and amphipods, as well as snails associated with the marine substrate. A comparison with studies of interior populations (Beal 1907, Welter 1935) in New York state and California suggest interesting differences between tidal- and freshwater -marsh populations that should be explored. Kale (1965) found that ants and spiders were larger components of the breeding season diet of the Marsh Wrens in saltmarshes than freshwater marshes, where the diet was dominated by Orthoptera, Odonata, and Lepidoptera.

Specialized forms of tidal-marsh mammals are restricted to a few rodents. Rodent granivores give way to herbivores in tidal marshes. Common tidal-marsh rodents include muskrats (*Ondatra zibethicus*), California voles (*Microtus californicus*), and meadow voles (*M. pennsylvanicus*) which are predominantly herbivorous (Thaeler 1961, Willner et al. 1980, Batzli 1986). Consistent with the dominance of herbivory in saltmarsh mammals, the salt marsh harvest mouse, which is endemic to tidal marshes, consumes considerably more plant material than does its closest relative, the western harvest mouse (*Reithrodontomys megalotis*) in uplands (Fisler 1965). Fisler (1965) demonstrated that the salt marsh harvest mouse has a substantially longer digestive tract than the upland species, which is likely an adaptation for digesting a greater proportion of plant material in the diet. Absent among rodents commonly found in tidal marshes are granivorous and frugivorous species (Greenberg and Maldonado, *this volume*).

Carnivory is also prevalent in small mammals in tidal marshes. Shrews, a common component of tidal-marsh faunas, are specialists on invertebrates, and indirect evidence suggests that at least one specialized tidal-marsh subspecies of the ornate shrew (*Sorex ornatus sinuosus*; Hays and Lidicker 2000) feeds on amphipods. The restriction in diet guilds to herbivores and carnivores may result from filtering of colonizing species (e.g., shrews are more likely to colonize marshes because they are carnivores), or it may result from niche shifts. Marsh rice rats (*Oryzomys palustris*), common in tidal marshes of the southeastern US (Daiber 1982, Wolf 1982) are known to depend heavily on invertebrates, particularly from the benthic substrate (Sharp 1967). Comparison of stomach contents between rats in tidal marsh and the immediately adjacent uplands (Kruchek 2004) showed that tidal-marsh rice rats consume far more invertebrates and less plant material than those in the adjacent old fields. Even in the grass-specialized herbivore, the California vole, stable-isotope analysis of the spring pelage of an individual captured in saltmarsh revealed a diet based completely on animal foods (Grenier 2004).

DORSAL COLOR SPECIALIZATION FOR FORAGING IN TIDAL MARSHES

Another less apparent feeding-related adaptation to a new habitat is cryptic coloration to reduce predation risk while foraging. Tidal-marsh sparrows and their close relatives in the upland forage mainly along the sediment and in low foliage, obtaining food from near to or on the ground (Wheelwright and Rising 1993, Greenlaw and Rising 1994, Post and Greenlaw 1994, Mowbray 1997, Arcese et al. 2002). Often they are observed foraging on the exposed surface of tidal channels and sloughs. Thus, dorsal coloration that matches the background color of the sediment may be important for reducing the risk of predation. For tidal-marsh sparrows, the periodic exposure by the tide of unvegetated sediment provides both a novel foraging substrate and a new predation risk, so selection for cryptic dorsal plumage is plausible.

Grinnell (1913) noted that tidal-marsh vertebrates tend to be darker than their upland relatives, and Greenberg and Droege (1990) reviewed this phenomenon, noting that the trend is sometimes more toward grayer hues rather than darker coloration. Saltmarsh melanism is clearly expressed within the tidal-marsh sparrows. Seaside Sparrows can be distinguished from other sparrow species in the field by their dark olive-gray dorsal coloration (Sibley 2000), and the recently extirpated Dusky Seaside Sparrow (*Ammodramus maritimus nigrescens*) exhibited distinctly melanistic plumage. Saltmarsh Sharp-tailed Sparrows are more heavily streaked with fewer white markings dorsally than Nelson's Sharp-tailed Sparrows, and *Ammodramus nelsoni subvirgatus* in tidal marshes of the Canadian maritime provinces are grayer than conspecifics from interior prairies (Ridgway and Friedmann 1901, Greenlaw and Rising 1994). The streaking of Belding's Savannah Sparrow is heavier and darker than typical Savannah Sparrows, but the large-billed subspecies are distinctly pale, although grayish in tone (Ridgway and Friedmann 1901, Wheelwright and Rising 1993). A quantitative study of variation in Song Sparrow dorsal plumage (Marshall 1948b), found upland *Melospiza melodia gouldii* to be reddish-brown, while tidal-marsh subspecies *maxillaris, samuelis* and *pusillula* were blackish-brown, blackish-olive, and yellowish-gray, respectively. Swamp Sparrows in tidal marshes also lose their colorful rusty tones dorsally, and rusty crown and nape plumage is replaced with black feathers (Greenberg and Droege 1990).

Saltmarsh melanism, or the tendency to be grayer or blacker, has been reported for several other birds as well (Table 1). The weaker plumage differentiation in Marsh Wrens is not surprising given that foraging in this species is more associated with water edge and vegetation rather than open sediment (Kroodsma and Verner 1997). In addition to these avian examples, saltmarsh melanism has been observed in various small mammals, namely cinereus shrew (*Sorex cinereus*), ornate shrew (*S. ornatus*), vagrant shrew (*S. vagrans*), California vole, meadow vole, and western harvest mouse (Grinnell 1913, Jackson 1928, Green 1932, Von Bloeker 1932, Rudd 1955, Hall and Kelson 1959, Thaeler 1961, Fisler 1965, Woods et al. 1982), and in one tidal-marsh snake (*Nerodia sipedon williamengelsi*; Conant et al. 1998).

It was long ago hypothesized (Grinnell 1913, Von Bloeker 1932) that the blackest forms of tidal-marsh species were found in brackish upper estuaries (such as Suisun Bay in the San Francisco estuary). This hypothesis is borne out by the following melanistic vertebrate taxa: Coastal Plain Swamp Sparrow, Suisun Song Sparrow (*Melospiza melodia maxillaris*), Dusky Seaside Sparrow, Suisun shrew (*Sorex ornatus sinuosis*), and tidal-marsh populations of the California vole.

The blacker and grayer dorsal coloration of tidal-marsh vertebrates may be an adaptation for blending in against the sediment background to reduce the risk of predation while foraging in the open. Relative to reddish sediment in

TABLE 1. COMPARISON OF PLUMAGE COLORATION OF TIDAL-MARSH-BIRD TAXA WITH THEIR CLOSEST NON-TIDAL-MARSH RELATIVES.

Tidal-marsh taxon	Taxon for comparison	Coloration in tidal marsh
Clapper Rail (East Coast; *Rallus longirostris crepitans* group)	King Rail (*Rallus elegans*)	Grayer.
Clapper Rail (West Coast; *R. l. obsoletus* group)	King Rail	Somewhat grayer.
Marsh Wren (*Cistothorus palustris griseus*) and other tidal-marsh populations	Marsh Wren, non-tidal marsh populations	Only *griseus* (southern Atlantic coast) is distinctly grayer.
Suisun Song Sparrow (*Melospiza melodia maxillaries*)	Modesto Song Sparrow (*M. m. mailliardi*)	Grayer with blacker markings.
Alameda Song Sparrow (*M. m. pusillula*)	Heermann's Song Sparrow (*M. m. heermanni*)	Grayer.
San Pablo Song Sparrow (*M. m. samuelis*)	Marin Song Sparrow (*M. m gouldii*)	Grayer.
Coastal Plain Swamp Sparrow (*M. georgiana nigrescens*)	Southern Swamp Sparrow (*M. g. georgiana*)	Grayer with blacker markings.
Belding's Savannah Sparrow (*Passerculus sandwichensis beldingi* group)	Savannah Sparrow (*P. sandwichensis*): interior subspecies	Grayer with blacker markings.
Large-billed Savannah Sparrow (*P. s. rostratus* group)	Savannah Sparrow (*P. sandwichensis*): interior subspecies	Grayer.
Saltmarsh Sharp-tailed Sparrow (*Ammodramus caudacutus*)	Nelson's Sharp-tailed Sparrow (*A. nelsoni*)	Grayer.
Acadian and James Bay Sharp-tailed sparrows (*A. n. subvirgatus* and *A. n. alterus*)	Nelson's Sharp-tailed Sparrow (*A. n. nelsoni*)	Grayer wings, less contrasting back.

terrestrial habitats, tidal marsh mud is grayer. This phenomenon occurs because sea water is rich in sulfates, and, in the anoxic conditions of tidal sediment, iron is reduced anaerobically to grayish iron sulfides rather than to the reddish iron oxides of aerobic soils (Mitsch and Gosselink 2000). The blacker coloration of birds and mammals in brackish marshes may be a result of background matching to blacker sediments. Upper estuary mud may have a greater quantity of dark organic materials relative to saltmarsh sediments, due to the closer proximity of riverine inputs (Odum 1988).

The Alameda Song Sparrow (*Melospiza melodia pusillula*) and the large-billed Savannah Sparrow group (*Passerculus sandwichensis rostratus* and related subspecies) are partial exceptions to the general trend of tidal-marsh melanism. Although both are relatively pale, they have grayish dorsal coloration, which is consistent with other tidal-marsh forms. Both sparrows inhabit the most saline and arid marshes of their respective regions. This correlation suggests background matching to a paler color for camouflage in drier, saltier marshes.

ENVIRONMENTAL AND GENETIC CONTRIBUTIONS TO TIDAL-MARSH ADAPTATIONS

Both environment and genes likely play a role in determining bill morphology. James (1983) found a surprisingly large environmental influence on bill shape in Red-winged Blackbirds (*Agelaius phoeniceus*) by cross fostering nestlings between geographic areas with morphologically distinct populations. Other studies show that bill size and shape are largely heritable (Boag and Grant 1981, Forstmeier et al. 2001). Given that Smith and Zach (1979) found Song Sparrow bill traits to be significantly heritable, we proceed with the assumption that the bill morphology of tidal-marsh sparrows is determined mainly by genetics. While heritability of bill morphology would make this trait potentially sensitive to selection, low heritability does not rule out the possibility that differentiation at the subspecies and species levels is genetically based (Merilä and Sheldon 2001).

The heritability of plumage coloration is not a simple question, because different feather colors are controlled by different mechanisms. A gene was recently identified that likely controls melanin deposition in Bananaquits (*Coereba flava*; Theron et al. 2001), and this gene or a number of other loci identified in poultry may contribute to increased melanin expression in tidal-marsh songbirds. Greenberg and Droege (1990) raised 11 Coastal Plain Swamp Sparrow nestlings taken from the wild at 4-6 d old in the lab under standardized conditions. When measured after the postjuvenal molt, these birds were similar in both dorsal color and bill size to adult Coastal Plain Swamp Sparrows and distinct from inland Swamp Sparrows. Although more experiments are required, this result suggests that bill morphology and plumage color have large heritable components in tidal-marsh sparrows.

EVIDENCE FOR A TAXON CYCLE IN TIDAL-MARSH SPARROWS

Generally only one or two sparrow species can be found breeding in the same tidal marsh. The exception occurs along the mid-Atlantic seaboard, primarily Chesapeake and Delaware bays. In these estuaries, three species cohabit tidal marshes—two saltmarsh specialists, Seaside Sparrow and Saltmarsh Sharp-tailed Sparrow, joined by the Coastal Plain Swamp Sparrow. Genetic data suggest a very recent colonization of the brackish upper estuary by Swamp Sparrows (Greenberg et al. 1998) in contrast to a much longer association with tidal marshes in the other species indicated by a deep divergence from upland relatives (Zink and Avise 1990). We suggest that coastal marshes existed along the gently sloping continental shelf of the eastern seaboard throughout the Pleistocene, allowing the more specialized *Ammodramus* sparrows to persist in saltmarshes (Malamud-Roam et al., *this volume*). The more recent development of estuarine marshes (Malamud-Roam et al., *this volume*) may have allowed Swamp Sparrows from small inland populations, which were expanding as the most recent continental glaciers receded (Greenberg et al. 1998), to colonize these brackish areas that were less suited to the adaptations of Seaside and sharp-tailed sparrows.

Support for this idea that adaptive specialization for the tidal-marsh environment accrues over evolutionary time is provided by the correlation between morphological divergence and genetic divergence in tidal-marsh sparrows (Fig 3; Grenier and Greenberg 2005). The four tidal-marsh sparrows with ancient divergence times from their upland counterparts (>200,000 yr BP) exhibited significantly more extreme bill elongation than recently diverged groups (<10,000 yr BP). We suspect that divergence will increase in the younger taxa as enough time passes for beneficial mutations to accumulate and be selected. The type of tidal-marsh habitat each taxon inhabits is not a confounding factor in this analysis; saltmarsh specialists are found in both the ancient (Seaside Sparrow, Savannah Sparrow, Saltmarsh Sharp-tailed Sparrow) and recent (San Pablo Song Sparrow and Alameda Song Sparrow [*Melospiza melodia pusillula*]) divergence groups. Brackish-marsh specialists are found only in the recent divergence group, either in sympatry with other tidal-marsh sparrows (Coastal Plain Swamp Sparrow), which fits with the proposed taxon cycle, or in relatively young marshes only a

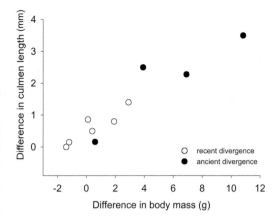

FIGURE 3. From Grenier and Greenberg (2005). Relationship between the difference in culmen length and the difference in body mass between sister taxa in different habitats. Differences are the mean for tidal-marsh birds minus the mean for the upland relative. Divergence is classified as either recent (<10,000 ybp) or ancient (>200,000 ybp).

few thousand years old (Suisun Song Sparrow; Atwater et al. 1979).

FUTURE RESEARCH

Although the patterns that we have discussed are compelling, rigorous field studies are required to empirically test if they reflect adaptation to tidal marshes. In variable environments like tidal marshes, natural selection acts in concentrated bursts to carve the evolutionary paths of populations (Benkman 1993), and for trophic adaptations the episodes of selective change are probably related to resource scarcity (Boag and Grant 1981, Schluter and Smith 1986). Therefore, studies are needed that track the distribution of feeding-related traits in a tidal-marsh population concurrent with the survival and reproductive success of individuals and the abundance of resources. Tidal-marsh passerines may be tractable systems for attempting this type of real-time measurement of natural selection, because populations are often dense enough to provide large sample sizes. More knowledge of the trophic ecology of tidal-marsh songbirds, particularly when and how they experience resource scarcity, would be a first step. Also, a better understanding of saltmarsh food resources, particularly invertebrate and seed distribution and abundance, is needed. Finally, cross-fostering experiments between upland and tidal-marsh subspecies would shed light on the importance of environment in creating the patterns we have reviewed.

ACKNOWLEDGMENTS

Thanks to S. R. Beissinger, E. Lacey, C. Benkman, and B. Eddleman for constructive reviews of earlier drafts of the chapter and to the American Museum of Natural History and the Museum of Vertebrate Zoology, University of California, Berkeley, for access to specimens. Financial support for field work on tidal-marsh birds and for manuscript preparation was provided by the Abbott Fund of the Smithsonian Institution, the Delaware Ornithological Society, the San Francisco Bay Fund, the Budweiser Conservation Scholarship, the P.E.O. Scholar Award, the Garden Club of America Award in Coastal Wetlands Studies, the Phi Beta Kappa Fellowship, and the Joseph Mailliard Fellowship in Ornithology.

REGIONAL STUDIES

Clapper Rail (*Rallus longirostris*) with fiddler crab (*Uca pugnax*)
Drawing by Julie Zickefoose

BREEDING BIRDS OF NORTHEAST SALTMARSHES: HABITAT USE AND CONSERVATION

ALAN R. HANSON AND W. GREGORY SHRIVER

Abstract. Saltmarshes and associated wildlife populations have been identified as priorities for restoration and conservation in northeastern North America. We compare results from a recent study on habitat requirements of saltmarsh-breeding birds in the Maritime Provinces of Canada to those from recently published studies for the New England Gulf of Maine, and the southern New England shore. Differences in geologic history, sedimentation rates, tidal amplitude, ice cover, sea-level rise, climate, and human activity have influenced the ecology, extent, and distribution of saltmarsh habitat among these regions. In Canada, Bay of Fundy saltmarshes studied were larger and less isolated compared to marshes in the Gulf of St. Lawrence or those along the Atlantic Coast of Nova Scotia. Saltmarshes in the Maritimes and the New England Gulf of Maine were large compared to those along the southern New England shore. In all study regions, species richness was greater in larger saltmarshes. In the Maritime Provinces, marsh area was an important determinant of the density of Nelson's Sharp-tailed Sparrows (*Ammodramus nelsoni*) and Savannah Sparrows (*Passerculus sandwichensis*). Willet (*Catoptrophorus semipalmatus*) density was not influenced by marsh area but was positively influenced by pond area. Proximity to other marshes, or the number of dwellings within 500 m of the study marsh did not affect any aspect of bird use. Nelson's Sharp-tailed Sparrow density was positively influenced by the presence of adjacent dike land. In the Maritimes, common reed (*Phragmites australis*) is not widespread and therefore not a useful predictor of avian habitat use in contrast to New England where studies have documented lower species richness where *Phragmites* is abundant. Based on findings from studies across the Northeast we conclude that: (1) habitat area is an important parameter for determining the occurrence of many species of saltmarsh-breeding birds, (2) habitat quality for saltmarsh-breeding birds is dependent on multiple spatial scales, and (3) wetland protection policies and conservation-restoration activities need to specifically address the collective habitat requirements and conservation concerns for individual bird species within locales.

Key Words: birds, Canada, conservation, isolation, Maritimes, New England, saltmarsh.

AVES REPRODUCTORAS DE MARISMAS SALADAS DEL NORESTE: UTILIZACIÓN DEL HABITAT Y CONSERVACIÓN

Resumen. Marismas saladas y poblaciones de vida Silvestre asociadas han sido identificadas como prioritarias para la restauración y conservación en el noreste de Norte América. Comparamos resultados de un estudio reciente sobre requerimientos del hábitat de aves reproductoras de marisma salada, en las Provincias Marítimas de Canadá, con aquellos estudios publicados recientemente para el Golfo de Nueva Inglaterra de Maine, y la costa sureña de Nueva Inglaterra. Diferencias en historia geológica, tasas de sedimentación, amplitud de marea, cubierta de hielo, levantamiento del nivel del mar, clima y actividad humana, han influenciado la ecología, el alcance y la distribución del hábitat de marisma salada entre estas regiones. En Canadá, las marismas saladas estudiados de la Bahía de Fundy fueron más grandes y menos aisladas, en comparación a las marismas en el Golfo de San Lawrence o a aquellas a lo largo de la Costa del Atlántico de Nova Scotia. Las marismas saladas marítimos en el Golfo de Nueva Inglaterra de Maine, fueron más grandes comparadas con aquellas a lo largo de la costa sureña de Nueva Inglaterra. En el área marítima, el área de marisma fue un importante determinante de la densidad de Gorriones Cola Aguda Nelson (*Ammodramus nelsoni*) y de Gorriones Sabana (*Passerculus sandwichensis*). La densidad del Playero Pihuiui (*Catoptrophorus semipalmatus*) no fue influenciada por el área de marisma, pero fue positivamente influenciada por el área del charco. La proximidad a otros marismas, o el número de viviendas dentro de los 500 m del estudio, la marisma no afecto ningún aspecto de la utilización del ave. La densidad del Gorrión Cola Aguda Nelson estuvo positivamente influenciada por la presencia de tierra del canal adyacente. En el área marítima, el carrizo (*Phragmites australis*) no es dispersado y por ello no es un vaticinador útil de la utilización del hábitat de aves, en contraste con Nueva Inglaterra donde estudios han documentado menor riqueza de la especie en donde *Phragmites* es abundante. Basándonos en hallazgos de estudios a través del Noreste, concluimos que: (1) el área del hábitat es un parámetro importante para determinar la aparición de muchas especies de aves reproductoras de marisma salada, (2) la calidad del hábitat para aves reproductoras de marisma salada depende en escalas espaciales múltiples, y (3) políticas de protección de humedales y actividades de conservación-restauración necesitan ser dirigidas específicamente a los requerimientos colectivos del hábitat y a preocupaciones de conservación para especies individuales de aves dentro de las locales.

Saltmarshes are unique ecosystems resulting from complex interactions between hydrology, sedimentation, salinity, tidal amplitude and periodicity, and primary productivity at the interface between terrestrial and marine ecosystems (Bertness 1999). The same physical and biological features that make saltmarshes some of the most productive ecosystems in the temperate zone, also supported European settlements during colonization of northeastern North America (hereafter Northeast). The history of human settlement patterns and use of saltmarsh ecosystems in the Northeast is an important factor in determining the present condition of saltmarshes. Human use of saltmarshes for agricultural purposes was widespread throughout the Northeast during the 1600–1900s. Ditching, draining, and infilling of saltmarshes occurred throughout the region and included diking in the Canadian Maritime Provinces (hereafter Maritimes). Since European settlement, increasing human populations and expanding cities and towns have resulted in the continued draining, infilling, and alteration of saltmarshes (Bertness et al. 2004). Loss of coastal wetlands in the US has been substantial, ranging from 30–40% (Horwitz 1978) with saltmarsh habitat in New England being particularly imperiled (Tiner 1984). In Canada, the amount of saltmarsh lost in some local areas is upwards of 85% (Reed and Smith 1972), although national statistics are not available (Glooschenko et al. 1988). Although much research has occurred on saltmarshes in the eastern US there have been few attempts to collectively assess saltmarsh forms, land-use histories, and wildlife communities in the Northeast (Bertness and Pennings 2000).

Despite the magnitude of habitat change, only recently have agencies concerned with wildlife conservation begun to systematically survey saltmarsh avifauna in the Northeast. Most of the research on the habitat function of saltmarshes in the Northeast has focused on fish (Weinstein and Kreeger 2000). Therefore, quantitative information about species occurrence, relative abundance, and density of key wildlife species is often unavailable. Inadequate information on the status and distribution of saltmarsh-bird populations limits the utility of North American Bird Conservation Initiative prioritization, and is the primary reason for many saltmarsh bird species being listed as species of high conservation concern (Pashley et al. 2000). In the Northeast, species such as Nelson's Sharp-tailed Sparrow (*Ammodramus nelsoni*), Saltmarsh Sharp-tailed Sparrow (*A. caudacutus*), Seaside Sparrow (*A. maritimus*), and Willet (*Catoptrophorus semipalmatus*) have been identified as species of concern by state, provincial, and federal agencies.

Saltmarshes are important landscape features for many bird species in the Northeast during all stages of the annual cycle (breeding, migration, and wintering). Despite their low floristic diversity, saltmarshes provide a continuum of habitat from terrestrial grassland to open water, heterogeneous distribution of micro-scale habitat features, and relatively high productivity. Habitat suitability studies have indicated that for wading and water birds the presence and configuration of open-water habitat is important (Burger and Shisler 1978, Hansen 1979) while many breeding passerines are sensitive to vegetation composition, structure, and configuration, as well as tidal inundation patterns (Marshall and Reinert 1990, Reinert and Mello 1995, DiQuinzio et al. 2002).

Understanding the conservation needs of saltmarsh-breeding birds requires knowledge of habitat requirements at multiple spatial scales, including within-patch habitat variables and the landscape configuration of patches. This knowledge is also critical in evaluating the effects of conservation and restoration activities (e.g., coastal land-use policies and regulations, habitat acquisition, and habitat restoration) as well as anticipating the potential negative impacts (e.g., infilling, drainage, and disturbance) on bird communities. Changes in the landscape configuration of saltmarsh patches is a likely outcome of increasing sea-level rise and may negatively effect the population viability of Seaside Sparrows in Connecticut (Shriver and Gibbs 2004). Understanding saltmarsh-bird habitat requirements is also critical to estimating the impacts of short-term habitat changes—weather and tidal cycles—on breeding bird distribution, abundance, and population trends. The effects of habitat and landscape features on saltmarsh-bird species richness in one region may not be the same in other regions, making large-scale, multi-region, coordinated studies and syntheses an important component in determining priorities for conservation and management options within regions, as shown for grassland birds by Johnson and Igl (2001).

Herein, we review information on the effects of habitat and landscape variables on saltmarsh bird communities in New England and present new information on these patterns in the Maritimes. We describe differences in saltmarsh distribution and land use among the regions, discuss patterns of saltmarsh-habitat area, isolation, human influence, and vegetative characteristics among distinct regions in the Northeast, and determine whether these variables influence saltmarsh-bird-species richness similarly among regions.

SALTMARSHES IN NORTHEASTERN NORTH AMERICA

Along the coastline of northeastern North America from Connecticut to Prince Edward Island (Fig. 1), five biophysical regions of saltmarshes can be recognized: southern New England shore, New England Gulf of Maine, Bay of Fundy, Atlantic Coast of Nova Scotia, and Gulf of St. Lawrence (Roberts and Robertson 1986, Wells and Hirvonen 1988, Mathieson et al. 1991, Shriver et al. 2004). We recognize that finer-scale spatial differentiation within regions is also possible (Kelley et al. 1988), however, the broad-scale regions we used differ substantially in geology, tidal amplitude, latitude, and human impacts on saltmarsh habitats (Table 1). An overview of climate, physical characteristics, and rocky shore ecology of regions from Cape Cod northward is provided by Mathieson et al. (1991).

SOUTHERN NEW ENGLAND SHORE

We define the southern New England shore as that area from the western edge of Long Island Sound to the southern shore of Cape Cod, Massachusetts. Marshes within this area share a similar geologic history, tidal range, and human land-use patterns. Long Island Sound is one of the largest estuaries along the Atlantic Coast of the US. It is a semi-enclosed, northeast–southwest trending basin which is 150 km long and 30 km across at its widest point. The mean water depth is 24 m with two outlets to the sea. The eastern end of the sound opens to the Atlantic Ocean whereas the western end is connected to New York Harbor through a narrow tidal strait. Fluvial input into the sound is dominated by the Connecticut River (Poppe and Polloni 1998). Long Island Sound is an estuary, with a watershed encompassing an area of more than 41,000 km^2 and reaching into portions of Massachusetts, New Hampshire, Vermont, New York, Rhode Island, and Canada. Long Island Sound is bordered by Connecticut and Westchester County, New York to the north, New York City to the west, and by Long Island, New York, to the south.

In this region, marshes have often formed in drowned river valleys and contain considerable deposits of peat. Most tidal wetlands along the southern New England shore are saltmarshes, and summer salinity averages about 20–30 ppt. Salinity in this region varies seasonally and with proximity to major sources of fresh water such as the Connecticut, Housatonic, and Thames

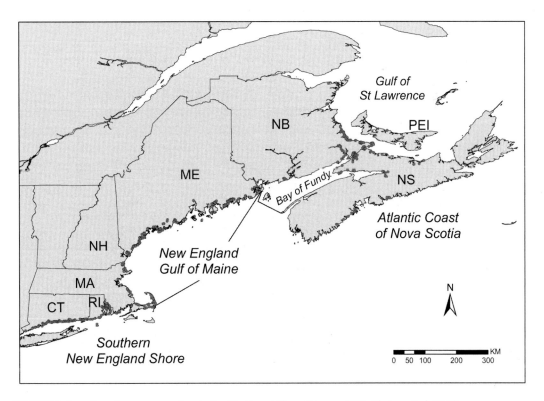

FIGURE 1. Location of survey marshes in the Northeast (from Hanson 2004, Shriver et al. 2004).

TABLE 1. BIOPHYSICAL CHARACTERISTICS OF SALTMARSHES IN THE NORTHEAST INDICATING THE IMPORTANCE OF DIFFERENT FEATURES AND FACTORS IN EACH REGION.

Feature/factor	Southern New England shore	NE Gulf of Maine	Bay of Fundy	Atlantic Coast of Nova Scotia	Gulf of Saint Lawrence
Agricultural diking	Low	Low	High	Low	Low
Tidal barriers	Moderate	Moderate	High	Moderate	High
Sea-level rise impact	Moderate	Moderate	Moderate	Moderate	High
Tidal amplitude	Low	Moderate	High	Moderate	Moderate
Sediment load	Low	Low	High	Low	Moderate
Ice	Low	Moderate	High	Moderate	High
Phragmites abundance	High	Moderate	Low	Low	Low
Low-marsh dominant vegetation	Smooth cordgrass (*Spartina alterniflora*)	Smooth cordgrass	Smooth cordgrass	Smooth cordgrass	Smooth cordgrass
High-marsh dominant vegetation	Saltmeadow cordgrass (*Spartina patens*)	Saltmeadow cordgrass	Saltmeadow cordgrass-saltmeadow rush (*Juncus gerardii*)	Saltmeadow cordgrass-saltmeadow rush	Saltmeadow cordgrass-saltmeadow rush
Upper high-marsh dominant vegetation	Marsh elder (*Iva frutescens*)	Baltic rush (*J. balticus*)	Prairie cordgrass (*S. pectinata*)	Prairie cordgrass-salt marsh sedge (*C. paleacea*)	Prairie cordgrass *C. paleacea*
Breeding bird species diversity	High	Moderate	Low	Low	Low
Breeding bird abundance	High	High	Moderate	Moderate	Moderate
Bird species designated at risk (AOU Alpha Code)	SESP, SSTS, WILL	SESP, NSTS, SSTS, WILL	None	None	None
Impoundment of tidal rivers	High	Low	Low	Low	High
Grazing	None	None	Low	Moderate	Moderate
Haying	Low	Low	Low	Moderate	Low
Agricultural ditching	None	Moderate	High	Moderate	Moderate
Mosquito ditching	High	Moderate	None	None	None
Infilling	High	Moderate	Low	Low	High
Fragmentation	Moderate	Moderate	Low	Low	High
Historical loss	Moderate-High	High	High	Low	Low
Saltmarsh remaining (hectares)	8,456	3,520	11,599	2,285	11,878
Restoration activity	High	High	Low	Low	Low
Wetland protection laws	High	High	Low	Low	Low

rivers, but is generally between 27 and 32 ppt. The Connecticut River contributes >70% of the fresh water influx, the Housatonic 12%, and the Thames 9% (Thomas et al. 2000). The tidal range of Long Island Sound increases from about 0.7 m in the east to about 2.2 m in the west, and its circulation is dominated by tidal currents (Koppelman et al. 1976). The water temperature of Long Island Sound fluctuates between ~0 C in the winter to >20 C in the summer (Thomas et al. 2000). The basic physical and biological structure of saltmarsh communities in Long Island Sound comes from smooth cordgrass (*Spartina alterniflora*) and saltmeadow cordgrass (*S. patens*).

The first European colonists in New England arrived in the early 1600s and used tidal marshes to graze livestock and provide fodder and bedding (Dreyer and Niering 1995). New England saltmarshes were both hayed and pastured continuously into the beginning of the twentieth century. Farmers began digging shallow ditches to drain standing water to increase yields of hay. By 1900, nearly 50% of the marshes in Connecticut were ditched and virtually all saltmarshes adjacent to the southern New England shore were ditched and altered by a variety of mosquito-control activities by the mid-1900s (Dreyer and Niering 1995). The effects of the ditching projects on saltmarsh ecosystem functions are difficult to determine as all but one marsh in New England has been ditched, leaving limited reference sites for comparison. The direct loss of saltmarsh habitat in New England occurred until the early 1900s through the practice of filling marshes with dredge spoil to create parking lots, industrial parks, airports, and shopping centers (Dreyer and Niering 1995).

Estimates of tidal wetland area presently occurring along Long Island Sound are just over 8,456 ha (Dreyer and Niering 1995), a 30% reduction from the pre-1880 estimates (Rosza 1995). In 1980, Connecticut began a tidal-marsh restoration program targeting systems degraded by tidal restrictions and impoundments (Rosza 1995). Such marshes became dominated by common reed (*Phragmites australis*) or cattail (*Typha angustifolia* and *T. latifolia*; Warren et al. 2002). These dense monocultures of reeds have been rapidly expanding in Connecticut's tidal wetlands with documented declines in avian diversity in plots associated with high density of reeds (Benoit and Askins 1999).

During 1960-1990, the human population along the southern New England shore increased by 40% (U.S. Census 2000). Not surprisingly, the expanding human population required increasing levels of infrastructure, particularly roads. Estimates in 1999 were that 13 roads/km^2 existed in coastal counties on Long Island Sound (Connecticut Department of Transportation, pers. comm.). Roads may reduce regional biodiversity by modifying wetland hydrology (Andrews 1990, Trombulak and Frissell 2000), facilitating invasive species (Cowie and Warner 1993, Lonsdale and Lane 1994, Greenberg et al. 1997), and increasing access to wildlife habitats by humans (Young 1994).

New England Gulf of Maine

The Gulf of Maine watershed encompasses land within Massachusetts, New Hampshire, Maine, Quebec, New Brunswick, and Nova Scotia, an area of 165,185 km^2. Over 5,000,000 people live around the Gulf of Maine. The entire population of Maine, 1,200,000 people, lives within the Gulf of Maine watershed and millions of tourists visit the Gulf of Maine every year. In New England, the Gulf of Maine extends from the tip of Cape Cod, Massachusetts (42°04'N, 70°15'W) to the St. Croix estuary in Calais, Maine (44°54'N, 66°59'W). The coast of Maine has 5,600 km of tidally influenced shoreline and is the third longest shoreline in the US. Tides along the Maine coast are semidiurnal and range from 2.6 m at Kittery to 5.6 m at Calais.

Generally, the amount of saltmarsh habitat in the New England Gulf of Maine decreases with increasing latitude. Saltmarsh alteration has a long history in all of New England including the Gulf of Maine. Saltmarsh habitat for the entire coast of Massachusetts was estimated at 12,600 ha in the 1990s (Koneff and Royle 2004; Koneff and Royle, unpubl. data). The majority of this saltmarsh habitat occurs within the Gulf of Maine. In New Hampshire and Maine, saltmarsh habitat in the 1990s was estimated at 1,900 and 5,200 ha, respectively (Koneff and Royle, unpubl. data). Jacobsen et al. (1987) estimated 7,980 ha of saltmarsh occurred in Maine based on 1960 aerial photographs. The Great Marsh in Massachusetts (>6,800 ha), is the largest marsh complex in New England and encompasses ~54% of saltmarsh habitat from Cape Ann Massachusetts to southern New Hampshire. Most of the other large (>100 ha) saltmarshes in New England occur on the south shore of Massachusetts. In southern Maine between Kittery and Cape Elizabeth, the two largest marshes (>1,000 ha) are at Webhannet Estuary and Scarborough Marsh. Cape Elizabeth, just south of Portland Maine, is a geologic division in coastal habitats for the Gulf of Maine. North of Cape Elizabeth, the Maine coast is dominated by rocky intertidal habitat with limited and patchily distributed saltmarsh habitat while south of Cape Elizabeth, where wave energy and tidal range are lower, the coast is dominated by sandy beaches and saltmarshes (Kelley et al. 1988).

BAY OF FUNDY

The Bay of Fundy is the northeast extension of the Gulf of Maine, located between the Canadian provinces of New Brunswick and Nova Scotia, and covers an area of 16,000 km^2. The Bay of Fundy is a macro-tidal system with a tidal range of 6 m in the outer bay and 16 m at the head of the Bay in Cumberland and Minas Basins (Desplanque and Mossman 2000, 2004). A single tidal flow into the Bay of Fundy involves 104 km^3 of water. During the day, therefore, the volume of water moving in and out of the Bay of Fundy is equivalent to four times the combined discharge of the world's rivers (Desplanque and Mossman 2004).

Higher elevations in Bay of Fundy saltmarshes are typically dominated by saltmeadow cordgrass (Ganong 1903, Chapman 1974, Van Zoost 1970, Morantz 1976, Thannheiser 1981, Thomas 1983, Chmura 1997, Van Proosdij et al. 1999). Only 3–4% of the tides per year for an average duration of 30 min, flood the high marsh in the upper Bay of Fundy (Palmer 1979, Gordon et al. 1985, Van Proosdij et al. 1999). Low marsh is dominated by smooth cordgrass and can be found at elevations between mean high water (MHW) and approximately 1.2 m below MHW (Van Proosdij et al. 1999). A mid-marsh zone which is a transitional zone between high marsh and low marsh has also been described (Wells and Hirvonen 1988, Van Proosdij et al. 1999) and can be dominated by goose tongue (*Plantago maritime*) in some marshes (Chmura et al. 1997). Another climatic-physical feature of Bay of Fundy saltmarshes is the role of ice in creating saltmarsh pannes, exporting detritus, and importing sediment (Bleakney and Meyer 1979, Gordon and Desplanque 1983, Gordon and Cranford 1994, Van Proosdij et al. 2000). The marshes in the upper Bay of Fundy differ from those in the other regions of the Northeast in that they are influenced by large amounts of available sediments. Water-column sediment concentrations typically range from 50–100 mg/L (Amos and Long 1987) and during fall storms, measurements of 6–7 g/L have been recorded (Amos and Tee 1989). Surface-water salinities are 30–33 ppt, with monthly mean water temperatures being affected by the Labrador Current and ranging from 0.6–13.0 C (Mathieson et al. 1991, Davis and Browne 1996a).

European settlement along the shores of the Bay of Fundy began in 1604. The process of diking and draining saltmarsh for conversion to agricultural fields was initiated in the 1630s along the Annapolis River and in the 1670s in the upper Bay of Fundy, with dikes being maintained to this day (Milligan 1987, Bleakney 2004). By 1920, 80% of all saltmarsh in the Maritimes was converted to agricultural land through diking (Reed and Smith 1972), a land use unique to the Bay of Fundy saltmarshes compared to other regions in the Northeast. The draining of wetlands through the use of dikes and water-control structures created 222,000 ha of agricultural land in Canada (Papadopoulus 1995). Currently 35,000 ha of dikeland are in the Bay of Fundy created through conversion from saltmarsh. In recent years most of the dikeland has been used for forage production or pasture (Collette 1995). This non-intensive agricultural use of the dikeland can provide habitat for grassland birds (Nocera et al. 2005).

Recently, dikeland has reverted back to saltmarsh in the upper Bay of Fundy when dikes and water-control structures failed and were not repaired or replaced. By the 1980s <65% of original saltmarsh area remained behind dikes compared to 80% in the 1920s (Milligan 1987, Austin-Smith 1998). Of New Brunswick's 141 Bay of Fundy saltmarshes, 35% were formerly diked (Roberts 1993). The Maritime Wetlands Inventory (Hanson and Calkins 1996) estimates that in the early 1980s, 7,793 ha of saltmarsh were in the Bay of Fundy (Table 2).

ATLANTIC COAST OF NOVA SCOTIA

The Atlantic Coast of Nova Scotia is a high-energy system, experiencing the effects of ocean swells, with a maximum tidal range of 2 m (Wells and Hirvonen 1988, Davis and Browne 1996b). Monthly mean surface-water temperatures are 0.9–15.0 C with salinities ranging from 32.0–33.5 ppt (Davis and Browne 1996a). The Atlantic Coast of Nova Scotia is a drowned coastline and has been subsiding for 7,000 yr (Fensome and Williams 2001) and is characterized by drumlins and terminal moraines (Roland 1982). Saltmarshes along this coastline are most often small wetlands protected by islands, or part of a few large complexes associated with estuaries (Scott 1980, Chagué-Goff et al. 2001). The vegetative zones in Atlantic Coast saltmarshes consist of smooth cordgrass in low marsh, and saltmeadow cordgrass, saltmeadow rush (*Juncus gerardii*), and Cyperaceae in the high marsh (MacKinnon and Scott 1984, Wells and Hirvonen 1988, Austin-Smith et al. 2000). Historically, little diking has occurred along the Atlantic Coast (Kuhn-Campbell 1979). In southwestern Nova Scotia where the coastal plain gradually slopes to below sea level, saltmarshes were hayed and

TABLE 2. LANDSCAPE LEVEL DESCRIPTORS OF SALTMARSHES IN THE ATLANTIC COAST OF NOVA SCOTIA, BAY OF FUNDY, AND GULF OF ST. LAWRENCE REGIONS.

Maritime wetland inventory data[a]	Atlantic	Bay of Fundy	Gulf of St. Lawrence
Number of saltmarshes	598	574	2,106
Total marsh area	6,091 ha	7,793 ha	11,880 ha
Median marsh size (hectares)	4.3 ha	5.9 ha	2.6 ha
Mean marsh size ± SE (hectares)	10.2 ± 0.83 ha	13.6 ± 0.91 ha	5.6 ± 0.21 ha
Number of marshes <5 ha	330	262	1463
Total area of marshes <5 ha	745 ha	587 ha	2,831 ha
Number of marshes 5–10 ha	125	116	343
Total area of marshes 5–10 ha	875 ha	843 ha	2,410 ha
Number of marshes 10–20 ha	74	90	183
Total area of marshes 10–20 ha	1,033 ha	1,260 ha	2,440 ha
Number of marshes 20–50 ha	47	69	98
Total area of marshes 20–50 ha	1,423 ha	2,214 ha	2,813 ha
Number of marshes >50 ha	22	37	19
Total area of marshes >50 ha	2,015 ha	2,889 ha	1,386 ha
Study marshes			
Number	16	72	72
Percent with adjacent dike land	0	29	6
Percent with old dikes in marsh	0	36	14
Percent with old ditches in marshes	0	47	37
Percent with ponds	69	61	85
Percent with reeds (Phragmites)	0.0	4.2	5.6
Percent with prairie cordgrass (Spartina pectinata)	47	64	74

[a] Maritime Wetland Inventory data obtained from Hanson and Calkins (1995).

grazed without the use of dikes. For much of the Atlantic Coast of Nova Scotia, the land rises steeply from the shoreline and there has been little infilling of saltmarsh for construction of human infrastructure. The Atlantic Coast of Nova Scotia is estimated to have 6,090 ha of saltmarsh (Table 2).

GULF OF SAINT LAWRENCE

The Gulf of St. Lawrence is a low-energy system compared to the Atlantic Coast of Nova Scotia and has a much smaller tidal range compared to the Bay of Fundy region (see Roland 1982). Tidal ranges are 1–4 m with mixed components of semidiurnal and diurnal influences. In the western section the tides are mainly diurnal with a period of 25 hr hence on some days tides can remain high for 12 hr (Davis and Browne 1996a). The shallow waters of the Gulf of St. Lawrence result in surface water temperatures ranging from 1.5–19.7 C, with maxima of >22 C being observed. In coastal areas, salinities of 25.2–28.0 ppt occur above the thermacline (Mathieson et al. 1991, Davis and Browne 1996a). The Gulf of St. Lawrence coast consists of a low-elevation plain (Fensome and Williams 2001) and is influenced by the transport of sandy materials, with many barrier islands, dunes, lagoons, and barachois ponds present (Reinson 1980). Residential development resulting in the infilling of saltmarshes and alteration of adjacent habitat, is the primary land use affecting saltmarsh habitat in the Gulf of St. Lawrence due to the presence of sandy beaches, warm water, and flat topography (Roberts 1993, Maillet 2000, Milewski et al. 2001). Gulf of St. Lawrence saltmarshes were not subject to the intense diking that Bay of Fundy marshes were, although some old hand-dug dikes can still be seen. Coastal marshes were, however, important to early agricultural activities (Hatvany 2001). Marshes were ditched to drain ponds and created drier soils for livestock and equipment as they were grazed and hayed.

The vegetative community of Gulf of St. Lawrence saltmarshes has been described as smooth cordgrass in the low marsh, saltmeadow cordgrass in the middle marsh and saltmeadow rush in the high marsh (Wells and Hirvonen 1988, Roberts 1989). Salt marsh sedge (Carex paleacea) and prairie cordgrass (Spartina pectinata) in the higher elevations of Gulf of St. Lawrence saltmarshes distinguishes them from New England saltmarshes (Gauvin 1979, Olsen et al. 2005). In comparison to Bay of Fundy or New England saltmarshes, the vegetative zones and ecology of Gulf of St. Lawrence marshes have received little study to date. The Gulf of St. Lawrence has 11,880 ha of saltmarsh (Table 2). The combination of relatively low land elevations, small tidal variation, intensive coastal-zone development, and erosive soils makes this area highly susceptible to sea-level-rise damage (Shaw et al. 1994). This seems to be confirmed by comparison of soil accretion rates to recent sea-level rise at some sites, but further study is necessary (Chmura and Hung 2004).

METHODS

Data to estimate avian-species richness, abundance, dominant vegetation, surface water area, previous human activity, adjacent land use, and proximity to adjacent saltmarsh were collected on saltmarshes in the Maritimes using techniques similar to those previously used throughout the Northeast (Benoit and Askins 1999, 2002; Shriver et al. 2004).

Maritime saltmarsh vegetative composition was characterized by estimating the percent areal cover of each macrophyte species in a 5-m radius centered on the survey point, and a 5-m-wide transect to the first survey point, between subsequent survey points, and from the last survey point to the marsh edge. The percent cover of salt-meadow vegetation was calculated by summing the percent cover of saltmeadow cordgrass, prairie cordgrass, salt marsh sedge, and saltmeadow rush. The presence of reeds, no longer maintained (old) ditches, or dikes were noted if they occurred within the marsh. Wetland inventory maps (Hanson and Calkins 1996), National Topographic Series maps (1: 50,000 scale) and the most recent aerial photographs were used to determine landscape level features. Marsh boundaries were determined by paved roads or water channels >100-m wide. These definitions of marsh boundaries ensure that the saltmarsh is a relatively homogeneous patch within the landscape matrix (Forman 1995). A proximity index, similar to Gustafson and Parker (1994), was estimated using wetland inventory maps and derived by summing the ratio of size (hectares) of an adjacent saltmarsh divided by its distance (kilometers) to the study marsh for all marshes within 1 km of the boundary of the study marsh (PI= $\sum$ [area in hectares] of nearby marsh i/[distance in kilometers] to nearby marsh i) for all marshes within 1 km of study marsh). This proximity index was based on the total area of an adjacent saltmarsh, not just the area within the 1 km buffer, and hence PI >10 was possible, unlike the proximity index used by Shriver et al. (2004). The number of buildings within a 500-m radius was determined as an index of human disturbance. The number and total area (hectares) of ponds in the marsh, the presence of dikes or ditches in the marsh, and the presence of dikelands within 250 m were determined based on aerial photographs.

To survey resident breeding-bird communities, 100-m-radius point counts (1–46 points/marsh) were established within each marsh and each point was visited at least twice from 10 June–30 July, 2000–2002, with at least 10 d between visits (Ralph et al. 1995). The number of points located in a marsh was determined by marsh size, with more points in larger marshes. We attempted complete coverage of the survey marsh. All point centers were >200 m from any other point center and at least 50 m from an upland edge. For small marshes, where the 100-m-radius point extended into adjoining upland habitat, only birds detected within the marsh were counted.

Observers, including volunteers, sampled for 10 min at each point and recorded all birds seen and heard within 100 m. Surveys were conducted from dawn to 1100 H on days with low wind (<10 km/hr) and clear visibility. All observers had experience in bird identification (by sight and sound) prior to this study, with additional training in identification of saltmarsh bird species if required. Differences among observers in ability to see and hear birds were not quantified.

Species richness in marshes was based on three guilds: (1) obligate wetland birds, (2) wading birds, and (3) passerines, similar to Shriver et al. (2004). Both the total number of species detected per marsh (total species) and the mean number of species detected per survey point in each marsh (species richness) were considered as response variables.

STATISTICAL ANALYSIS

General linear models (GLM) were used to determine which marsh-level and landscape-level features were significantly related to the mean number of birds or number of species observed per survey point in each marsh (SAS 2000). The mean number of individuals per survey point will simply be referred to as density. Separate models were developed for Nelson's Sharp-tailed Sparrow, Willet, and Savannah Sparrow densities, and species richness. Proportional data were arcsine-transformed prior to analysis, count data were square-root transformed, and other variables log-transformed prior to statistical analysis to improve normality, and reduce heterogeneity of variance (Zar 1999). W values indicated normal or near normal distributions (Proc UNIVARIATE, SAS 2000).

RESULTS

SALTMARSH CHARACTERISTICS

Surveys were conducted on 161 saltmarshes throughout the Maritimes. Saltmarshes in the previously described regions of the Maritimes differed in size distribution, the extent of human disturbance, and vegetative composition (Table 2). Saltmarshes surveyed in the Bay of Fundy were larger compared to saltmarshes surveyed along the Atlantic Coast, consistent with the size distribution of saltmarshes reported in the Maritime Wetlands Inventory (Table 2). Old dikes and ditching, adjacent dikeland, and reeds were not present in Atlantic Coast study marshes but were present in study marshes in the other regions (Table 2). Approximately three-quarters of the surveyed marshes in the Gulf of St. Lawrence had prairie cordgrass present (Table 3).

The mean number of buildings within 500 m for Gulf of St. Lawrence study marshes was 50 compared to 33 and 36 for saltmarshes along the Atlantic Coast and Bay of Fundy, respectively (Table 3). Gulf of St. Lawrence saltmarshes also had a greater number and greater total area of ponds compared to saltmarshes in the other two regions.

Nelson's Sharp-tailed Sparrow density was similar for study marshes among all three regions (Table 4). The density of Willets was higher for Atlantic Coast marshes compared to Gulf of St. Lawrence, and Bay of Fundy marshes (Table 4). Savannah Sparrow density was lower in Bay of Fundy marshes compared to those along the Gulf of St. Lawrence or Atlantic Coast.

LANDSCAPE AND PATCH-LEVEL EFFECTS

Individual bird species differed in their response to landscape and patch-level habitat characteristics. Marsh area was an important determinant of Nelson's Sharp-tailed and Savannah sparrow densities and avian-species richness (Table 4). Nelson's Sharp-tailed Sparrow density increased with marsh size up to 10 ha (Fig. 2, Table 5). Nelson's Sharp-tailed Sparrow density in marshes <5.0 ha (0.33 ± 0.07, $\bar{x} \pm$ SE) was less than that for marshes ≥5 ha (1.07 ± 0.09, $P < 0.001$), and was less in marshes ≤10.0 ha (0.50 ± 0.095, $\bar{x} \pm$ SE) compared to marshes >10.0 ha (1.2 ± 0.096, $P < 0.001$). Willet density was not influenced by marsh area but was positively influenced by pond area. In these study marshes a high correlation occurred between saltmarsh area and pond area ($r^2 = 0.73$) and density of Willets was positively associated with marsh area in models which included marsh area but not pond area (Hanson 2004). Savannah Sparrow density was negatively affected by pond area.

The average amount of saltmarsh-meadow vegetation positively influenced Willet density and species richness (Table 4). The number of dwellings within 500 m was positively correlated with species richness. The proximity index or the number of dwellings within 500 m of the study marsh did not affect any of the species habitat use response variables.

The presence of old dikes and old ditches on the marsh itself did not affect the density of Willets or Savannah Sparrows or species richness. Nelson's Sharp-tailed Sparrow, Willet

TABLE 3. SUMMARY STATISTICS OF STUDY SALTMARSHES IN THE ATLANTIC COAST OF NOVA SCOTIA, GULF OF ST. LAWRENCE, AND BAY OF FUNDY REGIONS.

	Atlantic Coast		Bay of Fundy		Gulf of St. Lawrence	
	Mean	SE	Mean	SE	Mean	SE
Marsh area (hectares)	19.98	4.78	52.75	10.97	24.12	4.68
N ponds	9.56	5.08	12.29	2.34	24.88	5.51
Pond area (hectares)	5.41	1.76	5.91	1.08	9.26	1.54
Proximity index	3.00	0.85	19.18	3.19	10.44	1.57
N dwellings <125 m	6.88	2.22	5.21	1.19	8.47	1.93
N dwellings 125–250m	7.88	1.90	8.38	2.67	10.56	2.51
N dwellings 250–500m	17.88	3.50	22.51	5.46	31.29	6.74
Total dwellings <500m	32.63	5.51	36.10	8.59	50.32	10.03
NSTS[a]/marsh	6.48	2.35	5.74	0.92	4.74	0.84
NSTS per survey point	0.97	0.19	0.85	0.13	0.90	0.09
WILL[a]/marsh	6.63	2.28	0.63	0.25	2.99	0.62
WILL per survey point	1.20	0.22	0.14	0.06	0.69	0.11
SAVS[a]/marsh	7.53	4.35	3.25	0.71	4.17	1.28
SAVS/survey point	0.80	0.37	0.42	0.08	0.89	0.12
Percent cover of salt-meadow vegetation	33.25	4.99	53.86	2.88	52.59	2.53
N passerine species	5.06	0.75	4.19	0.34	5.69	0.40
N wetland species	7.13	1.43	2.97	0.40	5.99	0.51
N wader species	0.94	0.11	0.49	0.07	0.76	0.07
N gull species	1.75	0.17	0.79	0.11	1.18	0.13
Total N of species/marsh	14.88	2.24	8.44	0.79	13.63	0.93
Mean N of species/survey point	7.32	0.85	3.92	0.29	8.40	0.53

[a] NSTS = Nelson's Sharp-tailed Sparrow; WILL = Willet; SAVS = Savannah Sparrow.

TABLE 4. RESULTS FROM GLMS FOR EVALUATING THE IMPORTANCE OF MARSH AND LANDSCAPE LEVEL HABITAT DESCRIPTORS ON THE MEAN NUMBER OF NELSON'S SHARP-TAILED SPARROWS, WILLETS, SAVANNAH SPARROWS AND SPECIES RICHNESS PER SURVEY POINT. RESULTS FROM MULTIPLE PAIR-WISE COMPARISONS AMONG REGIONS ALSO PRESENTED.

	Nelson's Sharp-tailed Sparrow		Willet		Savannah Sparrow		Species Richness	
	F	Pr > F	F	Pr > F	F	Pr > F	F	Pr > F
Model	35.20	<0.01	16.9	<0.01	17.38	<0.01	210.61	<0.01
Marsh area	20.04	<0.01	0.56	0.46	15.75	<0.01	4.22	0.42
Pond area	0.55	0.45	5.08	0.03	15.29	<0.01	0.61	0.44
Proximity index	0.03	0.87	0.55	0.46	0.42	0.52	0.86	0.35
Meadow cover	0.05	0.81	0.20	0.66	2.14	0.15	0.02	0.89
Old ditch	0.01	0.93	0.19	0.66	0.63	0.43	0.20	0.65
Old dykes	1.10	0.30	0.03	0.86	0.06	0.81	0.00	0.98
Dykeland nearby	12.17	<0.01	10.6	<0.01	2.61	0.11	6.91	0.01
Dwellings <500 m	0.04	0.84	0.56	0.45	1.67	0.20	4.01	0.05
Region	3.79	<0.01	26.25	<0.01	12.11	<0.01	32.31	<0.01
Model r^2	0.30		0.33		0.24		0.36	
BOF vs ATL [a]	NS		P < 0.05		NS		P < 0.05	
BOF vs GSL [a]	NS		P < 0.05		P < 0.05		P < 0.05	
ATL vs GSL [a]	NS		P < 0.05		NS		NS	

[a] Results from multiple pair-wise comparisons among regions: BOF = Bay of Fundy; ATL = Atlantic Coast of Nova Scotia; GSL = Gulf of St. Lawrence.

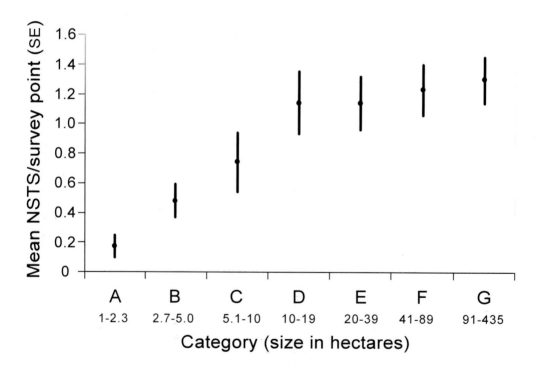

FIGURE 2. Mean (± SE) number of Nelson's Sharp-tailed Sparrows (NSTS) per survey point in relation to marsh size.

TABLE 5. SUMMARY STATISTICS OF STUDY SALTMARSHES ACCORDING TO MARSH SIZE.

Variable	Class A Mean	Class A SE	Class B Mean	Class B SE	Class C Mean	Class C SE	Class D Mean	Class D SE	Class E Mean	Class E SE	Class F Mean	Class F SE	Class G Mean	Class G SE
Pond area (hectares)	0.35	0.13	2.02	0.96	2.66	0.58	4.21	0.78	9.48	1.50	19.16	3.84	20.43	4.57
Marsh area (hectares)	1.55	0.13	3.72	0.19	7.47	0.27	13.96	0.60	28.85	0.85	60.72	3.14	222.85	31.48
Proximity index	4.28	2.15	4.51	1.85	8.19	2.45	16.14	5.48	18.50	4.34	22.00	5.57	25.05	5.65
N dwellings	42.75	11.60	44.43	15.17	43.41	18.36	36.21	12.67	37.62	10.88	27.11	6.46	75.79	35.62
Percent cover salt meadow	59.78	6.64	55.24	7.08	54.86	4.26	53.11	3.57	47.23	3.08	42.43	5.06	43.99	4.55
N NSTS[a]/marsh	0.17	0.08	0.79	0.25	2.00	0.62	3.87	0.88	7.13	1.18	11.85	2.59	16.56	2.36
N NSTS/point	0.17	0.08	0.48	0.11	0.74	0.20	1.14	0.21	1.14	0.18	1.23	0.17	1.30	0.16
N WILL[a]/marsh	0.42	0.19	1.05	0.46	1.36	0.52	1.76	0.35	3.16	1.19	4.87	1.98	4.26	1.69
N WILL/point	0.42	0.19	0.57	0.23	0.56	0.18	0.51	0.10	0.39	0.10	0.68	0.28	0.37	0.17
N SAVS[a]/marsh	0.28	0.13	1.16	0.37	1.88	0.66	1.75	0.53	5.07	1.03	6.51	3.78	17.07	5.92
N SAVS/point	0.28	0.13	0.94	0.31	0.64	0.14	0.62	0.22	0.73	0.13	0.50	0.21	1.06	0.26
N passerine species	3.20	0.56	3.19	0.63	4.17	0.46	5.00	0.65	6.50	0.57	5.22	0.64	7.57	0.76
N wetland species	2.10	0.62	1.62	0.37	4.52	0.89	4.33	0.66	6.38	0.76	7.39	1.19	7.00	1.24
N wader species	0.45	0.11	0.52	0.13	0.62	0.12	0.54	0.12	0.79	0.08	0.83	0.19	0.86	0.18
N gull species	0.60	0.18	0.71	0.18	1.07	0.24	0.75	0.15	1.38	0.19	1.17	0.22	1.86	0.25
Total N species/marsh	6.35	1.28	6.05	1.10	10.38	1.52	10.63	1.34	15.06	1.30	14.61	1.88	17.29	1.97
N Species/survey Point	6.18	1.26	5.19	0.88	6.86	0.82	6.14	0.74	7.19	0.78	5.41	0.76	5.96	0.69
Marsh size	1.0–2.3 ha		2.7–5.0 ha		5.1–10.0 ha		10.1–18.7 ha		20–39 ha		41–89 ha		91–435 ha	
Sample size	20		21		29		24		34		18		14	

[a] NSTS = Nelson's Sharp-tailed Sparrow; WILL = Willet; SAVS = Savannah Sparrow.

densities and species richness were positively associated with the presence of adjacent dike land (Table 4).

DISCUSSION

MARSH AREA

Marsh area had a consistent and positive influence on many breeding bird species occurrences and species richness in all regions of the Northeast. The occurrence of saltmarsh-obligate species was positively related to marsh area in the Maritimes, similar to previous findings for the New England Gulf of Maine and southern New England shore (Benoit and Askins 2002, Shriver et al. 2004). In the Maritimes, Nelson's Sharp-tailed Sparrow density was positively correlated with salt-marsh area. These findings are consistent with the findings for the effect of marsh area on the occurrence of Seaside Sparrows and Saltmarsh Sharp-tailed Sparrows in the New England Gulf of Maine and the southern New England shore (Benoit and Askins 2002, Shriver et al. 2004). Species richness was also greater on larger marshes than smaller marshes in the Maritimes and New England (Shriver et al. 2004). Shriver et al. (2004) observed that 13 of 14 species were more likely to be detected on larger marshes compared to smaller marshes. The number of species in the saltmarsh breeding bird communities declined with increasing latitude. The southern New England shore had the greatest species richness for saltmarsh breeding birds while species richness was lowest in the Bay of Fundy. Most of the wading bird species observed in U.S. saltmarshes are absent from the Maritimes, due to geographic range limits.

Both absolute and relative marsh size may influence bird distribution. Willets in Connecticut were absent in marshes <138 ha (Benoit and Askins 2002), whereas in the Maritimes, Willets were observed in smaller marshes, including a 2.0 ha saltmarsh that contained 0.40 ha of total pond area. Habitat use does not always equate with habitat quality (Van Horne 1983), and these different results may be due to the marsh patch-size distribution or low Willet populations (Benoit and Askins 2002). The affinity of Willets (and other shorebird species) for ponds highlights the importance of including measures of open water, rather than just marsh area, in analyses of habitat use as well as in conservation-restoration activities. Previous analyses indicate that Willets were more likely to occur on larger marshes than smaller marshes in all regions of the Northeast when pond area is not included in the model (Benoit and Askins 2002, Hanson 2004, Shriver et al. 2004). Erwin et al. (1994) observed the highest shorebird densities (including Willets) on ponds >0.10 ha. In the Maritimes, the density of Willets was positively correlated with pond area. In most marshes total pond area will be highly correlated with marsh area, and hence these findings collectively highlight the importance of large marshes as wildlife habitat.

In the Maritimes, marsh size was not important for facultative or opportunistic users of saltmarshes, such as Savannah Sparrow, perhaps because these species are also using several non-saltmarsh habitats, including upland grassland and dune ridges. Differences among species in the importance of marsh area are consistent with findings for grassland (Bakker et al. 2002) and forest birds (Mitchell et al. 2001) where individual species demonstrated scale-dependent differences in how they perceived habitat and landscape structure, and that no single scale was appropriate for assessing habitat. The importance of marsh size in different studies for different species in the Northeast suggests that large coastal marshes should remain intact and that the wildlife habitat benefits of several small saltmarsh restoration projects may not be as great as a single large project.

LANDSCAPE CONTEXT

The effect of marsh isolation on obligate saltmarsh breeding birds differed among regions and species. Marsh proximity was not an important variable in explaining Willet, Nelson's Sharp-tailed Sparrow, or Savannah Sparrow densities in the Maritimes and no difference was found in saltmarsh-breeding bird species richness among isolated or contiguous marshes in this region. This pattern was consistent with findings for Nelson's Sharp-tailed Sparrows in the New England Gulf of Maine where this species was not shown to be influenced by marsh isolation. The effect of marsh isolation on Willets was not consistent among regions. In all regions, except the New England Gulf of Maine, Willet occurrence or density was not influenced by marsh isolation. In the New England Gulf of Maine, Willets were more likely to occur on marshes in close proximity to other marshes (Shriver et al. 2004). The presence of Saltmarsh Sharp-tailed Sparrows, a species present along the southern New England shore and the New England Gulf of Maine, was also influenced by marsh isolation in the New England Gulf of Maine but not along the southern New England shore. Alternatively, Seaside Sparrows were positively associated with the proximity to other saltmarshes along the southern New England shore and only marginally in the New England Gulf of Maine (Shriver et al. 2004). The importance of marsh isolation likely depends on the distribution and characteristics of habitat patches within the landscape and the breeding ecology of the species in question. Patterns of marsh isolation on the saltmarsh breeding birds were not as consistent as the effects of marsh area. If patch size is large in relation to the home range of the species then, all other things being equal, proximity to other habitat patches may not be important. Correlations between the distribution of habitat patches across the landscape and within patch habitat quality will also influence the apparent importance of proximity indices in these analyses. Numerous and dispersed small saltmarshes in the New England Gulf of Maine, especially in northern Maine (Jacobsen et al. 1987), may also influence this relationship.

In comparison to eastern forested landscapes or western grassland landscapes, the tidal wetlands of the Northeast are relatively small, discrete habitats, unevenly distributed along the coast. Saltmarsh birds in many locales may be forced to use only one marsh because others are not available. The discrete, insular nature of saltmarshes may also explain why the number of dwellings near the marsh had no impact on densities of saltmarsh birds in the Maritimes. Shriver et al. (2004) did not observe an effect of road density on species richness in either the southern New England shore or the New England Gulf of Maine. The lack of correlation to these indices of human disturbance does not minimize the importance of the upland edge as nesting cover for species such as Willet and Nelson's Sharp-tailed Sparrow in the Maritimes (A. Hanson, pers. obs.).

Another unique finding from the Maritimes is that although proximity to adjacent saltmarsh habitat did not influence Nelson's Sharp-tailed Sparrow or Willet densities, the presence of adjacent dikeland habitat did. Nelson's Sharp-tailed Sparrows use tall-grass cover in agricultural areas, and riverine floodplains in the Maritimes (Townsend 1912, Conner 2002, Nocera et al. 2005). Willets will also nest in dikeland pasture (A. Hanson, pers. obs.) as well as considerable distances from estuarine feeding areas (Hansen 1979).

Within-Marsh Characteristics

Tidal flooding is an important proximate and ultimate determinant of nest success and hence nest-site selection by ground-nesting birds in saltmarshes (Reinert and Mello 1995, Shriver 2002). Singing male Nelson's Sharp-tailed Sparrows were associated with females who remain relatively close to the nesting area (Shriver 2002). This results in males using the higher elevations of the marsh, and they will use old fence posts, bushes, or spruce trees as singing perches if available adjacent to nesting areas (A. Hanson, pers. obs.). The species of plants associated with higher elevations of the marsh depend on absolute elevations of the marsh compared to water levels. Some marshes have only smooth cordgrass and saltmeadow cordgrass zones, whereas other marshes may also have zones of higher elevations that contain saltmeadow rush, salt marsh sedge, Baltic rush (*Juncus balticus*), or prairie cordgrass. Plant species associated with singing male sparrows may vary across marshes or regions depending on the plant species present in the highest elevations of the marsh. Habitat suitability and abundance of birds in Long Island Sound saltmarshes was largely influenced by birds using marshes with reeds less often than other marshes (Benoit and Askins 1999, Shriver et al. 2004). The limited distribution and abundance of reeds in the Maritimes presently precludes this relationship.

Implications for Bird Conservation

As described earlier, considerable differences exist in the nature of saltmarshes throughout the Northeast. Saltmarshes have been lost due to drainage or infilling and modified by activities such as ditching. The extent and intensity of such activities varies throughout the Northeast. Remaining saltmarshes may not be representative of past conditions and habitat use can only be based on habitat types currently available. Therefore, the observed differences in habitat use across regions may be due to differences in the amount of various habitat types available and not necessarily due to differences in habitat selection.

Nelson's Sharp-tailed Sparrows seem to be present in all moderately sized marshes in the Maritimes and 48% of saltmarshes in the New England Gulf of Maine. In the Maritimes, they use dykeland habitats and seem to be equally abundant in agricultural fields and floodplain grasslands (Conner 2002, Nocera et al. 2005). The data collected in the Maritimes support the recommendation that Nelson's Sharp-tailed Sparrows be designated as Not at Risk in Canada (Rompre et al. 1998).

Willets were hunted by market hunters almost to extirpation in the Northeast by 1910, with the Willet population north of Virginia reduced to a small breeding population in southern Nova Scotia (Tufts 1986). Willet populations increased throughout Nova Scotia after the 1920s and the passage of the Migratory Bird Convention Act, but were not reported to be nesting again in New Brunswick until 1966 and for Prince Edward Island not until 1974 (Erskine 1992). They returned to the southern New England shore and New England Gulf of Maine during the 1970s and 1980s (Lowther et al. 2001). The absence of Willets from many saltmarshes in the Maritimes, especially the Bay of Fundy, may reflect unsuitable or unused habitat. In Connecticut, it is thought that low population size results in much unused habitat (Benoit and Askins 2002). Low Willet populations or availability of suitable habitat are both cause for concern because of sensitivity to environmental catastrophe or habitat degradation. The lack of ponds on many saltmarshes in the upper Bay of Fundy may be due to vestigial dikes that preclude ice rafting and ditches that promote drainage of ponds. Without an understanding of natural pond-formation processes, direct human intervention to create ponds on saltmarsh by excavation may be considered habitat alteration and not restoration.

Saltmarsh Sharp-tailed Sparrows occur in this region from the Weskeag Marsh in Maine to the southern portion of the Southern New England Shore (Hodgman et al. 2002). This species is a high conservation priority because of its limited breeding distribution, the high proportion of its breeding population that occurs in the Northeast, and the threats to its coastal habitats (Pashley et al. 2000). Saltmarsh Sharp-tailed Sparrows were detected on 70% of the surveyed marshes along the southern New England shore and were less likely to occur on marshes invaded by reeds (Benoit and Askins 1999, Shriver et al. 2004). Conservation of this species in New England will likely be influenced by saltmarsh restoration projects that are designed to reduce invasive plant cover and remove tidal restrictions. The success of these projects in relation to increasing habitat quality for Saltmarsh Sharp-tailed Sparrows may be delayed, however, due to the time lag in vegetative response after initial flooding (DiQuinzio et al. 2002). Even though this species occurs on large percentage of marshes, its promiscuous mating system (Greenlaw and Rising 1994) and potentially male-biased sex ratio (Shriver, unpubl. data) may effectively reduce the number of source populations. Given that these sparrows are obligate saltmarsh birds, consideration of Saltmarsh Sharp-tailed Sparrow reproductive success should be incorporated into saltmarsh restoration projects that are designed to increase or restore marsh integrity.

Seaside Sparrows in the Northeast are distributed from southern New Hampshire (occasionally breeding in Maine) south to the southern portion of the southern New England shore. This species was very sensitive to marsh size along the southern shore of New England (Benoit and Askins 1999, Shriver et al. 2004) and only occurred on 15% of the surveyed marshes in this region (Shriver et al. 2004). Unlike Saltmarsh Sharp-tailed Sparrows, Seaside Sparrows are monogamous and territorial (Post and Greenlaw 1994), a contrast in behavioral strategies that may explain differences in the effects of marsh size on the occurrence and density of these two species. Seaside Sparrows tend to require larger marshes to establish breeding populations which are less common in the portion of the region where this species occurs. Shriver and Gibbs (2004) modeled the potential effects of sea level on the population viability of this species and found a significant increase in the probability of extinction given three estimates of sea-level rise. The ability of inland saltmarsh expansion with rising sea levels will be necessary in the coming decades if we are to conserve viable populations of saltmarsh breeding birds.

Saltmarsh conservation in New England has been facilitated through the enactment of various federal and state policies and regulations. In Canada, the Federal Policy on Wetland Conservation was implemented in 1991 and provincial governments in the Maritimes have recently passed wetland protection policies. Provincial regulations to protect coastal wetlands are forthcoming and much needed. In the Maritimes, coastal wetlands in the Gulf of St. Lawrence are most threatened due to high recreational use of the shoreline. Coastal wetlands may also be threatened in this region due to potential human responses to sea-level rise impacts on this low-elevation coastline. Studies in the Northeast have indicated the importance of large saltmarshes to bird diversity and density. Maintaining large saltmarshes intact without fragmentation should therefore be a conservation priority in the Northeast under current conditions and anticipated changes in coastal ecosystems due to rising sea levels.

ACKNOWLEDGMENTS

We would like to thank R. Greenberg, S. Droege, and J. Taylor for organizing and supporting the conference that lead to these proceedings. C. Elphick, G. Chumra, R. Greenberg, and L. Benoit provided valuable comments on earlier versions of this manuscript.

IMPACTS OF MARSH MANAGEMENT ON COASTAL-MARSH BIRD HABITATS

Laura R. Mitchell, Steven Gabrey, Peter P. Marra, and R. Michael Erwin

Abstract. The effects of habitat-management practices in coastal marshes have been poorly evaluated. We summarize the extant literature concerning whether these manipulations achieve their goals and the effects of these manipulations on target (i.e., waterfowl and waterfowl food plants) and non-target organisms (particularly coastal-marsh endemics). Although we focus on the effects of marsh management on birds, we also summarize the scant literature concerning the impacts of marsh manipulations on wildlife such as small mammals and invertebrates. We address three common forms of anthropogenic marsh disturbance: prescribed fire, structural marsh management, and open-marsh water management. We also address marsh perturbations by native and introduced vertebrates.

Key Words: Disturbance, impoundment, marsh endemic, marsh management, mosquito control, open-marsh water management, prescribed fire, structural marsh management.

IMPACTOS DEL MANEJO DE MARISMA EN HABITATS DE AVES DE COSTA-MARISMA

Resumen. Los efectos por las prácticas de manejo del hábitat en marismas de costa han sido pobremente evaluados. Resumimos la literatura existente que concierne a que ya sea si estas manipulaciones alcanzan sus metas, y los efectos de estas manipulaciones en organismos blanco (ej. Gallinas de agua y plantas de alimento de gallinas de agua) y en organismos no-blanco (particularmente en endémicos de marisma de costa). A pesar de que nos enfocamos en los efectos del manejo de marisma en aves, también resumimos la escasa literatura que concierne a los impactos de la manipulación de marisma en la vida silvestre, tales como pequeños mamíferos e invertebrados. Dirigimos tres formas comunes de perturbación antropogénica de marisma: quemas prescritas, manejo estructural de marisma, y manejo de agua en marisma-abierto. También dirigimos perturbaciones del marisma por vertebrados nativos e introducidos.

Nearly three-quarters of the 2,500,000 ha of coastal marshes of the US are located along the southeast Atlantic (North Carolina–Florida) and northern Gulf of Mexico (Florida–Texas) coasts (Alexander et al. 1986, Chabreck 1988). The extensive gulf and Atlantic marshes of the Southeast are intensely managed by federal and state land management agencies, conservation organizations, and private landowners. Managers disturb these ecosystems, often yearly, through prescribed burns, herbicide applications, ditching, and shallow pond construction. The rationale for these manipulations fall broadly under categories of: (1) wildlife enhancement, (2) flood control, (3) mosquito control, and (4) erosion mitigation (Daiber 1987, Chabreck 1988, Nyman and Chabreck 1995). These manipulations have occurred since historical times, but have been poorly, or only recently, evaluated in terms of their impact on wildlife.

In this chapter we address two primary questions: (1) are these manipulations achieving their wildlife management goals, and (2) what are the effects of these manipulations on non-target organisms, particularly coastal-marsh endemics? The majority of available data focuses on the effects of marsh management on birds. Much less is known about the impacts of marsh manipulations on small mammals, reptiles, and invertebrates. In this review, we address three common forms of anthropogenic marsh disturbance—prescribed fire, structural marsh management, open-marsh water management. Additionally, we address marsh perturbations by native and introduced vertebrates. Several other marsh-management actions we considered beyond the scope of this review including: insecticides targeting mosquitoes, herbicides for exotic or invasive species control, salt-hay cropping, and cattle grazing. These processes, particularly insecticide and herbicide use, often create impacts on nearly all of our marsh lands, but have been so little studied in the wildlife arena that little can be concluded, despite their clearly major impacts.

EFFECTS OF PRESCRIBED FIRE ON COASTAL-MARSH BIRDS

Prescribed fire is widely used to manipulate marsh vegetation. Although prescribed fire traditionally is used in gulf and southeast Atlantic Coast marshes, its application has spread throughout much of the eastern seaboard. For example, in 2002, the USDI Fish and Wildlife Service (USFWS), National Wildlife Refuge System burned approximately 9,500 ha of coastal marsh along the Texas coast; 9,300 ha of tidal and freshwater marshes along the North Carolina–Florida coast; 4,000 ha of coastal marsh in Louisiana; and 2,000 ha of salt and brackish tidal marshes in the Chesapeake Bay watershed (Dave Brownlie, USFWS Region 4, pers. comm.; Mark Kaib, USFWS Region 2 pers. comm.; Roger Boykin, USFWS Region 4 pers. comm.; Allen Carter, USFWS Region 5, pers. comm.). These figures exclude widespread prescribed burning by state agencies, National Park Service, or private individuals.

Prescribed fire in marshes gained initial support as a management tool for improving wintering waterfowl habitat in the 1930s and 1940s along the Gulf Coast (Lynch 1941, Nyman and Chabreck 1995). In the following decades, wildlife managers from the East and Gulf coasts embraced the recommendations of authors who advocated prescribed burning as a tool for conditioning marsh habitats (Lynch 1941, Hoffpauir 1961, Givens 1962, Hoffpauir 1968, Perkins 1968). Such conditioning includes removal of litter and dead vegetation, or vegetation considered to be of little or no value to gamebirds (e.g. cattails [*Typha* spp.] and cordgrasses [*Spartina* spp.]), and reduction of shrub cover. These burns were also purported to stimulate growth or seed production of food plants eaten by waterfowl such as bulrushes (*Schoenoplectus* spp. formerly *Scirpus* spp.), bristle grasses (*Setaria* spp.) and *Echinichloa* spp. Other purported benefits of marsh burning include: (1) maintaining a mixture of open-water and vegetated cover for resting, loafing, and breeding activities by waterfowl and other water birds, (2) facilitating trapping (primarily for muskrats [*Ondatra zibethicus*] and American alligators [*Alligator mississippiensis*]), (3) recycling dead plant material and increasing primary productivity through nutrient release, and (4) reducing the risk of unpredictable or uncontrollable fires, or fires that would damage the marsh system, e.g., peat fires (Nyman and Chabreck 1995, Foote 1996). Despite the long history of fire, and ongoing federal expenditures for prescribed fire programs, critical evaluations of the effects of fire on target and non-target species have been scarce.

In reviewing the available literature on fire in North American wetland ecosystems in 1988, Kirby et al. (1988:iii) declared that the science of using fire in natural and anthropogenic wetlands to perpetuate wildlife and plant communities was still in its infancy. Over a decade later, the predictive science of prescribed fire in wetlands remains weak; few analytical papers have documented fire's effects on marsh wildlife. Many coastal researchers consider marsh burning to be simply a "long-standing cultural practice...apparently done because of tradition or with poorly planned objectives" (Nyman and Chabreck 1995:134). We summarize extant information on the efficacy of prescribed burning on improving waterfowl habitat and populations and examine possible indirect effects of burning on non-target marsh birds.

Effects of Prescribed Burning on Waterfowl and their Habitat

Lynch (1941) advocated prescribed fires to enhance waterfowl wintering habitat. He reported that experimental burning on federal refuges in Louisiana attracted >500,000 geese and thousands of ducks to marshes that had previously held few waterfowl. Lynch (1941) states that these burns removed dense vegetation that interfered with growth of preferred waterfowl foods, increased nutritional quality of forage for cattle and geese, and increased waterfowl access to seeds and rhizomes. No details regarding counting methods, use of control marshes, or historical occurrence of fires or waterfowl in that area were provided. Interestingly, Lynch (1941) recommended that prescribed fires be used only on Gulf Coast marshes until experiments had been conducted in other coastal regions.

For 20 yr after Lynch, burning in coastal marshes continued without critical evaluation. In the 1960s, several authors began assessing marsh-burning efficacy, mostly based on observational and anecdotal data (Givens 1962, Hoffpauir 1968, Perkins 1968). Those authors focused on benefits to waterfowl by the favoring of preferred forage plants and maintaining shrub-free and otherwise open-marsh habitat.

As an example of the anecdotal nature of the evidence presented, Hoffpauir (1968:135) in coastal Louisiana, noted cover or wet burning 2–3 wk prior to arrival of Snow Geese (*Chen caerulescens*) provided fresh green vegetation and increased access to below-ground vegetation; however, geese used these areas for only 3–4 wk. Certain dabbling ducks appeared to use the burned areas extensively, feeding in potholes left behind by the activity of the Snow Geese. However, no information regarding use of controls, numbers of burned areas, or quantitative data were provided.

Despite these claims concerning waterfowl habitat improvement, we found only one study in which investigators surveyed waterfowl response to prescribed burns in coastal-marsh habitats using a standard methodology and comparing burned areas to controls (Gabrey et al. 1999). The authors conducted aerial surveys from December–February immediately following 14 prescribed burns on a 30,700 ha state wildlife refuge in coastal Louisiana. Gabrey et al. (1999) reported that 10 flocks of white geese (Snow Goose and Ross's Goose [*Chen rossii*]), ranging in size between 300 and 17,500 individuals, used recently burned marsh areas exclusively during the December–February period. However, the authors collected no behavioral or dietary data to assess goose activity or possible goose attractants in burned areas.

Habitat enhancement burns are intended to increase biomass and seed production of marsh plants preferred by migrating or wintering waterfowl (Lynch 1941), while reducing competition from less preferred plants such as inland saltgrass (*Distichlis spicata*), smooth cordgrass (*Spartina alterniflora*), or salt meadow cordgrass (*S. patens*) (Lynch 1941, Nyman and Chabreck 1995). DeSzalay and Resh (1997) evaluated late summer burns in inland saltgrass dominated coastal marshes in California and found that percent cover of inland saltgrass was reduced, while that of goosefoots (*Chenopodium* spp.) and purslanes (*Sesuvium* spp.) important in dabbling duck diets was increased, in burn treatments versus controls.

In brackish marshes in Chesapeake Bay, winter burning increased culm density and above ground biomass of live chairmaker's bulrush (*Schoenoplectus americanus*) in burned plots 1 yr post-fire, compared to plots that had not been burned for 2–3 yrs (Pendleton and Stevenson 1983, Stevenson et al. 2001). Biomass of dead bulrush was greater in unburned plots than burned plots. Burning did not affect biomass of plants other than bulrush. Pendleton and Stevenson (1983) concluded that the greater bulruish biomass produced following burning was a consequence of increased stem density rather than increased biomass of individual stems. Standing-dead material limited the total number of living culms in the unburned stands, and shaded new culms, therefore delaying the onset of spring growth.

Turner (1987) found that late-winter burning in smooth cordgrass marshes in Georgia reduced net aboveground primary production by 35%. Burning significantly reduced mean dry biomass of live rhizomes of smooth cordgrass in the top 10 cm of sediment. Burned plots exhibited a denser growth of smaller, finer smooth cordgrass plants than control plots.

Gabrey and Afton (2001) evaluated the effects of winter burning in 14 burned/unburned plot pairs in Louisiana saline, brackish, and intermediate marshes dominated by salt meadow cordgrass. Burns increased total live above-ground biomass but failed to increase bulrush species. Post-burn, species composition did not change, and post-burn flowering and seed production were nearly nonexistent, therefore, post-burn growth appeared to be from below ground rhizomes and roots of the burned plants (Gabrey and Afton 2001; S. W. Gabrey, pers. obs). Smooth cordgrass biomass in burned plots was lower compared to unburned plots; burning had no effect on inland saltgrass biomass (Gabrey and Afton 2001). The most notable and longest lasting effect of these burns was the dramatic reduction in dead above ground biomass, which remained below unburned levels for at least 3 yr post-burn.

Flores and Bounds (2001) studied six replicate marsh sites in the Chesapeake Bay of Maryland. All plots were burned in winter 1998 and treatment plots were burned again in winter 1999. Vegetation samples collected in the fall of 1999 (following treatments) showed that live above-ground biomass of inland saltgrass, chairmaker's bulrush, and saltmeadow cordgrass was greater in sites burned in 1999 than in those left unburned. Total biomass did not differ between treatments. Sites burned in 1999 had significantly higher mean stem densities than those left unburned. At 6 mo post-burn no significant difference was found in live aboveground biomass of black needlerush (*Juncus roemerianus*) or smooth cordgrass between burned and unburned treatments. The researchers report an overall increase in plant community stem density, but lack of increase in overall plant community biomass, in response to burning. Although burning increased biomass of bulrush, it did not reduce biomass of either cordgrass or saltgrass.

Some researchers report that burning coastal marshes enhances primary productivity. Spring, summer, and winter burns in Texas each increased live gulf cordgrass (*Spartina spartinae*) standing crop and the percentage of flowering plants by the end of the first growing season post-burn. The greatest growth response resulted from spring treatment, possibly because of post-burn rainfall (McAtee et al. 1979). Winter cover burns on the Mississippi coast increased net primary production (NPP) of above ground plant material by 56% and 49% in black needlerush and big cordgrass (*Spartina cynosuroides*) marsh communities, respectively (Hackney and de la Cruz 1981).

Season of burn and frequency of burn may explain in part the variability of vegetation response. In a greenhouse study using small buckets to simulate marshes, Chabreck (1981) showed that varying the season of burn altered the post-fire plant community. October burns appeared the most successful at promoting the growth of bulrush species. while burns between December and February promoted salt meadow cordgrass growth. In addition, O'Neil (1949) recommended 3-4 yr of repeated burning followed by periodic burning at 4-yr intervals to convert salt meadow cordgrass-inland saltgrass dominated marsh to sturdy bulrush (*Schoenoplectus robustus*)-saltmeadow cordgrass marsh in Louisiana. However, numerous other environmental variables, such as air or water temperature, salinity, pre-fire vegetation community, likely influence the composition of the post-burn plant community.

Another popular objective of habitat-enhancement burns in coastal marshes is to increase the nutritive quality of available plant foods. McAtee et al. (1979) report that digestible energy and crude protein content of gulf cordgrass was significantly increased on Texas coastal prairie in response to burning. Smith et al. (1984) conducted fall burning in a Utah alkali marsh; protein increased in inland saltgrass, tule bulrush (*Schoenoplectus acutus*), and cattail (*Typha* spp.), but not in alkali bulrush (*Bolboschoenus maritimus*). Schmalzer and Hinkle (1993) evaluated black needlerush and sand cordgrass (*Spartina bakeri*) marshes burned in December, at Merritt Island, Florida, and found that plant-tissue nutrient concentrations generally declined post-fire. One year after burning, nitrogen (N) content in live vegetation was lower than pre-burn content for all plant species. Phosphorous (P) concentrations increased in sand cordgrass, decreased in bulltongue arrowhead (*Sagittaria lancifolia*) and black needlerush in the black needlerush marsh, and were unchanged in other species. However, the P:N ratio increased in all live biomass types. Potassium (K) concentrations in live tissues declined or did not change significantly in all species whereas calcium (Ca) concentrations increased in black needlerush and sand cordgrass. Magnesium (Mg) concentrations decreased in live and dead black needlerush but increased in live bulltongue arrowhead and cordgrass species. Overall, biomass and nutrient content in these marshes did not return to pre-burn levels at 1 yr post-burn.

A potentially important, but poorly studied effect of prescribed fire is the possible impact on coastal-marsh invertebrates important in waterfowl diets. Some researchers have speculated that burning may reduce invertebrate populations in the short term by altering marsh surface temperature or exposing animals to greater predation risk (Hackney and de la Cruz 1981, DeSzalay and Resh 1997). Komarek (1984: 6) reported that following a single prescribed burn during the winter in a *Juncus-Spartina* marsh at St. George Island, Florida, three species of snail appeared to be more abundant in the burned section of the marsh. He observed higher populations of fiddler crabs (*Uca* spp.) in burned coastal marshes compared to unburned areas. However, because no data were provided to support these comments, it is unclear if such reported increases actually reflect increased invertebrate abundance, perhaps in response to greater nutrient availability, or greater invertebrate visibility in burned areas versus unburned sites.

A few studies demonstrate that various invertebrate taxa may respond differently to fire. On marsh islands in Virginia, Matta and Clouse (1972) collected invertebrates in sweep nets at 2-wk post-burn intervals from six sites representative of four burn treatments. The occurrence of most adult forms was not significantly affected by burning, although the principal insect herbivore, a meadow katydid (*Conocephalus* sp.) did show significant differences among sites, with fewer numbers at recently burned sites. Turner (1987) found that abundance of the periwinkle snail (*Littorarea irrorata*), an important winter food for American Black Ducks (*Anas rubripes*), was reduced by burning in smooth cordgrass in Georgia marshes. In the most extensive study to date, DeSzalay and Resh (1997) found densities of many invertebrates important in the diets of dabbling ducks in California wetlands (for example, *Chironomus* spp. and *Trichocorixa*. spp.) to be greater in burned compared to unburned control marshes. However, densities of other invertebrates, such as copepods and oligochaetes, were lower in open sections of burned marshes compared to unburned marshes. The researchers attributed lower densities and biomass of these invertebrates in burn areas to mortality due to vegetation removal, desiccation, or elevated soil temperatures.

The existing evidence supports the long-standing assumption that winter burning in coastal marshes does attract waterfowl; the evidence is strongest for geese. However, the mechanism for the attraction and the benefits accrued to the waterfowl populations remain unclear. Winter burning removes undesirable plants species and promotes growth of preferred waterfowl plant foods under some conditions (O'Neil 1949, Chabreck 1981, Pendleton and Stevenson 1983, Turner 1987, DeSzalay and Resh 1997, Stevenson et al. 2001) but not others (Flores and Bounds 2001, Gabrey and Afton 2001). Effects of burning on the nutritional quality of marsh vegetation appear ambiguous

(McAtee et al. 1979, Schmalzer and Hinkle 1993). The scant studies on marsh invertebrate community response are also inconclusive (Matta and Close 1972, Turner 1987, DeSzalay and Resh 1997). Plant and invertebrate community changes to burning are variable and likely depend on environmental factors such as season of burn, fire intensity, water depth and salinity, and post-burn rainfall. Although studies of vegetative productivity, plant nutritional quality, and invertebrate abundance are important, it is also necessary to determine if such changes indicate a change in habitat quality and benefit the organisms such as birds that forage on the vegetation or invertebrates. Habitat quality might be assessed by quantifying in these improved areas the: (1) activity or energy budgets, (2) foraging effort and behavior, (3) physiological indices such as sufficient energy stores for migration or breeding activities, (4) or movement among burned and unburned patches. We are unaware of rigorous fire studies that have been designed to answer these questions.

EFFECTS OF PRESCRIBED BURNING ON OTHER MARSH BIRDS

Given the lack of information on waterfowl response to burning in coastal marshes, it is not surprising that few quantitative studies address the effects of fire on other (non-game) birds (Rotenberry et al. 1995). Herein we summarize the few quantitative studies and include results of qualitative observational work on the effects of coastal-marsh fire on breeding and wintering coastal-marsh birds, including passerines and raptors.

Emberizidae (sparrows)

The Cape Sable Seaside Sparrow (*Ammodramus maritimus mirabilis*) is an endangered passerine whose relationship to fire has come under scrutiny due to recent population declines. Although now restricted to inland subtropical marshes and seasonally flooded prairies of southern Florida (Werner 1975), this subspecies of a primarily coastal-marsh species is one of the best researched passerines with respect to the effects of habitat burning, so we will review these studies in some detail. Werner (1975) tracked sparrow populations in two locations at Everglades National Park, Florida, for which historical fire data indicated that these areas had experienced wildfires in 1969 and 1972. A fire in 1974 also burned one of the two locations. The author reports that at each of these sites breeding densities of sparrows were sparse during the first year post-burn, but increased 3–4 yr post-burn. At one location, breeding densities declined during the fifth breeding season, post-burn. Werner (1975) suggests that sparrows decline in numbers immediately post-burn, then increase in density 3–4 yr after a fire, and may abandon a site after the sixth year after a fire as vegetation density increases. He speculates that optimum sparrow habitat could be maintained if marshes are burned every 4–5 yr. Werner (1975) based his conclusions on a very small sample size without control sites. Of additional interest in this study was the direct observation of individually marked sparrows fleeing the flaming front of a winter wildfire into adjacent unburned areas and flying in circles in areas of smoke and flames. Although sparrow density in the burned area returned to pre-burn levels by the next breeding season, none of the marked sparrows returned.

Taylor (1983) censused Cape Sable Seaside Sparrows at Taylor Slough, Florida. The study design was a randomized-block design, with three different prescribed burn treatments (annual, 3-yr rotation, 5-yr rotation) allocated to a set of three, 20-ha plots; a set of plots was located at each of three different marsh locations (blocks). In addition, a single control site had not been burned for 10 yr, and was not burned during the study. However, the season of prescribed fires differed between blocks, e.g., burns at one marsh were applied only in December (annually, every 3 yr, and every 5 yr), while at another marsh area, all burns were conducted in July. Therefore, the study had no true replication of treatments.

Taylor (1983) reported that on burned sites with deeper soils (>20 cm), vegetation recovery was more rapid and sparrow populations recovered and peaked earlier than on sites with shallower soils. The former populations re-colonized rapidly and began to decline 4 yr post-fire. In burned sites with shallow soils, plant biomass recovery was slower and sparrows did not even re-colonize these areas until about 4 yr post-fire and densities remained low for up to 10 yr. In addition, post-fire breeding territories were clumped, presumably because birds were forced to use marginal areas following large fires (Taylor 1983). Fires created long edges in which birds concentrated during the first breeding season post-fire and created a mosaic of unburned patches in which birds nested. Taylor (1983) concluded that fire regimes shorter than 8–10 yr could be detrimental to Cape Sable Seaside Sparrow populations.

Curnutt et al. (1998) provides the most comprehensive analysis of fire's effects on Cape Sable Seaside Sparrows. The authors analyzed

227 sites (as surveyed on a 1-km grid within Everglades National Park) that contained Cape Sable Seaside Sparrows between 1992-1996, and for which dates and spatial extent of fires from 1982-1996 were known. Sites had experienced fires caused by lightning strikes, unplanned human ignitions, or prescribed fire activities. The analysis did not control for likely differences beside fire frequency and time since last fire between sites (Walters et al. 2000).

For each site and for each sparrow survey, Curnutt et al. (1998) determined the frequency of fires, the number of days since the most recent fire, and whether the most recent fire occurred during the wet (1 June-31 October) or dry (1 November-May) season. They found that sparrow densities were lowest at sites that had a dry-season fire as their most recent fire occurrence. In contrast to Werner (1975) and Taylor (1983), Curnutt et al. (1998) found no evidence that sparrows abandon a site immediately post-burn and suggest that sparrows are able to occupy marsh sites immediately following a burn due to the patchy nature of natural fire in the Everglades. Curnutt et al. (1998) also found that sparrow populations increase in density with no evidence of eventual declines for up to 10 yr following a fire event. For those sites that held sparrows over the entire period of record, fire frequency ranged from one-seven fires/10 yr, with a mean of 2.97 fires/10 yr. Sparrow densities were highest where there had been one-two fires over the previous 10-yr period, lower where fire frequencies were greater or equal to three-six fires/10 yr, and absent from sites that were burned more than seven times in 10 yr. The authors' findings support those of Taylor (1983) — sparrows will use sites that had burned 10-12 yr previously — and contradict Werner's (1975) suggestions that sparrows decline in numbers and will abandon a site after the sixth year after a fire. The primary conclusion from this study is that frequent fires are harmful to Cape Sable Seaside Sparrows and may be the cause of declines in populations on the northeastern edge of the sparrow's range. Curnutt et al. (1998) also suggested that that the artificially drained nature of the coastal prairies in this region increased the flammability of these habitats, amplifying negative effects on Cape Sable Seaside Sparrows.

In summarizing past studies on the Cape Sable Seaside Sparrow, Walters et al. (2000: 1104) state that catastrophic sparrow population declines in the past decade cannot be directly attributed to fire. Nevertheless, authors of that paper concluded that fire has affected sparrow populations by altering habitat suitability, as demonstrated by direct evidence of immediate, negative effects of burning on sparrow populations, and the reported role of fire in periodically maintaining open habitats attractive to the sparrows (Werner 1975, Taylor 1983, Werner and Woolfenden 1983, Curnutt et al. 1998). Walters et al. (2000) analyzed population census data in Everglades National Park from 1981-1998 and concluded that two northeastern populations appear to have declined due to abnormally high fire frequencies, and that dry-season fires pose greater threats to breeding birds than wet-season fires. They cite evidence that increased fire frequency has been a direct result of anthropogenic water diversions, subsequent reduced hydroperiods, and exposure to human-caused dry-season fire. The authors speculate that occasional fire is necessary for continued occupancy of a marsh by Cape Sable Seaside Sparrows because it inhibits invasion by woody shrubs, including non-natives such as paper barked tea tree (*Melaleuca quinquenervia*; Curnutt et al. 1998), which can eliminate sparrow nesting habitat. Finally, Walters et al. (2000) stress the need to incorporate prescribed burning into rigorous experimental studies, including studies of the dispersal patterns of the birds through telemetry, to determine the direct effects of fire frequency on habitat and sparrow populations.

Effects of fire on other coastal sparrow populations, such as the Louisiana Seaside Sparrow (*Ammodramus maritimus fisheri*) and Nelson's Sharp-tailed Sparrow (*Ammodramus nelsoni*), have received recent attention (Gabrey et al 1999, Gabrey and Afton 2000, Gabrey et al. 2001). Gabrey et al. (1999) surveyed bird-species composition and abundance and vegetation structure on 14 pairs of winter-burned and unburned marshes in Louisiana. Winter bird surveys were conducted immediately following burns and again one full year post-fire. Immediately following burn treatment, plant community visual obstruction and percent cover were lower in burned plots; at 1 yr post-burn, vegetation structure was similar between treatment and control plots. Wintering Seaside Sparrows were absent immediately following burns, but were present in unburned marshes; Seaside Sparrows were present in burn-treatment plots 1 yr post-burn. Nelson's Sharp-tailed Sparrows, a migratory species that winters exclusively in coastal marshes (Greenlaw and Rising 1994), were present in burned marshes during the first winter but only in scattered patches of unburned vegetation; however, they were recorded frequently in unburned plots during both survey periods and in burn treatment plots 1 yr post-burn. The authors conclude that winter burning reduces the suitability of the marsh as winter habitat for these marsh-dependent sparrows, but only for a few months immediately following the burn.

Most studies of fire effects on birds rely on relative abundance as the response variable; rarely are demographic parameters such as nest success or survival addressed. Gabrey et al. (2002) used artificial nests to investigate effects of winter marsh burning on nest success of two coastal-marsh endemics—Louisiana Seaside Sparrows and Mottled Ducks (*Anas fuligula*). They recorded apparent nest success of nests containing quail eggs (to simulate sparrow nests) and chicken eggs (to simulate duck nests) in four pairs of burned and unburned marshes, during the breeding season prior to and following experimental burns. They found no difference in vegetation structure or success of either type of artificial nest in the post-burn breeding season. Although no effect of winter burns on artificial-nest success was detected, the authors caution that their study involved only four marshes and that the timing of burning may affect success of those birds that nest early in the season before sufficient vegetation re-growth has occurred (Gabrey et al. 2002).

Gabrey and Afton (2000) examined effects of winter marsh burning on Louisiana Seaside Sparrows nesting activity. Measurements were made during the breeding season (April–July) prior to experimental burns and during two breeding seasons post-burn. Male sparrows were absent from burned marshes during the start of the first breeding season after burns, but had reached abundances comparable to control marshes by June of that season. During the second breeding season post-burn, numbers of male sparrows were greater in burned marshes than in unburned marshes. Nesting activity indicators showed a similar but non-significant pattern in response to burning. The authors linked sparrow abundance and nesting activity to dead-vegetation cover, which was lower in burn plots during the first breeding season post-burn but recovered to pre-burn levels by the second breeding season post-burn. Gabrey and Afton (2000) speculated that reduced vegetation cover might provide less invertebrate prey and nest material for Louisiana Seaside Sparrows. During the study, the researchers recaptured birds banded as adults in unburned marshes during subsequent breeding seasons, but failed to recapture birds banded in burned marshes. The authors suggest that the sparrows move to nearby unburned marsh following a fire and that such displacement could affect short-term reproductive success by forcing dispersal into lesser quality habitats, increasing population density, interfering with pair bonds, and delaying territory establishment and nesting activities.

In other fire studies in the Chenier Coastal Plain in Louisiana, Gabrey et al. (2001) found that total abundance of sparrows (primarily Seaside Sparrows) did not differ between burned and unburned marshes during the first or third summers, post-burn, but were two times greater in burned than unburned marshes during the second summer post-burn. The peak in sparrow abundance coincided with the recovery of dead vegetation cover to pre-burn levels. Gabrey et al. (2001) concluded from both wintering and breeding season studies that periodic but infrequent fires that remove dense, dead vegetation benefit sparrow populations on the Chenier Coastal Plain.

Baker (1973) reported that two wildfires in December 1972 and January 1973 burned about 690 ha at St. Johns National Wildlife Refuge in Florida, leaving few patches of unburned sand cordgrass. Immediately following these fires, color-banded individuals of the now-extinct Dusky Seaside Sparrows (*Ammodramus maritimus nigrescens*) were displaced from burned areas. In early May, however, banded males reappeared and defended territories in burned areas. Baker (1973) speculates that rather than occupying small, unburned cordgrass patches within burn areas, the birds moved to nearby, unburned cover. Three birds banded on the area prior to burns were recaptured immediately after the burn in unburned cover, 900 m from their original locations.

Walters (1992) reported that in 1975, a fire intentionally set on private land escaped control lines and burned nearly 850 ha of Dusky Seaside Sparrow habitat on the St. Johns National Wildlife Refuge. Thirty-six male sparrows had occupied this area pre-fire; however, only seven were recorded post-fire. The refuge reported that six Dusky Seaside Sparrows escaped the fire to an adjacent private land area, which was subsequently burned by its owner. The sparrows then disappeared from the site.

Although difficult to quantify, fire may also have more direct effects on the survival of sparrows. Legare et al. (2000) captured and banded five Swamp Sparrows (*Melospiza georgiana*) on the St. Johns National Wildlife Refuge, in Florida. A sparrow banded on 20 March 1994 was recovered dead in burned sand cordgrass within 50 m of the original banding location on 5 January 1995, following a prescribed fire on the refuge. Although the authors report that the bird had most of its feathers burned, a conclusive cause of death was not reported.

Other passerines

In studies of wintering bird populations in coastal Louisiana marshes, Gabrey et al. (1999) found that several species of sparrows and wrens avoided recently burned marshes but reappeared one winter later. Common Yellowthroats (*Geothlypis trichas*) and Sedge Wrens (*Cistothorus platensis*) were absent from recently burned marshes, during the first winter, but present in unburned marshes. One year post-fire, Marsh Wrens (*Cistothorus palustris*) were found more frequently in unburned versus burned marshes. The authors concluded that winter habitat for several passerine species was reduced during the winter in which the burns occurred, particularly if a high proportion of the plot burned. In contrast, Boat-tailed Grackles (*Quiscalus major*) and Red-winged Blackbirds (*Agelaius phoenecius*) preferred recently burned plots, possibly because burns reduced visual obstruction and increased visual contact with conspecifics, and reduced ground cover, facilitating foraging for aquatic prey.

Gabrey et al. (2001) evaluated relative abundance of birds during the breeding season immediately following winter burns and for two consecutive breeding seasons thereafter. Structural vegetation characteristics (visual obstruction and percent cover) did not differ between burned and unburned plots by the first summer post-burn. Neither treatment affected bird species richness or species composition. Of the 10 most abundant bird species, only Sedge Wrens were absent from burned marshes but present in unburned marshes during the first post-burn breeding season. Sedge Wrens were present in burned marshes by the second breeding season post-burn. Total birds/survey for all species combined and for sparrows (primarily Seaside Sparrows) did not differ between burned and unburned marshes during the first or third summers post-burn, but were two times greater in burned than unburned marshes during the second summer post-burn, coinciding with the recovery of dead vegetation cover to pre-burn levels. The researchers concluded that managed burns for winter waterfowl foods appear compatible with maintaining populations of certain other marsh birds, provided that large contiguous marsh areas are not burned in any single winter, and >2 yr are allowed between burns.

Gabrey and Afton (2004) conducted multivariate analyses of breeding bird abundance in four pairs of burned and unburned marshes in the breeding season prior to experimental burns and in two breeding seasons post burn. Louisiana Seaside Sparrows, Red-winged Blackbirds, and Boat-tailed Grackles were the dominant species in these marshes. Winter burns dramatically lowered Seaside Sparrow abundance but increased blackbird and grackle abundance in the first breeding season post-burn. During the second breeding season post-burn, sparrow abundance increased and blackbird and grackle abundance decreased to the point where each variable was similar to pre-burn conditions (Gabrey and Afton 2004). Bird community changes were strongly correlated with percent cover of dead vegetation and live salt meadow cordgrass—plots with greater percent cover had greater sparrow densities and lower blackbird and grackle densities.

Raptors

Some research suggests that raptors use smoke and fire as a foraging cue, suggesting that raptors feed opportunistically upon prey either chased from cover by fire or left without cover by the burn (Baker 1940, Komarek 1969, Tewes 1984). Anecdotal evidence suggests that this occurs in marsh burns as well. Following two burns in gulf cordgrass communities at Aransas National Wildlife Refuge (ANWR), White-tailed Hawks (*Buteo albicaudatus*) reportedly dived through smoke to capture cotton rats (*Sigmodon hispidus*), pocket mice (*Perognathus* spp.), and grasshoppers (Acrididae) (Stevenson and Meitzen 1946). Tewes (1984) reported similar behavior during a 40-ha prescribed burn in gulf cordgrass at ANWR, when 14 White-tailed Hawks appeared near the fire, hovering near the ground and grasping prey in the ash. Other raptors noted soaring in the smoke column and hunting in the burned area included two Northern Harriers (*Circus cyanneus*), a White-tailed Kite (*Elanus leucurus*), an American Kestrel (*Falco sparverius*), and a Short-eared Owl (*Asio flammeus*). No raptors were noted during post-fire strip-transect counts, suggesting that the enhanced foraging opportunities afforded the raptors was extremely short lived. Tewes (1984) speculated this could be due to extensive and complete removal of vegetative cover forcing small mammals, snakes, and other prey species to abandon the site.

THE POSSIBLE ROLE OF LIGHTNING FIRES

Lightning-ignited fires are a common occurrence in coastal marshes, especially on the Gulf Coast and southeast Atlantic Coast. Such fires likely would have little detrimental impact on bird species endemic to these areas. Some evidence exists to support the idea that Seaside

Sparrow habitat, for example, in the Gulf Coast and southeast Atlantic Coast depends on periodic but relatively infrequent fires (Taylor 1983, Gabrey and Afton 2000); we are unaware of published studies that address effects of burning on Seaside Sparrows in the northern part of their range—habitats which naturally experience a lower frequency of lightning-ignited fires. In southern marshes in the absence of fire, vegetation density increases to a point at which the marsh is no longer suitable to Seaside Sparrows. Immediately post-fire, it appears that while numbers of breeding Seaside Sparrows and Marsh Wrens and wintering Nelson's Sharp-tailed Sparrows are reduced, these species may subsequently show a positive response for one or more years following the immediate post-burn season. However, as fire frequency increases (i.e., to every year), fires suppress vegetation density, rendering both breeding and wintering habitat unsuitable for several passerines (Common Yellowthroats and Sedge Wrens) including species dependent upon coastal-marsh habitats (Seaside and Nelson's Sharp-tailed sparrows). Frequent fires would likely also increase habitat availability for widespread habitat-generalist species such as blackbirds and grackles at the expense of habitat for endemic Seaside Sparrows or Nelson's Sharp-tailed Sparrows (Gabrey et al 1999, Gabrey and Afton 2004). Therefore, periodic but infrequent fires (Gabrey et al. 2001), possibly mimicking the historic fire regimes of these coastal habitats, are probably most likely to benefit sparrow and other passerine populations on the southeast coast. Whether such patterns occur in coastal marshes outside of the Southeast is unknown. Few studies have addressed effects on demographic parameters.

No studies to date have adequately addressed the likely effects of fire, either natural or prescribed, on other marsh-bird groups, such as raptors or colonial waterbirds. Research on population responses of these species to controlled fires in marsh habitats using standard methodologies and sound statistical design is needed to increase our understanding of the effects of prescribed burning on the entire coastal-marsh avian community.

MECHANISMS OF CHANGE IN COASTAL MARSH-BIRD COMMUNITIES IN RESPONSE TO FIRE

Because few scientific studies have focused on the effects of prescribed burns on marsh birds, the best we can do at present is to speculate about the potential effects of fire on various species and recommend that these potential relationships be investigated fully. This has been done based upon documented fire effects on coastal-marsh vegetation and known breeding or wintering habitat requirements of coastal-marsh birds. Prescribed burns may indirectly affect bird populations through a variety of pathways. Some of the more obvious mechanisms include direct or indirect effects on vegetation structure, changes in amount and distribution of open water, or changes in availability and quality of plant or animal food items. A summary of potential mechanisms is presented in Table 1. We emphasize that these are possible short-term impacts, based on the few quantitative studies that have been published. Long-term impacts have not yet been investigated. For example, Seaside Sparrow numbers may be temporarily reduced immediately following a fire but may increase for a period afterwards. In addition, many other variables such as water depth, salinity, and precipitation could influence vegetation responses to fire.

EFFECTS OF STRUCTURAL MARSH MANAGEMENT ON COASTAL-MARSH BIRDS

In addition to prescribed burns, marsh managers frequently alter marsh habitat by interrupting normal tidal cycles and manipulating the timing, depth, and duration of flooding, and salinity. Structural marsh management (SMM; Chabreck 1988, U.S. Environmental Protection Agency 1998) involves the use of weirs, dams, tide gates, canals, or other structures that alter the hydrology of coastal marshes. These structures allow managers to manipulate water depth, timing of flooding or drying, and salinity, to achieve the following objectives:

1. Prevent encroaching isohaline lines from changing the distribution of marsh types.
2. Encourage production of preferred waterfowl and muskrat foods while discouraging growth of plants with less waterfowl value (primarily cordgrass species).
3. Create or maintain shallow water or open water areas.
4. Reduce loss of existing marshes to erosion, sea-level rise, and saltwater intrusion.
5. Create new emergent wetlands from previously inundated areas.
6. Provide for ingress and egress of selected estuarine organisms (e.g., shrimp and larval fish).
7. Control biting insect populations (mosquitoes).

Although the scientific and management communities have begun to evaluate the effectiveness of SMM on coastal-marsh ecosystems,

TABLE 1. EFFECTS OF PRESCRIBED BURNS ON VEGETATION STRUCTURE AND POSSIBLE IMPACTS ON MARSH AVIFAUNA.

Effect of prescribed burn	Marsh appearance	Impact	Guild(s) potentially affected	Representative species
Cover burns reduce or remove emergent vegetation	Vegetation height and density temporarily reduced due to loss of standing vegetation	+	Waterfowl, gregarious, visual surface feeders	Snow Goose (*Chen caerulescens*), blackbirds (Icteridae).
Frequent fires inhibit woody vegetation establishment	Lack of shrubby cover, perches (live or dead) absent, standing dead vegetation cover removed, canopy removed	−	Species nesting in or foraging from strong perches; species nesting in shrubs or otherwise require woody structure	Colonial waterbirds, Belted Kingfisher (*Ceryle alcyon*), Peregrine Falcon (*Falco peregrinus*), Swamp Sparrow (*Melospiza georgiana*), American Bittern (*Botaurus lentiginosus*).
Cover burns reduce or remove surface litter	Marsh surface exposed, no decumbent vegetation above water level, cover for small mammals and tunnels removed	−	Species that nest in/on this material or that nest on mats of previous year's grasses; species that prey on rodents	Black Rail (*Laterallus jamaicensis*), Black Tern (*Chlidonias niger*), Forster's Tern (*Sterna forsteri*), Least Bittern (*Ixobrychus exilis*), Northern Harriers (*Circus cyaneus*), Short-eared owl (*Asio flammeus*).
		+	Species that forage on ground no longer inhibited by dense vegetation	Seaside Sparrow (*Ammodramus maritimus*)?
Peat or root burns expose underlying mineral soil	Open water more abundant; emergent vegetation replaced by shallow to moderately deep ponds; long-term inundation inhibits re-vegetation	−	Species using emergent vegetation in unbroken marsh; species that nest or forage on ground	Sedge Wren (*Cistothorus platensis*) and Marsh Wren (*Cistothorus palustris*), Red-winged Blackbird (*Agelaius phoeniceus*), Common Yellowthroat (*Geothlypis trichas*), Seaside Sparrow (*Ammodramus maritimus*), other sparrows (*Aimophila* spp.).
	Ponds remain as open water, submerged vegetation increases; or pond bottoms exposed due to tides or drought	+	Species using broken, hemi-marsh or mudflats for foraging or loafing	American Coot (*Fulica americana*), Pied-billed Grebes (*Podilymbus podiceps*), waterfowl, waders, shorebirds.
Fire suppression	Dense emergent vegetation and litter accumulation, woody shrubs colonize; small open-water ponds obscured	?		

few quantitative studies have been published. The greatest extent of SMM application to coastal marshes are in Louisiana and South Carolina (Day et al. 1990, U.S. Environmental Protection Agency 1998); consequently, most published studies of impacts on wetland vegetation and wildlife comes from these two states. We summarize below the current state of knowledge regarding effects of hydrology manipulations on coastal-marsh birds.

Bird Use of Impoundments During Winter

Waterfowl

As with prescribed burns, habitat management for waterfowl, particularly wintering habitat, has been a major justification for SMM. SMM became a common practice in the 1950s and the first evaluation of its effectiveness was presented by Chabreck (1960). He reported that prior to construction of impoundments in 1954, about 75,000 waterfowl wintered on Rockefeller State Wildlife Refuge in southwest Louisiana. In post-construction surveys, however, >320,000 waterfowl wintered in the new impoundments, with another 120,000 in surrounding areas within the refuge. He attributed the dramatic increase in numbers to increased food production and constant shallow water. Chabreck et al. (1974) later compared duck use of impoundments with that of control areas—unimpounded marshes and marshes that had been drained and converted to pasture. They reported that in general duck usage was highest in freshwater impoundments; numbers varied with vegetation type, water depth, and time of year.

Gordon et al. (1998) compared relative duck abundance between abandoned rice fields that were diked and managed for waterfowl and adjacent tidal (unimpounded) wetlands in South Carolina. Winter use of managed wetlands by seven dabbling duck species was greater than expected; winter use of unmanaged tidal marshes was less than expected for six of the seven species, American Black Duck (*Anas rubripes*) being the exception. The authors attributed these findings to differences in hydrology of the two types of marshes. Tidal marshes are flooded and drained daily; hence, availability of open water for foraging is unreliable. In addition, the intertidal period, in which the marsh surface is exposed, allows for denser vegetation growth that inhibits waterfowl access. In managed marshes, however, water level may remain relatively constant at a depth suitable for waterfowl foraging and the continuous flooding may prevent dense vegetation growth while maintaining large areas of open water suitable for dabbling duck foraging (Gordon et al. 1989). Finally, Gabrey et al. (1999) conducted five aerial counts of white geese (Snow Goose and Ross' Goose) wintering in managed marshes in southwestern Louisiana from December 1995 to February 1996 and found several flocks present in recently burned, unimpounded marshes or recently burned, impounded marshes; however, no description of goose behavior (e.g., foraging was reported). No geese were observed in unburned, unimpounded marshes.

Other bird species

Most assessments of the value of structural marsh management evaluate relative abundance of birds during winter or migration periods, and focus upon birds associated with ponds, mudflats, or open water, i.e., waterfowl, shorebirds, herons, egrets, gulls, and terns. Habitat within impounded marshes may supplement natural habitat in unmanaged marshes or provide protection from oil spills or other coastal catastrophes. Weber and Haig (1996) counted shorebirds and waterfowl in managed and unmanaged coastal marshes in South Carolina. Managed marshes were drawn down through April then re-flooded in June, July, or August. They found that throughout the winter and spring seasons (January–May), shorebird density at high tide (when natural marshes were flooded) was greater in managed exposed mudflats than in natural marshes. Even during low tides, shorebird density was generally greater in managed than in natural marshes. They concluded that managed marshes provide alternative or supplementary feeding or roosting habitats during high tides or adverse weather. Differences in shorebird density were attributed to consistently shallower water depth and greater invertebrate occurrence in managed marshes.

Impoundments in South Carolina are typically managed for production of wigeongrass (*Ruppia maritima*), spikerushes (*Eleocharis* spp.), bulrushes., and other waterfowl foods through spring drawdowns and summer re-flooding (Epstein and Joyner 1988). Consequently, vegetation composition differs between impounded marshes and natural tidally influenced marshes, which are dominated by big cordgrass (Epstein and Joyner 1988). Epstein and Joyner (1988) compared relative abundance of waterbirds in six managed (impounded) and two unmanaged (unimpounded) South Carolina marshes. Except for Clapper Rails (*Rallus longirostris*) and Northern Harriers, most bird species groups (particularly shorebirds, waterfowl, and waders) were more abundant in managed than in unmanaged marshes. The authors felt that the greater number of species and of individuals in managed marshes was due to moist soil conditions that increase access to invertebrates and seeds and to fish prey concentrated in progressively smaller ponds.

The above studies have addressed to some extent the question of whether waterbird use differs between impounded and natural or unimpounded marshes. However, birds that do not use open water or mudflat habitats but nest or forage in the emergent vegetation (e.g., passerines, rails, some herons, and egrets) have received less attention. Gabrey et al. (1999) addressed the issue of wintering passerines in impounded versus unimpounded coastal marshes in Louisiana. They found that some species (Seaside and Nelson's Sharp-tailed sparrows) were found almost exclusively in unimpounded marshes, possibly because of a preference for shorter vegetation and because ground-foraging behavior required exposed marsh surfaces. However, impoundment effects were confounded with salinity effects. This raises the question of the importance of vegetation variables in habitat selection. While most avian ecologists agree that vegetation structure is an important criterion, other factors such as invertebrate abundance, salinity, competitors, or predators may influence bird community composition differently in managed compared to unmanaged marshes.

It is interesting to note that three of the species listed above as being more abundant in unimpounded, natural marshes are coastal-marsh endemics (Clapper Rail, Seaside Sparrow, and Nelson's Sharp-tailed Sparrow). Consequently, although impounded marshes benefit a large suite of species, conservation of unimpounded coastal marshes is necessary for coastal-marsh endemics.

BIRD USE OF IMPOUNDMENTS DURING THE BREEDING SEASON

Waterfowl

Several waterfowl species breed in coastal marshes of the northeast Atlantic coast (e.g., Mallard [*Anas platyrhynchos*], American Black Duck, Blue-winged Teal [*A. discors*], and Gadwall [*A. strepera*] although most such populations are small (Bellrose 1976, Sauer et al. 2004). In southern marshes Mottled Ducks nest in large numbers (Moorman and Gray 1994). However, we are unaware of any published studies that address effects of marsh impoundment on any waterfowl nesting in coastal marshes.

Other bird species

Bird use of impounded and unimpounded marshes during the breeding season has received little attention. Brawley et al. (1998) compared breeding bird abundance in two restored (formerly impounded) marshes with that of three reference sites in Connecticut. They found that marsh specialists—those species that breed only in coastal marsh (Seaside Sparrow, Willet [*Catoptrophorus semipalmatus*], Marsh Wren, and Saltmarsh Sharp-tailed Sparrow [*Ammodramus caudacutus*])—were more abundant in the restored marshes than in the reference marshes. Three of these four species are listed as threatened (Willet) or of special concern (both sparrows), in the state. The authors state that these species were absent from the restored marshes prior to the re-establishment of tidal activity. Brawley et al. (1998) suggest that the frequent tidal inundation and exposure maintained the low-marsh community dominated by short-form smooth cordgrass in which Seaside and sharp-tailed sparrows prefer to nest. Marsh areas in the high-marsh zone are not flooded frequently enough or of sufficient duration to allow for establishment of short-form smooth cordgrass. Marsh areas below the low-marsh zone are permanently flooded and so also do not support smooth cordgrass.

Brawley et al.'s (1998) findings, that marsh impoundment benefits a diversity of bird species but limits habitat availability for marsh-specialist species, supports results from other regions. In New Jersey, Burger et al. (1982) found greater biomass and diversity of birds in impounded marshes compared to ditched or unimpounded marshes. However, species that nest exclusively in coastal marshes (Seaside and sharp-tailed sparrows, and Clapper Rails) were recorded only in unimpounded marshes. They (Burger et al. 1982) stated that while generalist or relatively abundant species used impounded marshes, maintaining natural unimpounded coastal marsh was necessary for the conservation of coastal-marsh specialists.

Gabrey et al. (2001) detected different bird communities present in impounded and unimpounded marshes in southwestern Louisiana. Red-winged Blackbirds and Boat-tailed Grackles were more abundant in impounded than in unimpounded marshes, whereas Seaside Sparrows were more abundant in unimpounded than in impounded marshes. The authors attributed these differences to vegetation structure and hydrology. Vegetation of impounded marshes included patches of cattails (*Typha* spp.) and common reed (*Phragmites australis*); blackbirds and grackles readily nested in these patches of tall vegetation. Unimpounded marshes, on the other hand, were dominated by low-growing salt meadow cordgrass and inland saltgrass. These two plant species form a low, densely interwoven canopy of relatively uniform height (<1 m), which presumably provides protection from predators for the ground-foraging Seaside Sparrow. In addition, the surface of impounded marshes is often continually flooded. The marsh surface of unimpounded marshes is exposed during part of the tidal cycle; hence sparrows are able to forage on the ground.

MECHANISMS OF CHANGE IN COASTAL-MARSH BIRD COMMUNITIES IN RESPONSE TO SMM

Structural marsh management influences coastal-marsh bird communities through its effects on open water or mudflat availability, timing and frequency of flooding, modification of the plant community, and salinity (Table 2). In general, SMM appears to benefit waterfowl and other species such as herons and blackbirds that are attracted to open water, exposed mudflats, lower salinity, or tall, dense vegetation. This likely is due to reduced diurnal variability in flooding due to the exclusion of tides. Thus, impoundments that are drawn down to moist soil conditions maintain those conditions until managers flood the impoundment. In contrast, unimpounded marshes flood at daily high tides; mudflats are then exposed for only about half a day. Disruption of tidal hydrology often increases the area of open water and decreases the amount of grass and short herbaceous vegetation. Consequently, although SMM provides habitat for a diversity of bird species, certain species such as Seaside Sparrows, sharp-tailed sparrows, and Clapper Rails, that require grassland-like conditions and alternating cycles of inundation and exposure of the marsh surface, likely do not benefit from impoundments.

EFFECTS OF MOSQUITO CONTROL AND OPEN-MARSH WATER MANAGEMENT ON COASTAL-MARSH BIRDS

The history of coastal-marsh alteration to control the mosquito as a human pest and disease vector, or for agriculture (livestock grazing and salt-hay farming), goes back centuries in the US (see Daiber 1987 for a review). During the early part of the 20th Century, the Old World notion of draining much of the high marsh was popular, and thus began an ambitious campaign of ditching both by hand and with horse or mule during the 1930s and 1940s (Daiber 1987,

Chabreck 1988). Ditches approximately 2–4 m wide and 1–2 m deep were dug in parallel fashion every 50 m in high-marsh areas from the upland-marsh ecotone bayward. The amount of Atlantic Coast marsh altered by this method has been estimated at about 90% and extends from Massachusetts to Florida (Tiner 1984, The Conservation Foundation 1988).

With the increasing awareness of the high productivity of coastal marshes in estuaries (e.g., the rise of the Odum school in ecology during the late 1950s and 1960s) and recognition of the importance of natural tidal flooding and hydrology to the integrity of marsh systems, improvements in marsh management were attempted. Experimentation began in New Jersey with a method that became known as open-marsh water management (OMWM) (Cottam 1938, Ferrigno and Jobbins 1968). This method substitutes biological control of mosquito larvae using predatory fish, and by altering mosquito egg-laying habitat, instead of drainage and pesticide applications. In short, mosquito depressions in the marsh not connected to existing ditches are either connected to ditches using new spurs or, if the depressions are very dense, a pond is constructed. The ponds originally were small (<0.05 ha), deep (often >60 cm), and had a deeper area or sump added to enable mummichog (*Fundulus* spp.) to survive during summer droughts. The material dredged to create the new ditches and ponds was spread thinly over the marsh surface to reduce the prospects that common reed (*Phragmites australis*) or woody vegetation such as marsh elder (*Iva frutescens*) might invade. Later, in other regions such as Delaware, the practice expanded but some modifications were added, such as adding sills to the ends of large ditches to retain ground water (Meredith et al. 1987). In spite of the popularity of the method in New Jersey and Delaware and its expansion to other states, little research on effects on wildlife has been performed and published in the peer-reviewed literature (Erwin et al. 1994 and references therein). In addition, Wolfe (1996) provided a summary of the effects of OMWM on birds, fish, mammals and other tidal resources in the Atlantic region. The practice remains somewhat controversial among wetland ecologists and federal and state resource managers because of concerns for converting and altering pristine marsh (Table 3).

Post (1974) was one of the earliest to demonstrate the behavioral and ecological effects of ditching on marsh birds, specifically Seaside Sparrows, revealing that ditches could alter the shape and sizes of territorial boundaries.

TABLE 2. EFFECTS OF STRUCTURAL MARSH MANAGEMENT (SMM) ON VEGETATION STRUCTURE AND POSSIBLE IMPACTS ON MARSH AVIFAUNA.

Effect of SMM	Marsh appearance	Impact	Guild(s) potentially affected	Representative species
Marsh ponds not subject to tides	Ponds retain water year-round; ratio of open water to emergent vegetation increases, submerged vegetation increases	+	Species that feed or loaf on shallow to moderately deep ponds	Waterfowl, coots, herons, egrets, terns.
	Marsh surface inundated year-round; access to invertebrates limited?	–	Ground-foraging or ground-nesting species	Seaside Sparrow (*Ammodramus maritimus*), Sedge Wren (*Cistothorus platensis*).
Timing of drawdowns controllable, not dependent on droughts; wetland manager actively conducts seasonal draw downs	Open mudflats may be available during spring or fall migration; summer drawdowns may enhance production of seed-producing annual plants; open water may be available throughout winter	+		Shorebirds, waterfowl, gulls.
	Marsh surface may be inundated year-round; access to invertebrates may be limited?	–	Ground-foraging or ground-nesting species	Seaside Sparrow, Sedge Wren?
	Availability of open water habitat may be limited during drawdowns	–	Species that feed or loaf on shallow to moderately deep ponds	Waterfowl, coots, herons, egrets, terns.
Flood-tolerant vegetation dominates; wetland manager favors out deep-water management regime	Cattail (*Typha*) or reeds (*Phragmites*) distribution increases	+	Species that nest in tall dense vegetation	Red-winged Blackbird (*Agelaius phoeniceus*), Boat-tailed Grackle (*Quiscalus major*), Marsh Wren (*Cistothorus palustris*), Common Yellowthroat (*Geothlypis trichas*).
	Short vegetation (*Distichlis* spp., salt meadow cordgrass) distribution decreases	–	Species associated with grassland-like habitats	Seaside Sparrow, Nelson's Sharp-tailed Sparrow (*Ammodramus nelsoni*).

TABLE 3. POTENTIAL ECOLOGICAL EFFECTS OF OPEN-MARSH WATER MANAGEMENT ON COASTAL EMERGENT MARSHES OF THE ATLANTIC COAST.

Negative effects	Positive effects
Loss of salt meadow cordgrass habitat for Seaside (*Ammodramus maritimus*) and sharp-tailed sparrows (*Ammodramus nelsoni* and *A. caudacutus*); loss of short-form smooth cordgrass (*Spartina alterniflora*),	Reduction of mosquito breeding sites.
Fragmentation of inner marsh with pools and radials; exacerbation of erosion and marsh loss in the face of sea-level rise	Increased forage fish populations and enhanced waterbird (wading birds, shorebirds) feeding habitats.
Compaction of emergent marsh due to operation of heavy equipment on marsh surface	Restoration of hydrology (with ditch plugging).
Invasion of shrubs, (*Iva* spp., *Baccharis* spp.), and reeds (*Phragmites australis*) due to slight elevation changes; change in vegetation community structure	Augmentation of perches and nesting substrates for passerines (marsh sparrows and wrens), wading birds.

For larger waterbirds, more recent studies in California have revealed that under certain circumstances, waterfowl use of marshes originally ditched and then diked for mosquito control can achieve positive results for both objectives (Batzer and Resh 1992). Along the Atlantic Coast, studies in New England demonstrated that draining of high marshes reduced their use by waterbirds because of the loss of ponds and pannes (Clarke et al. 1984, Brush et al. 1986, Wilson et al. 1987).

Other, early studies often examined marsh-alteration effects only at one local site and only on one or a few species, such as Herring Gulls (*Larus argentatus*; Burger and Shisler 1978) or Clapper Rails (Shisler and Schultze 1976). A more comprehensive analysis of OMWM effects on waterfowl at five New Jersey sites from 1959–1984 was attempted (Shisler and Ferrigno 1987); however, counting techniques and personnel changes rendered interpretation of the results difficult.

In a study of effects of OMWM on waterbirds in New Jersey, Erwin et al. (1994) determined year-round relative abundance of waterfowl, shorebirds, waders, gulls, and terns in ponds in OMWM-managed marshes, unmanaged tidal ponds, and managed impoundments (>400 ha). They found that spring and summer densities of American Black Ducks were greatest in the two large impoundments when compared to OMWM and tidal ponds.

In New England, several authors monitoring shorebirds, wading birds, and waterfowl have concluded that use of marsh sites treated with OMWM resulted in little difference when compared with sites with natural ponds. However, the method in New England did not include creation of new ponds (Clarke et al. 1984, Brush et al. 1986, Wilson et al. 1987). In Delaware, a 2-yr study by Meredith and Saveikis (1987) revealed that waterfowl use of OMWM ponds was only about one half that of natural ponds. The conclusions of that study are problematic however, because natural ponds were larger than were OMWM ponds. Walbeck (1989) conducted a study with limited information (only conducted for 1 yr) on the Eastern Shore of Maryland where large impoundment use by waterfowl was greater than use of OMWM ponds.

Several studies conducted in the mid-Atlantic region, examining many waterbird species, revealed that sizes of ponds and the water/marsh ratio of the study site were the most important determinant of use. Burger et al. (1982) examined six different marsh sites in New Jersey and found high use of larger ponds by some shorebird and wading bird species; however, they cautioned that adding ponds to the high marsh could adversely affect breeding Clapper Rails. Erwin et al. (1991) found among nine marsh sites in three states that the water/marsh ratio was positively correlated with use by waterfowl, and separately, American Black Ducks, but pond number was not. Larger ponds (>0.25 ha) tended to be used more than smaller ponds by most bird species, but no treatment effect (OMWM vs. natural pond) was found. In a later experimental study at Forsythe National Wildlife Refuge in New Jersey, Erwin et al. (1994) compared use by larger waterbirds of OMWM, small natural ponds, and nearby impoundments. They reported results that varied by guild and season. Higher densities were not always found in larger ponds for waterfowl, but this did seem to be the case for spring-summer shorebirds. When comparing small pond use (both OMWM and natural) with impoundment use, however, American Black Ducks and other waterfowl used impoundments in higher densities for both fall and winter feeding and nesting than they did small marsh ponds. They recommended that a smaller number of larger ponds be created in the high marsh if mosquito control is deemed necessary, and that ponds have shallow and sloping sides to accommodate shorebird, wading bird, and rail use. The authors also concluded that in areas near large impoundments, small water bodies would add little waterbird habitat value.

ANIMALS AS MARSH ARCHITECTS/MANAGERS (AND THEIR MANAGEMENT)

Although wildlife managers tend to think of marsh management as a strictly human endeavor, many animal species have demonstrated quite remarkable abilities to manipulate the structure, and hence functions, of marshes to differing degrees (Table 4). In some regions, as their populations have increased, some of these species have created conditions considered undesirable from the perspective of resource managers. Thus, managing the animal managers has become simultaneously a challenge and an ethical paradox, i.e., managing the marsh environment for human values is acceptable but for other animals to do so requires corrective (often lethal) measures (Table 5). We will explore and summarize some of the major aspects of animal architect activities in the US in the following sections.

BIRDS

The effect of marsh grazing by the Snow Goose can be significant in coastal marshes, because the birds typically pull up the aboveground stems to gain access to the rhizomes (Belanger and Bedard 1994, Jefferies and Rockwell 2002). In brackish marshes, geese tend to uproot primarily bulrush species while in saltmarshes along the Atlantic, the principal plant affected is smooth cordgrass. Larger patches of denuded marsh were referred to as eat-outs.

In the early years of study (1940s–1970s) of the Snow Goose, such goose eat-outs in winter were believed to be beneficial to wildlife, as apparently they opened up parts of the monotypic marshes and allowed access for feeding by a variety of other birds and fur bearers (Lynch et al. 1947, Chabreck 1988). In fact, small patches of eat-outs in the cordgrass marshes of New Jersey and Delaware, which make small fishes and invertebrates more available to predators, can attract over six species of feeding spring migrant shorebirds, as

TABLE 4. EXAMPLES OF MAJOR ANIMAL SPECIES THAT ACT AS MARSH ARCHITECTS IN MODIFYING THE STRUCTURE OF COASTAL MARSHES (PRIMARILY IN THE UNITED STATES).

Species	Area[a]	Effects on marsh (at high densities)	Effects on birds[b]
Snow Goose (*Chen caerulescens*)	East and Gulf coasts	Large eat outs of smooth cordgrass (*Spartina alterniflora*),	- for breeding rails, marsh sparrows, terns + for shorebirds and wading birds.
Muskrat (*Ondatra zibethicus*)	US wide	Eat outs especially Olney threesquare (*Schoenoplectus americanus*)	+ for waterfowl (roosting), open-water species, - for rails, marsh sparrows, waterfowl (feeding).
Nutria (*Myocastor coypus*)	LA, TX, MS, MD	Destruction of bulrushes (*Schoenoplectus* spp.); fragmentation of eat-outs	+ for waterfowl, open-water species, - for rails, marsh sparrows.
Horses (*Equus caballus*)	Southeast US islands (MD-GA), southern Europe	Reduced structure, trampling reduces cover and destroys nests	+ for total species richness of birds; - for nesting gulls, and total *numbers* of birds.
Cattle/sheep (*Bos taurus/Ovis aries*)	Atlantic and Gulf coasts in US; nearly global (cattle)	Trampling reduces both annual grasses and invertebrates, destroys nests	- for wintering waterfowl and shorebirds, rails due to reduced food.

[a] State names are abbreviated.
[b] + indicates a benefit for the target species/group, whereas a - sign indicates a negative effect.

TABLE 5. MANAGEMENT METHODS EMPLOYED TO CONTROL DENSITIES OF SPECIES THAT ACT AS MARSH ARCHITECTS IN MODIFYING THE STRUCTURE OF COASTAL MARSHES.

Species	Methods adopted	Outcome
Snow Goose (*Chen caerulescens*)	Special early season hunts, scare decoys and noisemakers, shooting on the breeding grounds (Canada)	Scaring and fall-winter hunts mostly ineffective; recent spring hunts in Canada under evaluation.
Muskrat (*Ondatra zibethicus*)	Trapping	Ineffective currently since market value is so low.
Nutria (*Myocastor coypus*)	Trapping; shooting, poisoning	Ineffective to date with low market values; shooting effective in some winters in Maryland.
Horse (*Equus caballus*)	Reducing size of herd by culling; use of exclosures, and sterilants	Sterilants costly and time consuming; exclosures only for local control. Roundups may be most effective (e.g., annual pony roundup in Chincoteague, Virginia).
Cattle/sheep (*Bos taurus/Ovis aries*)	Reducing size of herd by culling, exclosures, and pasture rotation	Pasture rotation (seasonal) and annual cull and sale most effective.

well as summering egrets, herons, and Glossy Ibis (*Plegadis falcinellus*; R. M. Erwin, unpubl. data). In the past few decades, however, Snow Goose populations have exploded, resulting in major marsh damage on the breeding grounds especially near St. James Bay, Canada, on staging areas along the St. Lawrence River, and on wintering areas from New Jersey to Maryland (see Batt 1998 for a summary of the goose problem). Intense grazing by large numbers of these social birds has resulted in rather large eat-outs that may require a decade or more for the vegetation to recover (Smith and Odum 1981, Young 1987). In some cases if the bare areas were extensive, they have lost their organic composition and became hypersaline; the marsh may have shifted to an alternative stable state (Jefferies and Rockwell 2002). Early attempts to remedy the goose problem relied on using hazing techniques and special extended season hunts during fall and winter on national wildlife refuges, initially at Forsythe National Wildlife Refuge, New Jersey (M. C. Perry, USGS, pers. comm.), Bombay Hook, and Prime Hook national wildlife refuges, Delaware (P. Daly, USFWS, pers. comm.). These proved largely unsuccessful however in reducing regional populations, and in recent years, the Canadian Wildlife Service is directing a special large-scale spring breeding season harvest (Batt 1998; Table 5).

The manager's challenge concerning the Snow Goose becomes one of partial suppression of a native species that is an important game species, a popular species among bird watchers and photographers, a charismatic species that precludes some types of draconian control methods (e.g., poisoning), and a species that has had a long co-evolutionary history with marsh-vegetation dynamics.

MAMMALS

Muskrats

The muskrat is a native species that, like the Snow Goose, has evolved in the marshes of North America. The role of muskrats and their management in marshes remains one of the classics in North American wildlife literature (Errington 1961). Without muskrats, fresh and brackish marshes may often become dominated by cattail although moderate muskrat densities control the cattail and keep the marsh open. Waterfowl managers speak of an ideal hemi-marsh with 40–50% open water in which muskrats are dense enough to control cattails and keep some open water, but are in turn kept under control by regular trapping (O'Neil 1949, Bishop et al. 1979). In coastal marshes along the Atlantic and Gulf coasts, the species that may benefit most from muskrat foraging activities and tunneling include migrant and wintering Blue-winged Teal, Green-winged Teal (*Anas crecca*), Mallard, American Widgeon (*A. americana*), and American Black Duck. During the breeding season, Coastal Plain Swamp Sparrows appear to achieve their highest densities in association with intense muskrat workings (B. Olsen and R. Greenberg, pers. comm.).

On occasion, muskrat population densities and associated tunneling activities may result in conflicts with wildlife management in marshes (Lynch et al. 1947). Examples include eroding the earthen plugs that marsh managers use in constructing OMWM sill ditches in Delaware (W. Meredith, Delaware Division of Fish and Wildlife, pers. comm.) and plugging old tidal ditches in New England (C. T. Roman, National Park Service, pers. comm.). Although poisons have been used on occasion as control, regular trapping remains the most widely acceptable method to control populations (Table 5); in recent decades however, with declining fur prices, reduced trapping has rendered population control ineffective (Chapman and Feldhammer 1982).

Nutria

The nutria (*Myocastor coypu*), a native of South America, was released in the Louisiana marshes in 1938 as part of a fur-bearing animal experiment and rapidly expanded throughout the Gulf Coast brackish marshes (Kinler et al. 1987, Chabreck 1988). Along the East Coast, nutria are found sporadically from Georgia north to the Blackwater National Wildlife Refuge in the Chesapeake Bay, Maryland, where they have created much controversy because of significant marsh losses on refuge lands (Chapman and Feldhammer 1982). As with the Snow Goose, small and localized nutria populations did not damage marshes, and it had been claimed that only for giant cordgrass (*Spartina cyanosuroides*) did nutria have any major impact (Harris and Webert 1962). In moderate numbers, nutria were felt to benefit some waterfowl, because the animals created open patches in otherwise dense marsh grass (Chabreck 1988). However, like the muskrat, the fur-trade decline has resulted in fewer trappers, and hence less control of local and region populations by trapping. As a result, large populations of these herbivores have caused very extensive eat-outs, resulting in marshes reverting to open water pools and lakes. In Maryland, the state natural resource agency is attempting to eradicate the species by trapping and shooting on all public lands (B. Eyler, Maryland Department of Natural Resources, pers. comm.).

Horses

In a relatively small number of regions today (e.g., southern France, Spain, and southeastern US), domestic or feral horses (*Equus caballus*) occur in coastal marshes, at times in high densities (Keiper 1985, Menard et al. 2002). As in previous examples, light to moderate grazing probably has little effect, but with more intense grazing impacts accumulate. In Georgia, significantly more periwinkle snails (*Littorea irrorata*), a potential waterbird prey, were found inside compared to outside of exclosures, and trampling by horses reduced above ground biomass of vegetation by 20–55% (Turner 1987). In southern Europe, horses reduced plants more than did cattle (*Bos taurus*), removed more vegetation per unit body mass, and maintained a mosaic of patches of short and tall grasses (Menard et al. 2002). This suggests potential indirect competition between horses and dabbling ducks (Menard et al. 2002). In the mid-Atlantic region of the US, horse grazing was thought to reduce the density of smooth cordgrass (Furbish and Albano 1994). In North Carolina, marshes subjected to moderate grazing by feral horses supported a higher diversity of foraging waterbirds, a higher density of crabs, but had less vegetation and a lower diversity and density of fishes than ungrazed marshes (Levin et al. 2002). Horse trampling of bird nests has occasionally been detected (I. Ailes, USFWS, pers. comm.) but is probably a minor factor in most locations.

The primary method of controlling feral horse numbers is simply reducing herd sizes and alternating the use of pasturage and wetland areas. On Assateague Island National Seashore and the Chincoteague National Wildlife Refuge in eastern Virginia, a fixed percentage of the annual foal production is removed from the herd during a July drive and managed roundup and are auctioned to the public in what has been a major tourist event (Keiper 1985). Experimentation with sterilization of horses has also been tried at several island locations along the Atlantic Coast, but with limited success (J. Schroer, USFWS, pers. comm.). Sterilization of dominant stallions without other control measures is unlikely to control feral horse populations.

Cattle and sheep

As with horses, light-to-moderate numbers of livestock (0.5–1.0 animal/ha) probably are not deleterious to marsh vegetation or to the associated bird life (Chabreck 1988). Cattle graze forbs and shrubs and may retard the invasion of woody vegetation into emergent marshes (Menard et al. 2002). In Europe, cattle grazing has been cited as benefiting grazing waterfowl as well as a common nesting shorebird (Redshank [*Tringa tetanus*]) by maintaining early successional stages and a diverse array of halophytic plant species, (Norris et al. 1997, Esselink et al. 2000). Along the US Gulf Coast, Chabreck (1968) mentioned that moderate cattle grazing in marshes might benefit the Snow Goose (Chabrek 1968) and, more interestingly, the Yellow Rail (*Coturnicops novaboracensis*; Mizell 1999). On the other hand, overgrazing by cattle reduces biomass of many annual seed-producing grasses and sedges, reducing food availability for wintering waterfowl, especially ducks (Chabreck 1968). In addition, in Germany an experiment conducted using three levels of grazing (0.5–2.0 animals/ha) over 9 yr demonstrated that grazing could depress population densities, species richness, and community diversity of invertebrates (Andresen et al. 1990); hence, many shorebird species could potentially be affected.

Sheep (*Ovis aries*) grazing in wetlands is most common in Europe. In general, as wetlands revert to upland pasture for sheep and cattle by drainage or diking into polders, potential wetland-dependent birds suffer habitat loss; such has occurred in The Netherlands (Hotker 1992) and elsewhere in Europe (Finlayson et al. 1992). In England, some attempts have been made to reduce the potential conflict between sheep grazing and wintering waterfowl use by imposing seasonal restrictions for sheep grazing from April–October in designated wet pastures (Cadwalladr and Morley 1973).

Management of potential deleterious effects of cattle and sheep involves reducing the herd periodically or alternating pasturage areas. Also, where significant waterfowl populations arrive in fall and winter, seasonal closures of some marshes from October through March may be appropriate to reduce disturbance.

SUMMARY AND FUTURE DIRECTIONS

Coastal marshes are subject to lightning-ignited fires that typically occur during the summer when thunderstorms are most frequent, and vegetation is actively growing. On the other hand, marsh managers typically burn marshes during late fall or winter, the time when migratory or wintering waterfowl are present, and vegetation, at least at higher latitudes, is generally dormant. Observational data provide limited evidence that these management burns attract some species of waterfowl (wintering Snow Goose in particular), at least occasionally. Unfortunately, lack of comparisons with unburned or control marshes limit inferences that can be made from these observations. We only can speculate as to what feature(s) of these burned marshes are attractive (e.g., food availability and nutritional content of vegetation, changes in predator communities, social interaction, and/or altered vegetation structure facilitating animal movement), or under what other environmental conditions waterfowl will use burned marshes (e.g., availability of food in the surrounding landscape).

Results of studies of vegetation responses indicate that prescribed burns sometimes, but not always, produce the desired results (i.e., changes in plant community composition, biomass, or seed production). Numerous environmental or other factors, including water depth or salinity, ambient or water temperature, humidity, fuel load, fire intensity, and season of burn likely strongly influence vegetation responses but have not been investigated. In particular, comparisons between biological responses to winter management burns and summer lightning fires could improve our understanding of the pre-management-era role of fire in these systems, and possible marsh community alterations caused by human-imposed fire regimes. Similarly, effects of prescribed fires on invertebrate foods are unclear.

Gulf Coast marshes in which Seaside Sparrows breed are prone to lightning-ignited fires; thus, these birds have likely evolved behavioral or other responses that allow their persistence in a frequently disturbed habitat. Prescribed fires appear beneficial to breeding sparrow populations, presumably because vegetation that inhibits the birds' movements along the ground is removed. Wrens and other small passerines apparently avoid recently burned marshes for about 1 yr, likely due to loss of vegetative cover. Burning marshes during the fall and winter reduces winter habitat quality for migratory species such as Sedge Wrens and Nelson's Sharp-tailed Sparrows, which winter almost exclusively in coastal marshes. Widespread and abundant species such as Red-winged Blackbirds and Boat-tailed Grackles seem to prefer recently burned marshes. Observations of fire effects on raptors and waterbirds are far too limited to make any significant inferences. Although these species do not necessarily nest in the marsh itself, they are important components of the marsh system as predators and vehicles of nutrient cycles; their responses should be investigated further.

Impounded marshes appear to attract waterbirds in greater numbers than do neighboring unmanaged, tidally influenced marshes and may contribute significantly to shorebird conservation because they provide supplemental feeding and roosting areas, particularly when natural marshes are inundated by high tide. However, passerines and other species that do not frequent large open-water ponds or mudflats may be negatively affected by conversion of tidal marshes into non-tidal marshes. Impounded marsh habitat differs sufficiently from unimpounded marsh habitat in that distinctive bird groups use one but generally avoid the other. Thus, managers are faced with a difficult task of integrating and improving management of impounded marshes with the management and preservation (in as natural a state as possible) of unimpounded marshes. Areas in which information appears to be lacking for coastal impounded marshes include effects of timing and duration of drawdown on wildlife use and invertebrate communities

Management values of impounded marshes during the breeding season are similar to those during winter and migration. Birds that benefit are typically those associated with open water ponds or mudflats; the specific nature of benefits for even these species have not been rigorously evaluated. At the same time, water levels within impoundments often are too deep to be suitable for ground-foraging passerines. These species appear to depend on periodic exposure of the marsh surface, possibly to facilitate foraging or because invertebrate prey are more vulnerable at low tides. Habitat structure may also play a role in the distribution and abundance of bird species because salinities and plant communities differ between impounded and unimpounded marshes. Invertebrate communities and availability may also differ between managed and unmanaged marshes.

Mosquito-control ditches drastically alter the hydrology, hence vegetation communities, of the marshes and set the stage for more dramatic marsh transformations. Since the 1960s in the mid-Atlantic region, OMWM has been developed to facilitate the biological control of mosquitoes. OMWM attempts to enhance fish populations while decreasing oviposition sites for mosquitoes by creating high-marsh pools and radial ditches isolated from daily tides. In spite of the appeal of depending upon biological rather than chemical means to control mosquitoes, the practice has proven controversial with some marsh ecologists remaining concerned about the mechanical alteration of marshes.

The effect of OMWM on waterbirds has been studied in several locations, but relatively little research has been done on a larger suite of potential ecological effects that might accrue due to OMWM treatment. Some of these potential effects are being addressed presently through research projects on six national wildlife refuges from Maine to Delaware (James-Pirri et al. 2004). Additional work is needed in other areas as well over longer time frames to determine the immediate versus longer-term effects of altering the hydrology and structure of the marsh. With the onset of sea-level rise, any additional interior fragmentation of marshes may prove inimical to a healthy marsh ecosystem.

In general, larger OMWM ponds (>0.1 ha) and pools attract more shorebirds and waterfowl than do small ones, although densities may not be greater. Several studies attempting to assess bird use of ponds were compromised due to either insufficient controls, or inappropriate survey methods. One experimental study in New Jersey indicated that, although larger ponds may be used by more birds than smaller ones, no treatment effect was detected (i.e., created versus natural pond use); also, at least for waterfowl, nearby large impoundments (100s of hectares in size) harbored both a larger number and higher density of birds than did the created ponds in fall and winter. Thus, the entire landscape surrounding the treatment areas of the marsh must be considered when addressing habitat use. Recent improvements in the design of small OMWM ponds include using very shallow, sloping perimeters to maximize shorebird use, and creating larger ponds unless dredged material deposition precludes that option.

Many animals other than humans have been marsh managers for years; however a limited amount of research has been conducted to evaluate effects of such activity. In general, removing animals (annual sales or culls) or rotating pasture lands have been effective in preventing overgrazing. In some cases, permanent fencing of selected areas may be necessary where critical species require increased protection (e.g., Piping Plovers [*Charadrius melodus*]) in the beach-marsh complexes of Chincoteague National Wildlife Refuge where feral horses are managed). Additional work is needed to assess the level of grazing and trampling that can be sustained by the local soil invertebrates and native grasses and sedges before community dynamics are altered.

Ironically, in light of the species' importance as an impetus of coastal-marsh management, recent increases in the Snow Goose in much of North America have been a major concern for state and federal wildlife managers and coastal wetland managers because of their potential to damage marshes and nearby crops. Special hunts have been used to attempt to reduce these populations; however, the effectiveness of these measures is unclear. Additional research and monitoring are necessary to determine the effectiveness of different levels of control in altering goose populations.

Medium-sized fur-bearing mammals also modify marshes considerably. The native muskrat, however is less cause for concern in its marsh plant consumption and tunneling than is the exotic nutria. Where population levels are moderate for each species, the opening of small pockets in the monoculture of marsh grasses may benefit waterfowl, rails, and other species. However, nutria have caused extensive marsh fragmentation and loss, especially in Louisiana and Maryland. Trapping no longer is viable economically nor is it effective in population control. An extermination program is underway in Maryland and Louisiana, and research efforts are underway to evaluate how population reduction rates are affecting declines and demographic aspects of the Maryland population.

Although some evidence suggests that we can improve marshes for waterfowl, herons, and, possibly in some cases, passerines, using certain marsh-management activities, success is often hit or miss. Additionally, the effects on non-target organisms, particularly those that depend on coastal marshes for at least part of their life cycle (e.g., endemic sparrows, rails, small mammals, snakes, and fish) are at best ambiguous and at worst harmful. As a result, many generally abundant and widespread species may benefit, whereas the few coastal-marsh specialists probably do not.

Nearly all studies of avian responses to coastal-marsh management document simple abundance or density measures that may not best reflect habitat quality (Van Horne 1983). Unknown are the effects of actions on biological parameters closely related to fitness, such as survival, nesting success, and physiological condition, or shifts in intrinsic (e.g., foraging behavior and social organization) or environmental factors (food availability and predator populations) that lead to changes in these parameters. In addition, most studies have attempted to relate bird responses local habitat features alone. Landscape scale variables such as area and extent of prescribed burns, proximity of other foraging areas, food sources, open water, or emergent vegetation, and habitat diversity and juxtaposition have also been largely ignored. Longer-term effects of changes in ecosystem processes (vertical accretion, compaction, sedimentation, and nutrient cycling) have also received comparatively little attention.

Finally, given the variable nature of coastal marshes, we should consider the merits of continuing to manage these habitats as we have historically (occasionally achieving some objectives) while risking potential irreversible ecosystem effects, such as the loss of a coastal-marsh endemic species. An alternative is to revise management goals and procedures to emphasize restoration of natural marsh processes (hydrology) and historic disturbances (fire). We suggest that scaling back the use of prescribed burning by reducing extent and frequency, particularly in areas in which fire is historically not a frequent disturbance, is certainly advisable, given the levels of uncertainty. In a similar vein, taking a go-slow approach on OMWM, especially in relatively pristine, unaltered coastal marshes is recommended. Coastal-marsh restoration, such as ditch plugging in the Northeast and opening up diked marshes (Cape Cod, Delaware Bay marshes; San Francisco Bay salt ponds; Merritt Island, Florida) should be encouraged.

A precautionary approach that uses adaptive resource management and attempts several experiments simultaneously to compare and evaluate model parameters is well advised. We encourage researchers and managers to work together to monitor and evaluate management activities while emphasizing an experimental approach (Ratti and Garton 1996). Such collaborations should emphasize well-designed long-term studies that document meaningful ecological responses (e.g., avian productivity or nutrient cycling). Only by treating each management activity, when possible, as a field experiment, complete with suitable control treatments and true replication, can significant advances in the science of coastal-wetland management be made. Information gleaned from these sound practices can be used to justify or alter coastal-marsh management activities with greater confidence.

ACKNOWLEDGMENTS

We would like to thank the USDI Fish and Wildlife Service, National Wildlife Refuge System Region 5, Northwestern State University, Smithsonian Environmental Research Center, USGS Patuxent Wildlife Research Center, and the University of Virginia. We also wish to thank S. Droege, R. Greenberg, and two anonymous reviewers for their discerning comments on previous drafts of this manuscript.

ENVIRONMENTAL THREATS TO TIDAL-MARSH VERTEBRATES OF THE SAN FRANCISCO BAY ESTUARY

John Y. Takekawa, Isa Woo, Hildie Spautz, Nadav Nur, J. Letitia Grenier, Karl Malamud-Roam, J. Cully Nordby, Andrew N. Cohen, Frances Malamud-Roam, and Susan E. Wainwright-De La Cruz

Abstract. The San Francisco Bay and delta system comprises the largest estuary along the Pacific Coast of the Americas and the largest remaining area for tidal-marsh vertebrates, yet tidal marshes have been dramatically altered since the middle of the 19th century. Although recent efforts to restore ecological functions are notable, numerous threats to both endemic and widespread marsh organisms, including habitat loss, are still present. The historic extent of wetlands in the estuary included 2,200 km^2 of tidal marshes, of which only 21% remain, but these tidal marshes comprise >90% of all remaining tidal marshes in California. In this paper, we present the most prominent environmental threats to tidal-marsh vertebrates including habitat loss (fragmentation, reductions in available sediment, and sea-level rise), habitat deterioration (contaminants, water quality, and human disturbance), and competitive interactions (invasive species, predation, mosquito and other vector control, and disease). We discuss these threats in light of the hundreds of proposed and ongoing projects to restore wetlands in the estuary and suggest research needs to support future decisions on restoration planning.

Key Words: Contaminants, disease, fragmentation, San Francisco Bay, sea-level rise, sediment supply, threats, tidal marsh, water quality, wetlands.

AMENAZAS AMBIENTALES PARA VERTEBRADOS DE MARISMA DE MAREA DEL ESTUARIO DE LA BAHÍA DE SAN FRANCISCO

Resumen. La Bahía de San Francisco y el sistema delta abarcan el estuario más grande a lo largo de la Costa Pacífico de las Américas y el área mas larga que aun queda para vertebrados de marisma de mar, a pesar de que los marismas de marea han sido dramáticamente alterados desde mediados del siglo 19. A pesar de que los esfuerzos recientes para restaurar las funciones ecológicas son notables, numerosas amenazas para ambos organismos de marea, endémicos y amplios, incluyendo pérdida del hábitat, están aun presentes. El alcance histórico de humedales en el estuario incluyeron 2,200 km^2 de marismas de marea, de los cuales solo el 21% permaneció, pero estos marismas de marea comprenden >90% de todos los marismas de marea que quedan en California. En este artículo, presentamos las amenazas ambientales más prominentes para los vertebrados de marisma de marea, incluyendo pérdida del hábitat (fragmentación, reducciones en el sedimento disponible, y aumento en el nivel del mar), deterioro del hábitat (contaminantes, calidad del agua, y disturbios humanos), e interacciones competitivas (especies invasoras, depredación, mosquitos y otro control vector, y enfermedes). Discutimos estas amenazas a luz de cientos de proyectos propuestos y llevados a cabo para restaurar humedales en el estuario, y las necesidades sugeridas por estudios para apoyar futuras decisiones en la planeación para la restauración.

Coastal and estuarine wetlands are resources of global importance to humans and wildlife, but they encompass <3% of the land surface in the Western Hemisphere and only 0.3% of the contiguous US (Tiner 1984). The most extensive regions of tidal marsh in the coterminous US are found along the Gulf Coast (9,880 km^2), southern Atlantic Coast (2,750 km^2), mid-Atlantic Coast (1,890 km^2), and New England and Maritime Coast (360 km^2; Greenberg and Maldonado, *this volume*). In contrast, much less tidal marsh is located on the West Coast, of which the largest extent is found in the San Francisco Bay and delta (SFBD; Fig. 1). SFBD is the largest estuary (4,140 km^2) on the Pacific Coast of the Americas, encompassing <7% of the land surface, draining >40% (155,400 km^2) of California (Nichols et al. 1986), and supporting 162 km^2 of remaining tidal marshes.

Saltmarsh plant communities along the California coast often form mosaic patches and are dominated by common pickleweed (*Salicornia virginica*, syn. *Sarocornia pacifica*) and Pacific cordgrass (*Spartina foliosa*). Common pickleweed occurs throughout the East and West coasts of the US (U.S. Department of Agriculture 2003), but Pacific cordgrass is traditionally found along the California coast from Bodega Bay (though it has been introduced to Del Norte County) in the north, to San Diego County in the south (Calflora 2003), extending into the Baja Peninsula of Mexico. In SFBD, relatively narrow strips (3–10 m) of Pacific cordgrass occur between mean tide level (MTL) and mean

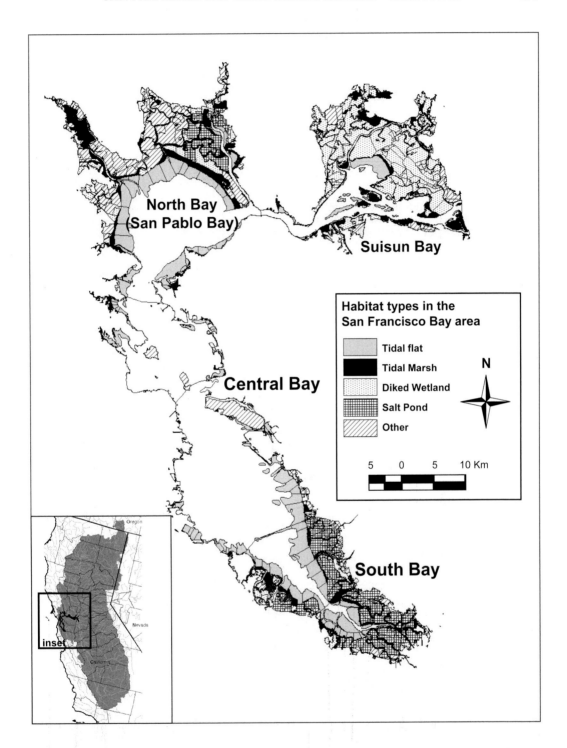

FIGURE 1. Area of California drained by the San Francisco Bay estuary and the Sacramento-San Joaquin River watershed (shaded), and distribution of tidal-marsh habitat within the estuary (San Francisco Estuary Institute 1998).

high water (MHW), and a wider band (up to a few kilometers) of common pickleweed ranges from mean high water (MHW) to mean higher high water (MHHW) (Josselyn 1983). In more brackish waters of the Sacramento and San Joaquin river delta, bulrushes (*Bolboschoenus* and *Schoenoplectus* spp.) are dominant. With the extensive losses of coastal wetlands in California, the SFBD now supports 90% of the remaining tidal wetlands (MacDonald 1977).

HUMAN DEVELOPMENT

The abundance of wildlife resources attracted the first humans to the estuary roughly 10,000 yr ago. Hunter-gatherer societies approached 25,000 inhabitants but likely posed little threat to most tidal-marsh wildlife (San Francisco Estuary Project 1991). In the last 200 yr, hunters and traders were attracted to the estuary by the abundant wildlife from as far away as Russia, causing the first notable decline of fur-bearing populations including sea otter (*Enhydra lutra*) and beaver (*Castor canadensis*) (San Francisco Estuary Project 1991). Spanish inhabitants set up missions and began grazing cattle and sheep in the 18th and 19th centuries. A rapid influx of humans occurred in 1848 when gold was discovered in the Sierra Nevada Mountains. Within 2 yr, the city of San Francisco grew from 400–25,000 individuals. Sierra Nevada hillsides were scoured by hydraulic mining and mercury (Hg) was used to extract gold. Roughly 389,000,000 m^3 of sediment, along with Hg-laden sediments, was transported downstream into the estuary from 1856–1983 (U.S. Geological Survey 2003).

The SFBD was home to over half of the state's population by 1860, and the population has steadily increased (Fig. 2). Population growth also stimulated rapid development and urbanization (Figs. 3, 4). Legislation for land reclamation (federal Arkansas Act of 1850, state Green Act of 1850) was enacted to encourage conversion of grasslands and wetlands to farmlands, and by the 1870s a network of levees had been constructed to protect low-lying fields. The deepwater harbor became a major shipping center, and by 1869, completion of the transcontinental railroad increased movement of food and goods from the region. Striped bass (*Morone saxatilis*) were intentionally introduced for a commercial fishery in 1879.

In the early 1900s, urban runoff polluted the bays, and thousands of hectares of wetlands were filled for development. Tidal wetlands in the South Bay were replaced by >5,000 ha of salt-evaporation ponds by the 1930s (Siegel and Bachand 2002). Dams and diversions on nearly every tributary prevented fish from spawning upstream, limited sediment transport downstream, and reduced freshwater

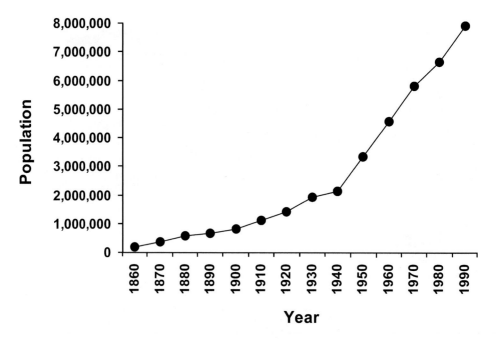

FIGURE 2. Increased human population in the San Francisco Bay estuary and delta from 1860–1990 (Bell et al. 1995).

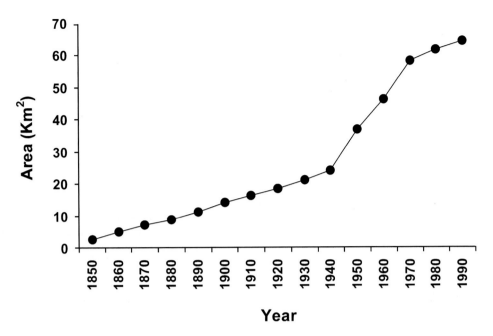

FIGURE 3. Expanding urban areas (in square kilometers) of the San Francisco Bay estuary and delta from 1850–1990 (Bell et al. 1995).

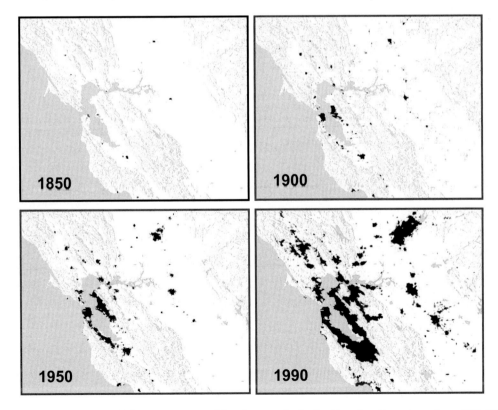

FIGURE 4. Distribution and changes in extent of development in the San Francisco Bay estuary from 1850, 1900, 1950, and 1990 (Pereira et al. 1999).

flows. The thousands of workers attracted to the area during World War II stimulated a housing and construction boom. Farmers became reliant on chemical practices increasing toxic runoff to estuary waters. The human population was nearly 6,500,000 in 1980, but it increased 34% to 8,700,000 by the end of the century (U.S. Census Bureau 2000). The historic (pre-1850) saltmarshes covered 2,200 km^2 or twice the extent of open water (Atwater et al. 1979), but the modern landscape has been altered by loss of 79% of the saltmarshes, 42% of the tidal flats, and construction of almost 14,000 ha of artificial salt-evaporation ponds (San Francisco Bay Area Wetlands Ecosystem Goals Project 1999).

THE MODERN ESTUARY

The diversity of the SFBD wetlands includes freshwater marshes in the eastern delta and upstream tributaries, brackish marshes in Suisun Bay and western delta, and saltmarshes (Fig. 1) in the South Bay, Central Bay, and San Pablo Bay (Harvey et al. 1992). That diversity supports 120 fishes, 255 birds, 81 mammals, 30 reptiles, and 14 amphibian species in the estuary, including 51 endangered plants and animals (Harvey et al. 1992). SFBD is a Western Hemisphere Shorebird Reserve Network site of hemispheric importance used by >1,000,000 shorebirds, supports >50% of wintering diving duck species counted on the Pacific flyway of the US during the midwinter, and is home to one of the largest wintering populations of Canvasbacks (*Aythya valisineria*) in North America (Accurso 1992).

Wetland conservation in the SFBD has evolved from slowing loss to preserving remnant wetlands to aggressively restoring areas. By 2002, 29 wetland restoration projects had been completed in the North Bay alone (S. Siegel, unpubl. data); however, these restoration projects were relatively small (≤12 ha). Large restoration projects, hundreds to thousands of hectares in size, have been proposed or initiated recently that will significantly change the regional landscape. Nearly 6,500 ha of salt ponds in the South Bay and 4,600 ha in the North Bay have been acquired (San Francisco Bay Area Wetlands Ecosystem Goals Project 1999, Steere and Schaefer 2001, Siegel and Bachand 2002). Some areas will be managed as ponds to provide habitat for thousands of shorebirds and waterfowl, but many areas are proposed for conversion to tidal marshes for the benefit of tidal-marsh species. Despite these wetland restoration efforts, tidal-marsh vertebrates still face many threats in the estuary.

In this paper, we summarize the major threats to tidal-marsh vertebrates including habitat loss (habitat fragmentation, sediment availability, and sea-level rise), deterioration (contaminants, water quality, and human disturbance), and competitive interactions (invasive species, predation, mosquito control, and disease). We describe how these threats affect tidal-marsh vertebrates, where proposed restoration projects may ameliorate their effects, and what studies would be helpful to support restoration planning.

HABITAT FRAGMENTATION

Habitat fragmentation results from changing large continuous areas to a pattern of smaller, more isolated patches of less total area within a matrix of altered habitats (Wilcove et al. 1986). Fragmentation usually also involves an increase in mean patch perimeter-to-area ratio and changes in patch configuration. Though habitat loss contributes directly to population decline, edge effects and habitat isolation may cause further reductions by an impact on dispersal and altering the ecological function within patches, especially near habitat edges (Andren 1994). These impacts may increase with time since isolation.

In the SFBD, loss of tidal-marsh habitat has resulted in many remnant marshes that are small and isolated from other marshes (San Francisco Bay Area Wetlands Ecosystem Goals Project 1999, San Francisco Bay Estuary Institute 2000). Roads, levees, urban development, and non-native vegetation have replaced upland edges and transition zones (Table 1), much to the detriment of animals that rely on upland areas as high-tide refugia. Many modern wetlands are

TABLE 1. HISTORIC (PRE-1820) AND PRESENT DISTRIBUTION OF TIDAL-MARSH PATCHES IN THE SAN FRANCISCO BAY ESTUARY (DERIVED FROM THE SAN FRANCISCO BAY ECOATLAS 1998).

Patch size (hectares)	Number of tidal and muted marsh patches[a]	
	Historic habitat (pre-1800)	Present habitat (1990s)
<10	190	370
10–50	46	97
51–100	21	23
101–500	33	38
501–1,000	18	3
1,001–3,000	20	1
>3,000	4	0
Total area of tidal marsh	77,530	16,996

[a] Patches derived by merging adjacent tidal (old and new) and muted marsh polygons. Muted marshes include wetlands without fully tidal flows.

patchy, linear or irregularly configured, confined to edges of large tidal creeks or sloughs, or on the bayside edges of levees where new marshes formed over sediments accreted during the 1850s. Levee networks, tide-control structures, and mosquito ditches not only fragmented wildlife habitat but also altered the hydrology and sediment dispersion patterns of outboard levee areas (Hood 2004). Although regulatory laws may prevent future losses of wetlands, increased urbanization and loss of native vegetation corridors may decrease the viability of populations (Andren 1994). Tidal-marsh species may be differentially affected depending on their level of habitat specialization (Andren 1994) and the scale of landscape heterogeneity to which they are sensitive (i.e., patch sensitivity, *sensu* Kotliar and Wiens 1990, Wiens 1994, Riitters et al. 1997, Haig et al. 1998).

Habitat fragmentation may act as an isolating mechanism resulting in higher extinction rates and lower colonization rates, lower species richness (MacArthur and Wilson 1963, 1967), higher nest-predation rates (Chalfoun et al. 2002), higher nest parasitization rates, and changes in ecological processes (Saunders et al. 1991). Numbers of avian marsh species in the prairie pothole region have been shown to vary with patch size and perimeter-to-area ratio as well as with vegetation and other local-scale factors (Brown and Dinsmore 1986, Fairbairn and Dinsmore 2001). The effects of fragmentation were also found to be variable, species-specific, and context-specific in a range of other habitat types (Bolger et al. 1997, Bergin et al. 2000, Chalfoun et al. 2002, Tewksbury et al. 2002).

Few studies have been done to assess the impact of fragmentation on tidal-marsh birds in the SFBD. Scollon (1993) mapped marsh patches and dispersal corridors for the Suisun Song Sparrow (*Melospiza melodia maxillaris*) based on theoretical estimates of dispersal distance and published population densities. Varying fragmentation effects on population size have also been predicted for San Pablo Song Sparrows (*M. m. samuelis*; Scollon 1993; Takekawa et al. chapter 16, *this volume*). The dispersal of San Pablo Song Sparrows from one fragmented marsh to another is thought to be rare since dispersal distance averages 180 m (Johnston 1956a). Thus, populations in smaller, more isolated fragments were more susceptible to local extinction. Recent empirical studies (Spautz et al., *this volume*; Point Reyes Bird Observatory, unpubl. data) have found that San Pablo, Suisun, and Alameda (*M. m. pusillula*) Song Sparrows, all California species of special concern, and California Black Rails (*Laterallus jamaicensis coturniculus*), a California state threatened species, respond to marsh size, configuration, isolation, and other landscape-scale factors as well as to local-scale factors such as vegetation composition and structure. However, the underlying processes contributing to these patterns, along with dispersal patterns across the estuary, are not well understood.

Restoration Concerns

Several large-scale wetland restoration projects are underway, including the Napa-Sonoma Marsh on San Pablo Bay (4,050 ha) and the former Cargill salt ponds in the South Bay (6,475 ha; Siegel and Bachand 2002). These projects involve restoration of areas that were historically tidal, but were converted to salt-evaporation ponds. The goals of these restoration projects are to provide large areas of contiguous habitat, increasing marsh area with minimal fragmentation. Previous restoration projects in the estuary have been relatively small and opportunistic with limited study of restoration effects at the landscape scale.

One of the most important considerations in restoration is determining the optimum configuration and size of the project. A single large expanse of habitat may be preferable for some species rather than several smaller, isolated habitat patches (particularly for those species requiring large territories). However, a population in a single large patch may be more vulnerable to extinction because of demographic stochasticity or catastrophic events (such as fires and disease; Carroll 1992). Highly vagile species, including most birds, are generally better able to disperse between isolated habitat patches than small mammals, reptiles, and amphibians. However, radio-telemetry studies of the endangered California Clapper Rail (*Rallus longirostrus obsoletus*) indicate low rates of movement between and within seasons (Albertson 1995) and habitat fragmentation is considered one of the main threats to the persistence of this subspecies (Albertson and Evens 2000).

REDUCTION IN SEDIMENT AVAILABILITY

Sediment deposition and tidal actions are the dynamic processes that sustain tidal-marsh wetlands. Rapid sediment accretion of tidal marshes in the SFBD extended for at least 20 yr after the start of hydraulic gold mining in the Sierra Nevada Mountains. San Pablo Bay received 300,000,000 m^3 of sediment, and by 1887 created 64.74 km^2 of new mudflats (Jaffe et al. 1998). The concentration of suspended sediments in the delta declined 50% by the 1950s with the cessation of hydraulic mining and the advent of dams on major tributaries from downstream reaches (Wright and Schoellhamer 2004). San Pablo Bay lost 7,000,000 m³ of sediment from 1951–1983 or an annual loss of 0.36 km² of mudflats (Jaffe et al. 1998).

Many wetland restoration project sites have subsided and require substantial sediment input to reach adequate levels for plant establishment. When sedimentation rates were studied in the south San Francisco Bay (Patrick and DeLaune 1990), one site (Alviso) was found to have subsided by >1 m based on records from 1934–1967 because of groundwater extraction (Patrick and DeLaune 1990). The large number of restoration projects occurring simultaneously may reduce predicted sediment availability (San Francisco Bay Area Wetlands Ecosystem Goals Project 1999). A shortage of sediment may result in a reduced turbidity, increased erosion, and a greater loss of mudflat and intertidal habitats (Jaffe et al. 1998).

Dredge material has been proposed for projects where natural sediment supply is inadequate. Annual yields from dredging operations produce an average of 6,120,000 m³ of sediment in the estuary (Gahagan and Bryant Associates et al. 1994). At current rates of sediment accretion, it would take 10–15 yr to raise elevations one meter in South Bay salt ponds (although actual rates will vary by pond) to a height appropriate for vegetation colonization (San Francisco Bay Area Wetlands Ecosystem Goals Project 1999). In recent years, regulatory agencies have included the potential use of dredge material for restoration to accelerate the process of restoration (U.S. Army Corps of Engineers 1987).

Restoration Concerns

Many contaminants such as Hg and PCBs are tightly bound to sediment particles. As a result, the transportation of contaminants is closely tied to the movement of sediments. Use of dredge material in restoration projects has the potential to transport and reintroduce buried contaminants to the soil surface where it may be biologically available (Schoellhamer et al. 2003). Another concern of dredge material use is the potential incompatibility with surrounding substrate conditions. Dredge material may not complement the fine particle size of naturally occurring tidal wetlands, and soils may not support vigorous plant growth (Zedler 2001). Dredge spoils may have coarser soils and less clay content (Lindau and Hossner 1981), and consequently less soil organic matter and microbial activity (Langis et al. 1991).

Coarser substrates may have a decreased ability to retain nutrients (Boyer and Zedler 1998) and may fail to support the vegetation structure and height required for target species (Zedler 1993). Despite the concerns of amending soils with dredge material, data in relation to its use in restorations are scarce. In the Sonoma Baylands restoration project, dredge materials were used to accelerate the restoration process. However, development of channels and vegetation has been slow, presumably because of limited tidal exchange. Additional studies would provide more detailed analyses of the benefits of dredge materials against the costs.

SEA-LEVEL RISE

Projections for future sea-level rise in the SFBD vary between 30 and 90 cm in the 21st century, depending upon which climate projection models are used (Dettinger et al. 2003). An estimated 10–20 cm of that rise is expected, regardless of anthropogenic global-warming effects based on the historic rate of 20 cm/100 yr seen during the course of the 20th century (Ryan et al. 1999). The remainder is due to the combined influence of thermal expansion as the ocean warms in response to global warming, and accelerated melting of glaciers and ice caps. Galbraith et al. (2002) predicted a conversion of 39% of the intertidal habitat in San Francisco Bay to subtidal habitat by 2100, as high as 70% in the South Bay. In addition to the projected long-term rise in global sea level, considerable short-term variability can be expected due to local factors including tides, increased storm surges and changes in upwelling along the coast of California, all of which act on a range of mechanisms and timescales (Table 2). The SFBD tidal marshes and their flora and fauna now face potentially severe threats associated with sea-level rise, because the magnitude of change and the accelerated rate of rise over the next few decades, and because human activities around the marshes have probably dramatically reduced the marshes' capacity for coping with sea-level changes.

Potential Impacts

Actual sea-level rise can be seen in the tidal data at the Golden Gate from 1897–1999 (Fig. 5), which have changed at different rates (Malamud-Roam 2000). For example, while mean sea level has increased by about 20 cm during the 20th century, the height of mean higher high water (MHHW) and the highest highs have increased by about 25 cm and 28 cm per century, respectively (Malamud-Roam

TABLE 2. MECHANISMS, VERTICAL RANGE, TEMPORAL PATTERNS, AND OVERALL EFFECTS OF VARIATION IN HYDRODYNAMICS ON SEA-LEVEL RISE IN THE SAN FRANCISCO BAY ESTUARY.

Mechanism	Vertical range	Temporal pattern	Effect
Wind waves, swash, and run-up	<1–60 cm	Oscillatory; frequency in seconds	Large waves may become more frequent if storms increase; Swash and run-up may increase where the shoreline is hardened.
Seiches	Few centimeters	Oscillatory; frequency in minutes	Increasing estuary depth may increase the frequency or amplitude; impact likely minor.
Tides	1–2 m	Oscillatory; dominant frequencies 12.5 and 25 hr	If tidal range continues to increase, the height of HW relative to MSL will increase, and the frequency and mean depth of over-marsh flooding will probably increase.
Storm surge	10s of centimeters	Episodic; winter	Frequency and amplitude will probably increase due to global warming.
Lunar modulation of the tides	Range ~ 120 cm; HW height ~ 60 cm; DHQ and DLQ ~ 30 cm	Dominant (spring/neap) = 14.6 d; Minor (lunar declination) = 13.6 d	Dredging can alter the fortnightly circulation patterns induced by lunar tidal modulation.
Annual cycles of solar radiation	Monthly MSL ~ 10 cm; Monthly MHHW ~20 cm	Annual cycle of rainfall and runoff (ppt high in winter, runoff high in spring, both low in fall)	Global warming could change amount and timing of precipitation and therefore runoff and water surface slope in the estuary.
Precession of lunar node	None	Daylight flooding occurs in winter and night flooding in the summer, varying slightly over a 18.6-yr period	Photosynthesis, predation, and dispersal may all respond to this pattern of flooding, temperature, and light.
Probably solar annual cycles, possibly beat frequencies	10–20 cm	MSL has annual cycle, low in April and high in September and MHHW has semi-annual cycle, high in January and July	Global warming induced changes in oceanic circulation may change seasonal tidal mean patterns, but it is not clear how.
El Nino/southern oscillation	15–30 cm rise	Episodic, occurring about every seventh year. Duration of rise in Estuary 6–12 mo	ENSO frequency apparently increasing. If this continues or accelerates, the frequency, duration, and height of extreme high water will increase.
Secular (multi-year) climatic cycles	100 m increase over Holocene, slowing about 7,000 yr ago.	Eustatic (absolute) rise ~ 1 mm/yr; relative (local) rise ~ 2 mm/yr	Global warming could increase the rate of relative rise in the bay to 3–9 mm/yr for the next 100 yr.

Note: The abbreviations include: DHQ = mean diurnal high water inequality (one-half the average difference between the two high waters of each tidal day observed over the National Tidal Datum Epoch); DLQ = mean diurnal low-water inequality; HW = high water; MSL = mean sea level; MHHW = mean higher high water; ENSO = El Niño southern oscillation.

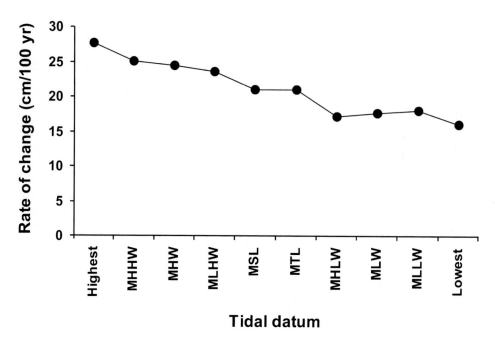

FIGURE 5. Rate of change in centemeters/100 yr in tidal data at the Golden Gate of the San Francisco Bay estuary from 1897–1999, including: MHHW = mean higher high water, MHW = mean high water, MLHW= mean lower high water, MSL = mean sea level, MTL = mean tide level, MHLW = mean higher low water, MLW = mean low water, MLLW = mean lower low water.

2000). Thus, for the animals living on the marsh surface, sea level has risen by an effective 25–28 cm in the last century. Should this pattern of a faster rate of change for the highest tides continue into the next century, projections of sea-level rise may underestimate the actual rise in sea level for the animals living in the tidal marshes. Furthermore, most global-warming scenarios for this region predict an increased frequency and severity of storm surges due to global warming (Cubasch and Meehl 2001).

Sea-level rise poses three major risks to wildlife that occupy the surrounding tidal marshes and mudflats.
1. Higher sea levels may drown the marshes and mudflats or increase storm surges causing greater shoreline erosion and habitat loss.
2. Increased frequency or duration of extreme high tides can to lead to higher mortality of songbird eggs and young birds during extreme high-tide floods (Erwin et al., *this volume*; Point Reyes Bird Observatory, unpubl. data). In addition amplified high tides may increase the vulnerability of the salt marsh harvest mouse (*Reithrodontomys raviventris*), California Clapper Rails (*Rallus longirostris obsoletus*), California Black Rails, and other marsh inhabitants to predation as they seek refuge upland.
3. Increased salinity intrusion into inland areas of the estuary can change plant assemblages and alter habitat for wildlife, such as the Common Moorhen (*Gallinula chloropus*).

CHANGING SALINITY

As sea level rises at the Golden Gate, salinity can be expected to intrude further up-estuary. In terms of salinity, the effect of rising sea level can be analogous to a decrease in fresh-water inflow, as the depth of the bay has remained fairly constant throughout the Holocene (Ingram et al. 1996). Vegetation patterns in the marshes have changed significantly during the last century, reflecting a significant increase in salinity (May 1999, Byrne et al. 2001, Malamud-Roam 2002). For example, pollen cores from the Petaluma River have shown dramatic shifts from brackish waters (characterized by tules) around 1,800 yr ago to saline conditions (characterized by pickleweed) around 1,400–800 years ago, and back again to brackish conditions around 750 yr ago (Byrne et al. 2001). Climate explains only a part of the change in estuarine salinity; water diversion for agriculture and human consumption explains the majority of the change (Peterson et al. 1995).

As sea level rises, the tidal-marsh habitats (fresh, brackish, and salt) can be expected to change as vegetation responds to higher salinity conditions. For some organisms, the changes may be favorable, with a greater area covered by saltmarsh habitat. Other organisms that rely on fresh water habitat, such as the Common Moorhen, may face significant loss of habitat, and a shift up-estuary in their range as the salt wedge approaches the delta. Strategies to compensate for the increase in marine influence due to sea-level rise could include increasing the area of marsh habitat within the delta or increased fresh water flows through the delta (requiring a decline in water diversion).

Restoration Concerns

A critical synergistic threat facing marsh inhabitants is the combined force of a sediment deficit and sea-level rise. The modern supply of sediments to the estuary has been significantly altered in historic times due to human modifications of the hydrologic system. However, future predictions based on historical rates of rise, and potential increases resulting from global warming, suggest significant losses of saltmarsh habitat in the South Bay and more significant losses of tidal mudflats, which are critical for shorebird species (Galbraith et al. 2002).

The North Bay, however, does not have a history of high subsidence as in the South Bay, though it is largely unknown whether the rates of sediment supply will be adequate to maintain marsh surface elevations in the future. Studies of long-term rates of sediment accretion in marshes in San Pablo Bay, the Carquinez Strait, and Suisun Bay indicate that for the last 3,000 yr the average rates of sediment accretion have matched sea-level rise (Goman and Wells 2000, Byrne et al. 2001, Malamud-Roam 2002). However, the long-term records also indicate that there were periods when sediment supply was clearly inadequate to maintain the marshes, resulting in a conversion of some areas to subtidal conditions (Goman and Wells 2000, Malamud-Roam 2002).

CONTAMINANTS

The SFBD has numerous sources of pollution, many of which are particle-bound, and their concentrations fluctuate with suspended sediment concentrations (Schoellhamer et al. 2003). Most restoration projects in the estuary are dependent on sediment inputs to elevate marsh plains; however, three sediment-associated contaminants including mercury (Hg), selenium (Se), and polychlorinated biphenyls (PCBs) are listed as priority pollutants in SFBD under section 303(d) of the Clean Water Act. Sediment-bound contaminants or those re-suspended by dredging operations or changes in sedimentation dynamics pose a potential threat to tidal-marsh vertebrates when transported into wetlands. Although numerous other pollutants occur in the bay, we focus on those suspected to have direct bearing on tidal marsh health and food webs.

One of the most prevalent contaminants is Hg. Mercury extracted from the Coast Range was used to recover gold and silver in Sierra Range mining operations during the Gold Rush era (Alpers and Hunerlach 2000). Between 1955 and 1990, the SFBD received an average of 6,030,000 m^3 of Hg-laden sediments annually (Krone 1996). Methylmercury (MeHg), a more toxic and readily bioaccumulated form (Marvin-DiPasquale et al. 2003), is magnified by passing through food webs with transformation initiated in phytoplankton. Acute toxicity in fish, birds, and mammals damages the central nervous system (Wolfe et al. 1998, Wiener et al. 2003), while lower-level exposure affects reproduction in vertebrates (Wiener and Spry 1996, Wolfe et al. 1998).

Human-health advisories for fish have been in effect since 1970, and Hg concentrations in striped bass have not changed significantly since that time (Fairey et al. 1997, Davis et al. 2002). Highest liver Hg concentrations for small mammals have been found in South Bay wetlands (Clark et al. 1992). A recent study documented that eggs of Forster's Tern (*Sterna forsteri*) and Caspian Tern (*Sterna caspia*) foraging in South Bay salt ponds and adjacent sloughs contained the highest concentrations (7.3 and 4 7 mg/kg total Hg) of any bird species in the estuary (Schwarzbach and Adelsbach 2002). Endangered California Clapper Rails exhibited depressed hatchability and embryo deformities with egg concentrations exceeding the lowest observed adverse effect concentration (LOAEC) of 0.5 µg/g (Schwarzbach et al. 2006; Novak et al., *this volume*).

Selenium (Se) is another persistent threat to tidal-wetland vertebrates. Se is substituted for sulfur in enzymes, resulting in reproductive failure and teratogenesis. Sources include oil refinery effluent and agricultural drainwater. Selenium loads from refineries have decreased (6.8 kg/d–1.4 kg/d) with recent regulation (San Francisco Bay Regional Water Quality Control Board 1992), but agricultural sources range between 20.4–53.2 kg/d (Luoma and Presser 2000). Dissolved Se concentrations are consistently <1 µg/L (Cutter and San Diego-McGlone 1990, San Francisco Estuary Institute 2003), below chronic criterion for aquatic life (2 µg/L). However, Se concentrations may be elevated because chemical speciation controls bioaccumulation (Luoma and Presser 2000). For example, Black-necked Stilts (*Himantopus mexicanus*) nesting at Chevron Marsh in Richmond, California, had eggs (20–30 µg/g dw) with similar concentrations found at Kesterson Reservoir (25–37 µg/g dw). Se in source water was 10% of the concentrations at Kesterson, but uptake was enhanced because the form of Se was selenite, the most bioavailable species.

Despite reduction in Se since 1998, concentrations in sturgeon and diving ducks remained elevated, possibly because of invasion of the Asian clam (*Potamocorbula amurensis*), a species that concentrates Se more than other clams (White et al. 1989, Linville et al. 2002, San Francisco Estuary Institute 2003). In a recent survey of wetland birds, Se concentrations in eggs ranged from 1.5 µg/g dw in Snowy Plover (*Charadrius alexandrinus*) eggs to 4.2 µg/g dw in Snowy Egret (*Egretta thula*) eggs, and 1.6 µg/g dw in failed California Clapper Rail eggs (Schwarzbach and Adelsbach 2002) compared with a threshold for reproductive problems of 6–10 µg/g dw (Heinz 1996). In birds, Hg and Se are toxicologically antagonistic, where exposure to one of these elements protects individuals from the toxic effects of the other (El-Begearmi et al. 1997); however, Heinz and Hoffman (1998) showed that the most environmentally realistic and most toxic forms of mercury (methylmercury) and selenium (selenomethionine) combined caused lower hatching success in Mallards (*Anas platyrhynchos*) than either contaminant alone.

Polychlorinated biphenyls (PCBs), synthetic chlorinated aromatic hydrocarbons with a wide variety of industrial uses, have been banned in the US since 1979. However, PCB concentrations in SFBD water remain high, and 83% of samples taken during 2001 by Regional Monitoring Program (San Francisco Estuary Institute 2003) exceeded water quality objectives. Sources to the estuary cannot be pinpointed, but are thought to be mainly historical. Runoff from creeks and tributaries has been identified as one significant source of PCBs (San Francisco Estuary Institute 2003).

PCBs are accumulated in fat and readily increase from trophic level to trophic level in estuarine food webs. Although PCB toxicity depends on the structure of individual congeners, general effects include thymic atrophy, immunotoxic effects, endocrine disruption, reproductive impairment, porphyria, and liver damage (Hoffman et al. 1996). Failed California Clapper Rail eggs collected during 1992 in the South Bay had total PCB concentrations between 0.65–5.01 µg g-1 (Schwarzbach et al. 2001). Rail sensitivity to PCB congeners is not known, but the authors indicate such concentrations may cause reduced hatching success.

Section 303(d) of the Clean Water Act lists elements copper (Cu) and nickel (Ni), as well as organic contaminants DDT, chlordanes, dieldrin, dioxins, and furans as priority pollutants. In addition, sediment monitoring showed that arsenic (As) and chromium (Cr) consistently exceeded guidelines at several sites (San Francisco Estuary Institute 2003). Sediment toxicity could threaten tidal-marsh vertebrates by decreasing their invertebrate prey base (San Francisco Estuary Institute 2003). Polybrominated diphenyl ethers (PBDEs), used as flame retardants, have been increasing in sediment and biota, and Caspian Tern eggs from SFBD contain the highest concentrations found in any bird species (T. Adelsbach, USDI Fish and Wildlife Service, pers. comm.). Tributyl tin (TBT), an anti-fouling additive in paint, is an endocrine disrupter that may pose a threat to wetland organisms by increasing through wetland food webs (Pereira et al. 1999).

Oil spills remain a threat, because California is the fourth largest oil-producing state and the third largest crude-oil-refining state in the nation. Six oil refineries are located in the bay and comprise nearly 40% of the state's total oil production capacity (California Energy Commission 2003). More than a thousand tanker ships pass through the estuary each year, along with countless container ships, recreational boats, and other vessels. The tanker Puerto Rican spilled 5,678,118 l of oil in 1984, killing approximately 5,000 birds, while a Shell Oil storage tank spilled 1,589,873 l in 1988. Over an extended period of time (from at least 1992–2002, though likely for many years earlier), heavy fuel oil leaked out of the freighter S.S. Jacob Luckenbach that sank southwest of the Golden Gate Bridge in 1953 (Hampton et al. 2003). Roughly 378,541 l of oil were removed from the wreck, but over 18,000 bird mortalities were estimated in 1997 and 1998 from this one source (Hampton et al. 2003). In 2002, of 6,867 oil spills reported in California, 445 were in the SFBD (Office of Spill Prevention and Response 2003).

RESTORATION CONCERNS

Projects in the bay that re-expose, accrete, or use dredged Hg-laden sediments may pose risks to fish and wildlife. This is especially true in tidal marshes where sulfate-reducing bacteria in anoxic sediments can transform inorganic Hg to MeHg. Boundary zones of oxic-anoxic areas in marshes have particularly high MeHg production. Site-specific parameters such as dissolved or organic carbon content, salinity, sulfate, and redox cycles also influence biotransformation rates and subsequent bioaccumulation in tidal wetland organisms (Barkay et al. 1997, Kelly et al. 1997, Gilmour et al. 1998).

WATER QUALITY

Water quality in the SFBD has changed dramatically due to human activity during the past 150 yr, often to the detriment of estuarine ecosystems. The impact of water quality change on tidal-marsh terrestrial vertebrates is virtually unstudied but could be severe if left unmanaged. In this section, we discuss water-quality threats exclusive of toxic contaminant and sediment budget issues addressed earlier. Changes in water quality probably affect vertebrate populations indirectly via long-term changes in the vegetative and invertebrate communities that inhabit the marsh and short-term cascading trophic effects starting with aquatic organisms low in the food web.

The SFBD is a highly variable environment commonly experiencing both wet and dry extremes within a year as well as wet and dry years. Humans have changed the timing and extent of these fluctuations as well as the chemistry of waters entering the estuary. Under current management regimes, river inflow and freshwater diversions can vary between years by as much as 25 times (Jassby et al. 1995). While native organisms are adapted to the seasonal fluctuations, recent and future extremes may tax their survival and reproductive capabilities.

Salinity

Salinity, a key variable determining tidal-marsh vegetative and invertebrate community composition, has changed significantly in recent decades. A decrease in salinity can convert a saltmarsh to fresh-water marsh, as in the tidal wetlands near the San Jose and Santa Clara Water Pollution Control Plant, where up to 567,811,800 l of treated fresh wastewater empty into south San Francisco Bay every day. Despite fresh-water flows from treatment plants, overall salinity in the estuary has increased because of a reduction in fresh-water flows from the delta to 40% of historical levels, due to water diversion for agriculture, municipal use, and local consumption (Nichols et al. 1986). Pollen and carbon isotope data from sediment cores from Rush Ranch in Suisun Bay clearly indicate a shift since 1930 in the dominance of marsh vegetation from freshwater to salt-tolerant plants as a response to increased salinity due to upstream storage and water diversion. This recent shift in the vegetative community was as extreme as the shift that occurred during the most severe drought of the past 3,000 yr (Byrne et al. 2001).

When freshwater flows are reduced, salt intrusion up-estuary can become a problem (Nichols et al. 1986), particularly in high-marsh areas that already become hyper-saline during certain times of the year. For example, historically large populations of the endangered San Francisco garter snake (*Thamnophis sirtalis tetrataenia*), have declined presumably due to the loss of several prey species from saltwater intrusion into less saline marsh habitats (San Francisco Bay Area Wetlands Ecosystem Goals Project 2000). Salt intrusion is a cause for concern particularly for tidal-marsh restoration projects in the Delta. When levees are breached, the total tidal prism increases, which allows saline waters to travel farther up-estuary during high tides. Change in salinity regimes may also facilitate the proliferation of invasive species. The Asian clam gained a foothold in Suisun Bay during a period of extreme salinity changes, with far-reaching consequences (Nichols et al. 1990).

Marsh vertebrates may be adapted to conditions that are specific to particular salinity ranges. For example, adaptations to kidney structures in the salt marsh harvest mouse may have allowed this species to use saline environments (MacMillen 1964). Zetterquist (1977) trapped the greatest numbers of salt marsh harvest mice in highly saline tidal marshes. Whether the high density of harvest mice in these marshes was due to an affinity for high salinity or lack of competitors is not known. In addition, Song Sparrow subspecies in San Pablo Bay saltmarshes and Suisun brackish marshes have different bill sizes and plumage colors that may reflect adaptation to local conditions (Marshall 1948). Because of the close association between vertebrate subspecies and tidal marshes of a specific salinity, long-term salinity changes could affect both the persistence and the evolution of endemic tidal-marsh vertebrates. Changes in salinity can affect the distribution of invertebrates as well, such as the winter salt marsh mosquito (*Aedes squamiger*), which is found not in freshwater, but in brackish and saline habitats.

PRODUCTIVITY

Water management can greatly affect primary productivity in the estuarine waters, which likely impacts marsh vertebrates that forage extensively on invertebrates and fish in the tidal channels. Phytoplankton forms the base of the food web that includes most of the wildlife in the estuary (Sobczak et al. 2002). Sometimes, productivity is low enough to affect fish growth and mortality, which may suggest that other vertebrates become food-limited as well (Jassby and Cloern 2000). Beneficial phytoplankton blooms occur when the null zone (the location where freshwater outflow balances saltwater inflow) is positioned across the broad shallows of Suisun Bay (Jassby et al. 1995). Management of delta inflows may alter the position of the null zone to facilitate optimal phytoplankton blooms and may be important for maintaining the food supply of vertebrates that forage in tidal channels.

EUTROPHICATION

Wastewater and runoff comprise a significant percentage of freshwater entering the estuary (Nichols et al. 1986). The main threat from these waters is toxic contaminants, but nutrient loading is also a concern. In past decades, summer die-offs due to eutrophication and ensuing oxygen depletion occurred in the South Bay. These events have not recurred following the institution of improved sewage treatment methods in 1979, although nitrogen concentrations remain high (Kockelman et al. 1982). Oxygen depletion continues to be a problem in areas of the delta when water retention times are long (B. Bergamaschi, U.S. Geological Survey, pers. comm.).

Eutrophication in the estuary is controlled by two factors. First, high turbidity causes algae to be light-limited, so nuisance blooms do not occur on the same scale as in other polluted bays (Jassby et al. 2002). Second, benthic bivalves consume much of the primary production and reduce the availability of nutrients in the water column (Cloern 1982). However, turbidity may decline in the near future due to retention of sediment behind dams and channel armoring (International Ecological Program 2003).

RESTORATION CONCERNS

Many agencies and hundreds of scientists are working to understand and manage water quality, although most of the focus is on open-water habitats rather than tidal wetlands. Delta inflow is currently managed to allow for movements of fish populations and keep summer primary productivity at optimal levels in the Suisun Bay. The Environmental Protection Agency suggested guideline maximum salinity levels for sensitive areas of the estuary. Investigators at the U. S. Geological Survey, California Department of Water Resources, Interagency Ecological Program, and Stanford University continue to monitor and model water quality in the estuary as well as conduct original research experiments.

HUMAN DISTURBANCE

Residents and visitors are drawn to the waters, shorelines, and wetlands in this highly urbanized estuary for aesthetic and recreational opportunities. The needs of >8,000,000 people in the SFBD estuary encroach upon the many wildlife populations dependent on estuarine habitats. For example, wetlands attract large numbers of visitors (10,000/yr, >75% of surveyed sites) for recreational activities such as bird watching and jogging (Josselyn et al. 1989). Although natural disturbances may create areas used by birds (Brawn et al. 2001), anthropogenic disturbances that elicit a metabolic or behavior response (Morton 1995) generally reduce the value of habitats for birds (Josselyn et al. 1989).

TYPES OF HUMAN DISTURBANCE

People traverse marshes on vehicles, bicycles, and on foot—jogging or walking through the areas along roads or trails. Automobiles on established roads may not greatly affect the behavior of wildlife, but boats and aircraft may be highly disruptive by causing animals to flush, exposing them to predators. The activity or noise of boats and personal watercraft may cause avoidance or flight behavior (Burger 1998), especially near waterways, or result in trampled vegetation, while wakes from boats may erode bank habitats. Low-flying aircraft may create large disturbances in estuaries (Koolhaas et al. 1993), especially near small airports or in agricultural areas where aerial spraying is used. Longer and more extensive use of areas by campers, fishermen, hunters, and researchers may have greater individual short-term effects, while cumulative long-term effects of visitors on trails or boardwalks may effectively decrease the size of the marsh, provide pathways for predators, and degrade the value of edge habitats.

TYPES OF EFFECTS

Knight and Cole (1991) described hierarchical levels of disturbance on wildlife species from death or behavior change (altered behavior, altered vigor, altered productivity, and death), to population change (abundance, distribution, and demographics), and finally community alteration (species composition, and interactions). In their studies of shorebirds in estuaries, Davidson and Rothwell (1993) described effects at local (movement) and estuary (emigration) levels, as well as impacts at estuary (mortality) and population (decline) scales. The adverse effects of human disturbance for waterbirds include loss of areas for feeding or roosting, elevated stress levels, and reduced reproduction including abandonment of nests or nestlings, as well as changes in behavior including avoidance of areas, reduction in foraging intensity, or feeding more at night (Pomerantz et al. 1988, Burger and Gochfeld 1991, Pfister et al. 1992, Burger 1993).

Human incursions into marshes may cause trampling of vegetation and soil compaction, reducing the quality of the habitats. Obligate saltmarsh species may be particularly sensitive to disturbance, especially because many are adapted to avoid predators by hiding within the vegetation and may avoid areas with repeated disturbance. Although habituation occurs for many species, birds displaced from coastal marshes were observed to fly to distant marshes rather than return to the same areas (Burger 1981), and disturbances displaced shorebirds from beaches (Pfister et al. 1992). Boats, especially small recreational boats, may be a particularly large disruption for waterbirds that flush at great distances (Dahlgren and Korschgen 1992).

DISTURBANCE LIMITS AND BUFFERS

Josselyn et al. (1989) found that long-legged waders in estuary marshes flushed when approached from 18–65 m, while waterfowl flushed from 5–35 m. Waterbirds avoided areas near paths where people traveled, and their behavioral responses were noted at distances of <50 m (Klein 1993). Green Heron (*Butorides striatus*) numbers were inversely proportional to the number of people at a site, and individuals that remained foraged less frequently (Kaiser and Fritzell 1984). Flushing distance was related to larger size and mixed-composition of flocks in waterfowl (Mori et al. 2001).

Stress responses are more difficult to detect. Hikers, joggers, and dogs, as well as avian predators, disturb the federally threatened Snowy Plover, prompting closure or fencing beach areas. Waterbirds avoided areas near paths where people traveled, and adverse behavioral responses were noted at distances of <50 m (Klein 1993). In addition, physiological monitoring studies have found elevated heart-rate levels in breeding marine birds (Jungius and Hirsch 1979) and wintering geese (Ackerman et al. 2004) when approached (<50 m). In response to these findings, many government agencies support regulatory buffer distances of 31 m including the California Coastal Act buffer between developments and wetlands. Studies have shown that buffers are effective if they are large enough; 35% of buffers <15 m had direct human effects (Castelle et al. 1992), but larger buffers up to 100 m were found to be effective for waterbirds (Rodgers and Smith 1997).

RESTORATION CONCERNS

Most tidal marshes in the SFBD are adjacent to urban development or levees. Unfortunately, tidal marshes are often fragmented by levees that have a narrow and steep transition from high marsh to upland. The salt marsh harvest mouse may be found in upland areas up to 100 m from the wetland edge during high tides (Botti et al. 1986, Bias and Morrison 1999). Other marsh inhabitants, such as the California Clapper Rail and the California Black Rail, use upland transitional areas as refuge from high tides; however the narrow width of these zones makes rails more vulnerable to predation. During prolonged flooding from high tides, Suisun shrews (*Sorex ornatus sinuosus*) utilize upland habitats for cover and food (Hays and Lidicker 2000). Without consideration of adjacent habitats, tidal-marsh restoration may fail to support target species. In addition, many restoration projects include public access to the marshes. How and where such access is allowed may greatly influence the value of the tidal marshes for some species.

INVASIVE SPECIES

The SFBD is perhaps the most highly invaded estuary in the world (Cohen and Carlton 1998). Many of the plants now found in the tidal marshes, most of the invertebrates in the marsh channels and adjacent tidal flats, and several terrestrial animals that forage in tidal marshes are not native to the Pacific Coast. Here, we focus on the exotic species that are most likely to harm the native vertebrates in these tidal marshes. An exotic disease, the West Nile virus (WNV), which is expected to soon appear in estuary tidal marshes, is discussed below.

PLANT INVASIONS

Grossinger et al. (1998) recommended that monitoring, research, and control efforts focus on the three exotic plants that have the widest distribution in tidal marshes: smooth cordgrass (*Spartina alterniflora*), dense-flowered cordgrass (*Spartina densiflora*) and broad-leaved peppergrass (*Lepidium latifolium*), and that four other exotic plants that as yet have a very limited distribution in these marshes be monitored: common cordgrass (*Spartina anglica*), saltmeadow cordgrass (*Spartina patens*), opposite leaf Russian thistle (*Salsola soda*), and oboe cane (*Arundo donax*). Of these species, smooth cordgrass and dense-flowered cordgrass are currently the most widespread and are likely to negatively impact native tidal-marsh vertebrates because they can become very abundant in the marsh plain (mid- to upper-marsh zones) (Ayres et al. 1999, Faber 2000). The marsh plain, which is naturally dominated by low-growing native cordgrass species provides key habitat for most of the resident tidal-marsh birds and mammals including the salt marsh harvest mouse (Shellhammer et al. 1982). Several native bird species including the federally endangered California Clapper Rail, the California Black Rail, and the three tidal-marsh Song Sparrow subspecies that are state species of special concern (Alameda, San Pablo, and Suisun), nest and forage in cordgrass (Johnston 1956a, b; San Francisco Bay Area Wetlands Ecosystem Goals Project 2000). Where smooth cordgrass and dense-flowered cordgrass become abundant, they can potentially alter marsh habitat by changing the vegetative structure (including canopy height, density, and complexity), subcanopy physical conditions, root density and soil texture, sediment deposition and erosion rates, and perhaps ultimately marsh elevation, marsh topography, and channel morphology (Callaway and Josselyn 1992, Daehler and Strong 1996, Faber 2000). However, few data are available on how these changes would affect tidal-marsh vertebrates.

Most research on the impacts of exotic plants on tidal-marsh vertebrates in the SFBD has focused on smooth cordgrass. This Atlantic cordgrass was introduced in the early 1970s and over the next decade began to spread and hybridize with the native California cordgrass (*Spartina foliosa*; Ayres et al. 1999). This exotic cordgrass (*Spartina alterniflora*, *S. alterniflora* × *foliosa*, or both) is highly productive and now occurs in >2,000 ha of marsh and tidal-flat habitats (Ayres et al. 1999). The dramatic alteration of tidal-marsh habitat by the tall, thick, exotic cordgrass will likely affect resident species the most. Native vertebrate species do occupy invaded marshes; however, it is unclear whether these subpopulations are sustainable (Guntenspergen and Nordby, *this volume*). For the California Clapper Rail, the biggest problem may be the loss of foraging habitat and food resources in invaded marsh channels where, during low tides, the rails do much of their foraging (Albertson and Evens 2000).

One ongoing study in the South Bay is investigating the impacts on Alameda Song Sparrows through changes in flooding regimes and interspecific interactions among native species. Preliminary analysis shows that Song Sparrow nests in exotic cordgrass are much more likely to flood than nests placed in native vegetation and so reproductive success may be lower in invaded marsh habitat (J. C. Nordby and A. N. Cohen, unpubl. data). The invasion may also be altering interactions among native species. Marsh Wrens (*Cistothorus palustris*), which are native to fresh- and brackish-water marshes, are now occupying invaded saltmarshes (J. C. Nordby and A. N. Cohen, unpubl. data). An increase in Marsh Wren density is potentially detrimental for Song Sparrows and other saltmarsh birds because Marsh Wrens are highly aggressive and are known to break the eggs of other species occupying adjacent nesting territories (Picman 1977, 1980). The addition of interference competition from Marsh Wrens could reduce the reproductive success and overall distribution of other saltmarsh-nesting bird species.

INVERTEBRATE INVASIONS

Exotic benthic invertebrates far outnumber natives in the marsh channels and mudflats, comprising >90% of the number of individuals and benthic biomass over most of the estuary (Cohen and Carlton 1995). Marsh birds and shorebirds must commonly feed on exotic invertebrates including clams, mussels, snails, and worms, and probably also ostracods, amphipods, and crabs (Carlton 1979). Moffitt (1941) found the Atlantic mussel (*Geukensia demissa*) abundant (57% of food by volume) and the Atlantic snail (*Ilyanassa obsoleta*) uncommon (2% of food) in the stomachs of 18 California Clapper Rails from South Bay. Williams (1929) observed California Clapper Rails feeding heavily on the western Atlantic clam (*Macoma petalum*), while ignoring the Atlantic snail. Overall, though, little specific information exists on what the tidal-marsh birds eat. Nor is it known what effect, if any, the replacement of native prey items by exotics has had on these birds—whether more or less food is available, whether it is more or less nutritious, or whether it is more or less contaminated by toxic pollutants than native prey.

De Groot (1927) reported in detail one impact of the Atlantic mussel. This mussel was first found in the SFBD in 1894, probably introduced in oyster shipments (Cohen and Carlton 1995). It quickly became abundant along channel banks and in the outer portions of cordgrass marshes where it typically lies partly buried so that the posterior margin of its shell protrudes just above the mud with the two valves slightly open. De Groot (1927) reported that the toes or probing beaks of rails were frequently caught and clamped between these valves. He estimated that at least 75% of adult rails had lost toes, others starved from having their beaks clamped shut or injured, and one–two nestlings per brood were caught by mussels and drowned by the incoming tide. Whether or not these injury and mortality estimates were valid, more recent observations confirm that California Clapper Rails in the SFBD are frequently missing one or more toes (Moffitt 1941, Josselyn 1983, Takekawa 1993), and Takekawa (1993) reported that a rail captured with a mussel clamped onto its bill subsequently lost part of its bill.

Exotic invertebrates may also have an indirect impact on tidal-marsh vertebrates by altering habitat. The southwestern Pacific isopod (*Sphaeroma quoyanum*) was first collected in the SFBD in 1893. It burrows abundantly in mud and clay banks along the channels and outer edges of saltmarshes, and has been credited with eroding substantial areas of marsh, though no direct studies or measurements have been made (Cohen and Carlton 1995). Two recently arrived crabs, the European green crab (*Carcinus maenas*, first seen in the estuary in 1989–1990) and the Chinese mitten crab (*Eriocheir sinensis*, first collected in the estuary in 1992), are also known to be common burrowers in saltmarshes or tidal channels (Cohen et al. 1995, Cohen and Carlton 1997). If these organisms do in fact contribute to the erosion and loss of marsh habitat, this would clearly have an impact on marsh vertebrates. The impacts could be greatest near marsh channels, since the slightly elevated areas alongside these channels are better drained and support taller vegetation, providing better nesting sites and habitat for saltmarsh Song Sparrows, and possibly for California Clapper Rails, and the salt marsh harvest mouse (Marshall 1948; Johnston 1956a, b; Shellhammer et al. 1982, Collins and Resh 1985, Albertson and Evens 2000).

VERTEBRATE INVASIONS

Red foxes (*Vulpes vulpes*) from Iowa or Minnesota were introduced into California in the last half of the 19th century either released by hunters or escaped from commercial fox farms. A wild population became established in the Sacramento Valley, and from this and other centers, red foxes spread to the East Bay region by the early 1970s. They were observed in the San Francisco Bay National Wildlife Refuge (SFBNWR) in the South Bay by 1986 and have continued to expand their range (Foerster and Takekawa 1991, Harvey et al. 1992, Cohen and Carlton 1995). Dens have been found in tidal saltmarshes and in adjacent levee banks. Red foxes prey on resident California Clapper Rails, Black-necked Stilts, American Avocets (*Recurvirostra americana*), and Snowy Egrets and on various other marsh and aquatic birds and mammals, including endangered endemic species (Forester and Takekawa 1991, Harvey et al. 1992, Albertson 1995).

The SFBNWR began a program of trapping and killing red foxes in 1991 (Foerster and Takekawa 1991, Cohen and Carlton 1995). Recent surveys show a strong recovery in local populations of California Clapper Rail following implementation of the red fox removal (Albertson and Evens 2000). In the early 1980s California Clapper Rail numbers in the South Bay were estimated at 400–500. The local population crashed to roughly 50–60 in 1991–1992 surveys, roughly 5 yr after the first detection of red foxes at the SFBNWR. In 1997–1998 winter surveys, rail numbers increased to 330. Because California Clapper Rails are year-long residents and have strong site tenacity, the variation between survey years is not thought to be from dispersal or migration.

Brown rats (*Rattus norvegicus*) became established in many parts of California by the 1880s. In the SFBD, brown rats are common in riparian areas, in fresh, brackish and saltwater tidal marshes, and in diked marshes (Josselyn 1983, Cohen and Carlton 1995). De Groot (1927) considered the brown rat to be the third most important factor in the decline of California Clapper Rail, after habitat destruction and hunting. More recent authorities (Harvey 1988, Foerster et al. 1990, Foerster and Takekawa 1991, Cohen and Carlton 1995) have also found substantial predation on California Clapper Rail eggs and chicks, with some estimating that brown rats take as many as a third of the California Clapper Rail eggs laid in the South Bay (Harvey 1988). Rats also prey on other marsh-nesting birds and their nest contents. Because brown rats are more likely in areas that abut urban development, habitat buffers might reduce their abundance in tidal marsh.

House cats (*Felis domesticus*) are widespread in California both as house pets and feral individuals. In the SFBD, house cats have frequently been seen foraging in saltmarshes, along salt-pond levees, and wading at the edge of tidal sloughs (Foerster and Takekawa 1991). House cats are known to have killed adult Light-footed Clapper Rails (*Rallus longirostris levipes*) in southern California (Foerster and Takekawa 1991) and at least one California Clapper Rail in the SFBD (Takekawa 1993), and presumably also prey on other marsh birds and mammals. The SFBNWR began a program of removing feral cats in 1991.

RESTORATION CONCERNS

Exotic cordgrass colonization is a threat for most South Bay tidal-marsh restoration projects due to its gross alteration of vegetative structure, sub-canopy physical conditions and root density, and potential alteration of soil texture, sediment deposition and erosion rates, marsh elevation, marsh topography and channel morphology, which could in turn affect native plant and invertebrate populations, as well as vertebrate populations. Although a regional exotic cordgrass control program began in 2004, complete control may take many years to achieve. It is likely to remain a significant issue in most restoration of this subregion and an imminent threat to the North Bay. In contrast, few areas are invaded by exotic cordgrass in the North Bay; thus, vigilant monitoring and removal of hybrid populations would be highly beneficial, and restoration of tidal marshes with native cordgrass may be more successful in this subregion. Control of nonnative predators will likely be an essential part of most tidal-marsh restoration projects to maintain native fauna. Nonnative predators will be of most concern in restoration areas adjacent to urban development.

PREDATION

Increased rates of predation on tidal-marsh vertebrates can result from three types of human-induced changes: (1) introduction of non-native predators, (2) changes in the distribution or abundance of native predators, and (3) alterations of habitat that influence predation effectiveness or avoidance. For birds and other vertebrates in tidal saltmarshes of the SFBD, as in most other ecosystems, predation is generally the dominant cause of adult and juvenile mortality and nest failure (Point Reyes Bird Observatory, unpubl. data). Although the primary cause of significant declines in populations of tidal-marsh vertebrates is habitat loss and degradation, other factors may be contributing to further population declines through increased predation: habitat fragmentation (Schneider 2001, Chalfoun et al. 2002), loss of vegetated upland edges for use as refugia from predators during high tides, establishment of boardwalks and power lines across marshes (the latter are used as perches by raptors), changes in marsh vegetation structure and the spread of urban-tolerant native predators (e.g., American Crow [*Corvus brachyrynchos*] Common Raven [*Corvus corax*], raccoon [*Procyon lotor*], and striped skunk [*Mephitis mephitis*]), feral animals (house cats) and other non-native predators (especially red fox), many of whom are human subsidized in urban and suburban areas (USDI Fish and Wildlife Service 1992). Sanitary landfills and riprap shorelines are also sources of predators (USDI Fish and Wildlife Service 1992).

In the SFBD, tidal-marsh fragmentation has resulted in an increased mean perimeter to edge ratio, and thus more edge per unit area. Predators are hypothesized to be more active at habitat edges. Current studies of the relationship between edge habitat and predation indicate that nest predators vary in activity and impact depending on the taxon and the surrounding land use (Chalfoun et al. 2002). Studies in estuarine marshes indicate that patterns of predation vary between sites (PRBO Conservation Science, unpubl. data). This variation is probably due to differences in the suite of predators, which itself may be dependent on variation in land use on adjacent uplands and vegetation type, and on variation in tidal flooding, channel and levee configuration, marsh vegetation structure and human disturbance patterns. Some changes in vegetation involving increases in vegetation density, such as that associated with the spread of invasive smooth cordgrass may actually result in decreased nest predation, but with ecological trade-offs (Guntenspergen and Nordby et al., *this volume*).

Several predators have been documented to depredate tidal-marsh birds, bird nests, reptiles, and mammals. They include upland mammal species that forage in marshes such as: raccoon, red fox, coyote (*Canis latrans*), striped skunk, house cat, domestic dog (*Canis familiaris*), house mouse (*Mus musculus*), brown rat, and black rat (*Rattus rattus*); wetland mammals such as the river otter (*Lutra canadensis*); snakes such as gopher snake (*Pituophis melanoleucus*) and garter snake (*Thamnophis* spp.) which have been observed swallowing nest contents (Point Reyes Bird Observatory, unpubl. data); and numerous wetland birds including Great Blue Heron (*Ardea herodeus*), Great Egret (*Casmerodius albus*), Snowy Egret, Black-crowned Night Heron (*Nycticorax nicticorax*), and gull species (USDI Fish and

Wildlife Service 1992, Albertson and Evens 2000; Point Reyes Bird Observatory, unpubl. data). Raptors, especially Northern Harrier (*Circus cyaneus*), White-tailed Kite (*Elanus leucurus*), and Red-tailed Hawk (*Buteo jamaicensis*) are also documented predators as are the Common Raven and American Crow. And finally, the nest parasite and egg predator, the Brown-headed Cowbird (*Moluthrus ater*) has been documented in SFBD tidal marshes, although rates of parasitism vary greatly among marshes (Greenberg et al., *this volume*). Some of these species, such as the Northern Harrier, nest in or near marshes and have probably always been part of the tidal-marsh food web. Other species, such as Common Ravens, American Crows, and raccoons, have adapted well to urban areas, and their large populations have resulted in increased predation in adjacent natural areas.

Restoration Concerns

Control of red foxes and other non-native predators in the south San Francisco Bay has contributed to a rebound in California Clapper Rail numbers. However, predator control is not a viable option in all parts of the estuary, and other measures to reduce predation, e.g., by modifying habitat, may have better long-term results (Schneider 2001). More studies are necessary to identify the primary predators of tidal-marsh birds and mammals in various parts of the estuary so that managers can decide which control measures, if any, are necessary.

MOSQUITOS AND OTHER VECTORS

Although wetlands, including tidal marshes, support high densities of many desirable species, they can also produce copious mosquitoes and potentially other disease vectors. These disease-carrying organisms can pose threats to tidal-marsh ecosystems because they can sicken and kill marsh animals as well as people, and because mosquito-control measures can have an adverse impact on marsh processes. Fortunately, neither traditional endemic vector-borne diseases nor current mosquito-control activities pose an imminent threat to existing marshlands; however, new diseases may have dramatic impacts on wildlife, particularly birds. The need to protect wildlife, as well as the public from diseases may pose serious challenges for wetland restoration proposals.

Marsh Mosquitoes and Mosquito-borne Diseases

Because some mosquito species transmit widespread and serious diseases to humans and other animals, they have been extensively studied over the last century (Durso 1996) and have been the subject of control programs in many areas, including SFBD. Mosquitoes are a diverse group of insects that share a common life history (egg, aquatic larvae, aquatic pupae, and flying adult), and a requirement for blood feeding by the adult females to produce eggs, with rare exceptions. In addition, all juvenile mosquitoes are weak swimmers and require habitats free from strong waves or currents or abundant predators. Thus, all mosquito species require shallow, still aquatic habitats for at least a few consecutive days.

Despite these similarities, mosquitoes vary considerably in their specific habitat requirements (several species may often coexist in close proximity) and in their potential for transmitting pathogens (public health significance of some species is higher than others). Mosquitoes are often distinguished by their specialized juvenile habitats, adult behavior, and vector status (Table 3 modified from Durso 1996, Maffei 2000). For example, *Culiseta incidens* is found in shaded, cool, clear fresh water, while *Aedes melanimon* prefers sunny, warm fresh water with dense grasses.

Some generalities are possible in this diverse assemblage. Larval habitat falls primarily along temperature (seasonal) and salinity (spatial) gradients, with only two truly salt-adapted mosquito species (*Aedes squamiger* and *A. dorsalis*) common in SFBD. Unlike freshwater genera that lay eggs in stagnant water, saltwater mosquitoes (*Aedes*), require an egg-conditioning period of at least a few days in which eggs cannot tolerate inundation. Thus, mosquito production is low in saltmarshes where dry periods are too short for egg conditioning (i.e., few impediments to drainage; Kramer et al. 1995). High flooding frequency is also beneficial for mosquito control because it is associated with currents sufficient to flush the larvae to unfavorable sites. Large populations of mosquitoes are almost invariably found where drainage is poor, whether the impounded water is saline (spring high tides that do not drain) or fresh (rain or seeps).

Although mosquito threats to human and animal health include disturbance, allergies, and infection secondary to scratching (Durso 1996), the most significant problems are the infectious pathogens carried between animals by mosquito blood feeding. West Nile virus (WNV) has killed hundreds of people, hundreds of thousands of birds (in almost all taxonomic groups), and smaller numbers of other vertebrate taxa, in the US over the last 4 yr (Center for Disease Control 2001, United States Geological Survey

TABLE 3. REPRODUCTIVE CONDITIONS, ADULT BEHAVIOR, AND VECTOR STATUS OF COMMON MOSQUITO SPECIES IN SAN FRANCISCO BAY ESTUARY MARSHES.

Common name	Scientific name	Eggs, larvae, and pupae	Adult behavior and vector status.[a]
Winter salt marsh mosquito (Maffei 2000)	Ochlerotatus squamiger (Aedes squamiger)	Egg conditioning on moist soil and plants. Simultaneous hatch following flooding. Highly salt tolerant. Cold water.	Flies 16–32 km. Bites humans day and dusk Localized pest. May transmit CE-like virus.
Summer salt marsh mosquito (Maffei 2000)	Ochlerotatus dorsalis (Aedes dorsalis)	Egg conditioning on moist soil and plants. Simultaneous hatch following flooding. Highly salt tolerant. Warm to hot water.	Flies up to 16 km. Bites humans and other large mammals day and night. Localized pest. Secondary vector of CE and WEE.
Winter Marsh mosquito (Maffei 2000)	Culiseta inornata	Eggs laid directly on cold standing water. Low salt tolerance.	Flies up to 8 km. Bites humans and other large mammals at night. Localized pest.
Washino's mosquito (Maffei 2000)	Ochlerotatus washinoi (Aedes washinoi)	Egg conditioning on moist soil and plants. Simultaneous hatch following flooding. Cool to cold water. Low salt tolerance.	Flies up to 1.6 km. Bites humans and other large mammals day and dusk. Localized pest. May transmit CE-like virus.
Western encephalitis mosquito (Durso 1996)	Culex tarsalis	Eggs laid directly on warm to hot standing water. Low to moderate salt tolerance.	Flies 16–24 km. Bites birds, and humans and other mammals at night. Primary vector of WEE, SLE. High vector competence for WNV in the lab.

[a] CE = California encephalitis; WEE = western equine encephalitis; SLE = St. Louis encephalitis; WNV = West Nile virus.

2003). WNV is the greatest immediate disease threat both to wetland organisms and humans. Lab research has demonstrated that it can infect *Culex tarsalis, Culex erythrothorax, Aedes dorsalis, A. melanimon, A. vexans,* and *Culiseta inornata,* and that all of these species can transmit the virus at some level, although the two *Culex* species were the most efficient vectors (Goddard et al. 2002). Field observations in states where *Culex tarsalis* occurs confirms that this species will probably pose the greatest threat in areas that have shallow fresh-water ponds (<5 ppt) that last until eggs, larvae, and pupae develop (5 d in the summer), but that become dry periodically to eliminate aquatic predators (Maffei 2000). While *Culex tarsalis* and its particular habitat types are the chief problems, other mosquito species and poorly drained marshes have been implicated in diseases in the past (Reisen et al. 1995, Durso 1996) and may also contribute to the establishment and spread of future pathogens (Center for Disease Control 1998, 2001).

RESTORATION CONCERNS

Because marsh mosquitoes have historically been recognized as a potential threat to animal health and human health and comfort, government agencies have acted to control marsh mosquito populations through a variety of activities, traditionally divided into physical control (habitat manipulation), biological control (stocking living predators or parasites), and chemical control (applications of biotic or chemical pesticides) (Durso 1996). Some of these strategies, such as widespread drainage of wetlands or extensive applications of DDT, clearly had substantial impacts on marshes and surrounding habitats in the past (Daiber 1986). Although these examples clearly indicate a need for continued monitoring and research, mosquito control activities have become more target-specific in recent decades and have not linked to significant adverse impacts on the marshes (Dale and Hulsman 1990; U.S. Environmental Protection Agency 1991, 1998, 2003; Dale et al. 1993, Contra Costa Mosquito and Vector Control District 1997, Center for Disease Control 2001).

In addition to mosquitoes, degraded tidal marshes can also provide habitat for brown and black rats (Breaux 2000), which are significant vectors of human disease, and for midges and other invertebrate pests (Maffei 2000). This combination of health threats and pests associated with marshes often has led to conflicts between wetland restoration proponents and neighbors, and the greatest threat to marshes associated with disease vectors may be the continuing development of residential areas nearby. Reducing this conflict will depend on good working relationships between wetland restoration advocates and mosquito control and other public health personnel (San Francisco Bay Area Wetlands Ecosystem Goals Project 1999).

DISEASE

Avian species are faced with the greatest known, or anticipated, threats from disease among all wildlife populations in the estuary. The three diseases of greatest concern and demonstrated mortality in the SFBD are West Nile virus, avian cholera, and avian botulism. Infectious diseases are currently on the rise for two probable reasons—the earth's climactic changes (Colwell 2004) and imbalance in biological systems, probably because of degraded habitat quality and diminished habitat quantity (Friend 1992).

For example, increased temperatures can boost the survival and growth of infectious diseases such as *Pasteurella multocida*, the bacteria that causes avian cholera (Bredy and Botzler 1989). Degraded habitat quality can lead to changes in microbial populations, and subsequently disease outbreaks (Friend 1992). A decrease in habitat size often results in a greater density of birds, increasing exposure, transmission, and spread of the disease to other locations (Friend 1992).

West Nile Virus

WNV is the most recent of the serious disease threats, first identified in the US in 1999. Reports of bird infections began in the eastern part of North America and have rapidly spread west. The first cases of WNV in the country were reported in New York City, New Jersey, and Connecticut. By December 2003, WNV had been identified all states except Oregon, Alaska, and Hawaii with 12,850 cases of human infection and 490 deaths (U.S. Geological Survey 2003). As of December 2003, California had two cases of human infection of WNV in Riverside and Imperial counties, and WNV was detected in the SFBD in 2004. WNV is transmitted by a variety of mosquitoes, but two species in particular appear to be primary vectors, *Culex tarsalis* and *C. erythrothorax*. At least 138 species of wild birds have been infected, with the family Corvidae demonstrating the highest prevalence of the disease (National Wildlife Health Center 2003a). American Crow and Blue Jay (*Cyanocitta cristata*) have most often been infected; however, this species composition may change as the virus moves westward. Coupled with the increasing number of corvids in the region, the threat to tidal-marsh birds is imminent. Other birds that inhabit wetlands of the estuary have been identified as hosts and vectors including cormorants, shorebirds, and Song Sparrows.

Transmission of WNV occurs when adult mosquitoes feed on the blood of an infected avian host followed by another vertebrate host. Mammals (humans and horses) do not appear to serve as intermediate hosts, though they can be infected (National Wildlife Health Center 2003a). The threat may be greatest for species of conservation concern, such as the three subspecies of Song Sparrows in the SFBD. The total estimated avian mortality due to the disease is over 100,000 individuals, though species breakdown is not available. Because the disease is so new, the mortality impacts, as well as the ecological interactions of the virus with its hosts, are virtually unknown (National Wildlife Health Center 2003a). Yet, if the Corvidae continues to be the principal group affected by the virus, WNV may have a beneficial impact on those bird species that compete with or are negatively affected by corvids.

Avian Cholera

AC is a highly infectious bacterial disease with the highest documented mortality rate of any disease for wetland birds in the estuary (USDI Fish and Wildlife Service 1992). Transmission is often direct and may involve surviving carrier birds; consequently, crowding is thought to increase incidence and mortality. California leads the nation in reported disease outbreaks, particularly in the delta. Waterfowl have been particularly affected, especially when concentrated on wintering areas or during spring migration. AC is of particular concern because roughly 50% of birds migrating along the Pacific flyway may pass through the SFBD. Outbreaks have occurred in a variety of habitats including freshwater wetlands, brackish marshes, and saltwater environments (National Wildlife Health Center 2003b). Since World War II, thousands of birds (mainly waterfowl) have been reported dead in each year. In one year, documented mortality was 70,000 birds for the state (USDI Fish and Wildlife Service 1992). The disease commonly affects more than 100 species of birds, though the Snow Goose (*Chen caerulescens*) has the greatest mortality (National Wildlife Health Center 2003b). Unlike botulism, AC often affects the same wetlands and the same avian populations year after year.

AVIAN BOTULISM

Avian botulism results from a neurotoxin produced by the bacterium, *Clostridium botulinum* type C (Friend 1987). The disease is caused by a bacterium that forms dormant spores in the presence of oxygen. The spores are resistant to heating and drying and can remain viable for years. Spores are widely distributed in wetland sediments and can also be found in the tissues of most wetland species, such as aquatic invertebrates and many vertebrates, including healthy birds. The botulism toxin is produced only when the bacterial spores germinate (Rocke and Friend 1999).

Although botulism is a more serious mortality causing factor than AC statewide, in the SFBD, the reverse is the case. Outbreaks of botulism in waterbirds are sporadic and unpredictable, occurring annually in some wetlands, but not in adjacent ones. In the past, mortalities from botulism have ranged from 0–1,000 in south San Francisco Bay. Botulism outbreaks caused 950 and 565 mortalities in 1998 and 2000, respectively, mostly ducks and gulls (C. Strong, San Francisco Bay Bird Observatory, pers. comm.). Botulism has been documented in the South Bay in Ruddy Ducks (*Oxyura jamaicensis*), Mallards, and Northern Shovelers (*Anus clypeata*; USDI Fish and Wildlife Service 1992).

Ecological factors that are thought to play a critical role in determining outbreaks include conditions that favor spore germination, the presence of a suitable energy source or substrate for bacterial growth and replication, and a means of transfer of toxin to the birds, presumably through invertebrate prey. Botulism outbreaks appear to be associated with moderately high pH (sediment pH 7.0–8.0) and low to moderate salinity (≤5 ppt). Botulism outbreaks are not specifically associated with shallow water and low dissolved oxygen (Rocke and Friend 1999).

RESTORATION CONCERNS

Tidal-marsh restoration will require careful management to avoid disease outbreaks. Increased density of vertebrate species in new restoration sites may encourage concentrations that result in disease outbreaks. Restoration projects adjacent to urban development may introduce potential disease sources. Finally, degraded environmental conditions may increase the effects of disease, impairing the species targeted for recovery under restoration efforts.

DISCUSSION

We have summarized some of the key threats to tidal-marsh vertebrates and have identified specific issues that are major concerns in tidal-marsh restoration projects in the SFBD. Unfortunately, it is difficult to compare the modern estuary to a historical period when the tidal marshes functioned naturally, because major changes to the system occurred before studies documented the importance of tidal marshes. For example, estuaries in SFBD and the arid Southwest are driven by snowpack conditions (Dettinger and Cayan 2003, Kruse et al. 2003), and water-user demands determine how closely the system follows the natural pattern of inflows. The climate results in two freshwater pulses—rainfall in the winter (November–February) and runoff in the spring (April–June). Most native vertebrate species are adapted to these wet periods, but changes in the environment have created a much different system than in the past. Flood protection and urbanization have resulted in a less dynamic estuary. The natural periodic, inter-annual, and annual flooding will be replaced by a static system that leaves little room for change. Under these conditions, the establishment and spread of exotic species may be facilitated. Exotic species continue to arrive in the SFBD at a rapid rate (Cohen and Carlton 1998), and the mechanisms introducing these species remain poorly regulated (Cohen 1997, Cohen and Foster 2000). Many important effects of exotic species may be indirect or subtle, such as ways in which exotic plants alter habitat for vertebrates.

In the modern estuary, combined threats may have the most detrimental consequences for many tidal-marsh vertebrate populations. Decreased water quality and increased contaminant loads may exacerbate the effects of vertebrate diseases. Human disturbance, fragmentation, and predation by species such as the red fox may reduce the carrying capacity of native vertebrate populations in remnant marshes. The loss of downstream sediments may greatly alter the sediment balance and the rate of marsh plain accretion in the SFBD. Dredge material may provide a beneficial solution to sediment deficits in some wetland restorations, but many estuarine contaminants are bound to sediments and the combined effects of dredging operations, dredge materials, and sediment-bound contaminants on vertebrates and their food webs are largely unknown.

In response to severe losses to wetland habitats, major efforts aim to regain and establish wetlands in SFBD. The current wave of restoration projects will alter the character of the estuary for the next century; however, with current rates of human development, the few remaining unprotected bay lands will no longer be restorable. Sea-level rise may eliminate tidal marshes squeezed between open water and urban development, rather than a mere relocation of marshes to higher elevations. Thus, identifying the critical environmental threats in advance may be the best way to guide restoration actions and management to ensure the conservation of tidal marshes into the future.

RESEARCH NEEDS

This paper represents the first step towards corrective action and management by identifying the major threats to tidal-marsh vertebrates in SFBD. Further efforts to comprehend the mechanisms and processes at work are listed below as the next step in our understanding of tidal-marsh ecosystems. This list is not intended to be a complete compilation of research needs, but to highlight some key issues that require immediate attention.

1. Conceptual models and data validation of the combined interaction of threats (i. e., interaction between water quality, contaminants, and disease) to tidal-marsh vertebrates is needed to understand system-wide processes.
2. Information about the effects of fragmentation on tidal-marsh function and demographic and population-level processes (dispersal, gene flow, survivorship, and predation) would improve prediction of those species' responses to habitat restoration.
3. A greater understanding of the hydrology and transport of bay sediments is needed to help determine local effects of restorations (sediment sinks) and their effect on flow patterns and sediment movement.
4. Detailed studies on the dredge-ameliorated wetlands including vegetation and structure, contaminant load, and vertebrate food webs may help resolve sediment concerns in marsh restoration.
5. A sediment-supply model based on recent empirical data would allow for better assessment of sediment changes and effects of sea-level rise on marshes that have a sharp upland transition zone (i. e., levee fringe marshes).
6. Research that would greatly improve understanding of contaminant threats in tidal marshes includes: (a) factors that control contaminant abundance and bioavailability within the wetlands, (b) relationship between foraging ecology and bioaccumulation in tidal-marsh vertebrates, and (c) prediction of wetland restoration activities on the concentrations, distribution, and bioavailability of contaminants.
7. Future research also should document how changes in water quality (i.e., salinity) may affect vertebrate distribution through a bottom-up control of vertebrate food webs. Increased salinity and low primary productivity have the potential to change the distribution and abundance of vertebrate species, but the magnitude of these effects and the mechanisms by which they act on vertebrates (e.g., via the food web or changes in the vegetative community) are not well understood.
8. The effect of disturbance on secretive species of saltmarshes is very difficult to study, because their responses are not easily observed. However, documenting human activities near tidal marshes and studying marked populations, including bioenergetic studies, may better quantify costs of disturbance and lead to specific management plans.
9. Studies that detail the direct and indirect effects of invasive species, as well as the mechanisms of exotic species arrival, establishment, and spread may lead to better regulation of introductory pathways and control options.
10. Predation studies would help to clarify the relationship between mortality and tidal-marsh fragment size, number, and distribution, and the potential effect of exotics.
11. Finally, mortality caused by disease, as well as degraded conditions under which they have the greatest effect, should be estimated within the context of overall annual mortality.

ACKNOWLEDGMENTS

We thank the editors, R. Greenberg, J. Maldonado, S. Droege, and M. McDonald for stimulating discussions that led to the preparation of this manuscript. We thank K. Phillips, D. Stralberg, N. Athearn, and K. Turner for helpful comments on the draft manuscript. The U.S. Geological Survey, Western Ecological Research Center, Wetland Restoration Program; National Science Foundation (DEB-0083583 to ANC), The Nature Conservancy, and the David H. Smith Conservation Research Fellowship (to JCN) provided partial support for preparing the manuscript.

ARE SOUTHERN CALIFORNIA'S FRAGMENTED SALTMARSHES CAPABLE OF SUSTAINING ENDEMIC BIRD POPULATIONS?

ABBY N. POWELL

Abstract. Loss of coastal saltmarshes in southern California has been estimated at 75–90% since presettlement times. The remaining wetlands are mostly fragmented and degraded, and most frequently have harsh edges adjacent to urban landscapes. Non-migratory Belding's Savannah Sparrows (*Passerculus sandwichensis beldingi*) and Light-footed Clapper Rails (*Rallus longirostris levipes*) are endemic to saltmarshes in southern California and Baja California, Mexico. Population sizes of Belding's Savannah Sparrows show a positive relationship with saltmarsh area, but few large wetland fragments remain within their range in California. Belding's Savannah Sparrows are sensitive to fragmentation and isolation, with small isolated marshes acting as population sinks. In addition, this subspecies shows low genetic variability, limited dispersal, and small effective population sizes. Light-footed Clapper Rails are habitat specialists, found in marshes with good tidal flushing that support California cordgrass (*Spartina foliosa*) habitats. Light-footed Clapper rails also show low genetic variability and limited dispersal and the remnant populations of clapper rails are relatively isolated from one another. Large wetland complexes may serve as population sources for both species, while small, isolated marshes may act as population sinks but more research is needed to estimate and model the dynamics of these two metapopulations. Mitigation for wetland loss and restoration projects should not be evaluated simply by presence of rare bird species alone, but instead efforts should be made to determine population sustainability.

Key Words: Belding's Savannah Sparrow, California, fragmentation, Light-footed Clapper Rail, metapopulation, saltmarsh.

SON CAPACES DE SOSTENER LAS MARISMAS DE MAREA FRAGMENTADOS DE CALIFORNIA POBLACIONES DE AVES ENDÉMICAS?

Resumen. La pérdida de las marismas de marea costeros en el sur de California ha sido estimada en un 75–90% a partir de los tiempos de pre-colonización. Los humedales que aun quedan se encuentran en su mayoría fragmentados y degradados, y con frecuencia sus bordes se encuentran adyacentes a paisajes urbanos. Los Gorriones Sabaneros No-migratorios (*Passerculus sandwichensis beldingi*) y el Rascón Picudo de Patas Ligeras (*Rallus longirostris levipes*) son endémicos en las marismas de marea en el sur de California y en Baja California, México. Los tamaños de las poblaciones de Gorriones Sabaneros muestran una relación positiva con el área de marisma salada, pero quedan pocos fragmentos largos de humedales dentro de su rango en California. Los Gorriones Sabaneros son sensibles a la fragmentación y al aislamiento, con pequeños marismas aisladas actuando como resumideros de población. Además, esta subespecie muestra variabilidad genética baja, limitada dispersión, y pequeños tamaños de población efectiva. Los Rascones Picudos de Patas Ligeras son especialistas del hábitat, encontrados en marismas con buena nivelación de marea, la cual mantiene habitats de pasto (*Spartina foliosa*). Los Rascones Picudos de Patas Ligeras también muestran baja variabilidad genética y limitada dispersión, y las poblaciones remanentes de Rascones Picudos se encuentran relativamente aisladas una de otra. Complejos de largos humedales quizás sirvan como fuentes de población para ambas especies, mientras que marismas pequeñas y aisladas quizás actúen como resumideros de población, pero se necesita más investigación para estimar y modelar las dinámicas de estas dos meta poblaciones. La mitigación para la pérdida de humedales y proyectos de restauración no deberían de ser evaluados simplemente por la presencia de aves raras por si solas, si no que los esfuerzos deberían hacerse para determinar la sustentabilidad de la población.

More than 16,000,000 people live along southern California's coast and the impact of a dense human population, coupled with high endemic biodiversity, has resulted in the listing of numerous species as threatened and endangered (Davis et al. 1995). Southern California's saltmarshes have suffered significant habitat degradation and loss of area. California has lost an estimated 91% of all wetlands and about 75% of its coastal wetlands since pre-settlement (Zedler 1982, Macdonald 1990). Estuarine systems in southern California have been highly altered by urban development, filling, river channelization, changes in freshwater flow, and invasion of exotic species. Marshes have become more and more isolated by the expansion of urban areas creating hostile environments for dispersing organisms. Isolation can hinder emigration, immigration, and gene flow (Shafer 1990, Andren 1994). Habitat fragments may become sinks ecological traps for some animal populations if production of young fails to exceed mortality, and local extinctions may occur unless immigration occurs from source habitats (Pulliam 1988, Howe et al. 1991).

Western saltmarshes provide nesting habitat for several rare species, including Belding's Savannah Sparrow (*Passerculus sandwichensis beldingi*) and Light-footed Clapper Rail (*Rallus longirostrus levipes*). Salt-pan habitats located within coastal marshes provide nesting sites for endangered California Least Terns (*Sterna antillarum browni*) and threatened Western Snowy Plovers (*Charadrius alexandrinus nivosus*), while channels and mudflats provide foraging habitat for these species. These marshes also provide important wintering grounds and foraging areas for migratory shorebirds and waterfowl.

The Belding's Savannah Sparrow was listed as endangered by the state of California in 1974 and the Light-footed Clapper Rail was listed as federally endangered in 1970 (USDI Fish and Wildlife Service 1979). Both subspecies are endemic to saltmarshes in southern California and Baja California, Mexico, and have suffered significant population declines due to wetland loss and degradation (Zembal et al. 1988, Massey and Palacios 1994).

Avian diversity within freshwater and brackish marshes has been attributed to marsh size, diversity of habitat types, amount of open water and degree of isolation from similar habitats (Kantrud and Stewart 1984, Brown and Dinsmore 1986, Peterson et al. 1995). Studies of avian abundance in coastal wetlands have typically focused on habitat use in the eastern US (Burger et al. 1982, Marshall and Reinert 1990, Erwin et al. 1995). Few studies have presented quantitative data on habitat use by birds of western coastal saltmarshes that have suffered considerable loss and degradation, particularly in coastal southern California. Even fewer studies exist on saltmarsh bird populations in adjacent Mexico. Here, I review the information available for two species endemic to southern California saltmarshes with respect to their sustainability within the US.

SOUTHERN CALIFORNIAN SALTMARSHES

Three littoral zones that have varying degrees of overlap in composition of vegetation types typically characterize saltmarshes in southern California. Low-marsh habitats occur in the lowest elevation and experience tidal inundation twice a day. California cordgrass (*Spartina foliosa*) is the dominant low-marsh species in marshes with full tidal flushing (access to tides has not been restricted by sedimentation or channelization) (Zedler 1982). In intermediate elevations, mid-marsh habitats have higher species diversity and are dominated by pickleweed (*Salicornia virginica*), which is tolerant of high soil salinities and inundation by salt and fresh water (Zedler 1982, Keer and Zedler 2002). Highest elevations in the marsh have the driest soils and highest soil salinities. The high-marsh zones are typically dominated by the Parish's pickleweed (*Salicornia subterminalis*; Zedler 1982). Loss of tidal circulation not only reduces the likelihood of cordgrass habitats, but also tends to decrease plant species diversity; monocultures of pickleweed are often found in these marshes (Zedler 1982). Considerable research has occurred on the restoration of these habitats in southern California saltmarshes (Zedler 1996, Zedler et al. 2001, Keer and Zedler 2002).

Belding's Savannah Sparrow

Belding's Savannah sparrows are non-migratory and endemic to southwestern saltmarshes, ranging from Goleta Slough in Santa Barbara County southward to Bahia de San Quintin, Baja California, Mexico (Fig. 1). Within the US, their southernmost local population occurs at Tijuana Estuary and they have been documented breeding in 30 marshes ranging from <1 ha to approximately 620 ha in size ($\bar{x}$ = 92.9 ± 136 ha; Fig. 2). This subspecies of Savannah Sparrow (*Passerculus sandwichensis*) is generally associated with *Salicornia* spp. habitats in mid- (dominated by pickleweed to high (dominated by Parish's pickleweed) littoral zones and avoids areas prone to frequent tidal inundation (Powell 1993, Powell and Collier 1998). In most remnant marshes, pickleweed habitats have been degraded by changes in tidal flow and freshwater inputs, invasion of non-indigenous plants, and fragmentation by trails and roads. Connections to native habitats beyond the high-marsh zone are rare in southern California and frequently this habitat type is adjacent to an urban interface; therefore use/importance of native uplands by these sparrows is unknown. Belding's Savannah Sparrows are rarely observed outside of saltmarsh habitats and are more frequently observed on adjacent beaches than uplands (Bradley 1973, Massey 1979; Powell, unpubl. data).

Volunteers have conducted censuses of Belding's Savannah Sparrows in southern California approximately every 5 yr since 1986. Counts occurred in 26–30 coastal saltmarshes and effort varied among wetlands and years. All 30 marshes were surveyed in 1986, 1991, 1996, and 2001, and the total estimated number of breeding pairs in California was 1,844–2,902

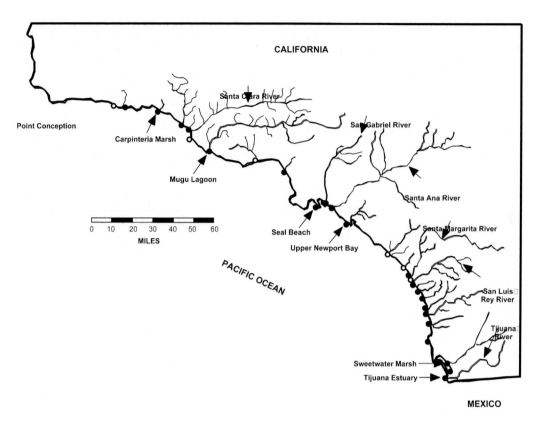

FIGURE 1. Map of coastal marshes in southern California. Dark circles are those marshes occupied by Belding's Savannah Sparrows.

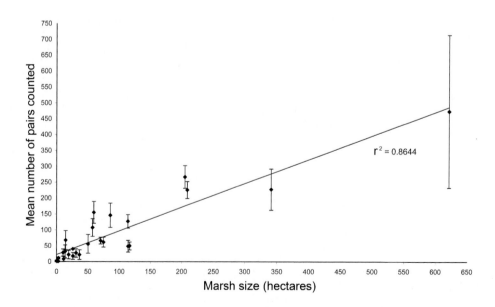

FIGURE 2. Relationship of Belding's Savannah Sparrow counts (1986–2001) to marsh area (hectares) in southern California.

(Zembal et al. 1988, Zembal and Hoffman 2002). Mugu Lagoon, the largest saltmarsh (620 ha) in southern California, consistently supported the largest local population of sparrows, followed by Tijuana Estuary (205 ha), Upper Newport Bay (208 ha), and Seal Beach (340 ha; Fig. 3). Data from these counts show that like grassland Savannah Sparrows, Belding's Savannah Sparrows are area sensitive; a positive relationship exists between size of wetland and indices of local population size, and sparrows are unlikely to occur in marshes <10 ha in size (Fig. 2; Powell and Collier 1998).

Work on breeding biology of Belding's Savannah Sparrows at Carpinteria Marsh, Santa Barbara County, California, indicated that effective population size is likely much smaller than the total population; <50% of males established territories and only 43% of those males managed to attract mates (Burnell 1996). The males that were unable to establish territories were considered to be floaters. In her 3-yr study, Burnell estimated that the effective population size ranged from 12–35% of the total population size during 1991–1993. She also determined that 33% of the males at Carpinteria were polygynous, with each male paired with two females within a territory. Within the Sweetwater Marsh complex, San Diego County, California, 93% of territorial males attracted mates and 9% of males were polygynous (Powell and Collier 1998). Powell and Collier (1998) did not know the total population of Sweetwater Marsh and therefore could not estimate the percentage of floaters. The discrepancy in the number of territory holders without mates between the two studies may partially be an artifact of sampling; the Powell and Collier (1998) was intensive (a total of 216 hr observing 54 territories during one breeding season), whereas Burnell's (1996) study was extensive (a total of 206 hr spread over 3 yr observing a total of 49 territories). Females are very difficult to observe because of their secretive behaviors, thus Burnell may have overestimated the number of males that failed to attract mates. Regardless, it should be noted that effective population size is likely to be a fraction of the number of sparrows present in a marsh.

Belding's Savannah Sparrow has limited dispersal and is a metapopulation with extirpation

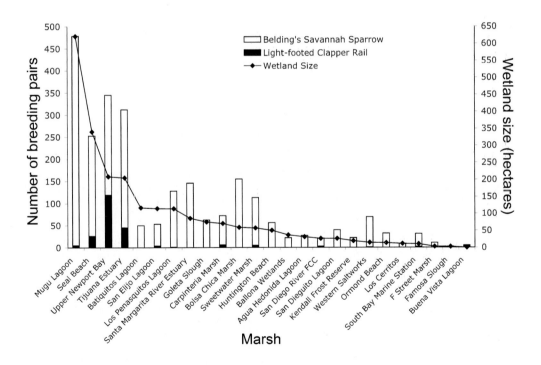

FIGURE 3. Relationships of Belding's Savannah Sparrow (average of four counts between 1986–2001) and Light-footed Clapper Rail (average of annual counts from 1980–2002) populations to marsh area (hectares) in southern California. Total marsh area decreases from left to right and overestimates the actual amount of saltmarsh habitat because areas were estimated using National Wetland Inventory E2 classification data (<http://www.nwi.fws.gov/> [31 July 2006]).

and recolonization of local populations (Zembal et al. 1988, Bradley 1994, Burnell 1996, Powell and Collier 1998). Bradley (1994) and Burnell (1996) found that local populations had distinct song dialects. In addition, Burnell (1996) found genetic evidence that little or no gene flow occurs among local populations of Belding's Savannah Sparrows and that allozyme differentiation among populations was most likely caused by genetic drift. Local populations of Belding's Savannah Sparrows in seven of the marshes she surveyed between Santa Barbara and San Diego counties had lower heterozygosity than expected in a Hardy-Weinberg equilibrium and there was also evidence of inbreeding within these populations (Burnell 1996).

In 1995, I established six study plots to examine reproductive success of Belding's Savannah Sparrows within the Sweetwater Marsh complex in San Diego County, California (Powell and Collier 1998). These marshes are highly fragmented and surrounded by urban and industrial landscapes. Plots were located in mid- and high-marsh habitats (Powell 1993). Comparisons of reproduction among plots showed significant differences between high reproductive success in high-marsh plots within the largest marsh (50.6 ha) and low reproductive success at the small isolated marsh (2.9 ha), where no fledglings were produced over the breeding season. The small marsh was isolated from other marsh habitats and surrounded completely by an urban landscape. In addition, despite the fact that this marsh was located only 0.5 km from the larger marsh, no movements of banded birds were observed between them, indicating exchange rates may be quite low.

I also examined the effects of habitat on reproductive success and found that areas with the tallest, densest vegetation and low quantities of bare ground were the best predictors of high-success (those that produced fledglings) territories. I found no relationship between territory size and reproductive success. Few of the study plots, with the exception of the small isolated marsh, had much space unoccupied by territories except for those areas with a high proportion of bare ground. Most of the vegetated areas consisting of mid- and high-marsh plant species were occupied by territories throughout the breeding season, suggesting that suitable nesting habitat was limited. Finally, I also banded 277 sparrows prior to the breeding season at Sweetwater Marsh in 1995 and did not see any of them at any other marsh within the San Diego Bay area in 1995 or following years (1995–1997). In addition, 45.5% of banded males within our plots occupied the same territories the following year, suggesting little emigration and high site fidelity. In summary, my research on reproductive success of Belding's Savannah Sparrows in different-sized wetlands within San Diego Bay suggested that small, isolated saltmarshes supported breeding birds but functioned as population sinks because they supported little or no productivity (Powell and Collier 1998).

LIGHT-FOOTED CLAPPER RAIL

The Light-footed clapper rail is a year-round resident of saltmarshes from Tijuana Estuary on the Mexican border, north to Santa Barbara County within the US, and like Belding's Savannah Sparrows, extends south to Bahia de San Quintin, Baja California, Mexico. Unlike Belding's Savannah Sparrows, Light-footed Clapper rails are closely associated with low-marsh habitats, particularly those consisting of cordgrass (Massey et al. 1984). This habitat type is associated with marshes with good tidal flushing and has disappeared from those marshes, such as Mugu Lagoon, with decreased tidal flow due to sedimentation, dredging, and river channelization. Only a small subset of southern California's saltmarshes currently supports cordgrass habitats: Tijuana Estuary, Sweetwater Marsh, Upper Newport Bay, and Seal Beach (Fig. 1; Zedler 1982).

Studies on the movements of Light-footed Clapper Rails indicate that they have strong site tenacity and rarely move >400 m; the farthest documented movement is 21.7 km (Zembal et al. 1989). In addition, genetic analysis of the subspecies indicated there was low genetic variability and reduced heterozygosity within Light-footed Clapper Rails (Fleischer et al. 1995). This subspecies shows a classic metapopulation structure, with local populations that experience extinction, recolonization, and limited dispersal (Fleischer et al. 1995, Zembal et al. 1998).

Light-footed Clapper Rails have been monitored annually in California since 1980. Marshes that potentially support clapper rails are visited in spring, and clapper rail calls are counted. Clapper rails use several distinct calls during the breeding season that can be used to distinguish single males, single females, and mated pairs (Massey and Zembal 1987). During a census, people walk slowly through the marsh at dawn or dusk and mark locations of calls on a map. In addition, taped calls may be played to elicit responses (Zembal 1998). Each year a breeding survey report is submitted to the California Department of Fish and Game. In 1980, the first year of the survey, the Light-footed Clapper

Rail metapopulation consisted of an estimated 203 breeding pairs; a high of 325 breeding pairs was counted in 1996 (Zembal et al. 1998). The number of estimated pairs has varied around the 22-yr mean of 231 pairs, but no overall pattern of decline or increase has occurred during this period. Upper Newport Bay consistently supported >50% of California's Light-footed Clapper Rails, and three sites combined (Upper Newport Bay, Tijuana Estuary, and Seal Beach) supported >80% of breeding pairs in any given year. These three estuaries are the second, third, and fourth largest in size within the range of the Clapper Rail (Fig. 3), Upper Newport Bay is relatively isolated from other saltmarshes in the region, and all three marshes are isolated from each other (Fig. 1). Mugu Lagoon, the largest wetland, supported on average only four pairs of rails, but this site has very little cordgrass habitat. Of the remaining 21 marshes where Light-footed Clapper Rails are found, none supported >4% of the metapopulation, and 15 each supported <1% of the metapopulation (usually one bird per wetland).

In addition to limited availability of cordgrass habitats in southern California, Light-footed Clapper Rail populations have been negatively impacted by predation. Removal of non-native red foxes (*Vulpes vulpes*) resulted in growth of the local population at Seal Beach from a low of five pairs in 1986 to a high of 65 pairs in 1993 (Zembal et al. 1998). In addition to predator management, nesting rafts were used at this site to increase nest site availability since the late 1980s. Recently however, the local population at Seal Beach appears to be lower (range = 10–24 pairs, $\bar{x}$ = 15.2; 1998–2002) than in the 1990s (range = 28–66 pairs, $\bar{x}$ = 43.9; 1990–1997; Zembal, unpubl. data). It is postulated that nesting rafts may actually increase rates of predation by raptors (Zembal et al. 1998).

DISCUSSION

It is clear that amounts of saltmarsh habitats, including pickleweed but in particular cordgrass, are currently limited in southern California. The remaining saltmarshes are mostly degraded to some extent and changes in tidal influence have eliminated the occurrence of cordgrass habitats in many marshes. In addition to degradation caused by changes in the hydrological regime, saltmarshes in southern California are typically surrounded by urban and/or industrial landscapes (Fig. 1). This creates a hostile environment for dispersal and likely causes naturally isolated wetlands to become functionally even more isolated. Neither Belding's Savannah Sparrows nor Light-footed Clapper Rails are thought to be good dispersers, and decreases in already low natural rates of immigration and emigration can have significant impacts on local population viability. Reduced dispersal can lead to local extinctions, reduced genetic variability, inbreeding depression, and decreased colonization rates (Pulliam 1988, Andren 1994).

Increased rates of predation related to human activity are another form of habitat degradation. Some predators, like red foxes, are not native to southern California but have become problematic as their populations have increased. Non-native red foxes have expanded their ranges and populations in California and impact coastal ecosystems, particularly Light-footed Clapper Rail populations (Zembal et al. 1998, Lewis et al. 1999). Common Ravens (*Corvus corax*), known predators of eggs, nestlings, and even adult birds, have increased substantially in California since the 1960s (Boarman and Berry 1995). Likewise, the proliferation of feral and domestic cats (*Felis catus*) in urban areas has a significant impact on native birds, and cats are frequently observed in these saltmarshes (Ogan and Jurek 1997). Increased rates of predation likely reduce survival rates of adults and young, and increase mortality during dispersal.

Although it is unlikely that new estuaries can or will be created in this region, it is possible to improve and expand cordgrass coverage in existing marshes. Foin and Brenchley-Jackson (1991) suggested that cordgrass habitat improvement within existing marshes could potentially triple the rail population, however restoration of saltmarsh vegetation can be a long and expensive process (Zedler et al. 2001). Despite the severe limitation of cordgrass habitat, captive breeding and reintroduction efforts have been initiated for Light-footed Clapper Rails (California Department of Fish and Game, unpubl. data). It has been well documented that the key predictors of successful translocations are habitat quality and the quality and number of animals released. In general, endangered species translocations are unsuccessful >50% of the time and if animals are released into habitats that are in poor condition or have insufficient area they are unlikely to persist (Griffith et al. 1989, Wolf et al. 1998). In addition, although Fleischer et al. (1995) suggested that translocations of Light-footed Clapper Rails could increase the genetic variability within local populations, they cautioned that documentation must first show that inbreeding depression is problematic for this species. Finally, more information is needed on natural recruitment into local populations of Light-footed Clapper

Rails. Given the small number of Light-footed Clapper Rails remaining in southern California and the limited and degraded condition of estuarine habitat, efforts to increase local populations should emphasize habitat creation and enhancement rather than costly translocations with low potential for success.

Restoration of high-marsh zones dominated by pickleweed is also possible. Degraded pickleweed habitats can be enhanced by restoring natural hydrological regimes, and the dominant species of pickleweed (*Salicornia virginica*), recruits readily (Zedler et al. 2001). Restoration of plant diversity in mid- to high-marsh zones increases vegetation structural diversity, which in turn may provide an increased prey base for Belding's Savannah Sparrows (Keer and Zedler 2002). Local populations of Belding's Savannah Sparrow have expanded after restoration efforts improved water flows at Mugu Lagoon (Zembal and Hoffman 2002).

Are the populations of Belding's Savannah Sparrows and Light-footed Clapper Rails in southern California sustainable? Given current information on low genetic variability, low dispersal rates, and low overall population sizes it seems questionable that either subspecies will persist unless more saltmarsh habitat is created and existing habitats are restored. More research is needed to determine and model the dynamics of these two metapopulations. Unfortunately, southern California's saltmarshes have not been characterized in relationship to habitat type, patch size and shape, connectivity, and isolation. Indeed, the amount of coverage by cordgrass and pickleweed-dominated habitats is unknown for most marshes. This information is critical, especially for the management of Light-footed Clapper Rails and Belding's Savannah Sparrows.

Both subspecies show metapopulation structure and should be managed as such. Planners and managers need to ask the following questions before designing habitat restoration and enhancement projects:

1. Is the existing wetland complex large enough to support self-sustaining local populations over time?
2. Are patches of specific habitat types (e.g., cordgrass or pickleweed) large enough to support self-sustaining local populations over time?
3. Are dispersers able to move between wetlands (will source or sink populations equilibrate over time)?
4. In a regional context, will the restoration benefit the metapopulation?

Finally, assessments of population size for each species need to consider that effective population size is likely a fraction of the total number of territorial birds counted.

We should take a regional approach to wetland restoration in order to enhance metapopulations of sparrows and rails. For example, unless overall wetland area is increased, creating cordgrass habitat to benefit Light-footed Clapper Rails may be at the expense of pickleweed habitats required for Belding's Savannah Sparrows and vice versa. Converting saltpan or dredged areas to saltmarsh may reduce the amount of habitat available to endangered California Least Terns, threatened Western Snowy Plovers, and other shorebirds that use these habitats. Considerations should be giving to the status of the target species, probability of success of habitat restoration, and overall ecosystem functioning.

ACKNOWLEDGMENTS

I thank C. Fritz and B. Peterson who worked with me in southern California's saltmarshes. I also thank J. Zedler as a mentor and for her tireless work to restore these wetlands. D. Zembal has graciously shared his data on Belding's Savannah Sparrows and Light-footed Clapper Rails, and it has been through his efforts that censuses of both species have continued through the years. B. Griffith provided valuable comments on this manuscript.

CONSERVATION BIOLOGY

Diamondback terrapin (*Malaclemys terrapin*)
Drawing by Julie Zickefoose

THE DIAMONDBACK TERRAPIN: THE BIOLOGY, ECOLOGY, CULTURAL HISTORY, AND CONSERVATION STATUS OF AN OBLIGATE ESTUARINE TURTLE

Kristen M. Hart and David S. Lee

Abstract. Ranging from Cape Cod to nearly the Texas-Mexico border, the diamondback terrapin (*Malaclemys terrapin*) is the only species of North American turtle restricted to estuarine systems. Despite this extensive distribution, its zone of occurrence is very linear, and in places fragmented, resulting in a relatively small total area of occupancy. On a global scale, excluding marine species, few turtles even venture into brackish water on a regular basis, and only two Asian species approach the North American terrapin's dependency on estuarine habitats. Here we describe some of the biological and behavioral adaptations of terrapins that allow them to live in the rather harsh estuarine environment. In this chapter we review the natural and cultural history of this turtle, discuss conservation issues, and provide information on the types of research needed to make sound management decisions for terrapin populations in peril.

Key Words: Adaptations, conservation, genetics, *Malaclemys terrapin,* population, management, saltmarsh.

LA TORTUGA DE AGUA DULCE: ESTATUS BIOLÓGICO, ECOLÓGICO, HISTORIA CULTURAL, Y ESTATUS DE CONSERVACIÓN DE UNA TORTUGA ESTUARINA OBLIGADA

Resumen. Extendiéndose desde el Cabo de Bacalao hasta casi la frontera entre Texas y México, la tortuga de agua dulce (*Malaclemys terrapin*) es la única especie de tortuga de Norte América restringida a sistemas de estuarios. A pesar de la extensiva distribución, su zona de ocurrencia es muy linear y en lugares fragmentados, lo qual resulta en una relativamente pequeña área total de ocupación. A escala global, excluyendo especies marinas, pocas tortugas se aventuran a aguas salobres en base regular, y solo dos especies asiáticas alcanzan la dependencia de la tortuga de agua dulce de Norte América en habitats de estuario. Aquí describimos algunas de las adaptaciones biológicas y de comportamiento de las tortugas de agua dulce que les permiten vivir incluso en el ambiente de estuario más duro. En este capítulo revisamos la historia natural y cultural de esta tortuga, discutimos asuntos de conservación, y proveemos información sobre los tipos de información que se necesita para tomar decisiones de manejo adecuadas para las poblaciones de tortuga de agua dulce en peligro.

Many freshwater fishes, mammals, and a variety of birds exploit marine and estuarine habitats so it is surprising that, even on a global basis, only a few reptiles occur regularly in salt and brackish marshes (Greenberg and Maldonado, *this volume*). The dearth of saltmarsh reptiles follows a general lack of reptilian species adapted to any marine environment. Crocodiles (two species), sea turtles (two families, seven species), sea snakes (about 50 species), and marine iguanas (one species) are the only truly marine reptiles. Most of the aforementioned species are strictly marine; the diversity of reptiles with a strong association with estuarine habitats, including saltmarshes, is lower. Although populations of various snakes and freshwater turtles have taken up residence in brackish habitats, few taxa (species or subspecies) are restricted to tidal marshes (Greenberg and Maldonado, *this volume*). In contrast, the diamondback terrapin appears to have a long evolutionary association with estuaries and their saltmarshes. The degree of divergence in terrapins is reflected in its status as representing a monotypic genus with time since divergence from non-estuarine taxa estimated as being in the neighborhood of 7–10,000,000 yr (Chan et al., *this volume*). If these estimates are correct, terrapins are the taxa with the longest estimated association with tidal marshes. Interestingly, two other species of turtle (also in monotypic genera) are largely restricted to estuarine habitats, but are found in tropical systems of southeast Asia (*Callagur* and *Orlitia*). Therefore, the diamondback terrapin, which occurs along the Atlantic and Gulf coasts of North America, is the only species of turtle specialized to saltmarsh and estuarine habitats in the temperate zone. In addition to its unique ecological evolutionary status, diamondback terrapins have achieved a level of economic and cultural importance that surpasses most of the saltmarsh vertebrates. Before populations were reduced, terrapins supported a multimillion dollar industry catering to the gourmet restaurant trade. Over-harvest and habitat modification virtually eliminated them in the late 1800s and early 1900s (Lazell 1979). All of these factors justify a focused look at diamondback terrapins as a saltmarsh endemic. We focus this chapter on terrapin adaptations to the salty estuarine environment, as well as cultural history or past human use of the terrapin as a food resource. We then discuss the management and conservation future of this well-known estuarine endemic.

RANGE OF THE DIAMONDBACK TERRAPIN

The diamondback terrapin has a range consisting of small, linearly distributed, and isolated populations in US coastal waters from Cape Cod, Massachusetts, to the Texas-Mexico border. Seven subspecies (Fig. 1) have been described, based primarily on differences in carapace morphology and skin coloring. Some of these smaller, regional subpopulations are extremely vulnerable to extinction.

CULTURAL HISTORY

Diamondback terrapins played an important role in the cultural history of colonial America. These turtles were an important food item of the Continental Army in the 1700s, and in later years were a major source of protein for slaves on tidewater plantations. In the late 1800s through the Great Depression terrapins were a highly sought-after item in exclusive restaurants as well as an important food source for families living in remote coastal settings. This high demand for terrapins resulted in a population crash and a major effort of the U.S. Federal Bureau of Fisheries to raise terrapins for restocking and commercial use. Because of their previous cultural and economic importance terrapins are arguably one of the most celebrated reptiles in North America.

At one time, slaves in tidewater plantations consumed a diet heavy in turtle meat, with terrapins reportedly served about two times a week. Then, for reasons difficult to explain terrapin meat became regarded as a gourmet item. Virtually overnight terrapins were sought with enthusiasm by the privileged. As early as 1830 the Prince of Canino tried to transplant terrapins to Italy. Later, the species was successfully established in Bermuda (D. Lee, pers. obs.) and at least two unsuccessful attempts were made to establish them in San Francisco Bay (Taft 1944, Hildebrand and Prytherch 1947). Eating terrapin became fashionable, in fact special terrapin bowls and terrapin forks became part of the flat and silverware of the affluent. Diamondbacks brought top dollar in markets and had the fashion continued, this turtle would likely be extinct today.

It is as difficult to explain the decline of the popularity of terrapin as it is to understand the appetite that developed for it; by the 1920s the species had been exploited to the extent that the industry could not sustain itself. Rather than conservation and economic concerns being responsible for the decline of terrapin harvest, it was Prohibition that made it difficult to obtain the various liquors in which the turtle meat was prepared. By the time of the Great Depression restaurants were no longer serving high-priced entrees and terrapin meat simply became just another seafood. The near collapse of terrapin populations in the wild kept the turtle meat market from rebounding, and the last restaurant to have terrapin on its menu closed its doors in Baltimore in the 1990s. The last possibility of the terrapin reclaiming its fame as a gourmet food item was in the Nixon presidency; once a year President Nixon threw a large formal affair in which diamondbacks were the main entree. For weeks before the event waterman throughout the Chesapeake Bay saved all the terrapins they could gather and every one was bought at top dollar for the affair. The increasing effort necessary to obtain enough terrapins for this annual dinner gave testament as to how uncommon the species had become. As an economic commodity

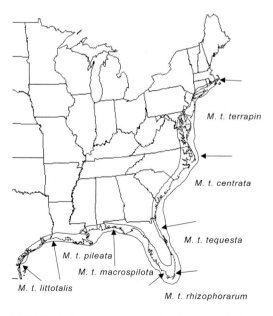

FIGURE 1. Current range and subspecies designations for the diamondback terrapin (modified from Carr 1952).

these turtles were a paradox, demanding top dollar in New York, Philadelphia, and Baltimore, and yet a staple food of the residents of remote places like Ocrocoke Island at least through the early 1940s.

In the early 1900s, considerable effort went into attempting to culture the species. The U.S. Bureau of Fisheries set up a number of terrapin pounds to study the feasibility of rearing and breeding captive diamondback terrapins. The bureau even made some unsuccessful attempts to improve the stocks by breeding the better tasting northern turtles with the larger ones found in Texas. The most prominent of these terrapin pounds was operated in Beaufort, North Carolina, between 1902 and 1948. The staff of the Beaufort lab published a number of studies centering on the propagation of terrapins and it is because of these studies that we have baseline information on the reproductive biology and growth of terrapins (Hildebrand and Hatsel 1926; Hildebrand 1929, 1932). The captive-breeding effort was extremely efficient and tens of thousands of hatchlings of various experimental stocks were released into the marshes and sounds of North Carolina, and other Atlantic and Gulf coast states when demand for the terrapin declined.

TERRAPIN BIOLOGY AND ECOLOGY

LIFE HISTORY

Although terrapins were raised in captivity and well studied in the early 1900s, relatively little is known about wild populations of terrapins. These medium-sized turtles (adults are 10–23 cm long) exhibit considerable sexual dimorphism, with females being three–four times larger by weight than males (Ernst et al. 1994). Diamondbacks are strong-jawed with a particular affinity for small mollusks and crustaceans. One of the major benefits to living in brackish water is the availability of a rich food supply—major food items include saltmarsh periwinkles (*Littorina irrorita*), small clams including blue and horse mussels (*Mytilus* and *Modiolus*), fiddler crabs (*Uca*), mud crabs (*Panopeus, Neopanopes,* and *Eurypanopeus*), and blue crabs (*Callinectes sapidus*). Less important foods include carrion, fish, and, on occasion, plant material (Tucker et al.1995). Although variations occur with latitude, male terrapins first reproduce after their fourth year whereas females reach sexual maturity after their seventh year (Hildebrand 1932, Montevecchi and Burger 1975, Auger 1989, Roosenburg 1990, Lovich and Gibbons 1990, Seigel 1994). Female terrapins lay one to several clutches of eggs, and this also varies from north to south throughout the range (Zimmerman 1992, Roosenburg and Dunham 1997).

MAKING A LIVING IN THE MARINE ENVIRONMENT

Although diamondbacks are seldom found in any of the freshwater habitats that adjoin the marshes and sounds in which they live, they can survive well in fresh water in captivity. Apparently it is the ability of terrapins to regulate osmotic pressures of brackish water that allows this turtle, one derived from freshwater ancestors, to survive in salty water. Species living in a marine environment must contend with maintaining osmotic balance. For the terrapin, a shell and scaled skin help to control dehydration, but other challenges exist with living in this environment. The turtle's total body weight decreases significantly (up to 0.32% per day) when exposed to pure (salinities of 34 ppt) seawater (Robinson and Dunson 1976). Whereas, most freshwater turtles have no tolerance for even brackish water, diamondback terrapins live in estuarine environments throughout their lives and survive through an interesting combination of physiological and behavioral adaptations.

Physiological adaptations

The saline environment presents a major adaptive challenge to life in a saltmarsh. Physiological regulation within blood, intercellular fluids, and various tissues plays a key role in maintaining osmotic balance. Red blood cells increase in number when terrapins are in water with high salt concentration, apparently in response to the need to remove ammonia and urea from the muscles where waste byproducts accumulate. The bladder and colon accumulate high concentrations of various compounds associated with exposure to seawater and much of it is excreted directly back into the water (Gilles-Baillien 1973). Like marine turtles and crocodiles, terrapins expel sodium through orbital glands near their eyes. Studies by Cowan (1969, 1971) showed the structure of these glands to be similar to other freshwater turtles. Although terrapins use these glands to secrete organic compounds, they are not specialized for increased salinity. Instead the lachrymal gland that is associated with the eye may have a more important role in maintaining salt balance—terrapins acclimatized to seawater show a 2.4-fold increase in sodium concentrations in the eye (Cowan 1969, 1971). Although this gland is adapted to minimize water loss as would be expected in an estuarine animal, it is clear that the gland is not dedicated to salt excretion, nor is it its primary purpose. Thus, modifications of the anatomy of terrapins contribute only in minor ways to terrapins' ability to exploit brackish water habitats.

Behavioral adaptations

The behavioral adaptations of terrapins to the harsh environment of the saltmarsh are perhaps the most interesting. Terrapins can discriminate between different salinities and much of their ability to cope with brackish water is the result of behavior (i.e., movements between salinity gradients and drinking freshwater from the surface after rains). Experimental animals retained in seawater for a week were able to rehydrate in less than 15 min when given access to freshwater (Davenport and Macedo 1990). When water is high in salt concentration (above 27.7 ppt) terrapins seem to avoid drinking it. At moderate salinities (13.6–20.0 ppt) terrapins drink small amounts of seawater, and when the salinity is low (<10 ppt) they drink large amounts (Robinson and Dunson 1976). When it rains, terrapins swim to the surface and drink from films of fresh water (D. Lee, unpubl. data). During rain they will stretch their necks above the surface and catch water in their open mouths. They also leave the water and drink rainwater that collects on the margins of their shells, and from their limb sockets, or from the sockets of other terrapins. While these observations were made on captive animals (Davenport and Macedo 1990), we have no reason to assume that captive turtles behaved differently from those in the wild. Nonetheless, although much seawater is taken orally when the turtles are feeding, even in extreme cases the turtles can quickly reverse osmotic imbalance (Davenport and Macedo 1990). Under normal conditions, of course, these turtles would only rarely be exposed to extremely low or high salinities.

SEASONAL MOVEMENTS

Terrapins are highly mobile, moving between water of different salinities in order to feed, mate, and brumate (brumation is a reptilian state analogous to hibernation), as well as to maintain proper osmotic balance. The pattern of these movements differs between age and gender classes of terrapins. In Maryland, Roosenburg et al. (1999) found that adult females in the Patuxent River moved more often and were found further from shore than adult males, juvenile males, and juvenile females. Their findings suggest that larger adult females move further and spend more time in deeper water while smaller males and all juveniles remain near the shore in shallower water.

In South Carolina, Tucker et al. (1995) found that large females spent more time in shallow portions of saltmarshes feeding on larger snails during tidal flooding and retreating with the ebbing tide or burying themselves in the mud. They also found juveniles and smaller males near the edges of marshes and channels where they foraged on smaller prey items.

During winter brumation, terrapins move into deep, fairly small creeks, select just the right bottom type and burrow into muddy substrate where the water is deepest. Brumation sites are far enough up creeks that salinities remain modest but tidal action keeps the water circulating. Terrapins brumate in the mud for several winter months and gradually increase their cellular osmotic pressure as sea water builds in their systems. Osmotic pressure also increases in their urea suggesting that they regulate salt to some degree through excretion (Gilles-Baillien 1973). Brumation locations are likely to be positioned so that during periods of heavy winter rain the saline nature of the water is periodically diluted. Precise locations of brumation sites is largely unknown, but it is generally thought that by late November terrapins settle in for the winter and many hundreds are often concentrated in a very small area. During the rest of the year the terrapins are more widely dispersed in creeks and sounds.

OTHER ASPECTS OF TERRAPIN LIFE HISTORY

Nesting ecology

Female terrapins require sandy upland substrate for egg laying. Narrow, sandy strips of land between the open estuarine water and marsh habitat provide ideal nesting habitat, and female terrapins congregate at such places in the summer to deposit one to several clutches per season. In many areas terrapins are forced to travel through a bay or marsh system each season—the prime feeding areas are not necessarily near their brumating quarters, and neither is likely to be in the proximity of nesting beaches. Studies on nest survivorship show that the turtles have a rather narrow spectrum of beaches on which a high percentage of the nests survive (M. Whilden, pers. comm.). Those nests isolated from terrestrial predators like skunks (*Mephitis* and *Spilogale*) and raccoons (*Procyon lotor*) do best. Terrapins, like many turtles, have temperature-dependent sex determination (TSD) whereby the sex of the turtle is determined as the embryo grows and develops within the egg in the nest chamber. Early July temperatures are the most influential on the gender of the developing embryo (Auger 1989), but specific characteristics of the nest and beach microhabitat cause different sex ratios in hatchlings—cooler, shady beaches produce mostly males and warmer, open sandy beaches produce mostly females. As a result of TSD it is important to have a number of nesting beaches available in any given area to ensure that enough turtles of each sex are produced each year.

Basking behavior

Terrapins are poikilothermic and, like other turtles, bask in the sun to elevate their internal temperature above that of the water. This elevated temperature accelerates the digestion of food and other metabolic processes. Terrapins generally bask on tidally exposed mud flats or while floating at the surface on calm days. By filling their lungs with air and extending their heads, necks, and hind limbs out of the water, they can absorb heat, a process facilitated by their dark integument, and quickly elevate their body temperatures.

TERRAPIN CONSERVATION

To date no range-wide evaluation of the population status of this turtle has been made, nor is much historical information available for comparison. Sites where long-term data are available, primarily small and isolated populations, suggest the species to be in peril (i.e., Florida, Seigel 1993; South Carolina, Gibbons 2001). This is consistent with the increasing combinations of factors that threaten terrapins. The USDI Fish and Wildlife Service listed this turtle as a status review species for decades and in the last few years various groups have initiated regional population assessments.

THREATS TO TERRAPINS

Despite limited protected status in some regions, populations of this long-lived turtle species generally have not recovered from past episodes of direct harvest (Seigel and Gibbons 1995). Only recently have scientists and policymakers recognized that the main threats to terrapin populations are linked to humans. Such threats include, but are not limited to, drowning in crab pots and entanglement in fishing gear (Bishop 1983, Roosenburg et al. 1997, Hoyle and Gibbons 2000), commercial harvest (Bishop 1983, Roosenburg et al. 1997), loss of critical nesting and basking habitat with accompanying effects on sex ratios (Lazell and Auger 1981), and incidental mortality by motorized vehicles (Lazell 1979, Roosenburg 1990, Wood and Herlands 1997). Turtle nests are depredated by raccoons (Seigel 1980, Feinberg 2003), Bald Eagles (*Haliaeetus leucocephalus*), and other predators whose populations have been enhanced by human activity. Nests are also disturbed by the rhizomes of grass roots (Lazell and Auger 1981) and nesting turtles suffer from competition for access to shoreline with developers and private property owners.

Interactions with gear designed to catch blue crabs

The incidental catch and subsequent drowning of diamondback terrapins in pots designed to catch blue crabs has become a major conservation issue along both the Atlantic and Gulf of Mexico coastlines. Crab pots deployed within the range of terrapin populations may directly threaten those populations (Wood, unpubl. data; Bishop 1983, Seigel and Gibbons 1995, Roosenburg et al. 1997), because terrapins of certain sizes are trapped in the pots and drown. Considerable mortality may also stem from terrapins getting lodged inside abandoned pots. In fact evidence of a crab-pot effect may be apparent in sex ratio data from Maryland (W. Roosenburg, pers. comm.) and North Carolina (K. Hart, unpubl. data). Sex ratios are consistently female-skewed in areas with intense commercial crabbing, which may be a result of differential mortality of males versus females in crab traps. However, until we know more about baseline terrapin sex ratios, population structure, mating systems, or vital rates, we cannot interpret skewed sex ratios as more than a predominance of females in the system.

Blue crabs support valuable commercial fisheries along the southeast and gulf coasts of the US, and today the majority of the total crab harvest is taken in crab pots. In North Carolina, for example, a 1998 estimate for the fishery places 1,063,331 crab pots in North Carolina waters, nearly doubling the number of pots set just 10 yr prior (North Carolina Marine Fisheries 1998). While these numbers are only estimates based on surveys, they indicate that potential accidental terrapin catch and mortality in crab pots in North Carolina can be highly detrimental to a species like terrapins that may be already declining. Interestingly, New Jersey and Maryland now require bycatch reduction devices (BRDs) on certain crab pots. BRDs are stiff, rectangular wire devices that are affixed to the funnel entrances of crab pots, reducing the size and height of the funnel opening. Recently, North Carolina outlined a requirement for BRDs for crab pots as a potential management option in the draft North Carolina Blue Crab Management Plan. However, it is currently unclear where, and when, such devices should

Table 1. State protection currently offered for the diamondback terrapin.

State	Protection
Massachusetts	Threatened.
Rhode Island	Endangered.
Connecticut	State regulated species.
New Jersey	Special concern, turtle excluder device (TED) on all crab pots.
Delaware	Species of state concern, regulated game species.
Maryland	Turtle excluder device (TED) on all noncommercial crab pots, harvest restricted to November through March, >15 cm plastron.
North Carolina	Species of special concern.
Georgia	Species of special concern.
Alabama	Species of special concern.
Mississippi	Species of special concern.
Louisiana	Species of special concern.

be required because of the lack of information on terrapin distributions and their overlap with blue crab fisheries. Furthermore, little is known about terrapin population structure and the extent or scope of terrapin mortality in crab pots. Characterizing the threat that commercial crab pots pose to terrapins, and quantifying terrapin movement and habitat use in a temperate estuarine system will help focus efforts to regulate the blue crab fishery towards the goals of continuing the valuable fishery and enhancing terrapin populations. Further, demonstrating economic benefits rather than losses from gear modifications appears to be an effective way to ensure implementation in commercial fisheries.

After conducting studies in Maryland, Roosenburg et al. (1997) concluded that between 10 and 78% of a local terrapin population might be captured annually in crab pots by recreational crabbing activity. Watermen on the Delaware Bay reported that during the warmer parts of the season a typical catch of 300 terrapins/day was normal (D. Lee, unpubl. data). Several Atlantic Coast state fisheries departments are now looking into requiring BRDs and changing harvest regulations.

Other threats

In addition to interactions with crab pots, terrapins are vulnerable to other anthropogenic disturbances at every phase of their life cycle. The list of threats to terrapins is long—from pollution to loss of wetlands, bycatch in fishing gear, loss of habitat to real estate developers, and predation by raccoons and bald eagles—and unfortunately diamondback terrapins often lose the battle against these pressures

Conservation Status

Currently, terrapins benefit from only limited protection (Table 1) yet their populations are declining or of unknown status in three-quarters of the states they occupy (Table 2) (Seigel and Gibbons 1995). Unfortunately, our current knowledge of terrapin ecology and population genetics is limited. Although we know that this long-lived turtle is much reduced from historical numbers, we do not know the scope and scale that either individual or collective threats pose at the population level (Roosenburg et al. 1997, Hart 1999).

POPULATION ASSESSMENTS AND MODELING

Researchers agree that terrapins are not nearly abundant as they once were (Ashton and Ashton 1991, Seigel and Gibbons 1995). Populations may be rebounding from severe harvest at the turn of the century (Conant and

Table 2. Status of regional diamondback terrapin populations.

Declining	Stable or increasing	Insufficient data
New York	Massachusetts	Delaware.
New Jersey	Rhode Island	Virginia.
Maryland	New York	Georgia.
North Carolina	Maryland	Florida (Gulf Coast).
South Carolina	Florida (Keys)	Alabama.
Florida (Atlantic Coast)		Texas.
Louisiana		
Mississippi		

Note: Data from Seigel and Gibbons 1995.

Collins 1991), however, relatively few surveys of terrapins have been published (Mann 1995). Wood (1992) recommended further surveys to establish baseline data for populations, but presently we do not have the information we need to delineate clear population trends for the species. Perhaps this is because short-term counts have been the primary criteria for gauging the size and health of such populations (Hurd et al. 1979). Multiple years of mark-recapture data are necessary to document population trends. Mark-recapture studies generate data to allow for eventual estimation of sex ratios, survival rates, age structure, and overall population size. Despite the efforts of several researchers in different study sites (Massachusetts, Auger 1989; Maryland, Roosenburg et al. 1999; Florida, Forstner et al. 2000; South Carolina, Bishop 1983; New Jersey, Wood 1992), we currently lack most critical vital demographic rates for terrapins. However, recent efforts by Hart (1999), Tucker et al. (2003), and Mitro (2004) to analyze long-term mark-recapture data sets from various locations revealed adult survival rates of 0.83, 0.84, and 0.95 for terrapins from sites in Massachusetts, South Carolina, and Rhode Island, respectively. These estimates are within the range of published survivorship rates for other emyid turtles (Iverson 1991) with similar age and size at maturity and longevity (40 yr, Hildebrand 1932).

The work by Dunham et al. (1989) on life-history modeling and Congdon et al. (1993) on Blanding's turtles (*Emydoidea blandingii*) focused attention on the life-history and demographic constraints of long-lived organisms. Recent work by Heppell (1998), Heppell et al. (2000), and Sæther and Bakke (2000) examined relationships among age at sexual maturity, adult survivorship, and juvenile survivorship within life histories of long-lived organisms. Results from their studies indicate that all long-lived vertebrates have coevolved life-history traits that limit their ability to respond to increased mortality imposed on any age group (Congdon et al. 1993). Understanding that long-lived vertebrates have a limited ability to respond to increases in mortality is particularly important in decisions related to populations that are subject to commercial harvest or bycatch of juveniles or adults (Crouse et al. 1987, Heppell and Crowder 1998).

ENFORCEMENT AND PROTECTION

Despite these limited state listings and the application of relatively new techniques in terrapin studies, multiple threats to diamondback terrapin populations exist in all states throughout their range and enforcement of harvest regulations is all but nonexistent. Unfortunately, protected status in a few states may not suffice to ensure the survival of the species. Nonetheless, the turtle is a potential candidate for listing—it has been a species under review for candidate 2 listing with the NOAA National Marine Fisheries Service for the last several decades, but new threats to terrapin existence continue to emerge. For example, human populations of Asian descent in the US and Canada have developed a dietary fondness for turtle meat and over the last several decades the market for terrapin has responded to their demand. However, terrapin is largely unregulated as a seafood and restrictions that are in effect were made long before we fully understood the turtles' habitat needs and well before current population modeling techniques were developed. Different states have different size limits for commercially harvested terrapins, but even a 10–13 cm size limit heavily favors collection of females. Because one male can fertilize dozens of females it is unclear how these regulations may influence what is needed to maintain reproductively viable populations.

CONSERVATION CONCERNS

Because of the rapid marketing that has developed for seafood, terrapins captured in the field one day often arrive in the markets of another state by the next morning to be sold. This makes it nearly impossible to track marketed terrapins, to learn of their origin, to enforce regulations of other states, and to obtain any statistical information on seasonal or even annual catch rates. As well, because terrapins were not an important seafood product for much of the middle part of the twentieth century, state agencies ceased collecting reports on terrapin landings and virtually no baseline information exists from which to establish regulations for commercial harvest. At this time, only scant information exists on the amount of bycatch of terrapins in crab traps and nets. Most of the turtles captured as bycatch drown and do not become part of the reported commercial harvest. Even if terrapins did not face problems in their coastal environments, slow-growing turtles with low annual reproductive output are not programmed to respond quickly to substantial harvest (Heppell 1998, Heppell et al. 1999).

SUMMARY AND FUTURE

Many turtle species worldwide are increasingly at risk of extinction (Eckert and Sarti 1997, Heppell et al. 1999). Given the general life-history characteristics of turtles, such as delayed sexual maturity, longevity, terrestrial nesting activity, and lack of parental care, they are particularly vulnerable to human-induced threats (Crouse et al. 1987, Congdon et al. 1993, Doak et al. 1994, Heppell 1998, Heppell et al. 1999). Despite annual reproduction schedules, turtles recover slowly from population declines because their populations require high juvenile and adult survival for stability (Congdon et al. 1993, Heppell 1998, Heppell et al. 1999). As such, increased mortality in the juvenile or adult stages will generally cause populations to decline. Threats that affect these life stages in particular need to be mitigated as soon as possible.

This species seems to have fallen through the cracks of local protection and state regulation. In general, fishery agencies base regulations on catch rates and, in most states, terrapin catch is not currently reported. In a number of areas adequate studies have been done to document the local decline of terrapins in the last several decades (Seigel 1993, Gibbons et al. 2001). These populations could benefit from immediate protective measures. Existing laws need to be enforced, harvest rates need to be reported, and the extent and nature of bycatch and other mortality sources needs to be documented on a region-by-region basis. Although more research is necessary, management decisions need not be put off any longer. Concerted efforts to synthesize available data and protect the terrapin should be initiated.

Many isolated populations will be lost if we wait until the last pieces of research are analyzed and incorporated into management plans and regulations. However, terrapin conservation faces real challenges because development of coastal habitats carries on, direct exploitation of terrapins is again expanding, unregulated crabbing continues, interstate traffic of terrapins continues to be facilitated by members of the coastal seafood industry, and enforcement of existing regulations is minimal to non-existent. States that do provide various levels of protection to terrapins have different size limits and seasons, and most watermen are often not informed of these regulations and even fewer watermen report annual catch results consistently. Additionally, crabbers are likely to resist gear modifications such as BRDs, despite findings that their crab-catch rates would not likely decrease with such devices, and the general public is largely unaware of terrapins, their decline, or their modest needs.

Despite these challenges, we have hope for this resilient turtle. Practical, general measures like protecting saltmarsh habitat and specific management actions like installing temporary fences along roads where terrapin road kill is high, affixing BRDs to crab pots, and halting direct harvest would work to protect many terrapins throughout their range. While much research remains to be completed in order to make long-range decisions regarding management regulations, a number of local conservation efforts could be initiated immediately to protect declining populations. Maintaining the integrity of saltmarsh ecosystems is tantamount to ensuring the long-term protection of terrapins.

The future might be bleak for the terrapin if real protection is not afforded to the species soon. The time to address the pressing threats is upon us. Management and protection strategies can be fine-tuned as more information becomes available and the turtles, over time, respond to these efforts. But waiting for completion of long-term studies is not a viable option for a vulnerable, slow-growing species with limited reproductive output, confined to habitats that are under heavy use and continued development.

ACKNOWLEDGMENTS

We thank the editors and three anonymous reviewers for their comments on an earlier draft of this chapter. KMH acknowledges the U.S. Geological Survey and Duke University for support and thank the North Carolina Sea Grant Fishery Resource Grant Program for funding.

HIGH TIDES AND RISING SEAS: POTENTIAL EFFECTS ON ESTUARINE WATERBIRDS

R. Michael Erwin, Geoffrey M. Sanders, Diann J. Prosser, and Donald R. Cahoon

Abstract. Coastal waterbirds are vulnerable to water-level changes especially under predictions of accelerating sea-level rise and increased storm frequency in the next century. Tidal and wind-driven fluctuations in water levels affecting marshes, their invertebrate communities, and their dependent waterbirds are manifested in daily, monthly, seasonal, annual, and supra-annual (e.g., decadal or 18.6-yr) periodicities. Superimposed on these cyclic patterns is a long-term (50–80 yr) increase in relative sea-level rise that varies from about 2–4 + mm/yr along the Atlantic coastline. At five study sites selected on marsh islands from Cape Cod, Massachusetts to coastal Virginia, we monitored marsh elevation changes and flooding, tide variations over time, and waterbird use. We found from long-term marsh core data that marsh elevations at three of five sites may not be sufficient to maintain pace with current sea-level rise. Results of the short-term (3–4 yr) measures using surface elevation tables suggest a more dramatic difference, with marsh elevation change at four of five sites falling below relative sea-level rise. In addition, we have found a significant increase (in three of four cases) in the rate of surface marsh flooding in New Jersey and Virginia over the past 70–80 yr during May–July when waterbirds are nesting on or near the marsh surface. Short-term, immediate effects of flooding will jeopardize annual fecundity of many species of concern to federal and state agencies, most notably American Black Duck (*Anas rubripes*), Nelson's Sharp-tailed Sparrow (*Ammodramus nelsoni*), Saltmarsh Sharp-tailed Sparrow (*A. caudacutus*), Seaside Sparrow (*A. maritima*), Coastal Plain Swamp Sparrow (*Melospiza georgiana nigrescens*), Black Rail (*Laterallus jamaicensis*), Forster's Tern (*Sterna forsteri*). Gull-billed Tern (*S. nilotica*), Black Skimmer (*Rynchops niger*), and American Oystercatcher (*Haematopus palliatus*). Forster's Terns are probably most at risk given the large proportion of their breeding range in the mid-Atlantic and their saltmarsh specialization. At a scale of 1–2 decades, vegetation changes (saltmeadow cordgrass [*Spartina patens*] and salt grass [*Distichlis spicata*] converting to smooth cordgrass [*Spartina alterniflora*]), interior pond expansion and erosion of marshes will reduce nesting habitat for many of these species, but may enhance feeding habitat of migrant shorebirds and/or migrant or wintering waterfowl. At scales of 50–100 yr, reversion of marsh island complexes to open water may enhance populations of open-bay waterfowl, e.g., Bufflehead (*Bucephala albeola*) and Canvasback (*Aythya valisneria*), but reduce nesting habitats dramatically for the above named marsh-nesting species, may reduce estuarine productivity by loss of the detrital food web and nursery habitat for fish and invertebrates, and cause redistribution of waterfowl, shorebirds, and other species. Such scenarios are more likely to occur in the mid- and north Atlantic regions since these estuaries are lower in sediment delivery on average than those in the Southeast. A simple hypothetical example from New Jersey is presented where waterbirds are forced to shift from submerged natural marshes to nearby impoundments, resulting in roughly a 10-fold increase in density. Whether prey fauna are sufficiently abundant to support this level of increase remains an open question, but extreme densities in confined habitats would exacerbate competition, increase disease risk, and possibly increase predation.

Key Words: Atlantic coast, breeding habitat, marsh flooding, marsh surface, sea-level rise, tidal fluctuations, waterbirds.

MAREAS ALTAS Y MARES QUE ASCIENDEN: EFECTOS POTENCIALES EN AVES ACUÁTICAS DE ESTUARIO

Resumen. Las aves acuáticas de costa son vulnerables a los cambios en el nivel del agua, especialmente bajo las predicciones acerca del levantamiento acelerado del nivel del mar y el aumento en la frecuencia de tormentas durante el siguiente siglo. Las fluctuaciones causadas por marea y viento en los niveles del mar que afectan a las marismas, sus comunidades de invertebrados, y sus aves acuáticas dependientes son manifestadas en periodicidades diarias, mensuales, estacionales, anual y supra-anuales (ej. en décadas o 18.6 años). Super impuestos en estos patrones cíclicos hay un incremento de largo plazo (50–80 años) en el levantamiento del nivel del mar que varia de cerca de 2–4 + mm/años a lo largo de la línea costera del atlántico. En cinco sitios de estudio seleccionados en islas de marisma de Cabo de Bacalao, Massachussets hasta la costa de Virginia, monitoreamos los cambios en la elevación de la marisma e inundaciones, las variaciones de la marea en el tiempo, y utilización de aves acuáticas. De datos centrales de marisma de largo plazo encontramos que las elevaciones de marisma en tres de los cinco sitios quizás no son suficientes para mantener el ritmo con las actuales elevaciones en el nivel del mar. Resultados de las medidas de corto plazo (3–4 años) utilizando tablas de elevación de la superficie, sugieren una diferencia más dramática, con un cambio en la elevación de marisma de cuatro sitios cayendo por debajo de la elevación del nivel del mar. Además, hemos encontrado un incremento significativo (en tres de los cuatro casos) en la proporción de la superficie de marisma en inundación en Nueva Jersey y Virginia durante los últimos 70-80 años durante Mayo-Julio cuando las aves acuáticas están anidando en o cerca de la superficie de la marisma. A corto plazo, efectos inmediatos de inundación ponen en peligro la fecundidad anual de muchas especies del interés de agencias federales y estatales, más notablemente en el Pato Negro Americano (*Anas rubripes*), el Gorrión Cola Aguda Nelson (*Ammodramus nelsoni*), el Gorrion de Marisma Salado Cola Aguda (*A. caudacutus*), el Gorrion Costero (*A. maritima*), el Gorrion Pantanero (*Melospiza georgiana nigrescens*), la Polluela Negra (*Laterallus jamaicensis*), el Charran de Foster (*Sterna forsteri*). El Charran Picogrueso (*S. nilotica*), el Rayador Americano (*Rynchops niger*), y el Osterero Americano (*Haemotopus palliatus*). Los Charranes de Foster se encuentran probablemente más en riesgo, dada la gran proporción de su rango reproductivo en el Atlántico medio y su especialización a la marisma salada. A la escala de 1-2 décadas, los cambios en la vegetación (*Spartina patens* y *Distichlis spicata* convirtiéndose a *S. alterniflora*), la expansión interior de charcos y la erosión de marismas reducirán el hábitat de anidamiento para muchas de estas especies, pero quizás mejoren el hábitat de alimento de aves migrantes de playa y/o de gallinas de agua migrantes o de invierno. A escalas de 50-100 años la reversión de los complejos de islas de marisma para abrir el agua quizás mejoren las poblaciones de gallinas de agua de bahía abierta, ej. Pato Monja (*Bucephala albeola*) y Pato Coacoxtle (*Aythya valisneria*), pero reduzcan dramáticamente habitats de anidación para las especies de marisma de anidacion nombradas anteriormente, quizás se reduzca la productividad de la marisma por la pérdida de la cadena alimenticia detrital y el hábitat de criadero para los peces e invertebrados, y causa la redistribución de las gallinas de agua, aves de playa y otras especies. Tales escenarios son más susceptibles a suceder en las regiones medias y del norte del Atlántico, ya que estos estuarios son más bajos en la repartición de sedimento en proporción a aquellos en el sureste. Un simple ejemplo hipotético de Nueva Jersey es presentado donde las aves acuáticas son forzadas a cambiar de marismas naturales sumergidas a encharcamientos cercanos, resultando en aproximadamente un incremento en densidad de 10 pliegues. Si la fauna de presa es suficientemente abundante para soportar este nivel, sigue siendo una pregunta abierta, pero densidades extremas en habitats confinados exacerbaría la competencia, incrementaría el riesgo de enfermedades, y posiblemente incrementaría la depredación.

Concern is growing in many areas of the US and throughout the world that as sea levels continue to rise, coastal-marsh elevations may not be able to keep pace (Titus 1988, Emery and Aubrey 1991, Warrick et al. 1993, Brinson et al. 1995, Nicholls and Leatherman 1996, Nerem et al. 1998). This will have large implications not only to human infrastructure (Titus 1991, Titus et al.1991) but also to many rare and imperiled species of animals and plants (Reid and Trexler 1992). The obvious first victims of accelerated sea-level rise will be those plants and animals that are obligate saltmarsh residents or coastal-dependent migratory species such as waterbirds. Even though the total area of the coastal estuarine zone is a fraction of the upland areas of the US, the large number of migratory waterbirds (nearly 100 species in the US) and threatened-to-endangered species using the coastal fringe is disproportionately high (Reid and Trexler 1992, Daniels et al. 1993).

Global sea levels are predicted to rise from 10-90 cm during the next 100 yr, with a median model estimate of 48 cm (Intergovernmental Panel on Climate Change 2001). Of course, the degree to which marsh surface dynamics in a given estuary departs from global average predictions depends on complex interactions of marsh shallow subsidence (e.g., compaction, decomposition, and water storage and extraction; Cahoon et al. 1999), plate-tectonic movements (glacial or isostatic rebound; Emery and Aubrey 1991), landscape position relative to sediment source (Kearney et al. 1994, Roman et al. 1997), storm frequency (Giorgi et al. 2001), and biotic factors such as grazing or trampling by herbivores (Chabreck 1988; Mitchell et al., *this volume*).

Variation in the effects of large-scale phenomena such as sea-level rise occurs at all spatial scales. Within a coastal embayment, the position of the marsh may strongly influence its ability to maintain elevation. Lagoonal marshes, in the middle of a large bay, may have insufficient sources of sediments to maintain their elevations compared with marshes near barrier islands (storm-driven inorganic sand) or those close to the mainland (high organics) (Hayden et al. 1991, Roman et al. 1997). In addition to local variation at the sub-estuary and estuary levels, regional differences occur in marsh accretion and relative sea-level rise (RSLR), i.e., that due to both water-level changes as well as change in land-surface elevation. In part, this is caused by post-glacial crustal uplift in New England and northward, but down warping in the mid-Atlantic region (National Academy of Science 1987, Emery and Aubrey 1991).

In parts of New England and the Carolinas along the U.S. Atlantic Coast, average marsh-accretion rates seem to be much greater than RSLR; however, in parts of Georgia, the

Chesapeake Bay of Maryland, and Connecticut, RSLR is greater than average marsh accretion (National Academy of Science 1987, Warren and Niering 1993). Examining accretion rates, however, only reveals part of the dynamics. Subsidence can be significant in areas of the Chesapeake Bay (Nerem et al. 1998), in Virginia, and especially in Louisiana (Boesch et al. 1983). Recognizing the importance of measuring both components has led to the evolution of the surface elevation table (SET), a mechanical device that allows for monitoring of surface elevation changes over time from a permanent benchmark (Cahoon et al. 2002 and references therein).

Tide and sea-level variations also have frequency components associated with daily, monthly, yearly, and decadal and 18.6-yr periods (Kaye and Stuckey 1973, Pugh 1987, Stumpf and Haines 1998; Fig. 1). Regular fluctuations in the timing of high- and low-water events can have profound effects on the evolution and life histories of the myriad of organisms associated with saltmarshes (Bertness 1999). These predictable temporal variations must be factored into the ecological and evolutionary responses of organisms that are also facing steadily increasing sea levels that may also be associated with more frequent but unpredictable storm events (Intergovernmental Panel on Climate Change 2001).

One of the end results of rising sea levels and marsh submergence is declining quality or loss of wildlife habitat. Global warming with concomitant storm increases and sea-level rise has sparked major concern among ornithologists and coastal managers with disturbances or loss of both intertidal habitats of international importance as well as adjacent emergent marshes (Titus et al. 1991, Peters and Lovejoy 1992, Ens et al. 1995, Fenger et al. 2001). Special concerns have been voiced for loss of shorebird habitats (Myers and Lester 1992, Galbraith et al. 2002). Using a coarse-simulation model of changes in water levels and coastal elevations at five critical shorebird staging areas in the US, Galbraith et al. (2002) estimated intertidal habitat losses of 20–70%.

In addition to shorebirds, a number of other species and groups of waterbirds are potentially vulnerable to coastal storms and sea-level rise (Table 1). Along the Atlantic Coast over the past 50 yr, habitat quality and quantity have declined as human density and disturbances have increased. A number of waterbird species that use Atlantic marshes during some part of the year has been listed by federal and/or state agencies as being at risk or of special concern because of population declines in certain regions; these include American Black Ducks (*Anas rubripes*), Forster's Terns, (*Sterna forsteri*). Gull-billed Terns (*S. nilotica*), Common Terns (*S. hirundo*), Black Skimmers (*Rynchops niger*), American Oystercatchers (*Haematopus palliates*), Black Rails (*Laterallus jamaicensis*), and marsh-nesting passerines such as Seaside Sparrows (*Ammodramus maritimus*), Nelson's Sharp-tailed Sparrows (*A. nelsoni*), Saltmarsh Sharp-tailed Sparrows (*A. caudacutus*), and Coastal Plain Swamp Sparrows (*Melospiza georgiana nigrescens*) (see appendices in the Waterbird Conservation for the Americas Plan <http://www.waterbirdconservation.org> [6 July 2006]).

Because global climate-change and sea-level rise scenarios have major implications to coastal habitats, especially federally owned parks and national wildlife refuges, the U.S. Geological Survey has provided major funding within its research programs directed at these topics since 1998. Our study has focused on monitoring marsh changes, tide levels, and waterbird use of selected coastal sites that are known to be important for one or more guilds of waterbird and where coastal managers and scientists have voiced concerns over marsh changes.

The questions we pose are:
1. Is sea-level rise occurring at consistent rates at all mid- and north Atlantic locations? How do these rates compare with marsh elevation changes in lagoons over the short and long terms?

TABLE 1. NUMBER OF AVIAN SPECIES USING EMERGENT SALTMARSHES FOR NESTING, FEEDING, AND/OR RESTING IN THE MID-ATLANTIC COASTAL REGION.

Group	Breeding	Migration	Wintering
Waterfowl[a]	6	26	24
Shorebirds[b]	5	29	12
Seabirds[c]	12	13	4
Wading birds[d]	10	10	4
Marsh birds[d]	3	8	1

[a] Bellrose (1976), Palmer (1976).
[b] Bent (1962a, b).
[c] Bent (1963a).
[d] Bent (1963b).

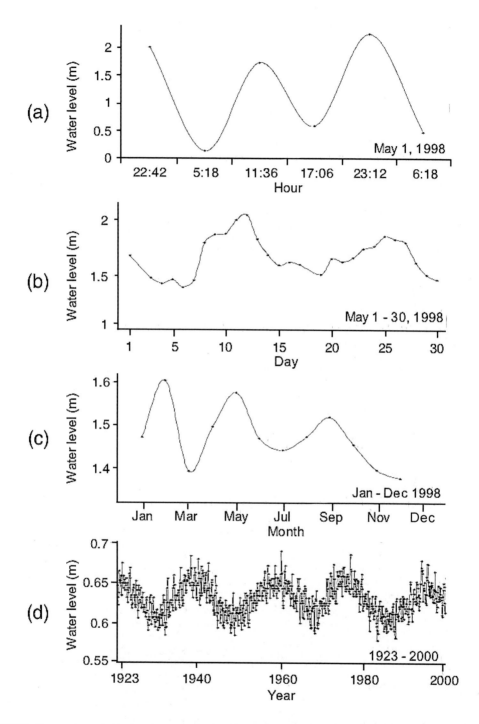

FIGURE 1. Comparison of simultaneous temporal cycles associated with tidal waters. Raw data obtained from National Oceanic and Atmospheric Administration's (NOAA) National Ocean Service (<http://www.co-ops.nos.noaa.gov/> [31 July 2006]). All data were obtained for the Atlantic City tide gauge with mean low low water as the datum. (a) Low and high tides based on 24 hours, 1 May 1998; (b) Daily high high tides for the month of May 1998 (full moon and new moons were 11 May and 25 May, respectively); (c) Monthly mean high tides for 1998; and (d) Monthly mean sea level minus monthly mean high tide calculated to show the approximate 18.6-yr cycle known as the metonic cycle or lunar nodal cycle (Kaye and Stuckey 1973).

2. What are flooding rate patterns in marshes and how might these affect nesting species?
3. Which waterbird species are affected most by marsh inundation over short, intermediate and long terms?
4. What are some implications of habitat shifts required of affected waterbird species?

METHODS

SITE SELECTION

Three regional study areas were established in saltmarshes along the Atlantic Coast of the US to determine the latitudinal variations in sea-level rise, marsh-elevation change, and bird use. The study areas in New Jersey and Virginia each consisted of two separate sites (Fig. 2), resulting in a total of five sites from Cape Cod to southern Virginia. Sites were chosen based on their importance to waterbirds and the presence of ongoing management (especially on federal lands such as Cape Cod National Seashore and E. B. Forsythe National Wildlife Refuge [NWR]) and research activities. The Virginia sites are within the Virginia Coast Reserve, an area of ongoing estuarine research supported by the University of Virginia's Long Term Ecological Research program (<http://www.vcrlter.virginia.edu/> [7 July 2006]). We selected lagonal marsh complexes, i.e., islands in the middle of embayments, because for most species of waterbirds these are the most important. Each site included randomly selected marsh elevation sampling plots (N = 15; located in the high marsh at least 15 m from the main marsh channel) as well as a waterbird survey plot containing tidal ponds, pannes, and/or mudflats to document waterbird usage. Predominant vegetation included short-form smooth cordgrass (*Spartina alterniflora*), glassworts (*Salicornia* spp.), and beach salt grass (*Distichlis spicata*) with one exception being a site at Mockhorn Island, Virginia, where the dominant vegetation type was tall-form smooth cordgrass The plots to be sampled for marsh surface elevation change were replicated at each site (two–six replicates) with at least 100 m separating the marsh surface plots.

MARSH ELEVATION

Surface elevation table

Changes in marsh elevation were measured using a surface-elevation table (SET after Boumans and Day 1993, Cahoon et al. 2002), a portable device that attaches to a permanent stable benchmark pipe driven into the substrate until the point of refusal, roughly 5–6 m (Fig. 3). The device is designed to detect changes in marsh elevation at a high resolution (+/- 0.7–1.8 mm; Cahoon et al. 2002) by repeatedly measuring the same position on the marsh surface over time. We took measurements every 6 mo in the spring and fall and rates of elevation change were determined by linear regression analysis of the cumulative elevation change (i.e., $time_t - time_0$).

Accretion (feldspar)

The SET measures total marsh elevation change. To determine the influence of shallow subsidence or compaction, the accretion component must be separated from total elevation change. Sediment accretion was measured on the same time interval as the SET. Three feldspar marker horizons (N = 45) were positioned around each SET plot at the time of the initial SET reading (Cahoon and Turner 1989). A liquid nitrogen cryocorer was used to obtain a frozen sediment core in a manner that does not cause compaction of the sample (Cahoon et al. 1996). The amount of accretion was determined by measuring the amount of material (both organic and inorganic) above the white feldspar layer to the nearest 0.1 mm using calipers.

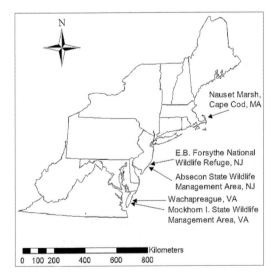

FIGURE 2. General locations of marsh study sites along the East Coast of the US: Nauset Marsh, Cape Cod National Seashore, Massachusetts; E. B. Forsythe National Wildlife Refuge, and Absecon State Wildlife Management Area, New Jersey; Wachapreague and Mockhorn Island State Wildlife Management Area, Virginia.

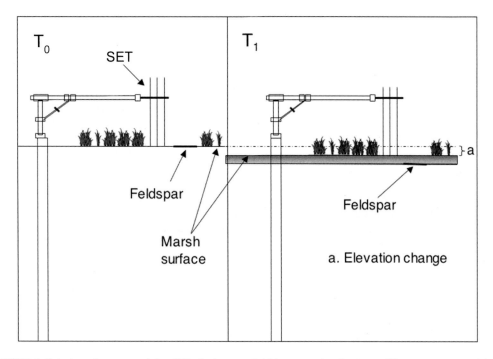

FIGURE 3. Relative placement of the SET platform and feldspar marker horizons (diagram not to scale). At time $_0$ the baseline elevation readings are taken and the feldspar marker horizons are placed on the marsh surface surrounding the SET. At time $_1$ (~6 mo later), marsh elevation is measured again along with vertical accretion. The example above depicts a scenario where vertical accretion was present (evidenced by the gray shaded area) but shallow subsidence was greater yielding a net loss of elevation. This is just one example and not necessarily representative of all results.

Long-term accretion (^{210}Pb)

The SET and cryogenic coring data present a short-term picture of marsh-elevation dynamics; however, radiometric dating using radioisotopes in the substrate allows one to establish historic accretion rates (ca. 100 yr; Kastler and Wiberg 1996). Two sediment cores were taken from the high marsh at each of our study sites in Virginia and New Jersey using a piston corer, approximately 1 m long and 10.25 cm in diameter (C. Holmes, USGS, unpubl. data). Long-term sedimentation rates for Nauset Marsh were obtained from previously published results (Roman et al. 1997). The cores were sampled at 1-cm intervals for the first 10 cm and at 2-cm intervals thereafter. Using the radioactive isotopes ^{210}Pb and ^{137}Cs, the age of the sample was determined by comparing the original isotopic concentrations to the percent remaining in the sample. A constant-flux:constant-sedimentation model was deemed most appropriate to analyze the sedimentation rates for the New Jersey and Virginia cores. This model assumes a constant flux of unsupported ^{210}Pb and a constant dry-mass sedimentation rate. (C. Holmes, USGS unpubl. data; Robbins et al. 2000).

SEA-LEVEL TRENDS AND MARSH FLOODING

Water-level data were obtained from the National Oceanic and Atmospheric Administration's (NOAA) National Ocean Service (<http://www.co-ops.nos.noaa.gov/> [7 July 2006]) to show geographic differences in rates of sea-level rise for each general location along the East Coast (Boston, Massachusetts; Montauk, New York; Atlantic City, New Jersey; and Chesapeake Bay Bridge Tunnel, Virginia). Data from the Boston and Atlantic City tide gauges span roughly the past 80 yr; however, the New York and Virginia data sets date back approximately 50 and 25 yr, respectively. Because the Wachapreague, Virginia, local tide-gauge station only dated from 1980, those water-level data were not used; instead we used the tidal data from Sandy Hook, New Jersey, because this is the standard long-term reference for coastal Virginia. Monthly mean data were used to calculate sea-level rise trends and linear regression analysis was used to determine the rate of sea-level rise for each location.

During the nesting season (May–July), extreme spring tides often destroy the nests of and cause mortality to marsh-nesting waterbirds such as Forster's and Common Terns (Storey 1987, Burger and Gochfeld 1991a). Based on previous observations, we estimated that a tide that reached at least 20 cm above the surface of the marsh would be sufficient to flood the majority of nests in most colonies. To determine the frequency of such tides, we installed data loggers at our study sites at E. B. Forsythe NWR, New Jersey, and at Wachapreague, Virginia, in spring and summer 2001 to monitor the height of the water above the marsh surface. We used both the Solinst Leveloggers®, which use pressure sensors to detect changes in water level, and powdered cork, which was placed inside clear acrylic tubes. When the water rose inside the tube, the cork adhered to the sides revealing the maximum height of the previous tides.

The data loggers provided readings of water level, date, and time (5-min intervals over 3–7 d) and the cork gauges corroborated the level of the tide event. Comparing these data to that from local NOAA tide gauges, we determined the NOAA tide-gauge reading at the time when the marsh reached our designated flooding point (20 cm above the marsh surface). Tide gauges in Atlantic City and Sandy Hook, New Jersey, were used to reference past flooding events for the New Jersey and Virginia sites, respectively. We hindcasted through the tidal record to determine the frequency of marsh flooding events (>20 cm over the marsh surface) during the breeding season (mid-May–mid-July) at E. B. Forsythe NWR, New Jersey, and Wachapreague, Virginia.

Two methods of hindcasting were used. The first approach assumes steady-state conditions of the marsh surface wherein historic tide heights were compared to current marsh sediment elevations. It is apparent, however, given the long-term accretion rates revealed from the ^{210}Pb data, that material was deposited on the marsh surface, suggesting an increase in marsh elevation over time. Neglecting to account for long-term accretion may in effect reduce the number of flooding events predicted in earlier years, resulting in bias toward an increasing flooding trend over time. In an attempt to correct for this potential bias, a second test was performed where the historic water-level data were compared to marsh elevations adjusted for vertical accretion over the duration of the tidal record.

Making the rather large assumption that accretion was constant over time, we determined an annual accretion rate by averaging the annual rates from each of the two cores taken at each site. This rate was multiplied by the number of years included in the tidal record to provide a total amount of accretion over the given time period. Because we did not know the actual present or historical elevation of the marsh surface relative to mean sea level, we had to adjust the water-level data in order to compensate for marsh elevation change. To do this, we increased the water levels for the initial year in the tidal record by the cumulative amount of accretion gained over the entire time series. In effect, this lowered the elevation of the marsh relative to sea level. The amount added to the water levels was reduced annually by the calculated rate of sediment accretion occurring from the beginning of the tidal record to the year in question.

Both of these methods of documenting the frequency of marsh-flooding events have flaws. Assuming a constant rate of accretion is probably somewhat unrealistic, given the stochasticity of storm events. However, we feel that this exercise represents both conservative and liberal estimates of historic flooding frequencies.

HYPOTHETICAL SEA-LEVEL-RISE SCENARIO: ATLANTIC COUNTY, NEW JERSEY

Relative sea-level rise could have a profound effect on waterbird habitat, especially the intertidal pannes and pools that are especially important to shorebirds using these areas for feeding during migration. We developed a hypothetical scenario for Atlantic County, New Jersey, to estimate the numbers of shorebirds displaced at high tide if a major portion of their natural feeding habitat was eliminated due to sea-level rise and to identify possible refugia capable of providing suitable feeding habitat.

We calculated an estimate of total saltmarsh area and shorebird feeding habitat for Atlantic County, New Jersey using a geographic information system (GIS) and National Wetland Inventory (NWI) maps obtained from the USDI Fish and Wildlife Service (<http://wetlands.fws.gov/nwi> [7 July 2006]). Of the 12 quadrangles used, all were originally classified using aerial photography from March 1977; however, nine of the 12 were recently updated and reclassified based on aerial photography from April 1995. All geoprocessing and analyses were done using ArcGIS 8.2 (Environmental Systems Research Institute 2002). Maps were imported into ArcGIS 8.2 and the areas located in Atlantic County, New Jersey, representing intertidal as well as ponds and pannes, as these are the primary shorebird feeding habitat, were selected.

NWI maps are often not detailed enough to show features such as interior pools or pannes. Because of that, many such features were digitized from digital orthophotos. Four marsh areas were selected based on randomly generated points. A 1,000-m buffer was established around each point and any marsh body that intersected the buffer was selected and all of the pannes and pools in that marsh were digitized. It should be noted that the digital orthophotos we used for digitizing habitat features only covered roughly 50% of the coastal Atlantic County marshes. As a result, when the random points were generated they were only generated for the area covered by the photos. Even so, we are confident that they marshes sampled in this analysis accurately represent the marshes throughout the county. Upon completing the digitizing, the areas of the pannes and pools were calculated for the four randomly selected marshes. Any polygon not having an area of at least 100 m^2 was eliminated to ensure consistency. The proportion of panne and/or pool area to marsh area for the four randomly selected subsets was used to help estimate the total area of pannes and pools in the entire county.

Assuming a large portion of shorebird feeding habitat such as the pannes and pools in Atlantic County, New Jersey, would no longer be available, we identified a likely suitable alternative. E. B. Forsythe NWR, which has a complex of three large impoundments that already provides feeding habitat for many migrant shorebirds and would presumably attract birds displaced from their natural feeding areas by sea-level rise.

An empirical example of the number of shorebirds using the marsh pannes and pools in the county for feeding was estimated based on surveys (N = 17) we conducted during the fall (August–October) of 2000 at our study site (28 ha) located on a marsh island in the county. Surveys were conducted from a 3-m platform and divided into four consecutive 30-min intervals that began 2 hr prior to the peak tide. Shorebird density was based on the combined area of the main feeding habitats (pannes and pools) and the peak survey for shorebird species (15August 2000; 93 birds). This figure was then extrapolated to yield an estimate of shorebirds using similar feeding habitat in the county based on the estimated area of suitable feeding habitat derived above from the GIS analysis.

Shorebird data for the impoundments (540 ha) at E. B. Forsythe NWR for fall 2000 were collected on a weekly basis by refuge volunteers (N = 10 surveys). As with the previous data set, density was calculated based on the known areas of the impoundments and the maximum number of shorebirds seen on a single survey during that season (31 August 2000; 9,445 birds). This density was then recalculated to account for the additional number of shorebirds that would be using the impoundments due to the loss of their natural habitat.

RESULTS

Changes in RSLR and Marsh Surfaces

Rising water levels are apparent from the long-term tide station records from Boston, Massachusetts, to the mouth of the Chesapeake Bay, Virginia, with estimates ranging from 2.6 mm/yr for Boston to 5.9 mm/yr for the Chesapeake Bay Bridge-Tunnel (Fig. 4).

When comparing the increases in RSLR to long-term (ca. 100 yr) accretion rates, it appears that marshes are maintaining pace with RSLR based on five of nine cores (Table 2). However,

TABLE 2. Long-term trends (ca. 100 yr) of marsh elevation (accretion) change based on core analyses compared with relative sea-level rise (RSLR).

Location	Number of cored sites	Long-term accretion (^{210}Pb) (mm/yr)	Relative sea-level rise (RSLR) (mm/yr)
Nauset Marsh, Massachusetts[a]	1	4.2 ± 0.6	2.4[b]
Absecon, New Jersey	2	3.9 ± 0.0	4.1[a, c]
		3.7 ± 0.0	
E. B. Forsythe NWR, New Jersey	2	3.4 ± 0.1	4.1[a, c]
		4.1 ± 0.0	
Wachapreague, Virginia	2	3.1 ± 0.0	3.9[a, d]
		3.9 ± 0.0	
Mockhorn Island, Virginia	2	5.4 ± 0.0	3.9[a, d]
		16.0 ± 0.0	

[a] Data from Roman et al. (1997) RSLR figure based on NOAA tide gauge data from 1921-1993 (Boston, MA station number 8443970; core data only from the Fort Hill site).
[b] P < 0.01, regression analysis H$_0$: sea-level rise = 0.
[c] Calculated from 77 yr (1923-2000) of NOAA tide gauge data (Atlantic City, New Jersey, station number 8534720).
[d] Calculated from 67 yr (1933-2000) of NOAA tide gauge data (Sandy Hook, New Jersey station number 8531680, which is a reference gauge for Wachapreague station number 8631044).

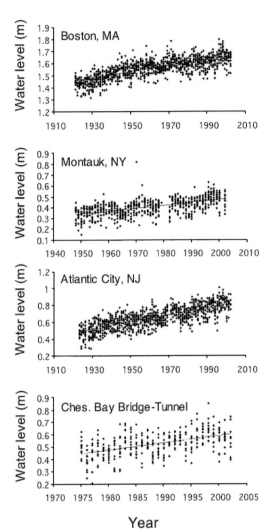

FIGURE 4. Comparison of relative sea-level rise rates along the Atlantic Coast based on monthly mean tide heights (datum mean low low water). Raw data obtained from National Oceanic and Atmospheric Administration's (NOAA) National Ocean Service (<http://www.co-ops.nos.noaa.gov/> (31 July 2006]). Annual RSLR rates, regression equations, r^2 values, and P-values are: Boston: rate = 0.26 cm/yr, y = -3.48 + 0.00257x, r^2 = 0.55, P < 0.01; Montauk: 0.26 cm/yr, y = -4.76 + 0.0026x, r^2 = 0.31, P < 0.01; Atlantic City: 0.41 cm/yr, y = -7.4 + 0.0041x, r^2 = 0.58, P < 0.01; Chesapeake Bay Bridge-Tunnel: 0.59 cm/yr, y = -11.2 + 0.0059x, r^2 = 0.23, P < 0.01.

considering the current conditions, the SET results (past 3–4 yr) suggest that, except for Nauset Marsh, sediment elevations are not keeping pace with the current rates of RSLR, with deficits varying from ca. 2.5–4.1 mm/yr (Table 3). All values of RSLR and marsh elevation change (except E. B. Forsythe NWR) indicated significant (P < 0.05) positive levels of change using least-squares regression analyses.

Estimates of flooding frequencies at the New Jersey and Virginia sites, based on our temporary tide-gauge recordings suggest a significant increase in marsh flooding (>20 cm above marsh surface) in both areas during the May–July period over the past ca. 70–90 yr (Fig. 5). This analysis however, accounts only for water-level changes, assuming a steady-state in marsh surface. Correcting the marsh elevation changes back to the early 1900s based on our core accretion accumulation rates is crude at best, but we considered it necessary to attempt to estimate both marsh and water surface changes, even if a subsidence component was not estimable; the results (Fig. 5) still reveal a significant increase in flooding in Virginia, but a non-significant (P > 0.05) increase in New Jersey.

WATERBIRD IMPACTS

The effects of marsh inundation and flooding frequency for the waterbirds in these regions are numerous. In the short term, the effects will be expressed most dramatically for the species that are marsh nesters (Table 4). For those species with reasonable breeding estimates available, we calculated the number and percentages of the Atlantic Coast populations that reside in the most vulnerable region from New Jersey to Virginia (Table 5). For both Forster's Terns and Laughing Gulls (*Larus atricilla*), a large majority of their populations nest in this region and, thus, their populations are highly vulnerable.

At the intermediate time scale (1–2 decades), changes in marsh morphology and vegetation are expected with negative effects on populations of nesting species, but some positive effects on feeding waterfowl and shorebirds (Table 6). Also, at longer (>50 yr) time scales, with loss of most lagoonal marshes, waterfowl and shorebirds may enjoy some benefit, however, nesting habitat loss will be dramatic, and large changes in estuarine productivity may result (Table 7).

HYPOTHETICAL SEA-LEVEL-RISE SCENARIO: A CASE STUDY AT ATLANTIC COUNTY, NEW JERSEY

We digitized a total of 2,402.5 ha of saltmarsh habitat from the four randomly selected marsh locations in Atlantic County, New Jersey and, of the total area digitized, roughly 10% was categorized as pannes or pools, i.e., shorebird feeding habitat. According to the National Wetland Inventory maps, the total area of saltmarsh for Atlantic County was approximately 17,974.5 ha. Hence, if we assume that the four randomly selected subsamples are accurate

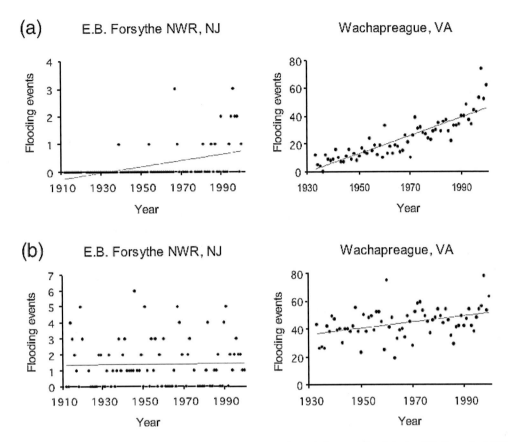

FIGURE 5. Frequency of marsh flooding events (>20 cm inundation) during May–July for E.B. Forsythe NWR, New Jersey, and Wachapreague, Virginia. The first scenario (a) represents annual marsh flooding frequency for each site assuming a static marsh elevation measure (based on year 2000) in relation to NOAA water levels. The second scenario (b) represents flooding frequencies calculated using a marsh elevation estimate hindcasted by subtracting annual accretion for each year. Averaged accretion rates were 3.8 and 3.5 mm/yr for E. B. Forsythe NWR (N = 2) and Wachapreague (N = 2), respectively. Forsythe flooding frequencies were calculated using 89 yr of NOAA tide gauge data (1912–2000); 68 yr (1933–2000) of data were used for Wachapreague. Regression equations, r^2 values, and P-values are: (a) Forsythe: y = -20.9 + 0.010x, r^2 = 0.19, P < 0.001; Wachapreague: y = -1260.0 + 0.653x, r^2 = 0.77, P < 0.001; and (b) Forsythe: y = -1.3 + 0.001x, r^2 = 0.001, P > 0.05; Wachapreague: y = -388.3 + 0.220x, r^2 = 0.16, P < 0.001.

TABLE 3. SHORT-TERM TRENDS (3–4 YR) IN MARSH ELEVATION CHANGE BASED ON SURFACE ELEVATION TABLE (SET) SITES COMPARED WITH RELATIVE SEA-LEVEL RISE (RSLR).

Location	Number of sites	Elevation change (mm/yr)	Relative sea-level rise (RSLR) (mm/yr)[a]
Nauset Marsh, Massachusetts	4	3.4[b]	2.6[b]
Absecon, New Jersey	3	1.7[b]	4.1[b]
E. B. Forsythe NWR, New Jersey	3	-1.1	4.1[b]
Wachapreague, Virginia	3	2.1[b]	3.9[b]
Mockhorn Island, Virginia	2	1.4[c]	3.9[b]

[a] RSLR rates same as those in Table 2, except for Nauset Marsh where seven additional years were added since the Roman et al. (1997) report using NOAA tide gauge data (Boston, MA station number 8443970).
[b] P < 0.01, regression analysis H_0: if no superscript, assume change = 0.
[c] P < 0.05, regression analysis H_0: if no superscript, assume change = 0.

TABLE 4. SHORT-TERM (SEASONAL) EFFECTS OF SEA-LEVEL RISE ON ATLANTIC COAST WATERBIRDS AND HABITAT.

Effect	Biological effect on waterbirds	
	Direct	Indirect
Storm flooding	Nest losses[a]: AMOY, BLSK, CLRA, COTE, FOTE, GBTE, LAGU, SESP, STSP	A. Renesting energetics costs (adult survival?). B. Shift nest sites. C. Fitness of late season fledglings (immature survival?).

[a] Species codes: AMOY = American Oystercatcher; BLSK = Black Skimmer; CLRA = Clapper Rail; COTE = Common Tern; FOTE = Forster's Tern; GBTE = Gull-billed Tern; LAGU = Laughing Gull; SESP = Seaside Sparrow; STSP = sharp-tailed sparrow.

TABLE 5. ESTIMATES OF NUMBER OF NESTING WATERBIRDS ALONG THE ATLANTIC COAST[a].

Species	States (N)	Total[a]	N (and %) in mid-Atlantic[b]	N (and %) in marshes[b, c]
Laughing Gull (*Larus atricilla*)	10	136,774	97,032 (0.71)	97,032 (1.00)
Common Tern (*Sterna hirundo*)	13	51,389	4,702 (0.09)	1,657 (0.35)
Forster's Tern (*Sterna forsteri*)	6	6,449	5,255 (0.81)	5,255 (1.00)
Gull-billed Tern (*Larus nilotica*)	5	1,127	631 (0.56)	61 (0.09)

[a] Maine–Georgia (13 states). Colony estimates (in pairs) based on 1993–1995 inventories based on unpublished censuses coordinated by state biologists.
[b] New Jersey–Virginia.
[c] Percent of total mid-Atlantic (New Jersey–Virginia) populations located in marshes.

TABLE 6. INTERMEDIATE (1–2 DECADES) EFFECT OF SEA-LEVEL RISE ON ATLANTIC COAST WATERBIRDS AND HABITATS (+ INDICATES POSITIVE EFFECTS, - INDICATES NEGATIVE).

Effect	Ecological change	Effects on waterbirds[a]
Changes in marsh morphology and/or vegetation	A. Vegetation community shift (*Spartina patens* to *S. alterniflora*)	SESA, STSP nesting habitat (-).
	B. Enhanced ponding in interior marshes	Shorebirds (15+ species) ABDU, BWTE feeding habitat (+).
	C. Erosion of marsh island perimeters	AMOY, BLSK, COTE, FOTE, GBTE nesting habitat (-).

[a] Species codes: ABDU = American Black Duck; AMOY = American Oystercatcher; BLSK = Black Skimmer; BWTE = Blue-winged Teal; COTE =

TABLE 7. LONG-TERM (>50 YR) EFFECTS OF SEA-LEVEL RISE ON ATLANTIC COAST WATERBIRDS AND HABITAT.

Effect	Ecological Change	Effects on waterbirds[a]
Inundation of marsh Islands	A. Emergent marshes convert to tidal flats	Migrant shorebirds, wintering waterfowl feeding habitat (+); Tern, LAGU, sparrow, rail nesting habitat (-); move to alternative habitat?
	B. Potential loss of nursery area for forage fish, shellfish	Reduced K for waterbirds; shifts to impoundments (increase in competition)?
	C. Redistribution of marshes latitudinally	Winter redistributions of ABDU, other waterfowl, shorebirds.

[a] Species codes: ABDU = American Black Duck; LAGU = Laughing Gull.

representations of the habitat in the County, the area of pannes and pools should be roughly 10% of the entire saltmarsh area, or 1,797.5 ha.

We calculated a maximum density of 51.7 shorebirds/ha during autumn 2000 at our survey location in Atlantic County (see Appendix 1 for species lists). Assuming that this number is also representative of the feeding densities throughout the county, we can estimate that 92,928 shorebirds (51.7 shorebirds/ha × 1,797.5 ha) are using the pannes and pools as feeding habitat at any given time during peak autumn migration. Supposing that this habitat is no longer available due to sea-level rise, E. B. Forsythe NWR offers one of the only local viable alternatives, with three impoundments totaling ca. 541 ha. The maximum density of shorebirds in the impoundments during autumn 2000 was 17.5 shorebirds/ha (J. Coppen, E. B. Forsythe NWR, unpubl. data). Presuming no yearly population changes of shorebirds, that density would increase 10-fold to 189 shorebirds/ha, assuming that shorebirds were unable to utilize their natural feeding habitat due to sea-level rise and all shifted to impoundments.

DISCUSSION

Relative Sea Levels and Saltmarshes

As reported earlier (National Academy of Science 1987), we showed a gradient in RSLR from New England to the mid-Atlantic region. A number of factors may account for this geographic variation, including deep subsidence associated with crustal dynamics and isostatic rebound differences (Kaye and Stuckey 1973, Douglas 1991). Local variation may be pronounced as well; to wit, our findings of major accretion estimate differences from cores only about 1 km apart at our Mockhorn Island site (Table 2). At another lagoonal marsh site within 20 km of our Wachapreague, Virginia site, an earlier estimate of 1.5–2.1 mm/yr was reported from two cores at Chimney Pole Marsh (Kastler and Wiberg 1995), putting that estimate far below the 5.4 and 16.0 values we obtained. Altogether, these data and others suggest marsh inundation has been occurring for some time, at least in Virginia. Knowlton (1971) found a 17% loss of cordgrass marsh from the period 1865–1965 comparing coastal geodetic survey maps. This variation also suggests that it may be quite problematic to model large-scale marsh changes without having a large number of additional samples.

In contrast to the Virginia data, Roman et al. (1997) claimed that, at least on Cape Cod, marshes were easily keeping pace with sea-level rise; however, our SET data suggest that marsh elevation change from 1998–2002 is more modest than the core data taken by Roman et al. (1997) suggest. One factor that might account for this difference was mentioned by Roman et al. (1997); storm-driven sand from the nearby barrier spit may provide infrequent but significant sources of surface accretion that was reflected in the longer-term cores but not in our SET results. In general, the northern New England marshes may be safer than marshes further south because recent tide-gauge data reveal a slower rate of RSLR over the most recent 40 yr compared to the entire record (ca. 80 yr) for the Boston, Portland, and Eastport, Maine stations (Zervas 2001). No other tide gauge (N = 60 examined) showed any other significant change comparing the recent data with the entire record (Zervas 2001). However, the greater tidal amplitudes in New England, relative to the mid-Atlantic, may offset any differences in RSLR, resulting in similar flooding vulnerabilities and decreasing their attractiveness as nesting sites on the marsh surface.

Morris et al. (2002) underscore the complications of marsh responses to sea-level rise along the southeastern US coast. They present an optimality model showing sea level as a constantly moving target (subject to interannual, annual, decadal, and longer temporal cycles). In this model, marsh vegetation is always adjusting toward a new equilibrium, and lag times are important. Thus, comparing short-term SET rates with long-term core data is fraught with peril since subsurface processes such as root production, decay, and compaction have not had time to occur. They further point out large geographic differences, with southeastern marshes being sediment-rich compared to mid- and north Atlantic marshes. Absolute elevation differences of the marshes are also critical in assessing vulnerability to RSLR (Morris et al. 2002:2876).

The increases in flooding frequency we documented in Virginia during the breeding season of waterbirds raises concerns about the future ability of some species to sustain their populations. Even though the small increase in flooding we suggest in New Jersey was not significant by one analysis, we feel that throughout the mid-Atlantic, waterbird habitat will become increasingly flooded as average global sea-level rise is expected to accelerate in the next 100 yr. The best estimate of an ca. 48-cm increase (Intergovernmental Panel on Climate Change 2001) representing a much higher rate than the global average of ca. 15 cm over the past century ± our estimates of 39–41 cm (3.9–4.1 mm/yr rates) for Virginia and New Jersey, respectively, over the past century (Table 2). In addition to mean water-level changes, storm frequencies have been predicted to increase by 5–10% along the Atlantic, with possible increases in intensity of storms as well (Giorgi et al. 2001). In addition to strong directional winds and normal lunar-seasonal cycles (Fig. 1), a long-term metonic or nodal cycle of 18.6 yr may also play a role in coastal flooding of marshes (Kaye and Stuckey 1973). Stumpf and Haines (1998) report that, with metonic, or lunar-nodal, cycles coupled with an annual tidal signal, some Gulf Coast wetlands may experience discrepancies of up to 5 cm from predicted tidal charts. In the mid-Atlantic region, this suggests that increases in water levels, and hence marsh flooding, probably occurred more often in the mid-1970s and again the early 1990s (Panel D, Fig. 1). Unfortunately, we were unable to locate any long-term data on breeding success by any marsh-nesting birds to confirm this prediction. However, Eyler et al. (1999) did find high nest flooding losses of marsh-nesting Gull-billed Terns during the 1994–1996 breeding seasons in Virginia. For long-lived marsh specialists such as Laughing Gulls that might breed for up to a decade, an intriguing question arises: does evidence exist for any long periodicities in reproductive effort that might be associated with these tidal cues?

WATERBIRD IMPACTS

The effects of slow inundation of lagoonal marshes along the Atlantic Coast are widespread. For waterbirds, the most vulnerable group is the marsh-nesting species, most notably those species that are saltmarsh-habitat specialists such as Laughing Gulls, Forster's Terns, Clapper Rails, and Seaside Sparrows, sharp-tailed sparrows, and Coastal Plain Swamp Sparrows. Forster's Terns appear to be the most vulnerable at present given their saltmarsh specialization (Storey 1987) and their large proportion of breeding range within the mid-Atlantic. Although we have some recent estimates of nesting populations of some of the colonially nesting species (Table 5), no large-scale surveys are available for most nesting species of cryptic passerines or rails. The status of the two species of sharp-tailed sparrows and the Coastal Plain Swamp Sparrow is currently under investigation by a number of biologists along the mid- and north Atlantic coasts (J. Taylor and R. Dettmers, USDI Fish and Wildlife Service, pers. comm.; Greenberg and Droege 1990, Greenberg et al. 1998). At present then, we have little idea of the extent to which populations of these cryptic species of sparrows and rails are threatened at the regional or Atlantic Coast scale. Nest losses of all these affected species will no doubt result in frequent renesting efforts that may often be relatively unsuccessful (e.g., Roseate Terns, [*Sterna dougallii*], Burger et al. 1996; Common Terns, Nisbet et al. 2002). Numerous anecdotal reports are also available of entire season nesting failure of Laughing Gulls, Gull-billed Terns, and Black Skimmers in coastal Virginia due to repeated flooding washout events (R. M. Erwin, unpubl. data; B. R. Truitt, The Nature Conservancy, unpubl. data). This occurred in most colonies during the mid-1990s and again in 2001–2002, with almost no young-of-the-year terns or skimmers seen in August roosts (B. R. Truitt, unpubl. data). Increased reproductive effort in a given season may prove costly to adult survival as well (Cam et al. 1998 and references therein); thus, both fecundity and survival may be adversely affected.

Abandoning marsh islands is another option. However, even if marshes are able to expand along the mainland or the backside of the barrier islands as the sea rises (Brinson et al. 1995), the ability of these marsh-nesters to shift to new breeding areas may be limited. First, in many developed coastal areas, the extent of bulkheads, sea walls, and other human structures may severely restrict the area of any significant long-term marsh transgression (Titus et al.1991). Second, even with marsh development, these newer marshes may become ecological traps as mammalian predators including red foxes (*Vulpes vulpes*), raccoons (*Procyon lotor*), and feral cats (*Felis domesticus*) will have easier access to nesting birds than when they nested on islands (Erwin et al. 2001). All of these carnivorous mammals have increased in the Atlantic coastal region as humans continue to increase along the coast and barrier island regions (Erwin 1980, Burger and Gochfeld 1991)

At the intermediate-time scale, changes in vegetation with smooth cordgrass invading areas now dominated by the higher marsh saltmeadow cordgrass (*Spartina patens*) and beach salt grass will presumably result in loss of breeding habitat for the marsh sparrows that depend on high marsh grasses and low marsh shrubs such as marsh elder (*Iva frutescens*). As sea levels encroach into marshes and cause more frequent flooding, many species will simply suffer reduced breeding habitat or shift to alternative habitats. Some colonial species such as Common Terns, Black Skimmers, and Gull-billed Terns have shifted to manmade habitats, e.g., an abandoned parking lot at the Hampton Roads, Virginia, bridge-tunnel complex (Erwin et al. 2003) but such sites are obviously limited. Specialized species such as rails, Forster's Terns,

Laughing Gulls, and Seaside Sparrows may be strongly affected. For shorebirds and waterfowl especially, habitats that are flooded too deeply for feeding in large marshes may cause a larger proportion to shift to agricultural fields nearby if conditions are proper (Rottenborn 1996), to manmade public impoundments or, may force them to shift to other regions. Very large concentration of birds within impoundments raises questions about both the capacity of the prey base to sustain the populations, as well as the potential for disease under artificially high densities (Combs and Botzler 1991). In any event, either being forced into much higher densities in smaller habitat areas, or shifting to more distant regions will probably take a toll in the population levels of a number of species.

Finally, over the long term, with more unvegetated flats and open water replacing marsh islands, wintering waterfowl and migrant shorebirds may benefit by having more feeding habitat available. Especially for diving species such as bay and sea ducks, the newly created shallow water should provide good additional foraging habitat (Perry and Deller 1996). However, less marsh may force changes in distribution during breeding, migration, and even wintering seasons. American Black Ducks may be forced to winter further north as mid-Atlantic marshes decrease in area, especially if marsh loss is not as severe or rapid in coastal regions north of New Jersey. Again, this probably comes with a cost in terms of survival and population size.

EFFECTS ON FAUNAL COMMUNITIES AND THE ESTUARY

Although we have focused our attention here on the potential effects of sea-level rise on the waterbird community, numerous ecological impacts are expected to occur along many coastal regions of the US and elsewhere. As examples, Reid and Trexler (1992) argue that, in just five southeastern US states, almost 500 rare and imperiled species use the coastal zone below the 3-m contour, and that human density is increasing in the coastal zone of all of these states. Similarly, Daniels et al. (1993) reveal the potential effects of sea-level rise in South Carolina, where 52 endangered or threatened plant and animal species were found in the 3-m contour. They argue that species as varied as American alligator (*Alligator mississippiensis*), loggerhead turtle (*Caretta caretta*), shortnose sturgeon (*Acipenser brevirostrum*), and Wood Stork (*Mycteria americana*) will be threatened by salt-water intrusion, flooding, inundation or erosion of feeding or nesting habitat if predictions prevail.

At the ecosystem level, major state changes would be expected to occur as estuaries once dominated by emergent cordgrass marsh become dominated by open water where primary productivity is dominated by macroalgae, submerged aquatic plants, or phytoplankton (Hayden et al. 1991, Valiela et al. 1997). Not only would major changes in primary productivity occur, but secondary production may also be greatly affected. The loss of the detrital food web within emergent marshes would have major implications for all trophic levels, to nearshore coastal production, and might severely jeopardize nursery areas for commercially important fisheries (Bertness 1999).

RESEARCH NEEDS

As has been pointed out above, we need much more information on the densities and distributions within saltmarshes of most waterbird species, especially those depending on the marshes for breeding. We have some estimates for the large, conspicuous colonial waterbirds, but very little large-scale data exist for the cryptic rails and marsh sparrows. As in most wildlife studies, few long-term monitoring efforts have been directed at assessing any demographic parameters, especially fecundity and survival; at best, annual population estimates are obtained for breeding adults for some species. In addition, we need more accurate global positioning system data to describe and model marsh elevations and topography, especially as they relate to tidal means and ranges. Having these data may allow us to construct more accurate models that will project various sea-level rise scenarios to landscapes that are relevant to bird habitat use (Galbraith et al. 2002). Or conversely, as Morris et al. (2002) suggest, marshes in sub-estuaries may have their own characteristic pattern of surface and sub-surface dynamics that strongly limit extrapolation.

ACKNOWLEDGMENTS

We thank the organizers, especially S. Droege and R. Greenberg, for inviting us to participate in the symposium. Many individuals assisted in establishing the SETs and helping with measurements, but we especially thank C. Roman, E. Gwilliams, and J. Lynch for their help. The staffs of Cape Cod National Seashore, E. B. Forsythe NWR, the Virginia Institute of Marine Sciences, and the University of Virginia's Virginia Coast Reserve Long-Term Ecological Research project (NSF Grant # DEB 0080381) provided much logistical support for the project. Funding was provided by the Global Change Program within the US. Geological Survey. Waterbird breeding estimates were kindly provided by state wildlife biologists in Maine, Massachusetts, Rhode Island, Connecticut, New York, New Jersey, Delaware, Maryland, Virginia, North Carolina, South Carolina, and Georgia. We also thank S. Kress of the National Audubon Society for northern New England data and B.R. Truitt, The Nature Conservancy, and B. Williams for records from coastal Virginia. S. Droege and two anonymous reviewers provided constructive comments on an earlier draft.

APPENDIX 1. RANKING OF SHOREBIRD SPECIES OBSERVED AT OUR SURVEY LOCATION (LBNJ) AND THE IMPOUNDMENTS AT E. B. FORSYTHE NATIONAL WILDLIFE REFUGE, NEW JERSEY ON THE PEAK SURVEY DATE FOR EACH LOCATION (15 AUGUST 2000 AND 31 AUGUST 2000, RESPECTIVELY). SPECIES ARE LISTED IN DECREASING ORDER.

LBNJ bird survey site		E. B. Forsythe impoundments	
Common name	Scientific name	Common name	Scientific name
Semipalmated Sandpiper	Calidris pusilla	Semipalmated Sandpiper	Calidris pusilla
Semipalmated Plover	Charadrius semipalmatus	Semipalmated Plover	Charadrius semipalmatus
Short-billed Dowitcher	Limnodromus griseus	Black-bellied Plover	Pluvialis squatarola
Black-bellied Plover	Pluvialis squatarola	Least Sandpiper	Calidris minutilla
Willet	Catoptrophorus semipalmatus	Short-billed Dowitcher	Limnodromus griseus
American Oystercatcher	Haematopus palliatus	Western Sandpiper	Calidris mauri
Greater Yellowlegs	Tringa melanoleuca	Greater Yellowlegs	Tringa melanoleuca
Whimbrel	Numenius phaeopus	Willet	Catoptrophorus semipalmatus
		Lesser Yellowlegs	Tringa flavipes
		Stilt Sandpiper	Calidris himantopus
		Killdeer	Charadrius vociferus
		Ruddy Turnstone	Arenaria interpres
		American Oystercatcher	Haematopus palliatus
		Wilson's Phalarope	Phalaropus tricolor

THE IMPACT OF INVASIVE PLANTS ON TIDAL-MARSH VERTEBRATE SPECIES: COMMON REED (*PHRAGMITES AUSTRALIS*) AND SMOOTH CORDGRASS (*SPARTINA ALTERNIFLORA*) AS CASE STUDIES

GLENN R. GUNTENSPERGEN AND J. CULLY NORDBY

Abstract. Large areas of tidal marsh in the contiguous US and the Maritime Provinces of Canada are threatened by invasive plant species. Our understanding of the impact these invasions have on tidal-marsh vertebrates is sparse. In this paper, we focus on two successful invasive plant taxa that have spread outside their native range—common reed (*Phragmites australis*) and smooth cordgrass (*Spartina alterniflora*). A cryptic haplotype of common reed has expanded its range in Atlantic Coast tidal marshes and smooth cordgrass, a native dominant plant of Atlantic Coast low-marsh habitat, has expanded its range and invaded intertidal-marsh habitats of the Pacific Coast. The invasions of common reed in Atlantic Coast tidal marshes and smooth cordgrass in Pacific Coast tidal marshes appear to have similar impacts. The structure and composition of these habitats has been altered and invasion and dominance by these two taxa can lead to profound changes in geomorphological processes, altering the vertical relief and potentially affecting invertebrate communities and the entire trophic structure of these systems. Few studies have documented impacts of invasive plant taxa on tidal-marsh vertebrate species in North America. However, habitat specialists that are already considered threatened or endangered are most likely to be affected. Extensive experimental studies are needed to examine the direct impact of invasive plant species on native vertebrate species. Careful monitoring of sites during the initial stages of plant invasion and tracking ecosystem changes through time are essential. Since tidal marshes are the foci for invasion by numerous species, we also need to understand the indirect impacts of invasion of these habitats on the vertebrate community. We also suggest the initiation of studies to determine if vertebrate species can compensate behaviorally for alterations in their habitat caused by invasive plant species, as well as the potential for adaptation via rapid evolution. Finally, we urge natural-resource managers to consider the impact various invasive plant control strategies will have on native vertebrate communities.

Key Words: food webs, geomorphology, invasive plants, marsh birds, North America, *Phragmites*, *Spartina alterniflora*, saltmarsh, vertebrates.

EL IMPACTO DE PLANTAS INVASORAS EN ESPECIES DE VERTEBRADOS EN MARISMA DE MAREA: EL CARRIZO (*PHRAGMITES AUSTRALIS*) Y EL PASTO (*SPARTINA ALTERNIFLORA*) COMO CASOS DE ESTUDIO

Resumen. Grandes áreas de marisma de marea a lo largo EU así como y en las Provincias de Marisma de Canadá, se encuentran amenazadas por especies de plantas invasoras. Nuestro entendimiento acerca del impacto que estas especies tienen en los vertebrados de marismas de marea es escaso. En este artículo nos enfocamos en dos taxa de especies de plantas exitosas que se han dispersado fuera de su rango nativo—el carrizo (*Phragmites australis*) y el pasto (*Spartina alterniflora*). Un haplotipo críptico de carrizo ha expandido su rango en las marismas de marea en la Costa del Atlántico. El pasto (*Spartina alterniflora*), nativa y dominante del hábitat de marisma baja de la Costa del Atlántico, ha expandido su rango e invadido hábitats de marisma intermareal en la Costa Pacífico. Las invasiones del carrizo en las marismas de marea de la Costa Atlántica y el pasto en marismas de marea de la Costa del Pacífico parece que tienen impactos similares. La estructura y composición de estos hábitats ha sido alterada y la invasión y dominancia por estas dos taxa, pueden derivar en cambios profundos en los procesos geomorfológicos, alterando la mitigación vertical y pueden potencialmente afectar las comunidades de invertebrados y toda la estructura trófica de estos ecosistemas. Pocos estudios han documentado impactos de taxa de plantas invasoras en especies de vertebrados de marisma de marea en Norte América. Sin embargo, especialistas del hábitat, los cuales ya están considerados como en peligro, son los que están siendo más afectados. Se necesitan estudios extensivos experimentales para examinar el impacto directo de especies de plantas invasoras en especies nativas de vertebrados. El monitoreo cauteloso de los sitios durante los estados iniciales de la invasión de plantas y el rastreo de los cambios del ecosistema en el tiempo son esenciales. Debido a que las marismas de marea son el foci para la invasión por numerosas especies, también necesitamos los impactos indirectos de invasión de estos hábitats en la comunidad de vertebrados. También sugerimos el inicio de estudios para determinar si especies de vertebrados pueden compensarse en términos de comportamiento por las alteraciones en su hábitat causado por especies de plantas invasoras, como también el potencial para la adaptación vía evolución rápida. Finalmente, recomendamos a los manejadores de recursos naturales a que consideren el impacto que tendrán las estrategias de control de varias plantas invasoras en comunidades nativas de vertebrados.

Invasive plant taxa are profoundly changing North American saltmarshes, but this is not an isolated phenomenon. The introduction of non-indigenous plants in diverse habitats represents some of the most dramatic examples of biological invasions. Their impact on natural habitats, and the biodiversity of those habitats, is a pervasive threat and one of the most daunting ecological challenges facing twenty-first century natural-resource managers.

Considerable attention has been devoted to understanding the attributes of successful invaders and the characteristics of invaded regions and habitats, as well as to documenting the patterns and history of invasion (Mooney and Drake 1986, Drake et al. 1989, Cronk and Fuller 1995, Pysek et al. 1995). The consequences of invasive taxa on biological communities and ecosystem processes have been documented more recently (D'Antonio and Vitousek 1992, Mack et al. 2000). MacDonald et al. (1989) calculated that of the 941 vertebrate species thought to be in danger of extinction worldwide, 18.4% are threatened in some way by introduced species. In North America, they calculated that >13.3% of the native avifauna is threatened by invasive species. In another study focusing on threats to biodiversity in the US, Wilcove et al. (1998) found that 49% of all imperiled species (plants and animals) were threatened by invasive species.

The study of biological invasions has only recently focused on coastal and estuarine habitats (Grosholz 2002). Large numbers of non-indigenous species have been identified in U.S. coastal estuaries >200 non-indigenous species from San Francisco Bay alone (Cohen and Carlton 1998). Most research has concentrated on non-native aquatic invasive species including crustaceans, clams, crabs, and hydrozoans (Cordell and Morrison 1996, Crooks 1998, Bagley and Geller 2000, Byers 2000). In contrast, efforts to examine the ecological effects of non-native emergent wetland plant taxa in saltmarshes have lagged behind (except Weinstein et al. 2003). In this chapter, we focus on two of the more problematic invasive marsh-plant taxa in North American saltmarshes — common reed (*Phragmites australis*) and smooth cordgrass (*Spartina alterniflora*).

ATLANTIC AND PACIFIC TIDAL SALTMARSHES

In the contiguous US and Maritime Provinces of Canada saltmarshes occur in three distinct regions: the Northeastern Atlantic Coast from the Hudson River north to the St. Lawrence estuary, the coastal plain of the United States from New Jersey south along the southeastern US Atlantic Coast to the northern Gulf of Mexico, and the western US along the Pacific coast (Dame et al. 2000, Emmett et al. 2000, Roman et al. 2000). In this chapter, we are interested in two areas — the Atlantic Coast, with an emphasis on the northeastern Atlantic region, and the Pacific Coast.

Northeast-coast saltmarshes are formed largely by reworked marine sediments and in situ peat formation. These marshes are largely limited to small, narrow fringing systems because the physiography of the region and broad expanses of rocky coast limit their areal extent (Nixon 1982). Farther south, more extensive saltmarshes occur in the drowned valley estuaries of Delaware Bay and Chesapeake Bay (Teal 1986). Pacific Coast saltmarshes occur in a geologically young region structured by tectonic and volcanic forces (Emmett et al. 2000). Because of the rocky and unfeatured wave-dominated shoreline along much of the Pacific Coast, extensive areas of saltmarsh are restricted to large estuaries such as those associated with the San Francisco Bay and the Columbia River or behind sheltering bay-mouth bars.

Strong physical gradients of salinity and tidal inundation contribute to the characteristic patterns of tidal-height zonation. The two main marsh zones include low marsh and high marsh. Low marsh is lower in elevation relative to mean low water and is regularly flooded by tides. High marsh occupies the higher elevations in the intertidal zone, and is less influenced by tidal forces.

The organization of tidal-marsh vegetation communities varies in the different regions. A short-statured grass, saltmeadow cordgrass (*Spartina patens*), dominates the high marsh along the northeast coast — often intermixed with the short form of smooth cordgrass and black needlerush (*Juncus gerardi*) at the upland border of the high marsh. Smooth cordgrass persists as a dominant species in low-marshes, in this region reaching heights as tall as 1.25–2 m (Teal 1986). Open tidal mudflats characterize the lower intertidal zone of Pacific Coast estuaries. The mid-intertidal zone is dominated by California cordgrass (*Spartina foliosa*), which forms a narrow band of vegetation along the outer edges of the native terrestrial vegetative zone and along tidal channels (Mahall and Park 1976, Ayers et al. 2003). California cordgrass's range extends from Baja California north to

Humboldt Bay (Josselyn 1983). California cordgrass, the only species of *Spartina* native to the Pacific Coast, grows sparsely and is relatively short (usually <1 m tall; Ayres et al. 2003). The Pacific Coast high marshes, with high salinity and saturated soils, are dominated by low-growing (<0.5 m) cordgrass species (Baye et al. 2000). A transition in vegetative composition of saline marshes occurs in the Pacific Northwest and these marshes are dominated by *Salicornia* spp. and beach salt grass (*Distichlis spicata*) in the high marsh and seaside arrowgrass (*Triglochin maritima*) in the low marsh (Seliskar and Gallagher 1983) but much of the intertidal area in the Pacific Northwest remains un-vegetated (Simenstad et al. 1997).

Over the last 200 yr, marsh and tidal flats have been lost to or degraded by human development activities including diking, draining, dredging, or filling for agriculture or urbanization, and conversion to salt-production ponds. A substantial portion of U.S. tidal wetlands has been destroyed (Tiner 1984) and unaltered coastal saltmarshes are rare (Roman et al. 2000). Over 80% of the saltmarshes that once occurred in New England have already been lost (Teal 1986). Originally, New England saltmarshes had networks of salt ponds, pannes, potholes, and channels in the high marsh where the water was semi-permanent. Roads and other obstacles have cut off or reduced tidal flow into these habitats (Roman et al. 1984, Burdick et al. 1997). Most saltmarshes along the Atlantic Coast have also been ditched to remove standing water and pools and prevent mosquito breeding, resulting in lowered water tables, vegetation changes, and associated trophic impacts on fish and waterbirds (Roman et al. 2000). Pacific Coast marshes have suffered the same fate as those along the Atlantic Coast. For example, a 70% loss of tidal wetlands has occurred in the Puget Sound estuary in Washington with localized loss being virtually complete in heavily urbanized areas (Washington Division of Natural Resources 1998). At the turn of the 19th century, the San Francisco Bay estuary included approximately 76,900 ha of tidal marshes and 20,400 ha of open tidal flat. Today, only about 16,300 ha (21%) of tidal marshes and 11,800 ha (58%) of tidal flats remain (Goals Project 1999).

Northeast coast saltmarsh vegetation patterns have changed dramatically over the past 50 yr. Surveys of southern New England saltmarshes suggest that increases in sea levels leading to increased waterlogging of upland marsh soils and plants has in turn led to the replacement of black needlerush in the upper high marsh by seaside arrowgrass and the replacement of saltmeadow cordgrass by the short form of smooth cordgrass (Niering and Warren 1980, Warren and Niering 1993). Cultural eutrophication leading to higher loadings of nitrogen to northeast tidal marshes is also hypothesized to have resulted in changes in tidal-marsh vegetation patterns (Bertness et al. 2002). Nitrogen fertilization experiments in nitrogen-limited New England tidal marshes resulted in increased abundance of smooth cordgrass in high-marsh plots while marsh hay decreased (Levine et al. 1998, Emery et al. 2001).

Invasive Tidal Saltmarsh Plant Species

Atlantic and Pacific coast tidal saltmarshes are characterized by a few dominant emergent plant species organized in characteristic zones resulting from both physical stress and competition, leading to distinct plant communities at specific elevations (Bertness and Ellison 1987). But, because of habitat degradation, they may be among the most susceptible to invasive plant species. Shoreline development, tidal restriction, and habitat destruction result in disturbed conditions including bare soil, high nutrient inputs, altered hydrology, and high light levels which are thought to be among the conditions that promote successful plant invasions. The colonization and spread of common reed in Atlantic Coast marshes and cordgrass in Pacific Coast marshes has been rapid and follows a pattern often typical of plant invasions. Windham (1999) describes the typical invasion sequence of reeds in Atlantic Coast saltmarshes initiated by the first appearance of isolated small patches, the continued initiation of numerous other isolated patches over time, the coalescence of these patches and eventual dominance of an area. She cited an average annual rate of spread >20% at a site in southern New Jersey from 1972–1991.

Common reed

Common reed is found worldwide. It tolerates a range of abiotic conditions and is found in both freshwater and coastal habitats, although its establishment and growth is limited by flooding duration and high salinity and sulfide levels (Chambers 1997). Reeds have been shown to form extensive stands in tidal marshes with salinities <15 ppt. Small, more recently established plants grow well at salinities from 0–5 ppt, exhibit some reduction in growth up to 35 ppt, and have difficulty persisting when salinities exceed 35 ppt (Chambers et al. 1999). In North America, the range of reeds has expanded dramatically since the late 19th century, and in some areas reeds have formed extensive monocultures displacing native species (Chambers et al. 1999). Reeds now occupy many tidal habitats in Maritime Canada, New England, the mid-Atlantic, and the northern Gulf of Mexico. Reeds form dense monocultures following establishment (Meyerson et al. 2000) and are thought to be a robust competitor relative to other saltmarsh species. They grow to 3–5 m tall and can form solid stands with stem densities ranging from 50–125 shoots/m^2 (Meyerson et al. 2000).

Wrack accumulation, erosion, ice scour that promotes bare soil, ditching and other hydrologic disturbances, and nutrient enrichment associated with shoreline development provide reeds with opportunities to become established (Chambers et al. 1999). Dispersal and burial of large rhizome fragments into well-drained and low-salinity sites improve the chances of successful establishment (Bart and Hartman 2003). Once established, poorly drained areas and sites with high salinity and sulfide levels tend to be invaded by clonal spread (Chambers et al. 2003).

Many explanations have been invoked for the recent change in the relative abundance and distribution of reeds in North American tidal marshes (Chambers et al. 1999, Orson 1999). Recent advances in genomics, including the ability to examine nucleotide sequences in chloroplast DNA, have shed considerable light on this question. Comprehensive genetic analyses of herbarium specimens collected before and after 1910 reveal significant changes in the haplotype frequency of North American reed populations (Saltonstall 2002, 2003). Today one distinct haplotype derived from an introduced Eurasian lineage (Type M) is the dominant type found in the tidal marshes of the Northeast and mid-Atlantic Coast although populations of native haplotypes still persist in the region. Although native haplotypes still dominate along the Pacific coast, haplotype M has been identified in urban areas in the western US (Saltonstall 2003).

It is currently not known why haplotype M has become the dominant reed lineage and has increased its distribution throughout Atlantic Coast tidal habitats. Type M may be a superior competitor or environmental conditions may have changed and played a role in the expansion of its range (Silliman and Bertness 2004). New experiments evaluating the growth and persistence of native and invasive haplotypes along salinity and hydrologic gradients as well as competition experiments with other saltmarsh dominants are currently underway (Vasquez et al., 2005).

Smooth cordgrass

Among the invasive plants in Pacific Coast marshes, several cordgrass species have been particularly successful because they are among the most abundant and aggressive intertidal plants in North America (Adam 1990). For instance, four of the 12 non-native plant species identified as introduced species of concern in the San Francisco Bay estuary are cordgrass species—smooth cordgrass, saltmeadow cordgrass, dense-flowered cordgrass (*S. densiflora*) and common cordgrass (*S. anglica*) (Grossinger et al. 1998). The introduction and spread of smooth cordgrass, however, is arguably the most devastating of the tidal-marsh-plant invasions on the Pacific Coast. The dominant plant species of low marsh along the Atlantic Coast of the US, smooth cordgrass has become established in open tidal mudflats of the Pacific Coast and has extended its range up through the high-marsh zone as well.

Multiple intentional and accidental introductions of smooth cordgrass have occurred in Pacific estuaries. In Washington, smooth cordgrass was accidentally introduced to Willapa Bay sometime before 1911 (Scheffer 1945) and by 1988 it had spread to occupy >445 ha of tidal flat (Aberle 1993). More recently, the rate of spread appears to be accelerating. In 1997, the area solidly covered by smooth cordgrass was estimated at >1,315 ha, and by 2002 it was estimated at >2,500 ha equaling nearly 47% of the tidalflat habitat in Willapa Bay (Washington Division of Natural Resources 2000, Buchanan 2003). In the 1930s and 1940s, smooth cordgrass was intentionally introduced in four areas of Puget Sound for duck-habitat enhancement, but the spread there has been minor compared to that in Willapa Bay (Frenkel 1987). Smooth cordgrass was also introduced in the late 1970s into one area in Oregon, the Siuslaw estuary (Aberle 1993). In California, the Army Corps of Engineers brought smooth cordgrass plants into the South Bay of San Francisco Bay in 1973 for a marsh-restoration project and over the next decade it was transplanted to at least two other sites, and likely others, within the South Bay (Ayres et al. 2004, Grossinger et al. 1998). Over the next 10 yr, smooth cordgrass spread slowly to other areas and began to hybridize with the native California cordgrass. This hybrid (smooth cordgrass × California cordgrass) is highly productive, out competes both parental species, and is the form that is now aggressively spreading throughout San Francisco Bay (Daehler and Strong 1997, Ayres et al. 2004). The latest surveys show that smooth cordgrass/hybrid now occupies approximately 2,030 ha which equals approximately 17% of the tidal flat and marsh habitat in south San Francisco Bay where the invasion is concentrated (Ayres et al. 2004). Ayres et al. (2003) predicted that, if unchecked, invasive smooth cordgrass has the potential to spread throughout the San Francisco Bay estuary and beyond such that it would cause the global extinction of the native California cordgrass. Indeed, recent surveys have confirmed the presence of smooth cordgrass in two other California estuaries north of San Francisco Bay (Bolinas Lagoon and Drakes Estero) indicating the potential for widespread colonization of other Pacific Coast estuaries (Ayres et al., 2003). Daehler and Strong (1996) estimate, that along the Pacific Coast of the US, the final distribution for smooth cordgrass will stretch from Puget Sound, Washington, south through the Tijuana River Estuary, California.

Effects of Invasive Reeds and Cordgrass on Tidal-marsh Habitat

The expansion of reeds into high-marsh areas along the Atlantic Coast of the US can result in important changes in plant community structure and potential declines in the vertebrate species dependent on these habitats. In New England marshes, the impacts of human development and cultural eutrophication are affecting the distribution of plant species (Bertness et al. 2002). Shoreline development and enhanced nitrogen supplies appear to be associated with the expansion of common reed populations into the high marsh. Rooth et al. (2003) documented increased rates of sediment accretion following invasion by reeds in oligohaline tidal marshes of the Chesapeake Bay. The high productivity of reeds and accumulation of litter on the marsh surface, coupled with high stem density and high inorganic sediment loading, appears to be the mechanism resulting in the higher rates of sediment accretion. The enhanced rates of sediment accumulation in reeds stands can alter the physical structure of tidal marshes by building up the marsh surface and filling in topographic depressions and first order tidal channels, resulting in a loss of microtopographic variation (Lathrop et al. 2003).

Similar habitat alteration occurs in Pacific Coast estuaries that have been invaded by non-native smooth cordgrass (and/or the hybrid form smooth cordgrass × California cordgrass). The non-native cordgrass often grows to heights of 2 m or more and the above- and below-ground biomass is much denser than any of the native plant species (Callaway and Josselyn 1992). Smooth cordgrass is able to occupy a much larger portion of the tidal gradient than any of the native marsh plants and has been dubbed an ecosystem engineer because of its ability to alter habitat through increased sediment accretion (Ayres et al. 1999). When invaded by smooth cordgrass, marshes can ultimately be transformed into solid non-native cordgrass meadows (Daehler and Strong 1996). In San Francisco Bay, this non-native cordgrass colonizes open intertidal mudflats and clogs tidal channels (growing as low as 73 cm above the lower limit of the intertidal zone), and grows throughout the marsh plain up to the high marsh (as high as 15 cm below the maximum elevation of tidal-marsh vegetation) where it appears to be displacing native plant species (Ayres et al 1999, Collins 2002). Based on estimates of smooth cordgrass tidal inundation toleration rates, current water levels, and tidal regimes, Stralberg et al. (2004) predicted that approximately 33% of intertidal mudflat habitat could be encroached upon by smooth cordgrass and its hybrids. In addition, the upward spread of smooth cordgrass could be accelerated by future sea-level rise (Donnelly and Bertness 2001).

Changes in habitat structure and composition that accompany the smooth cordgrass invasion on the Pacific Coast and the common reed invasion on the Atlantic Coast, lead to alterations in geomorphological processes in tidal marshes and have implications for many aspects of the tidal-marsh ecosystem including basic hydrologic function (e.g., altering flow regimes in marshes by clogging tidal channels). Thus the effects of the invasion and dominance of tidal wetlands by common reed and smooth cordgrass could cascade throughout the tidal-marsh system and alter the trophic structure of the marsh ecosystem as well, although little is currently known about these effects. For example, the increase in sediment accretion (e.g., 1–2 cm/yr in Willapa Bay [Sayce 1998]), coupled with the increase in mass and density of aboveground biomass of smooth cordgrass invasions in Pacific Coast estuaries, could potentially change the invertebrate community composition of intertidal zones, reducing benthic invertebrate densities (Capehart and Hackney 1989), while increasing insects and arachnids of the cordgrass canopy.

Potential Impacts on Tidal-marsh Vertebrates

The impact of the introduction and spread of non-native reeds and smooth cordgrass on tidal-marsh vertebrate populations remains largely unstudied. For instance, few data correlate the distribution of these invasive plant species with the distribution and abundance of native tidal-marsh bird or mammal species.

One of the most widely recognized values of saltmarshes is their support of migrant and resident avian species. Fundamental changes in habitat structure, shifts in primary productivity, and the potential modification of trophic pathways that accompany the invasion will likely have their biggest impacts on resident, non-migratory species that are dependant year-round on tidal marshes.

POTENTIAL IMPACTS OF COMMON REED

Several studies provide evidence that many species of vertebrates use marshes dominated by reeds (Kiviat et al., pers. comm.), which appears to be more important to wildlife as shelter than as food. Wildlife species tend to use the edges of stands, mixed-reed stands, and smaller patchy stands than the dense extensive interiors of larger stands. Interestingly, colonial-nesting long-legged wading birds may benefit from the proximity of reed stands. In certain sites in Delaware, reeds provide critical habitat for nesting colonial wading birds by offering substrate and material for nesting, and serves as a buffer from human disturbance (Parsons 2003).

Reed-dominated marshes support more species of coastal marsh-breeding birds than commonly believed (Kiviat et al., pers. comm.). Kiviat et al.'s literature review documented 24 species of birds that utilized reed stands located in either estuarine tidal marshes and creeks or saltmarsh habitat. Although dense populations of reeds appear to have little value for birds, stands interspersed with tidal creeks and open water and mixed stands or habitat on the edge of reed stands do support some bird species (Swift 1989, Brawley 1994, Holt and Buchsbaum 2000).

Holt and Buchsbaum (2000) suggested that factors other than the dominant plant species also have a major role in determining the distribution of bird species in tidal marshes. They found that the presence of reeds in northern Massachusetts's coastal marshes appeared to have little effect on the numbers of Red-winged Blackbirds (*Agelaius phoeniceus*), Marsh Wrens (*Cistothorus palustris*), Virginia Rails (*Rallus limicola*), or Saltmarsh Sharp-tailed Sparrows (*Ammodramus caudacutus*). Benoit and Askins (1999) conducted one of the few direct comparisons of bird use of reeds and unaltered saltmeadow cordgrass habitat in Atlantic Coast saltmarshes. They found significantly fewer species of birds in reed-dominated stands than in high-marsh saltmeadow cordgrass stands. The Seaside Sparrow (*Ammodramus maritima*), Saltmarsh Sharp-tailed Sparrow, and Willet (*Catoptrophorus semipalmatus*), three tidal-marsh specialists adapted to nesting in the short high-marsh vegetation, had low frequencies in stands dominated by reeds. The Marsh Wren, Swamp Sparrow (*Melospiza georgiana*), and Red-winged Blackbird (marsh generalists that prefer tall reedy vegetation) were found in sites dominated by reeds. Even within these examples of species that maintain populations in reeds, more detailed study is required. Olsen and R. Greenberg (pers. comm.) report that in Delaware, Swamp Sparrows require saltmeadow cordgrass for nest cover. Clumps of this vegetation can be found along the edge of reed beds, but not in the interior of large stands. Therefore, it is unlikely that Swamp Sparrows can maintain nesting populations in larger stands of reeds. In fact, Benoit and Askins (1999) also found that homogeneous stands of reeds did not provide sustainable habitat for many wetland bird species. Wading birds, shorebirds, and waterfowl were absent from surveyed reed stands. By contrast, the high-marsh stands dominated by short-stature grasses included a wide variety of generalists: waders, shorebirds, ducks, and aerial insectivores as well as high-marsh specialists. A phenomenon less well documented was the use of reed stands embedded in a larger more heterogeneous landscape. Benoit (1997) reported Virginia Rails and King Rails (*Rallus elegans*) using patches of reeds interspersed with areas of open brackish marsh.

In Atlantic Coast tidal marshes where reeds have recently established, the availability of prey resources (snails, amphipods, and isopods) to adult mummichogs (*Fundulus heteroclitus*) may be no different than in non-invaded tidal marshes (Fell et al. 1998). However, as the hydrology of these sites change and marsh surface heterogeneity and topographic depressions disappear, there is evidence that fish recruitment and utilization may change in reed-dominated stands (Weinstein and Balletto 1999, Osgood et al. 2003). A growing body of research suggests that mummichogs may exhibit reduced feeding and reproduction in response to the structural changes that occur as tidal-marsh sites naturally dominated by cordgrass species become dominated by common reed (Able et al. 2003, Raichel et al. 2003). This suggests that prey for larger wading birds may not be accessible within dense stands of reeds but these large wading birds may forage on the edges of reed stands intermixed with more typical low or high marsh.

POTENTIAL IMPACTS OF SMOOTH CORDGRASS

On the Pacific Coast, the smooth cordgrass invasion may have negative effects on native vertebrate species, but as yet few data are available.

The salt marsh harvest mouse (*Reithrodontomys raviventris*), a federally endangered species endemic to San Francisco Bay saltmarshes, prefers the mid- and upper-tidal areas that are largely dominated by pickleweed (*Salicornia virginica*; Shellhammer et al. 1982). Shellhammer et al. (1982) found very few mice in pure stands of salt-marsh bulrush (*Schoenoplectus maritimus*), a tall, reedy bulrush with structural characteristics more similar to the non-native smooth cordgrass than to the preferred *Salicornia* spp. Shellhammer et al. speculated that the value of pickleweed was higher for the saltmarsh harvest mouse than was bulrushes because pickleweed provides denser cover and more horizontal branching. While smooth cordgrass may provide fairly dense cover, it provides little horizontal structure. Other mammals that occur in San Francisco Bay tidal marshes that may be affected by habitat alteration associated with smooth cordgrass include the saltmarsh wandering shrew (*Sorex vagrans haliocoetes*), the Suisun shrew (*Sorex ornatus sinuosis*), and the California vole (*Microtus californicus*). All three of these species are known to occur in the middle and upper intertidal zones of salt or brackish marshes (Lidicker 2000, MacKay 2000, Shellhammer 2000).

Relatively few bird species are restricted year-round to tidal saltmarshes in San Francisco Bay. Resident bird species that breed in the tidal marshes include the federally endangered California Clapper Rail (*Rallus longirostris obsoletus*) and three subspecies of tidal-marsh Song Sparrow (*Melospiza melodia pusillula, M. m. samuelis, and M. m. maxillaris*) listed as California species of special concern because of habitat loss and because they have extremely restricted ranges and adaptations for nesting in Pacific Coast saltmarshes (Marshall 1948a, Johnston 1954). While California Black Rails (*Laterallus jamaicensis coturniculus*) and Salt Marsh Common Yellowthroats (*Geothlypis trichas sinuosa*), also species listed in California, typically breed in brackish or freshwater marshes, they do occur in saltmarshes during winter and so may also be affected by the smooth cordgrass invasion.

Although California Clapper Rails do occur and nest in areas that have been invaded by non-native smooth cordgrass (S. Bobzien, pers. comm.; J. C. Nordby, pers. obs.), it is unclear whether these sub-populations are sustainable. Clapper Rails forage mainly at low tide when the mud substrate in tidal channels and tidal flats is exposed and their preferred foods (clams, mussels, snails, and crabs) are more available (Williams 1929, Moffit 1941, Albertson and Evens 2000). By colonizing tidal flats and clogging tidal channels, smooth cordgrass may reduce the foraging habitat of rails as well as alter what food items are available. Also, Clapper Rails do occasionally nest in native California cordgrass (Zucca 1954) but no studies have yet examined the success of Clapper Rail nests placed in either exotic or native cordgrass.

Like the California Clapper Rail, the Alameda Song Sparrow (*Melospiza melodia pusillula*) does occupy marshes that have been invaded by smooth cordgrass (J. C. Nordby and A. N. Cohen, unpubl. data). In a native marsh, Song Sparrow breeding territories are typically arrayed in a tight linear fashion in the taller plants (gumweed [*Grindelia stricta*] and Virginia picklewood) that occur along tidal channels (Marshall 1948a, Johnston 1954). Preliminary analyses from an ongoing study of how saltmarsh Song Sparrows are responding behaviorally to the rapid alteration of habitat by smooth cordgrass have shown that Song Sparrows do include the non-native cordgrass habitat in their territories and use those areas for foraging as well as for nesting. However, no observed Song Sparrow territories have been composed entirely of smooth cordgrass, and nests that were placed in smooth cordgrass were somewhat less successful and much more likely to fail due to tidal flooding than were nests placed in native vegetation (J. C. Nordby and A. N. Cohen, unpubl. data). It is possible that the Song Sparrows are drawn to inappropriate nesting sites in smooth cordgrass that are too low in elevation relative to the tides. Whether smooth cordgrass is acting as an ecological trap for Song Sparrows, where overall reproductive success is reduced, remains to be tested

The impact of the smooth cordgrass invasion is not restricted to resident species because the open tidal flats of Pacific estuaries provide crucial habitat for migrating shorebirds. San Francisco Bay is designated as a Western Hemisphere Shorebird Reserve Network that provides breeding habitat or critical migratory stopover sites for >1,000,000 waterfowl and shorebirds each year (Kjelmyr et al. 1991), more than any other wetland along the Pacific Coast of the contiguous US (Page et al. 1999). Most of these bird species forage extensively on benthic organisms found in the vast tidal mudflats that rim the bay (Takekawa et al. 2000). In a study of the affect of the spread of common cordgrass (a close relative of smooth cordgrass) on shorebird populations in the British Isles, Goss-Custard and Moser (1988) found the largest reduction in Dunlin (*Calidris alpina*) in estuaries where the cordgrass had replaced much of the intertidal mudflat foraging habitat. Ayres et al. (2004)

predicted that in San Francisco Bay the loss of tidal mudflat habitat to smooth cordgrass colonization could be extensive if the invasion goes unchecked over the next two centuries. Stralberg et al. (2004) estimated that 33% of shorebird habitat value (range 9–80%) could be lost under realistic spread scenarios.

In Willapa Bay, Washington, where smooth cordgrass increased three-fold between 1994 and 2002 (Buchanan 2003), aerial surveys conducted in 2000–2001 suggest a reduction in shorebird numbers by approximately 60% and foraging time by as much as 50% in the southern portions of the bay as compared with data from the 1991–1995 surveys (Jaques 2002).

In addition to the direct alteration of habitat, invasive plants may be altering competitive interactions among native species as well. Pacific Coast Marsh Wrens (*Cistothorus palustris paludicola*), which normally nest in freshwater or brackish marshes and not in open saltmarshes (Verner 1965), have begun to establish breeding territories in the newly available smooth cordgrass habitat in San Francisco Bay as well as in other smooth cordgrass-invaded marshes such as those in Willapa Bay (Williamson 1994; J. C. Nordby and A. N. Cohen, unpubl. data). Marsh Wrens are highly territorial and will defend their nesting areas by breaking the eggs of other species that attempt to nest nearby (Picman 1977). They can control the distribution and alter the behavior and reproductive strategy of much larger and aggressive birds, such as Red-winged Blackbirds and Yellow-headed Blackbirds (*Xanthocephalus xanthocephalus*; Picman 1980, Picman and Isabelle 1995). Preliminary analyses of birds in smooth cordgrass-invaded marshes in San Francisco Bay have shown that Song Sparrows and Marsh Wrens have segregated territories with little overlap and that marsh wren territories are more highly correlated with the non-native cordgrass habitat than are Song Sparrow territories (J. C. Nordby and A. N. Cohen, unpubl. data). It is not yet known, however, whether Marsh Wrens are excluding Song Sparrows from the smooth cordgrass habitat or if song sparrows are selecting against those areas for other reasons (e.g., nesting habitat or food resources are limited).

FUTURE RESEARCH NEEDS

It is clear that the replacement and dominance of tidal-marsh communities in North America by invasive non-native reeds and cordgrasses can have important and perhaps severe consequences. These taxa may alter geomorphological processes, hydrologic regimes, and habitat structure. It is presumed that invasion by these taxa can affect the trophic structure and vertebrate species composition of tidal marshes. However, we know of no experimental studies of vertebrate species that provide quantitative estimates of these effects. These studies are needed to examine the impact of habitat alteration by invasive plant species on the structure and function of tidal-marsh communities in settings that allow for rigorous comparisons with appropriate controls.

Additional studies are needed that determine the current distribution, abundance, and population trends of native vertebrate species and their correlation with the presence of different species of invasive plants, as well as the effects of invasive species on important demographic parameters such as reproductive success and survival. We also need to assess the landscape-scale consequences of plant invasions in tidal marshes and whether a relationship exists between vertebrate community structure and landscape patchiness. Small isolated stands characteristic of the early stages of invasion may not negatively impact native vertebrate populations and may even provide additional edge habitat for certain species. As patches coalesce, however, and a threshold is reached in the invasive cover of an area, we may only then see detrimental effects as dense interior areas occupy a greater share of the landscape and intact native habitat becomes increasingly rare. Because the spread of exotic species is an ongoing process, we often have unique opportunities to establish baseline data in areas that are not yet invaded and also to track changes over time in areas where invasions are actively spreading. The development of predictive theoretical and empirical models that incorporate metapopulation dynamics of vertebrate species would enhance our understanding of the potential future impacts of these invasions.

It is also important that we assess the behavioral and genetic responses of native species to the exotic-species invasions. Because the alteration of habitat can occur so rapidly, we need to understand whether, or to what extent, native species can alter their behavior to compensate for changes in their environment. A high level of behavioral plasticity would be beneficial as it could also buy species more time to evolve adaptations to their rapidly changing habitat.

Not only must we examine the direct impact of non-native, invasive species, we also need to expand our understanding of the indirect impact of habitat alteration that can be associated with invasions such as trophic cascades in which the entire food web is altered, facilitation of further exotic invasions as newly altered habitat attracts additional non-native species or even alters interspecific interactions among native species. Only by examining both the direct and indirect effects of habitat alteration and the effects of other invasive species on tidal-marsh vertebrates will we be able to determine the full extent of the impacts of invasive plant species.

Millions of dollars are being spent on controlling the cordgrass and reed invasions and natural-resource managers have been making decisions about invasive-plant control measures without knowing the appropriateness of the different control programs currently in place. These control measures (e.g., large-scale glyphosate spraying, fire, mechanical removal, or tarps), as well as the seasonal timing of application, may well have unintended consequences for native species and should be balanced with careful monitoring of vertebrate communities. A mechanistic understanding of the impacts of invasive species on vertebrate communities is an essential step in determining if suitable alternative management strategies are needed.

ACKNOWLEDGMENTS

We thank S. Droege and R. Greenberg for their encouragement and endless patience and E. Kiviat and D. Stralberg for valuable comments that improved the manuscript. JCN was supported by a David H. Smith Conservation Research Fellowship from The Nature Conservancy. GRG received support from the U.S. Geological Survey Eastern Region State Partnership Program and the U.S. Geological Survey-Fish and Wildlife Service Science Support Program to examine the impact of *Phragmites* invasion in tidal marshes.

TIDAL SALTMARSH FRAGMENTATION AND PERSISTENCE OF SAN PABLO SONG SPARROWS (*MELOSPIZA MELODIA SAMUELIS*): ASSESSING BENEFITS OF WETLAND RESTORATION IN SAN FRANCISCO BAY

JOHN Y. TAKEKAWA, BENJAMIN N. SACKS, ISA WOO, MICHAEL L. JOHNSON, AND GLENN D. WYLIE

Abstract. The San Pablo Song Sparrow (*Melospiza melodia samuelis*) is one of three morphologically distinct Song Sparrow subspecies in tidal marshes of the San Francisco Bay estuary. These subspecies are rare, because as the human population has grown, diking and development have resulted in loss of 79% of the historic tidal marshes. Hundreds of projects have been proposed in the past decade to restore tidal marshes and benefit endemic populations. To evaluate the value of these restoration projects for Song Sparrows, we developed a population viability analysis (PVA) model to examine persistence of *samuelis* subspecies in relation to parcel size, connectivity, and catastrophe in San Pablo Bay. A total of 101 wetland parcels were identified from coverages of modern and historic tidal marshes. Parcels were grouped into eight fragments in the historical landscape and 10 in the present landscape. Fragments were defined as a group of parcels separated by >1 km, a distance that precluded regular interchange. Simulations indicated that the historic (circa 1850) *samuelis* population was three times larger than the modern population. However, only very high levels (>70% mortality) of catastrophe would threaten their persistence. Persistence of populations was sensitive to parcel size at a carrying capacity of <10 pairs, but connectivity of parcels was found to have little importance because habitats were dominated by a few large parcels. Our analysis indicates little risk of extinction of the *samuelis* subspecies with the current extent of tidal marshes, but the vulnerability of the smallest parcels suggests that restoration should create larger continuous tracts. Thus, PVA models may be useful tools for balancing the costs and benefits of restoring habitats for threatened tidal-marsh populations in wetland restoration planning.

Key Words: fragmentation, *Melospiza melodia samuelis*, population viability analysis, salt ponds, San Francisco Bay, San Pablo Song Sparrow, wetlands.

EVALUANDO LOS BENEFICIOS DE LA RESTAURACIÓN DE HUMEDALES EN LA BAHÍA DE SAN FRANCISCO

Resumen. El Gorrión Cantor de San Pablo (*Melospiza melodia samuelis*) es una de las tres subespecies de Gorriones Cantores morfológicamente distintas en marismas de marea del estuario de la Bahía de San Francisco. Estas subespecies son raras, ya que la población humana ha crecido, El dragar y el desarrollo han resultado en una pérdida del 79% de las marismas de marea históricas. Cientos de proyectos han sido propuestos en la década pasada con el fin de restaurar las marismas de marea, así como para beneficiar poblaciones endémicas. Para poder evaluar el valor de estos proyectos de restauración para los Gorriones Cantores, desarrollamos un modelo de análisis de viabilidad de población (AVP) para examinar la persistencia de subespecies samuelis en relación al tamaño de la parcela, conectividad y a la catástrofe en la Bahía de San Pablo. Un total de 101 parcelas de humedal fueron identificadas de coberturas de marismas de marea modernas e históricas. Las parcelas fueron agrupadas en ocho fragmentos en el paisaje histórico y 10 en el paisaje actual. Los fragmentos fueron definidos como un grupo de parcelas separadas por >1 km, una distancia que impedía intercambio regular. Las simulaciones indicaron que la población histórica (circa 1850) samuelis era tres veces mas grande que la población moderna. Sin embargo, solamente altos niveles de catástrofe (>70% mortandad) pondrían en peligro su persistencia. La persistencia de las poblaciones fue sensitiva al tamaño de la parcela con una capacidad de carga de <10 pares, pero la conectividad de las parcelas se encontró que tenía poca importancia porque los habitats estaban dominados por unas pocas parcelas. Nuestro análisis indica que hay poco peligro de extinción de la subespecie samuelis con el actual alcance de las marismas de marea, pero la vulnerabilidad de las parcelas más pequeñas sugieren que la restauración debería de crear tramos contiguos más largos. Es por ello que modelos AVP quizás sean herramientas útiles para balancear los costos y beneficios de habitats en restauración par alas poblaciones en peligro de marisma de marea en las plantaciones para la restauración de humedales.

Predicting how birds use habitat patches is a fundamental requirement in being able to identify functions and structures of landscapes critical to a bird's life cycle (Wiens 1994, 1996; Walters 1998). For birds of tidal marshes, the size, shape, and orientation of wetland patches may determine their value for local populations (Benoit and Askins 2002). Area-sensitive species respond to the size of habitats and may decline or fail to find or use small habitat patches with increased fragmentation. In planning wetland restoration projects, larger patches are typically thought to be more valuable for most species (Dramstad et al. 1996, Goals Project 1999), and corridors between tidal-marsh patches have often been considered valuable for maintaining populations. However, in urbanized areas, restoring small patches may be more cost effective than developing movement corridors (Beier and Noss 1998).

Habitats in the San Francisco Bay estuary have been reduced, modified, and fragmented by loss of 79% of its tidal marshes, 42% of its tidal flats, and construction of >13,000 ha of artificial salt evaporation ponds (Goals Project 1999). Hundreds of recent wetland restoration projects will create significant changes to the landscape, including conversion of thousands of hectares of salt ponds to tidally influenced marshes (Goals Project 1999, Steere and Schaefer 2001). Salt evaporation ponds have been part of the estuary for decades (Ver Plank 1958), and they now support a rich community of migratory birds during the migration and wintering periods (Takekawa et al. 2001), as well as breeding populations during the summer. Unfortunately, limited information is available to predict how the proposed changes will affect population viability of the target tidal-marsh species.

The San Pablo Song Sparrow (*Melospiza melodia samuelis*), hereafter referred to as *samuelis*, is one of three subspecies of Song Sparrows found in the San Francisco Bay estuary (Fig. 1). These include the Suisun Song Sparrow (*M. m. maxillaris*) in the eastern reach of the estuary, and the Alameda Song Sparrow (*M. m. pusillula*) in the southern reach. These three subspecies are listed as species of concern by California (Laundenslayer et al. 1991). The viability of *samuelis* is threatened because of increased fragmentation and reduction of tidal-marsh habitat around San Pablo Bay (Walton 1978, Marshall and Dedrick 1994, Nur et al. 1997). Of concern are the persistence of *samuelis* in tidal marshes of San Pablo Bay, and genetic integrity of the subspecies.

Our objective was to examine the relationship between extinction and tidal-marsh wetland parcel size. We used existing information on vital rates (Marshall 1948a, b; Johnston 1956a, b; Walton 1978, Collins and Resh 1985, Marshall and Dedrick 1994, Nur et al. 1997) to develop a population viability analysis (PVA) for *samuelis*. We used the model to estimate current and historical population size from modern and historic availability of habitats and determined the risk of extinction for the population given the current fragmented habitats. Finally, we evaluated PVA modeling as a tool in wetland restoration planning to establish the benefits of restoring bay lands to tidal marshes.

METHODS

POPULATION RANGE

The *samuelis* subspecies is found in the remaining tidal marshes surrounding the San Pablo Bay sub-region (Fig. 1). From Richmond to the southeast, the range of *samuelis* extends around the northern edge of San Pablo Bay to Tiburon in the southwest. The highest densities are found in Petaluma Marsh (Nur et al. 1997) and along the Petaluma River, the largest continuous tract of *samuelis* habitat (Marshall and Dedrick 1994). This area is connected to the maze of sloughs, levees, and ditches that comprise the baylands at the northern end of San Pablo Bay. To the east, isolated patches of tidal marsh south of the Carquinez Strait contained small numbers of breeding *samuelis* in the mid 1970s (Walton 1978) that were still present in the late 1990s (B. N. Sacks, unpubl. data). Southwest of the Petaluma River, *samuelis* is currently found in patches of tidal marsh including a large area north of San Rafael and smaller areas to the south (Nur et al. 1997).

MODEL DEVELOPMENT

The PVA was based on a modified Leslie matrix model where elements contained functions instead of constants. Simulations were carried out for 50 time steps (t), each corresponding to a single generation (year). Parameters N, n, p, m, and K were defined respectively as total female population in the spring just prior to breeding, number of females in age class x, age-specific survivorship, age-specific fecundity expressed as the average number of female fledglings produced by each female (sex-ratios were assumed to be 50:50) and carrying capacity (see POPULATION PARAMETERS below for details). N was calculated as:

$$N_{t+1} = \sum_{x=1 \text{ to } 7} n_{x(t+1)} \quad (A1)$$

Sizes of age groups were calculated for age class x = 1,

$$n_{x(t+1)} = \sum_{x=1 \text{ to } 7} n_{x(t)} m_{x(t)} p_{0(t)} \quad (A2)$$

and for age classes x = 2–7,

$$n_{x(t+1)} = n_{x-1(t)} p_{x-1(t)} + n_{x-1(t)} SD_1 RV_1 \quad (A3)$$

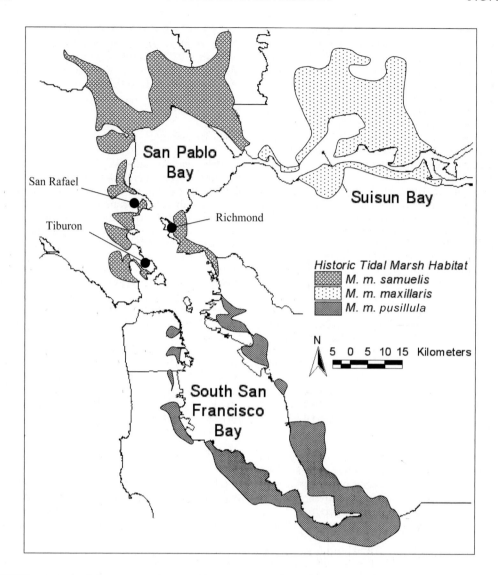

FIGURE 1. Ranges of three Song Sparrow subspecies in the San Francisco Bay estuary of California (adapted from Walton 1978).

where,

$$p_{0(t)} = 1 - [(0.304N_{(t)}/K + 0.4176) + SD_2RV_{2(adult\ =\ juvenile\ px)}] \quad (A4)$$

for age classes x = 2–7,

$$p_{x(t)} = [\text{mean estimated } p_x] + SD_3RV_{3(adult\ =\ juvenile\ px)} \quad (A5)$$

and,

$$m_{x(t)} = [\text{mean estimated } m_x] + SD_4RV_3 \quad (A6)$$

We incorporated demographic stochasticity into (A3) by calculating the binomial SD (Burgman et al. 1993) as

$$SD_1 = \sqrt{[(1 - p_{x(t)})p_{x(t)}/n_{x(t)}]} \quad (A7)$$

and multiplying by an approximately normally distributed and standardized random variable (RV_1), where

$$\text{pre-standardized } RV_1 = \arcsin \sqrt{X},$$

and X was the average of two randomly selected numbers between 0 and 1. Values were then standardized such that average $RV_1 = 0$ and SD = 1. A different RV_1 was chosen for each age class to reflect independence.

Environmental stochasticity was also incorporated into the models (A4, A5, A6). RV_2 was sampled from the negative of an approximately lognormal and standardized distribution (Burgman et al. 1993), to minimize effects of constraints (see below), and RV_3 was sampled from an approximately standard normal distribution as follows:

$$\text{pre-standardized } RV_2 = -(e^{RV_3}),$$

where RV_3 was the standardized arcsine of the square root of the average of two randomly selected numbers between 0 and 1. Separate random variables were selected for m_x (A6) and p_x (A4, A5), because Arcese et al. (1992) reported no correlation between residuals from these two vital rates when regressed on density. Carrying capacity, defined as the number of pairs in a parcel, was not determined explicitly from empirical data, but was used as a variable to examine the population response across a range of values for K. Projections were based on 100 stochastic simulations.

POPULATION PARAMETERS

We used data from previous studies (Johnston 1956a, b; Arcese et al. 1992) to estimate population parameters for the modeling. Three constraints were imposed on population numbers and vital rates. First, only whole numbers were used in modeling, such that when equations produced fractions of individuals, numbers were rounded to the nearest integer. Second, survival was constrained to fall between 0 and 1 (adults) or 0 and 0.4 (juveniles); juvenile survival was constrained to a maximum of 0.4 to be consistent with empirical data (Johnston 1956a, b), and to minimize the chance of juvenile survival exceeding adult survival. Third, annual fecundity rate was constrained between 0.5 and 3 daughters per female to be consistent with empirical data (Johnston 1956a, b).

A density-dependent function for p_0 (A4) was derived from regressions on data from Song Sparrows (*M. melodia morphna* and upland subspecies) on Mandarte Island, British Columbia (Arcese et al. 1992), but the intercept term was modified to be consistent with the larger clutch sizes for *samuelis* at Point San Pablo (Johnston 1956a, b) by adding 1.1 (based on the density-specific difference in expectations). Parameters are shown in equation (A4). Adult p_x ($\bar{x} = 0.43$) in (A5) and m_x ($\bar{x} = 2.2$ daughters) in (A6) were from Johnston's (1956a, b) data. Estimates of SD were based on standard deviations for density-independent parameters (A5), and root-mean-square errors for density-dependent (A4) parameters; specifically, SD for juvenile (0.08; A4) and adult (0.09; A5) p_x were calculated from Mandarte Island data (Arcese et al. 1992). SD for m_x (0.70; A6) was taken directly from Point San Pablo data (Johnston 1956a, b).

HABITAT PARCELS AND FRAGMENTS

We treat habitat units hierarchically, where a parcel is defined as a contiguous tract of tidal marsh, and a fragment is defined as a group of parcels separated from other fragments by >1 km. We determined the modern extent of the tidal-marsh habitat available for *samuelis* by intersecting the modern ecoatlas coverage (San Francisco Estuary Institute 2000) with the reported range for the subspecies (Fig. 1). We identified wetland parcels and fragments in the San Pablo Bay sub-region (Figs. 2a, b) from detailed geographic information system (GIS) coverages of habitats in the estuary known as the San Francisco Bay Area Ecoatlas (San Francisco Estuary Institute 2000). The ecoatlas coverages included modern (1997) habitats in the estuary (Fig. 2a) and historic (1770-1867; Fig. 2b) wetland parcels delineated from an extensive collection of eighteenth- and nineteenth-century maps and other sources. We define parcels as contiguous tracts of tidal marsh from wetland polygons in the ecoatlas with a buffer distance of >50 m, a separation distance reported to be rarely crossed by Song Sparrows (Marshall 1948a, Collins and Resh 1985, Scollon 1993).

Little is known about dispersal rates among subpopulations of the Song Sparrow. Adults generally do not disperse and first-year birds disperse only an average of approximately 180 m (Johnston 1956a, b). Fragments separated by >1 km seemed poorly connected by dispersal on the basis of findings by Nur et al. (1997). This is also consistent with observations of Smith et al. (1996) in British Columbia where subpopulations on several small islands separated by distances >1 km were primarily driven by within-population dynamics. Most of the smaller eastern and western fragments of tidal marsh on San Pablo Bay are separated by 1-5 km, suggesting that dispersal among these patches is rare.

PARCEL CONNECTIVITY

Although habitat loss has likely reduced the historic *samuelis* population by decreasing carrying capacity (Marshall and Dedrick 1994),

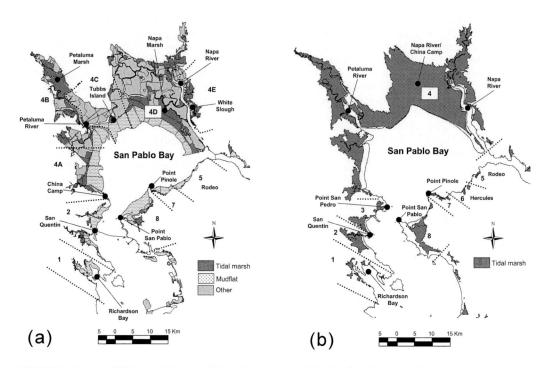

FIGURE 2. Modern (1997; a) and historical distribution (circa 1850; b) of tidal-saltmarsh parcels in the northern Central Bay and San Pablo Bay sub-regions. Figure modified from the San Francisco Bay Area Ecoatlas (San Francisco Estuary Institute 2000). Dashed lines indicate large habitat fragments separated by >1 km of terrestrial or aquatic barriers and numbers correspond to definitions in Table 1.

we wished to evaluate the added importance of loss of connectivity among wetland parcels. Rather than try to model the complex and poorly understood process of dispersal and re-colonization, we examined the most extreme assumptions, 0% and 100% connectivity among parcels within fragments, bracketing the range of possibilities. Specifically, we used equations estimating the probability of extinction as a function of parcel size under two sets of assumptions: (1) 100% connectivity of all parcels made by summing K across parcels within fragments and then calculating the expected number of sparrows in this combined area, and (2) 0% connectivity of parcels made by calculating the expected number of sparrows for each parcel and then summing these numbers across parcels to arrive at a projection for the fragment. In both cases, the expected number of sparrows (in parcel or fragment, respectively) was the product of K+1 (where K was assumed to be 2.75 pairs/ha; Johnston 1956b) minus the probability of extinction—a function of K determined from the simulation model described above. Assuming that the probability of extinction is a concave function of K, the expected number of sparrows in a fragment with 100% connectivity among parcels will always be greater than the expected number of sparrows in a fragment with 0% connectivity among its fragments (Jensen's inequality). This modeling was done for the habitat configuration during historical and modern times. Extinction either occurs or does not occur and is thus a binomial variable with respect to each isolated parcel. We assumed that parcels were filled to capacity when not extinct, such that projected estimates for the fragment subpopulation were calculated as the sum over all parcels of the probability of extinction multiplied by the carrying capacity.

CATASTROPHIC EVENTS

We did not have any a priori expectation about catastrophic events because of the paucity of data on the frequency and extent of catastrophes on the *samuelis* population. Therefore, we ran a sensitivity analysis on these parameters at carrying capacities varying from 20–3,000. We ran simulations for three different extents of catastrophes (50, 70, and 90% mortality in each parcel over winter) that occurred in 10% of years, and 90% mortality at 5% frequency. At K <50, the outcome was extremely sensitive to the extent of catastrophe, but at K >100, the outcome was highly sensitive when extent was >50%, but relatively insensitive when catastrophic extent was 0–50%. Frequency made little difference relative to extent at 90% mortality. The analysis assumed 0% connectivity of parcels and was then used to calculate expected population numbers (2.75 females/ha; Johnston 1956b) for both historical and modern habitat maps.

RESULTS

Parcels and Fragments

The area of tidal marsh included large fragments (>500 ha) in the Petaluma and Napa marshes at the center of the range, surrounded by medium-sized fragments (>350 ha) at China Camp, Tubbs Island, and White Slough, and the smallest fragments at the southwestern and southeastern extent of the range (Fig. 2a). The modern tidal-marsh area was estimated to be 8,076 ha, composed of ten fragments (1–33 parcels each, mean parcel size = 80 ha). Cogswell (2000) reported a similar number of fragments, but he defined different areas. The mean fragment size ranged from 12–3,887 ha (Table 1).

Tidal-marsh area differed greatly from the historic to the modern landscape. In the past, fragments ranging from 30–23,225 ha represented 3.2 times more tidal marsh than in the present (Table 2). Nearly 90% of the historic tidal wetlands were encompassed within a single large fragment that extended from China Camp in the southwest to Napa River in the northeast (Fig. 2b). Two other fragments to the southwest exceeded 500 ha, and the area of Point San Pablo was >1,000 ha. Each fragment was composed of 3–39 tidal-marsh parcels with a mean size of 259 ha.

Population Size Estimates

The number of pairs estimated to occupy the 10 fragments ranged from 29 pairs in a 12 ha fragment to 10,648 pairs in a 3,887 ha fragment (Table 1). We estimated the current population of *samuelis* as 22,079 pairs. This population estimate is higher than earlier figures by Walton (4,600 pairs in 1978) and Marshall and Dedrick (15,000 pairs in 1994); however, our model results compared well with a more recent

Table 1. Habitat fragments, parcels, areas, and population projections by level of connectivity (0%, 100%) for pairs of San Pablo Song Sparrows in the modern baylands determined from the Ecoatlas (San Francisco Estuary Institute 2000).

				Connectivity	
Fragment	Name	Number of parcels	Area (ha)	0%	100%
1	Richardson Bay	14	107	269	294
2	San Quentin	15	187	488	514
4A	China Camp	4	475	1,304	1,305
4B	Petaluma Marsh	11	2,210	6,072	6,079
4C	Tubbs Island	5	477	1,306	1,313
4D	Napa Marsh	33	3,887	10,648	10,688
4E	White Slough	1	384	1,057	1,057
5	Rodeo	3	12	29	32
7	Point Pinole	9	69	176	190
8	Point San Pablo	6	268	730	738
Total		101	8,076	22,079	22,210

Table 2. Habitat fragments, parcels, areas, and population projections by level of connectivity (0%, 100%) for pairs of San Pablo Song Sparrows in the historic baylands circa 1850 determined from the Ecoatlas (San Francisco Estuary Institute 2000).

				Connectivity	
Fragment	Name	Number of parcels	Area (ha)	0%	100%
1	Richardson Bay	22	276	731	759
2	San Quentin	4	726	1,992	1,996
3	San Pedro	5	586	1,602	1,612
4	Napa River-China Camp	39	23,225	63,841	63,870
5	Rodeo	6	30	72	81
6	Hercules	5	58	155	160
7	Point Pinole	3	83	227	228
8	Point San Pablo	16	1,008	2,762	2,772
Total		100	25,992	71,382	71,478

estimate (25,000 pairs) based on extensive surveys in the sub-region (Nur et al. 1997).

In the historic landscape, tidal-marsh habitats surrounding San Pablo Bay were more than three times larger and the model estimated a population of 71,400 pairs, a number very similar to the 71,000 pairs calculated by Marshall and Dedrick (1994). We estimated abundance ranging from 72–63,841 pairs in the eight fragments, with most pairs (89%) found in the fragment spanning Napa River to China Camp (Table 2; Fig. 2b).

CONNECTIVITY AND CATASTROPHE

Modern population estimates of *samuelis* (22,210) differed <1% compared to the number predicted (22,079) if connectivity of populations within parcels was 100% or 0% (Table 1). Similarly in the historic landscape, we found little difference (<1%) in estimated *samuelis* numbers with 0% and 100% connectivity of parcels (Table 2). For parcels in which K <10 pairs, populations were likely to go extinct at any catastrophe level (Figs. 3a, b). Low catastrophe levels (0–50%) were related to high frequency of extinction only in the parcels where K ≤10 pairs. The predicted frequency of extinction was only affected in the smallest parcels with carrying capacities of ≤10 pairs by catastrophic rates of 0–50% in 50 yr (Fig. 3a).

DISCUSSION

PVA models may be valuable as decision tools to assess risks in reaching proposed management goals for target species (Reed et al. 2002). For example, PVAs have been used to find the best management options to reduce the chance of catastrophe and save species such as the threatened Florida Scrub Jay (*Aphelocoma coerulescens*) from extinction (Root 1998). Proposed restoration of up to 8,900 ha of tidal marsh in San Pablo Bay (Goals Project 1999) may result in an increase in *samuelis* numbers. However, our PVA model suggests that *samuelis* is not in imminent danger of extinction with the current extent of tidal-marsh habitats. Restoring more tidal marsh will increase the population size of this subspecies, but it will not make its long-term persistence more likely. Other factors, including habitat quality and predator control, may become much more important determinants of population size.

Most salt-marsh parcels surrounding San Pablo Bay were relatively small ($\bar{x}$ = 16 ha), and the largest 15% of the parcels comprised the majority (89%) of the total area. Even when no inter-parcel dispersal occurred (0%

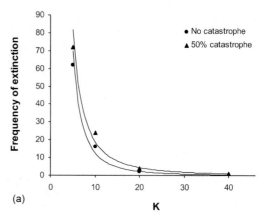

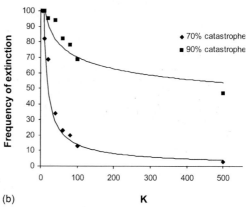

FIGURE 3. Frequency of extinction predicted as a function of carrying capacity (number of pairs per parcel) at (a) low (0, 50%) and (b) high (70, 90%) catastrophe rates within a 50-yr period as a function of K.

connectivity), population projections for *samuelis* were very similar to results assuming 100% connectivity, except when catastrophic extent was >70% mortality. Thus, although we concluded that connectivity among parcels was not a critical issue in the current landscape, our finding that small parcels had greater extinction probabilities indicates that it will be important to prevent increased fragmentation in the future. In addition, we did not examine patch shape, a factor that may affect habitat value for tidal-marsh species (Benoit and Askins 2002). Parcel shape may be an important consideration in restoring parcels for species like *samuelis* that typically use linear channel habitats (Collins and Resh 1985).

We did not incorporate habitat quality in our model, but it is known to affect population persistence in birds (Root 1998). Nur et al. (1997) found that densities of sparrows varied as a function of patch size, and smaller patches may be sinks where populations are not maintaining themselves. Densities were much lower in the southeast part of their range compared with the northwest (Nur et al. 1997), suggesting that habitat quality in the sub-region did vary. *Samuelis* densities are greater within tall native vegetation such as gumplant (*Grindelia stricta*) and coyote bush (*Baccharis pilularis*) that are used for nesting and cover (Collins and Resh 1985). Song Sparrows seem to have greater reproductive success (Johnston 1956a, b; Nur et al. 1997) in habitats dominated by pickleweed (*Salicornia virginica*), as opposed to those dominated by native California cordgrass (*Spartina foliosa*), probably because pickleweed is associated with higher elevation marsh zones with taller vegetation and less potential for nest flooding.

CATASTROPHIC EVENTS

Although Johnston (1956a, b) found unusually high tides to be an important cause of nest failure from flooding, recent studies (Nur et al. 1997) were unable to corroborate high tides as an important source of nestling mortality (N. Nur, PRBO Conservation Science, pers. comm.). Compared with Mandarte Island, British Columbia, where cold, rainy winter conditions decimated the population in two of 17 yr (Arcese et al. 1992, Smith et al. 1996), winter storms are likely not major catastrophic events for Song Sparrows in the warmer climate of the San Francisco Bay area. In >18 yr of mist netting an upland subspecies of Song Sparrow (*M. m. gouldii*) at nearby Point Reyes National Seashore, no evidence showed any catastrophic decline (N. Nur, PRBO Conservation Science, pers. comm.). However, oscillations in Sierra Nevada stream flows may lead to periodic flood and drought events and salinity changes in the estuary (Dettinger and Cayan 2003), conditions that may alter tidal-marsh plant composition and habitat structure (Zedler et al. 1986). Effects of this type of catastrophe, delayed by time lags of more than a year, may be difficult to identify (Knopf and Sedgwick 1987).

Risks of sea-level rise may exacerbate nest inundation, especially during highest high tides. Although sea-level rise of 30–90 cm is predicted to occur in the next 100 yr, extreme high tides may increase at a higher rate. For example, one scenario indicated that sea-level rise of 20 cm may produce a 28 cm increase in extreme high tides (Malamud-Roam 2000). With many of the bayland wetlands adjacent to cities or behind levees, loss of wetland habitats may result in catastrophic losses of tidal-marsh populations. Though actual effects of higher high tides are marsh specific (due to differences in elevation and geomorphology), preliminary analysis shows that Song Sparrow nests in exotic cordgrass are much more likely to flood than nests placed in native vegetation (J. C. Nordby and A. Cohen, unpubl. data). The risks of sea-level rise in combination with the invasion of smooth cordgrass (*Spartina alterniflora*) and *S. foliosa* × *S. alterniflora* hybrids may create catastrophes at the level where persistence of the subspecies may be threatened (Takekawa et al., chapter 11, *this volume*). Smooth cordgrass may invade future restoration projects, leading to reduced food resources and foraging habitat for Song Sparrows and creating more favorable habitat for Marsh Wrens (*Cistothorus palustris*) that may displace Song Sparrows (J. C. Nordby, unpubl. data).

PVAs IN RESTORATION PLANNING

Wetland conservation in the San Francisco Bay region has evolved from a period of preservation to an era of aggressive restoration. Marshall and Dedrick (1994) declared, "priceless tidal marshes have become monotonous salt-evaporation ponds, pastures, cities, factories, and game refuges for fresh-water ducks." However, in this highly urbanized ecosystem, blanket condemnation of artificial habitats, or conversely, a belief that restoring a few wetlands in a vast, yet greatly degraded landscape may return function to more natural or diverse communities is an oversimplification.

Numerous migratory and native species use artificial habitats such as salt-evaporation ponds (Takekawa et al. 2001, Warnock et al. 2002), and mosquito ditches that provide channel habitat for as many as 2,000 *samuelis* in the Petaluma Marsh (Collins and Resh 1985). Conversely, many wetland restoration projects have failed to create marshes with values and functions of older marshes. For example, cordgrass plants in a created wetland of southern California may be less vigorous, and did not provide the height structure needed for the endangered Light-footed Clapper Rail (*Rallus longirostris levipes*; Zedler 1993). Similarly, Song Sparrows in restored terrestrial habitats with less structure were less productive and prone to predation because of lack of cover (Larison et al. 2001).

Multi-species management in complex ecosystems such as this highly urbanized estuary has become a difficult balancing act that requires simultaneously weighing costs and benefits of alternatives for several species. Recent restoration planning has included efforts to determine what comprises the best landscape for the most diverse community with an emphasis on tidal-marsh species (Goals Report 1999). However, increasing populations of threatened tidal-marsh species may require actions that benefit them at the expense of other, less threatened species (Takekawa et al. 2000). With PVA analyses, the benefits of converting habitats for threatened tidal-marsh species may be compared with predicted population losses of other species (Stralberg et al. 2005; N. Warnock, PRBO Conservation Science, unpubl. data), providing for better balance in restoration decisions.

ACKNOWLEDGMENTS

The U.S. Geological Survey (National Biological Survey) and the Species-at-Risk Program provided funding for the project. W. M. Perry and J. Lanser provided assistance with geographic information system coverage and background mapping based on excellent San Francisco Estuary Institute Ecoatlas coverage. N. Nur (PRBO Conservation Science) provided recent data for comparison with older population parameters. We thank K. J. Phillips, N. D. Athearn, J. Yee, J. E. Maldonado, and two anonymous reviewers for helpful comments on the draft manuscript.

MULTIPLE-SCALE HABITAT RELATIONSHIPS OF TIDAL-MARSH BREEDING BIRDS IN THE SAN FRANCISCO BAY ESTUARY

HILDIE SPAUTZ, NADAV NUR, DIANA STRALBERG, AND YVONNE CHAN

Abstract. We modeled the abundance or probability of occurrence of several tidal-marsh-dependent birds found in the San Francisco Bay estuary—the San Pablo Song Sparrow (*Melospiza melodia samuelis*), Alameda Song Sparrow (*M. m. pusillula*), Suisun Song Sparrow (*M. m. maxillaris*), Salt Marsh Common Yellowthroat (*Geothlypis trichas sinuosa*), California Black Rail (*Laterallus jamaicensis coturniculus*), and Marsh Wren (*Cistothorus palustris*)—based on marsh characteristics at several scales. Local habitat variables included vegetation type, structure, and height, and tidal-channel characteristics. Landscape variables included marsh size and configuration, distance to edge, and type of surrounding land use. For each species considered, both landscape and local habitat factors were significant predictors in multi-variable, multi-scale, linear or logistic regression models. While the best models contained both local and landscape variables, all four bird species were also well predicted by local habitat or landscape variables alone. Predictor variables differed by species, but each species responded strongly to vegetation composition (specific plant species) as well as the overall structure (height or complexity) of the vegetation. Scale effects also differed by species. For Song Sparrows, land-use variables were most important at a relatively small spatial scale (500 m) while for Marsh Wrens and Common Yellowthroats they were important at the largest scale examined (2,000 m). Certain elements of vegetation type and structure, as well as marsh size and configuration (perimeter to area ratio) and surrounding land use, were important across several species, suggesting a suite of habitat and landscape characteristics that may be useful in identifying sites important to multiple bird species.

Key Words: Cistothorus palustris, Geothlypis trichas sinuosa, habitat selection, *Laterallus jamaicensis coturniculus, Melospiza melodia,* San Francisco Bay, tidal marsh.

RELACIONES DE HABITAT A ESCALAS MULTIPLES DE AVES REPRODUCTORAS DE MARISMA DE MAREA EN EL ESTUARIO DE LA BAHÍA DE SAN FRANCISCO

Resumen. Modelamos la abundancia o la probabilidad de ocurrencia de varias especies de aves dependientes de marisma de marea, encontradas en el estuario de la Bahía de San Francisco—el Gorrión Cantor de San Pablo (*Melospiza melodia samuelis*), el Gorrión Cantor de Alameda (*M. m. pusillula*), el Gorrión Cantor Suisun (*M. m. maxillaris*), la Mascarita Común de Marisma Salado (*Geothlypis trichas sinuosa*), la Polluela Negra de California (*Laterallus jamaicensis coturniculus*), y el Chivirin Pantanero (*Cistothorus palustris*)—basados en las características de la marisma a diferentes escalas. Las variables locales incluyeron el tipo de vegetación, estructura y altura, y las características del canal de la marea. Variables del paisaje incluyeron el tamaño de la marisma y su configuración, la distancia a la orilla, y tipo de uso del suelo de los alrededores. Para cada especie considerada, tanto el paisaje como los factores locales del hábitat fueron vaticinadores significativos en los modelos de multi-variable, multi-escala, linear o de regresión logística. Mientras que los mejores modelos contenían tanto variables locales como de paisaje, las cuatro especies fueron también bien pronosticadas por el hábitat local o las variables de paisaje solas. Las variables de predicción se diferenciaron por especies, pero cada especie respondió fuertemente a la composición de la vegetación (especies de planta específicas) como también a la estructura total (altura o complejidad) de la vegetación. Efectos de escala también difirieron por las especies. Para los Gorriones Cantores, las variables del uso del suelo fueron más importantes a una escala espacial relativamente pequeña (500 m), mientras para los Chivirines Pantaneros y las Mascaritas Comunes de Marisma Salada fueron más importantes a la escala mayor examinada (2,000 m). Ciertos elementos del tipo y de la estructura de la vegetación, como también el tamaño y la configuración de la marisma (perímetro al radio del área) y el uso del suelo de los alrededores, fueron importantes a través de algunas especies, sugiriendo un juego de características del hábitat y el paisaje que quizás sea utilizado para identificar sitios importantes para múltiples especies de aves.

Tidal marsh, formerly the dominant habitat type in the San Francisco Bay estuary (hereafter the estuary), has been reduced to <20% of its original extent as a result of human activities, such as diking, dredging, and urban development (Goals Project 1999). In addition, many remaining tidal marshes have been hydrologically altered and subdivided by levees, mosquito-control ditches, boardwalks, and power lines. Many have also been degraded by contaminants, invasive species, and recreational use (Takekawa et al., chapter 11, *this volume*). This habitat loss and degradation has adversely affected a unique assemblage of marsh-dependent plants, animals, and invertebrates, many of which are specifically adapted to the range of salinity and tidal regimes in the estuary's marshes.

Tidal-marsh passerine birds, including three endemic subspecies of Song Sparrow (San Pablo Song Sparrow [*Melospiza melodia samuelis*], Alameda Song Sparrow [*M. m. pusillula*], and Suisun Song Sparrow [*M. m. maxillaries*] hereafter referred to as Song Sparrows or tidal-marsh Song Sparrows), the endemic Salt Marsh Common Yellowthroat (*Geothlypis trichas sinuosa*), and the Marsh Wren (*Cistothorus palustris*), have experienced a severe habitat loss, and have been restricted in many areas to isolated and degraded marsh fragments with extensive urban upland edges. All but the Marsh Wren are considered species of special concern by the state of California. The California Black Rail (*Laterallus jamaicensis coturniculus*), a state of California threatened species and a federal species of management concern, is a tidal-marsh-dependent species that is now absent from many estuary marshes, and its small population size raises concerns about its long-term persistence in the estuary (Evens et al. 1991, Nur et al. 1997). All of these species merit special attention due to their limited distributions and relatively small population sizes, but they may also serve as habitat indicators for other tidal-marsh-dependent plant and animal species, several of which have state and/or federal threatened or endangered status.

Important earlier studies of the three focal songbird species in the San Francisco Bay estuary, primarily concerning the Song Sparrow (Johnson 1956a, b; Collins and Resh 1985, Marshall and Dedrick 1994), were based on field data limited in scale and extent. Until recently, data sufficient for analyzing regional and landscape-level habitat associations have not been available. Studies that published data on estuary-wide songbird distributions (Hobson et al. 1986, Nur et al. 1997) did not generally contain corresponding information on critical habitat and landscape characteristics. Black Rail distribution patterns in the estuary have been more systematically identified due to their special conservation status (Evens et al. 1991, Evens and Nur 2002), but landscape-level habitat associations of this species other than relationship with marsh size have not been previously analyzed.

In 1996, we began conducting annual surveys of breeding songbirds and Black Rails in 21 San Francisco Bay estuary tidal marshes, adding new sites each year to result in a total of 79 marshes surveyed at least once between 1996 and 2003. This comprehensive dataset provides a unique opportunity to examine regional distribution and abundance patterns and, most importantly, to assess the effects of local habitat characteristics, landscape composition, and habitat fragmentation on these distribution and abundance patterns. Knowledge of specific habitat requirements of these tidal-marsh birds will improve the ability of land managers and wildlife agencies to plan restoration, management, and acquisition activities.

All bird species display some degree of specificity in terms of the habitat types in which they choose to set up territories, forage, seek shelter, and breed; habitat relationship models seek to quantify and clarify these apparent preferences (Cody 1985). Birds tend to respond to particular characteristics of vegetation structure and patchiness, often at several scales (Rotenberry and Wiens 1980, Wiens and Rotenberry 1981, Saab 1999). Habitat characteristics found to be important for wetland birds include various aspects of vegetation structure and density (Collins and Resh 1985, Leonard and Picman 1987, Weller 1994, Benoit and Askins 1999, Whitt et al. 1999, Poulin et al. 2002), water depth and cover (Leonard and Picman 1987, Craig and Beal 1992), and tidal-channel characteristics (Collins and Resh 1985). However, many bird species may also respond to the landscape context of a habitat patch, as well as its size and shape. Numerous studies over a range of habitat types have demonstrated a significant effect of surrounding landscape at various scales on species richness, relative abundance, and nest success of breeding passerines (Flather and Sauer 1996, Bolger et al. 1997, Bergin et al. 2000, Fairbairn and Dinsmore 2001, Naugle et al. 2001, Tewksbury et al. 2002), as well as scale-dependent responses to habitat characteristics (Pribil and Picman 1997, Naugle et al. 1999).

Bird relationships to patch size and shape have been studied in other habitats, especially with regard to the process of habitat fragmentation. Many researchers have evaluated island biogeography (MacArthur and Wilson 1967) principles for habitats ranging from eastern deciduous forest (Ambuel and Temple 1983, Robbins et al. 1989) to southern California chaparral (Soulé et al. 1988, Bolger et al. 1991) to wetlands (Brown and Dinsmore 1986, Naugle et al. 2001). Others have focused on fragmentation as a process occurring along a gradient, recognizing the intermediate stages between contiguous habitat and isolated fragments (Wiens 1994) and the potential for differential effects on wildlife along that fragmentation gradient (Andrén 1994). More recent reviews and meta-analyses have suggested that, for most species, habitat fragmentation may actually have little demonstrable effect beyond the direct effects of habitat loss and degradation (Bender et al. 1998, Harrison and Bruna 1999). In addition, fragmentation effects on breeding birds appear to be scale-dependent (Chalfoun et al. 2002, Stephens et al. 2004). However, few studies have evaluated the effects of tidal-marsh fragmentation on breeding songbirds or rails (but see Benoit and Askins 2002).

The Baylands Ecosystem Habitat Goals Report (Goals Project 1999) recommended the creation and maintenance of large, interconnected blocks of tidal marsh with a minimum of upland intrusions and urban edge interface. But these recommendations were based largely on expert opinion, rather than empirical evidence. The Goals Report also summarized the best available information at the time regarding the habitat preferences of the Song Sparrow (Cogswell 2000), Salt Marsh Common Yellowthroat (Terrill 2000), and Black Rail (Trulio and Evens 2000), including qualitative analyses of habitat requirements; but at the time, no one had attempted to develop quantitative, predictive multiple scale models for the habitat requirements of these taxa.

For this study we developed models predicting breeding songbird responses to differences in landscape patterns and local habitat characteristics, in order to provide information about an ecosystem that has been increasingly fragmented and degraded by human activities.

The specific objectives of this study were: (1) to identify elements of marsh-vegetation composition and structure that affect Song Sparrow, Salt Marsh Common Yellowthroat, Marsh Wren, and Black Rail abundance or probability of occurrence during the breeding season; (2) to identify the importance of surrounding land use, marsh size, and landscape-scale habitat configuration on abundance or probability of occurrence; (3) to identify the spatial scale at which landscape influences on marsh-bird distribution and abundance are most strongly expressed; (4) to compare the relative influence of local habitat- and landscape-level factors on each species evaluated; (5) to contrast the patterns observed among the four species; (6) to evaluate the variation in relative abundance across the San Francisco Bay estuary; and (7) to consider implications of these results for monitoring programs, restoration projects, and land and wildlife managers.

METHODS

STUDY AREA

Study sites were located in tidal marshes throughout the San Francisco Bay estuary in San Francisco, San Pablo, and Suisun bays (Fig. 1). Although access limited marshes available for bird surveys, efforts were made to select sites that encompassed a range of habitat conditions over a broad geographic area. A special effort was made to identify and survey marshes in a range of sizes from the smallest fragments to larger areas of contiguous marsh (Table 1).

The data used in these analyses were obtained from bird surveys conducted during the spring and summer of 2000 and 2001. Point count surveys (Ralph et al. 1993) were conducted twice per year and Black Rail surveys were conducted only in 2001.

POINT-COUNT SURVEY METHODS

We conducted point-count surveys at 421 locations in 54 fully tidal and muted tidal marshes — marshes that receive less than full tidal flow due to physical impediments (Goals Project 1999) — distributed fairly evenly across the estuary (Table 1). Surveys were conducted within 4 hr of sunrise, one or two times between 20 March and 31 May in 2000 and twice between 20 March and 29 May in 2001. Successive survey rounds were conducted at least 3 wk apart.

We placed survey points 150–200 m apart along transects, with a randomly chosen start location and one to 20 points per site, depending on marsh size. In the smallest marsh fragments there was only enough room for one survey point (N = 5). Points were often placed along levees or boardwalks to decrease impact to marsh habitat, but where possible they were placed within the marsh vegetation to reduce the bias of sampling from habitat edges. At each point, a trained observer recorded all birds detected by sight and sound for 5 min. For detections within 100 m from the observer, distance was estimated within 10-m bands; detection type (visual or auditory) was also recorded for each bird.

We calculated an abundance index (number of birds detected per hectare) for each passerine species at each survey point within a 50-m radius of the observer to correspond with the area in which we collected vegetation measurements (see below). Because some surveys were conducted from habitat edges, we adjusted this index for area surveyed by dividing by the actual area of marsh habitat surveyed, calculated from geographic information system (GIS)

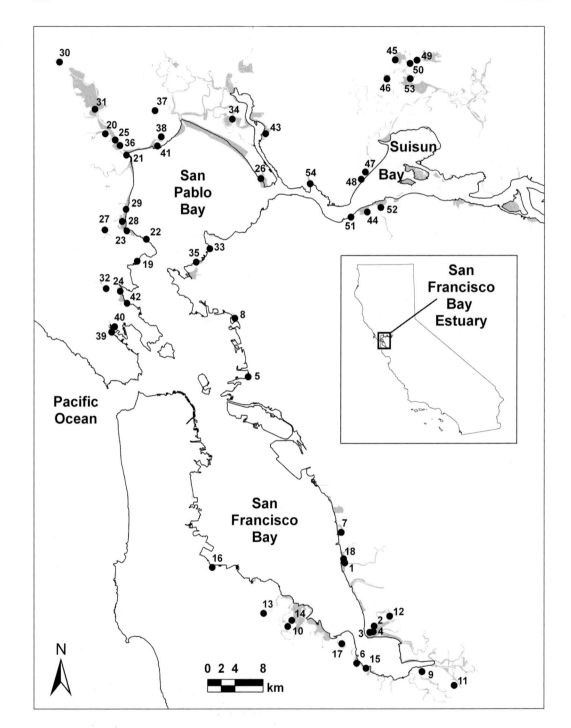

FIGURE 1. San Francisco Bay estuary tidal marsh study sites used in analyses. See Table 1 for corresponding study site names. Tidal-marsh habitat is shown with gray shading.

TABLE 1. TIDAL-MARSH BIRD SURVEY SITES IN THE SAN FRANCISCO BAY ESTUARY SURVEYED BETWEEN MARCH 2000 AND MAY 2001.

Site Name	Number of survey points	Perimeter/area ratio (meters/hectares)	Patch size (hectares)
San Francisco Bay			
1. Old Alameda Creek	6	133.4	234.5
2. Hetch-Hetchy east	5	64.3	446.2
3. Hetch-Hetchy west	7	64.3	446.2
4. Dumbarton Marsh	14	64.3	446.2
5. Emeryville Crescent	5	223.6	20.1
6. Faber-Laumeister Tract, east Palo Alto	7	68.3	124.5
7. Hayward regional shoreline	11	116.3	100.8
8. Hoffman Marsh, El Cerrito	5	209.6	14.8
9. Mouth of Alviso Slough	6	292.4	10.0
10. Middle Bair Island west	5	20.1	1,283.8
11. New Chicago Marsh	6	96.7	1,768.9
12. Newark Slough	7	64.3	446.2
13. Oral B fragment	1	468.9	6.5
14. Outer Bair Island west	3	20.1	1,283.8
15. Palo Alto baylands	9	68.3	124.5
16. Park Plaza fragment	1	396.9	2.0
17. Ravenswood Slough	8	233.2	35.7
18. Whalestail marsh	12	133.4	234.5
Total number of survey points	118		
San Pablo Bay			
19. Beach fragment	1	387.9	1.3
20. Black John Slough	20	34.2	1,806.5
21. Day Island	8	70.6	1,132.8
22. China Camp fragments	2	929.4	0.4
23. China Camp State Park	16	70.6	1,132.8
24. Corte Madera Ecological Reserve	10	96.9	104.6
25. Green Point Centennial Marsh	7	34.2	1,806.5
26. Mare Island	20	37.5	1,428.7
27. Mitchell fragment	3	155.0	11.8
28. McInnis Marsh	10	70.6	1,132.8
29. Hamilton south / McInnis north	10	70.6	1,132.8
30. Petaluma Dog Park	4	98.78	36.7
31. Petaluma Ancient Marsh	9	34.2	1806.5
32. Piper Park	5	221.4	58.8
33. Point Pinole south	3	256.5	9.3
34. Pond 2A restoration	10	12.1	5,767.8
35. San Pablo Creek	9	97.1	60.6
36. Petaluma River Mouth (Carl's Marsh)	10	67.1	393.0
37. Sears Point	10	164.7	123.2
38. Tolay Creek	11	67.1	393.0
39. Tam High School (Richardson Bay)	5	156.0	38.5
40. Travelodge fragment	1	344.2	2.4
41. Lower Tubbs Island (muted marsh)	8	67.1	393.0
42. Triangle/MCDS fragment	1	204.7	5.0
43. White Slough Marsh	5	71.6	265.2
Total number of survey points	198		
Suisun Bay			
44. Bullhead Marsh	10	65.2	205.8
45. Cordelia fragment	3	13.1	6,658.5
46. Grey Goose	6	13.1	6,658.5
47. Goodyear Slough north	10	13.1	6,658.5
48. Goodyear Slough south	10	13.1	6,658.5
49. Hill Slough east	6	133.5	28.3
50. Hill Slough west	5	171.6	12.8
51. Martinez Regional Shoreline	10	137.5	40.8
52. Point Edith	10	23.1	1,034.5
53. Rush Ranch	10	42.7	557.9
54. Southampton Bay/ Benicia State Park	10	112.7	71.3
Total number of survey points	90		

Note: Numbered site locations are shown in Fig. 1.

data (San Francisco Estuary Institute 2000) and verified in the field. For analysis, the area-adjusted abundance index was averaged over all surveys for that point (see below).

BLACK RAIL SURVEY METHODS

Black Rail surveys were conducted at 216 points in 28 San Pablo and Suisun bay marshes (Table 1) between 18 April and 29 May during the breeding season of 2001. We did not survey San Francisco Bay sites because Black Rails are not usually found there during the breeding season. We established one to 20 survey points in each marsh, depending on marsh size. In several marshes we surveyed from rail survey points previously established by Evens et al. (1991), but most marshes were surveyed from points that we established for point-count surveys. Survey points were placed at least 100 m apart but at most sites they were 200 m apart, as was the case for point counts.

Surveys were conducted following a standardized taped-call-response protocol (Evens et al. 1991, Nur et al. 1997). The observer listened passively for 1 min after arriving at the survey point, and then broadcasted tape-recorded black rail vocalizations consisting of 1 min of "grr" calls followed by 0.5 min of "ki-ki-krr" calls. The observer then listened for another 3.5 min for a total of 6 min per point. At each point, rails heard calling <30° apart were considered the same bird (unless the calls were simultaneous), and those >30° apart were considered different birds. We summarized the data by counting the number of rails detected within 50 m of the observer to correspond with the point count and vegetation data; this is also the maximum distance at which Black Rails can be reliably counted (Spear et al. 1999).

We determined whether rails were present during any rail survey or breeding season point count survey in either year (i.e., a point was coded absent for Black Rails if none was detected at any survey in 2000 or 2001) and included in our analysis only the points where rail taped-call-response surveys were conducted in 2001.

VEGETATION SURVEY METHODS

At each survey point, vegetation and other local habitat data were collected in the field in 2000 or 2001 by trained observers (Table 2). These data were limited to the habitat within 50 m of each point. By walking through the habitat along perpendicular, randomly selected transects we estimated visually the percent of marsh habitat, percent cover of tidal channels, shrub and non-woody vegetation (and of each individual plant species), and pans or ponds. We scored cover for each habitat variable as proportion of total cover, measured on a 0–1 scale and scored cover of each plant species as proportion of total vegetation cover, also measured on a 0–1 scale. We measured vegetation density by counting the number of times vegetation hit a 6 mm-diameter pole at 10 cm intervals from the ground at five sample points on the transects. We summed all hits, and also summed those under and over 30 cm, a height previously determined to be important for marsh birds and grassland birds (Rotenberry and Wiens 1980, Collins and Resh 1985). For analysis we calculated mean hits for each density-height category. We also measured the distance from the center of the survey point to the nearest tidal channel and that channel's width; and developed a channel index by counting the number of channels of several width categories (<1 m and <2 m) crossed by the transects.

GIS METHODS

For each survey-point location, we used ArcView GIS 3.2a (Environmental Systems Research Institute 2000) and extensions to derive a set of landscape parameters characterizing that point and the surrounding marsh. GIS data for bayland habitats were obtained from the EcoAtlas modern baylands GIS layer (San Francisco Estuary Institute 2000). To characterize upland habitats, we derived a composite land-use layer for the San Francisco Bay region consisting of the most recent 1:24,000 land-use GIS layers from the California State Department of Water Resources (Department of Water Resources 1993–1999) where available, and 1:24,000 land-use GIS layers from the U.S. Geological Survey (USGS) Midcontinent Ecological Science Center (1985). We generated three general classes of landscape metrics (Table 3): edge proximity metrics, habitat configuration metrics, and landscape composition metrics.

Edge proximity metrics were calculated for each point-count location using the Alaska Pak extension for ArcView 3.x (National Park Service 2002). Habitat configuration (marsh size and shape) metrics were calculated for the marsh patch underlying each point count using the Patch Analyst extension for ArcView 3.x (Elkie et al. 1999). Marsh patches were defined as contiguous areas of tidal marsh, muted marsh, tidal channels <60 m across, diked baylands, ruderal baylands, managed marsh, and inactive salt ponds (San Francisco Estuary Institute 2000). Landscape composition metrics were calculated for each point-count location

TABLE 2. LOCAL-HABITAT VARIABLES EXAMINED.

Variable	Description
Proportion of cover of dominant native and non-native plant species:	Relative proportion of vegetated area (if >0.01).
Salt grass (*Distichlis spicata*)	Short dense grass found in saline soils of upper marsh.
Gumplant (*Grindelia stricta*)	Leafy, composite woody shrub with many stems; found on channel banks in more saline marshes.
Rushes (*Juncus* spp)	Short rush found in brackish to fresh water areas; most typically Baltic rush (*J. balticus*).
Pepperweed (*Lepidium latifolium*)	Tall perennial non-native herb (>1 m tall) found in brackish to fresh areas, along channel banks and in the upper marsh; forms dense tangled canopy mid-season; falls to near-horizontal when foliage is densest.
Common reed (*Phragmites australis*)	Tall grass up to 2 m high; forms dense stands; found in brackish to fresh areas; may be non-native.
Pickleweed (*Salicornia virginica* syn. *Sarcocornia pacifica*)	Short often dense perennial, found in upper marsh, saline soils; dominant in San Pablo Bay and San Francisco Bay; typically 30–40 cm tall but can grow taller.
All sedge species and alkali bulrush (*Schoenoplectus* spp. and *Bolboschoenus maritimus*)	
Common tule and California bulrush (*Schoenoplectus acutus-S. californicus*)	Tall, rounded perennial sedge (>2 m tall) found in brackish to fresh areas; often on channel banks, often submerged.
Olney's bulrush (*Schoenoplectus americanus*)	Short- to medium-height perennial sedge found in saltier areas than common tule and California bulrush; old stems form dense structure used for nesting.
Alkali bulrush (*Bolboschoenus maritimus*)	Medium height triangular perennial sedge found in saltier areas than Olney's bulrush; old stems form dense structure used for nesting.
Smooth cordgrass (*Spartina alterniflora*; non-native invasive)	Perennial cordgrass forms taller (>1 m), denser stands in lower and higher elevations than native California cordgrass; interbreeds with and outcompetes native; focus of invasion in the estuary is San Francisco Bay.
California cordgrass (*Spartina foliosa*)	Native perennial cordgrass (~1 m) found in narrow band in low marsh and in channels.
All *Spartina* spp.	
Cattails (*Typha* spp.)	Tall (>1 m) perennial in fresh water areas.
Vegetation species richness	Total number of plant species counted within 50 m.
Vegetation species diversity	Shannon diversity index[a].
Ground cover proportion	Estimated ground cover proportion within 50 m of survey point.
Marsh habitat proportion	Estimated proportion of marsh habitat, including internal levees, within 50 m of survey point.
Shrub cover proportion	Shrubs including gumplant and coyote brush (*Baccharis pilularis*).
Vegetation cover proportion	All herbaceous and woody marsh vegetation
Pond/pan cover proportion	Estimated proportion of tidal or non-tidal open water or dry pans within 50 m of survey point.
Channel cover proportion	Estimated proportion of tidal channels or sloughs within 50 m of survey point.
Distance to closest channel (meters)	Distance to closest channel >0.2 m in width.
Width of closest channel (meters)	Width of the closest channel >0.2 m in width.
Channel density; channels <1 m in width	Number of channels of less than 1-m width crossed on two 100-m transects centered on survey point and set at right angles; divided by total length of transects.
Channel density; channels <2 m in width	As above but using channels of <2-m width.
Number of stems at height: <10 cm, 10–20 cm, 20–30 cm, <30 cm, >30 cm	Mean count of stems touching a 6-mm dowel placed at five sample points (predetermined distances but randomly selected directions from center survey point).
Total number of stems	Sum of all stems counted.

[a] (Krebs 1989).

Note: All variables were measured within a 50-m radius circle of survey points. Only variables that were significantly correlated with bird abundance or probability of occurrence ($P < 0.05$ for passerine species; $P < 0.20$ for Black Rail [*Laterallus jamaicensis*]; see text) were considered in the model selection procedure.

TABLE 3. LANDSCAPE METRICS CALCULATED FROM GIS DATA LAYERS.

Landscape metric	Type	Data source[a]
Edge proximity		
Distance to nearest water edge (meters)	Point	EcoAtlas.
Distance to nearest non-marsh edge (meters)	Point	EcoAtlas.
Distance to nearest upland edge (meters)	Point	EcoAtlas.
Distance to nearest urban edge (meters)	Point	EcoAtlas, DWR, USGS.
Habitat configuration[b]		
Marsh patch size (hectares), Log [marsh patch size, hectares]	Patch	EcoAtlas.
Distance to nearest marsh patch (meters)	Patch	EcoAtlas.
Marsh patch perimeter/area ratio (meters/hectare)	Patch	EcoAtlas.
Fractal dimension: [2 × log [patch perimeter (meters)]] / [patch area (meters^2)]	Patch	EcoAtlas.
Landscape composition		
Tidal and muted marsh proportion within circles of radius 500 m/1,000 m/2000 m	Point	EcoAtlas.
Non-tidal wetland proportion within circles of radius 500 m/1,000 m/2,000 m	Point	EcoAtlas, DWR, USGS.
Urbanization proportion within circles of radius 500 m/1,000 m/2,000 m	Point	EcoAtlas, DWR, USGS.
Agriculture proportion within circles of radius 500 m/1,000 m/2,000 m	Point	EcoAtlas, DWR, USGS.
Salt pond proportion within circles of radius 500 m/1,000 m/2,000 m	Point	EcoAtlas, DWR, USGS.
Agriculture proportion within circles of radius 500 m/1,000 m/2,000 m	Point	EcoAtlas, DWR, USGS.

[a] DWR = California Department of Water Resources (1993–1999), USGS = U.S. Geological Survey (1996), EcoAtlas (SFEI 2000).
[b] Marsh patches were defined as contiguous areas of tidal marsh, muted marsh, tidal channels <60 m across, diked baylands, ruderal baylands, managed marsh, and inactive salt ponds (San Francisco Estuary Institute 1998).
Note: Point-level metrics were calculated from the center of the point count station. Patch-level metrics were calculated for the entire marsh patch, which generally included several point-count locations.

by creating circular buffers of different widths (500, 1000, and 2000 m) and using ArcView's Spatial Analyst extension (Environmental Systems Research Institute 1999) to calculate the area of each land use category within each buffer area.

STATISTICAL METHODS

For analysis of Song Sparrow abundance, we used a square-root transformation to improve the normality of regression model residuals. Relationships between this variable and the habitat variables were analyzed using linear models (Neter et al. 1990) in Stata 8 (StataCorp. 2003). For the less abundant Common Yellowthroat, Marsh Wren, and Black Rail, we evaluated presence/absence per survey point with logistic regression analysis (Hosmer and Lemeshow 1989) using the logit command in Stata 8 (StataCorp. 2003).

For each species, we constructed linear or logistic regression models (Neter et al. 1990) to assess the separate and combined effects of local habitat and landscape-level variables, and to develop models with maximum explanatory power and predictive ability. We were not testing specific hypotheses about the determinants of bird abundance or presence, but rather attempting to characterize the suite of variables that were important for each species and identify specific habitat and landscape variables of predictive value.

To select local-habitat variables for analysis (from a potential list of 32 variables, Table 2) we first looked at the Pearson correlation coefficient (r) between the bird abundance or presence variable and each of the habitat variables. For Song Sparrows, Marsh Wrens, and Common Yellowthroats, we selected the variables for which the pairwise Pearson correlation coefficient was statistically significant ($P < 0.05$); this produced a set of 14–19 candidate local habitat variables for each passerine species. For Black Rails, we used a less stringent significance criterion ($P < 0.20$) because of the reduced sample size (less than half the number of survey stations), but corresponding to a comparable strength of association criterion ($|r| \geq 0.1$ for all four species).

For each bird species and each surrounding landscape-composition variable (Table 3), we first chose the most appropriate scale of measurement (500-, 1000-, or 2000-m radius), selecting the scale that resulted in the highest r^2 value in a separate multiple variable regression analysis, thus reducing the total number of landscape variables to 15. We also compared the predictive ability of log-transformed marsh area to marsh area untransformed (while controlling for a bay main effect), and selected the best variable for each species.

We used two variable-selection approaches in order to identify two sets of variables: first, a more concise core set of variables (ideally, 5–10 variables) that were the most predictive with respect to abundance or presence of each species, and second, a more complete set of variables that included the concise set of variables as a subset, but also included variables of lesser importance, which nonetheless could improve the predictive ability to characterize abundance or presence/absence. To compare predictive abilities of the concise models and the inclusive models we used r^2 or, for logistic regression, its analogue, pseudo r^2.

After reducing the number of candidate variables to be considered, as described above, we constructed local-habitat models using stepwise regression or logistic regression analysis (backward elimination, P < 0.05) on our local-habitat variables of interest (Table 2), thus producing a single local-habitat model for each species. Then we repeated this process with landscape variables (Table 3) to generate a landscape model for each species, using the same elimination procedure. We included all landscape variables (for a given spatial scale) in the starting model for each species. In this approach, a habitat or landscape variable was retained only if its retention reduced the deviance of that model by 3.84 units, i.e., reduced AIC by 1.84 units (Lebreton et al. 1992), compared to the comparable model without the specified variable; note that AIC = deviance + 2 × k, where k = number of parameters in the model (Burnham and Anderson 2002).

We compared these models, which we termed concise models, to models obtained through an AIC-minimizing backward stepwise process (Catchpole et al. 2004), which we termed inclusive models. Beginning with the full model, variables resulting in the greatest reduction (and thus improvement) in AIC were removed sequentially until the AIC value could no longer be reduced further and removing the remaining variables would result in an increase in AIC. Thus, the AIC-minimization approach was more inclusive; a variable was retained if so doing reduced the deviance by at least 2.00 units, i.e., reduced AIC by any amount at all. The AIC-minimization criterion is asymptotically equivalent to using a P-value of P < 0.157 to retain a variable when using a likelihood ratio test with 1 df (Lebreton et al. 1992).

To carry out the stepwise AIC-minimization procedure, we used the swaic command for Stata 8.0 (Z. Wang, unpubl. Stata extension). For linear regression models (i.e., for Song Sparrows only), we carried out the stepwise procedure using the comparable P-value to decide whether to retain the specified variable (i.e., P < 0.157).

To evaluate the relative contribution of local-habitat and landscape variables as predictors of bird abundance/presence, we entered the variables from each final inclusive model (local habitat and landscape) into a single backwards elimination stepwise regression analysis to obtain a combined scale model. We used both selection criteria (P < 0.05 and aic minimization) to arrive at competing final multi-scale models. Finally, we examined each of the remaining significant variables for differences in their effects across bays by testing each variable individually for significant (P < 0.05) variable × bay interactions, while adjusting for the remaining variables in the model. Where it was possible to test for interactions across all bays (three bays for the passerine species, two bays for the Black Rail), we report any variable whose effect differed significantly across bay regions.

RESULTS

SONG SPARROW

Song Sparrow relative abundance (hereafter abundance) was significantly higher in San Pablo Bay than in Suisun or San Francisco bays ($F_{2,419}$ = 8.99, P < 0.001; Table 4). Controlling for differences among bays, Song Sparrow abundance was positively associated with the cover of the shrubs gumplant (*Grindelia stricta*) and coyote brush (*Baccharis pilularis*) within a 50-m radius of each survey point (Table 5a; Fig. 2). A negative association was found between the cover of rushes (*Juncus* spp.) and ponds and

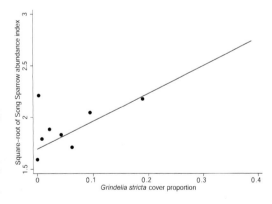

FIGURE 2. Tidal-marsh Song Sparrow (*Melospiza melodia*) abundance versus gumplant (*Grindelia stricta*) cover within 50 m. The linear regression line represents the effect of gumplant cover on Song Sparrow abundance after controlling for all other variables in the final multi-scale model (Table 8a). Points represent the mean abundance for each decile of gumplant cover values (i.e., 10% of the observations contained in each category, with 36% equal to zero). These mean values are shown for illustration purposes only and were not used to calculate regression line.

TABLE 4. MEAN VALUES BY BAY OF DEPENDENT VARIABLES AND SELECTED INDEPENDENT VARIABLES.

Variable	San Francisco Bay	SD (N)	San Pablo Bay	SD (N)	Suisun Bay	SD (N)
Dependent variables						
Song Sparrow (*Melospiza melodia*) abundance index (birds/hectare)	3.39	2.37 (122)	4.15	2.37 (199)	3.01	1.66 (100)
Common Yellowthroat (*Geothlypis trichas*) presence proportion	0.123	0.330 (122)	0.171	0.377 (199)	0.760	0.429 (100)
Marsh Wren (*Cistothorus palustris*) presence proportion	0.451	0.500 (122)	0.482	0.501 (199)	0.940	0.239 (100)
Black Rail (*Laterallus jamaicensis*) presence proportion	–	–	0.438	0.498 (146)	0.381	0.490 (63)
Independent variables						
Salt grass (*Distichlis spicata*) cover	0.027	0.061 (120)	0.056	0.116 (199)	0.075	0.094 (91)
Gumplant (*Grindelia stricta*) cover	0.041	0.065 (120)	0.049	0.063 (199)	0.021	0.034 (91)
Pickleweed (*Salicornia virginica*) cover	0.675	0.212 (120)	0.660	0.266 (199)	0.141	0.228 (91)
All sedge species cover	0.017	0.079 (120)	0.118	0.217 (199)	0.203	0.163 (91)
Alkali bulrush (*Bolboschoenus maritimus*) cover	0.016	0.079 (120)	0.112	0.204 (199)	0.036	0.088 (91)
Vegetation cover	0.810	0.327 (120)	0.888	0.117 (198)	0.883	0.110 (91)
Channel cover	0.140	0.165 (120)	0.057	0.058 (198)	0.055	0.068 (91)
Distance to closest channel (meters)	16.8	17.1 (96)	19.6	32.6 (181)	24.3	29.3 (90)
Number of stems <30 cm	7.40	3.84 (79)	6.86	3.06 (199)	7.90	4.09 (90)
Number of stems >30 cm	0.80	1.760 (84)	1.57	1.47 (199)	5.24	3.44 (90)
Total number of stems	7.40	3.84 (79)	6.86	3.06 (199)	7.90	4.09 (90)
Marsh patch size (hectares)	451	504 (116)	1,159	1214 (214)	2,263	2,966 (94)
Marsh patch perimeter/area ratio (meters/hectare)	110	82 (116)	86.5	103.9 (214)	66	54 (94)
Marsh patch fractal dimension [2 × log [patch perimeter (meter)]]/ [patch area (meters²)]	1.15	0.567 (115)	1.61	0.478 (214)	1.11	0.028 (94)
Tidal and muted marsh proportion within 2,000 m	0.176	0.116 (120)	0.176	0.114 (210)	0.165	0.123 (95)
Non-tidal-marsh proportion within 2,000 m	0.048	0.051 (120)	0.033	0.043 (210)	0.214	0.188 (95)
Agriculture proportion within 2,000 m	0.005	0.017 (120)	0.150	0.161 (210)	0.011	0.029 (95)
Natural upland proportion within 2,000 m	0.009	0.023 (120)	0.130	0.142 (210)	0.167	0.119 (95)
Urbanization proportion within 2,000 m	0.164	0.195 (120)	0.158	0.193 (210)	0.166	0.159 (95)
Salt Pond proportion within 2,000 m	0.284	0.216 (120)	0.038	0.133 (210)	0	0 (95)
Distance to nearest non-marsh edge (meters)	82	115 (120)	148	125.5 (210)	84	80.4 (95)
Distance to nearest water edge (meters)	351	328 (120)	401	494 (210)	405	317 (95)
Distance to nearest urban edge (meters)	998	905 (120)	737	674 (210)	637	515 (95)
Distance to nearest upland edge (meters)	819	938 (120)	401	557 (210)	256	233 (95)

Note: The bird abundance index represents the number of birds detected per hectare of tidal-marsh habitat surveyed (within a 50-m radius). Presence proportion is the proportion of survey stations with one or more detections of a species over all surveys at that station.

TABLE 5. REGRESSION COEFFICIENTS AND MODEL STATISTICS FOR LOCAL HABITAT REGRESSION MODELS.

(a). Song Sparrow (*Melospiza melodia*)

$r^2 = 0.176$[a]	B + SE	P	Partial r^2
Coyote brush (*Baccharis pilularis*) cover	3.46 + 1.05	0.001	0.033
Gumplant (*Grindelia stricta*) cover	2.57 + 0.490	<0.001	0.069
Rushes (*Juncus* spp). cover	-2.49 + 0.653	<0.001	0.041
All sedge species cover		NS	
Olney's bulrush (*Schoenoplectus americanus*) cover		NS	
Alkali bulrush (*Bolboschoenus maritimus*) cover		NS	
Smooth cordgrass (*Spartina alterniflora*) cover	(-)	NS	
Shrub cover		NS	
Vegetation cover		NS	
Pond/pan cover	-0.0145 + 3.87e-2	<0.001	0.044
Distance to closest channel (meters)		NS	
Width of closest channel (meters)		NS	
Channel density <1 m		NS	
Channel density <2 m		NS	
Bay main effect		0.030	0.023
N	401		

[a] AIC minimization model $r^2 = 0.183$.

(b). Common Yellowthroat (*Geothlypis trichas*)

Pseudo $r^2 = 0.427$[a]	B + SE	P	Partial r^2
Coyote brush (*Baccharis pilularis*) cover		NS	
Rushes (*Juncus* spp.) cover	(+)	NS	
Pepperweed (*Lepidium latifolium*) cover	8.84 + 2.79	<0.001	0.028
Common reed (*Phragmites australis*) cover		NS	
Pickleweed (*Salicornia virginica*) cover	(+)	NS	
Common tule and California bulrush (*Schoenoplectus acutus-S. californicus*) cover	(+)	NS	
Olney's bulrush (*Schoenoplectus americanus*) cover	(+)	NS	
Alkali bulrush (*Bolboschoenus maritimus*) cover	3.03 + 0.889	<0.001	0.027
California cordgrass (*Spartina foliosa*) cover		NS	
All *Spartina* spp. cover		NS	
Cattails (*Typha* spp.) cover	(+)	NS	
Vegetation species richness	(-)	NS	
Vegetation species diversity	(+)	NS	
Marsh habitat proportion		NS	
Shrub cover	0.126 + 0.0308	<0.001	0.045
Vegetation cover		NS	
Channel cover	-0.0531 + 0.0267	0.029	0.011
Distance to closest channel (meters)	-0.0222 + 8.88e-3	<0.001	0.049
Width of closest channel (meters)		NS	
Number of stems between 20–30 cm		NS	
Number of stems >30 cm	0.211 + 0.0747	0.004	0.019
Total number of stems		NS	
Bay main effect		<0.001	0.138
N	330		

[a] AIC minimization model pseudo $r^2 = 0.489$, N = 329.

Note: NS = not significant and dropped during backwards stepwise process. Signs in parentheses indicate direction of relationship for additional variables retained in an alternative inclusive final model developed using Akaike information criteria (AIC) minimization. For (b), (c), and (d), P values refer to likelihood ratio tests.

pans within 50 m. Using AIC minimization criteria, Song Sparrow abundance was also negatively associated with smooth cordgrass (*Spartina alterniflora*) cover (Table 5a). The local-habitat model (including bay) explained 17.6% of the variance in Song Sparrow abundance (Table 5a).

Landscape-level characteristics were also significant predictors of Song Sparrow abundance. Song Sparrows responded to land use most strongly at the smallest scale examined, 500 m (Table 6). The final landscape model (including bay) explained 18.8% of Song Sparrow abundance (Table 7a). Abundance was positively associated with log-transformed marsh-patch size (Fig. 3), with the proportion of natural uplands within 500 m, and with the distance of a survey point from the nearest water edge.

TABLE 5. CONTINUED.

(c). MARSH WREN (*Cistothorus palustris*)			
Pseudo r^2 = 0.405[a]	B + SE	P	Partial r^2
Coyote brush (*Baccharis pilularis*) cover	60.6 + 23.8	0.002	0.028
Salt grass (*Distichlis spicata*) cover	-6.36 + 1.83	<0.001	0.030
Gumplant (*Grindelia stricta*) cover		NS	
Rushes (*Juncus* spp.) cover		NS	
Pepperweed (*Lepidium latifolium*) cover		NS	
Common reed (*Phragmites australis*) cover		NS	
Pickleweed (*Salicornia virginica*) cover		NS	
Common tule and California bulrush (*Schoenoplectus acutus-S. californicus*) cover	(+)	NS	
Olney's bulrush (*Schoenoplectus americanus*) cover	(+)	NS	
Alkali bulrush (*Bolboschoenus maritimus*) cover	17.4 + 3.38	<0.001	0.142
Cattails (*Typha* spp.) cover		NS	
Vegetation species richness		NS	
Vegetation species diversity		NS	
Marsh habitat proportion	0.0191 + 8.92e-3	0.027	0.010
Vegetation cover	(+)	NS	
Distance to closest channel (m)	-0.0191 + 6.51e-3	0.002	0.030
Width of closest channel		NS	
Number of stems below 10 cm		NS	
Number of stems between 10–20 cm		NS	
Number of stems <30 cm		NS	
Number of stems above >cm		NS	
Bay main effect		<0.001	0.223
N	361		

[a] AIC minimization model pseudo r^2 = 0.428, N = 361.

(d). CALIFORNIA BLACK RAIL (*Laterallus jamaicensis*)			
Pseudo r^2 = 0.102[a]	B + SE	P	Partial r^2
Salt grass (*Distichlis spicata*) cover	-4.53 + 1.79	0.006	0.032
Gumplant (*Grindelia stricta*) cover	(+)	NS	
Common reed (*Phragmites australis*) cover		NS	
Pickleweed (*Salicornia virginica*) cover	(-)	NS	
Common tule and California bulrush (*Schoenoplectus acutus-S. californicus*) cover	-16.0 + 5.07	<0.001	0.066
Alkali bulrush (*Bolboschoenus maritimus*) cover		NS	
California cordgrass (*Spartina foliosa*) cover	(-)	NS	
All *Spartina* spp. cover		NS	
Vegetation species richness		NS	
Vegetation cover	(+)	NS	
Distance to closest channel (meters)	(-)	NS	
Channel density <1 m	36.9 + 17.4	0.030	0.020
Channel density <2 m		NS	
Bay main effect		0.023	0.023
N	176		

[a] AIC minimization model r^2 = 0.168, N = 168.

Note: NS = not significant and dropped during backwards stepwise process. Signs in parentheses indicate direction of relationship for additional variables retained in an alternative inclusive final model developed using Akaike information criteria (AIC) minimization. For (b), (c), and (d), P values refer to likelihood ratio tests.

Negative associations occurred with the proportion of agriculture as well as tidal and non-tidal marsh within 500 m (Table 7). No additional landscape variables were retained using the AIC minimization procedure.

When local habitat and landscape variables were combined in one model, all vegetation and landscape variables remained highly significant except tidal-marsh proportion (Table 8a). The combined model's explanatory power was 32.2%—close to the summed combined power of the individual local and landscape level models (36.4%). Smooth cordgrass cover was the only additional significant variable retained using the AIC minimization procedure. Controlling for all variables in the final model

TABLE 6. COMPARISON OF RELATIVE SIGNIFICANCE OF THREE DIFFERENT SCALES OF SURROUNDING LAND USE.

Scale	Tidal Marsh Song Sparrow (Melospiza melodia) r^2	Salt Marsh Common Yellowthroat (Geothlypis trichas) pseudo r^2	Marsh Wren (Cistothorus palustris) pseudo r^2	California Black Rail (Laterallus jamaicensis) pseudo r^2
500 m	**0.12**	0.33	0.23	0.07
1,000 m	0.11	0.35	0.24	**0.09**
2,000 m	0.09	**0.39**	**0.26**	0.07

Note: For each species, all surrounding land-use variables of a particular scale were put into a model along with bay. The scale of the model with the highest r^2 or pseudo r^2 (values in **bold**) were used for the multi-variable landscape model.

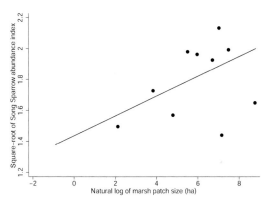

FIGURE 3. Tidal-marsh Song Sparrow (Melospiza melodia) abundance index versus log-transformed marsh patch size (hectare). The linear regression line represents the effect of marsh patch size on Song Sparrow abundance after controlling for all other variables in the final multi-scale model (Table 8a). Points represent the mean abundance for each decile of log-transformed patch size (hectare) values (i.e., 10% of the observations contained in each category). These mean values are shown for illustration purposes only and were not used to calculate regression line.

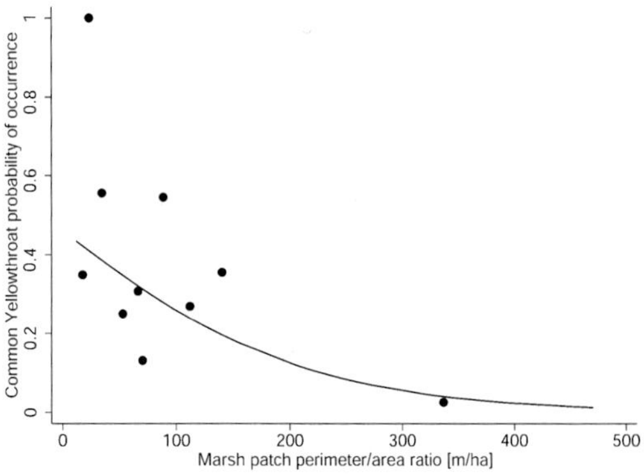

FIGURE 4. Common Yellowthroat (Geothlypis trichas) probability of occurrence versus marsh patch perimeter/area ratio (meters/hectare). The logistic regression line represents the effect of marsh patch perimeter/area ratio on Common Yellowthroat probability of occurrence without controlling for any other variables in the final multi-scale model. Points represent the mean probability of occurrence for each decile of perimeter/area ratio values (i.e., 10% of the observations contained in each category). These mean values are shown for illustration purposes only and were not used to calculate regression line.

left a significant interaction between bay and the proportion of non-tidal marsh (significantly negative in San Francisco and Suisun bays only) and patch size (significant only in San Francisco and San Pablo bays; Table 9).

SALT MARSH COMMON YELLOWTHROAT

The proportion of points with Common Yellowthroats was significantly higher in Suisun Bay than in San Pablo and San Francisco bays; it was lowest in San Francisco Bay ($F_{2,419} = 99.26$, $P < 0.001$; Table 4). For the Common Yellowthroat, local habitat variables (while controlling for variation among bays) predicted 42.7% of the variation in probability of occurrence (Table 5b). Common Yellowthroats were more likely to be found at sites with more stems above 30 cm in height and with higher shrub cover (primarily gumplant and coyote brush), pepperweed (Lepidium latifolium) cover and alkali bulrush (Bolboschoenus maritimus) cover (Table 5b). A negative relationship existed with distance to channel and channel cover proportion within 50 m (Table 5b). Using AIC minimization methods, additional positive relationships occurred with cover of rushes, pickleweed (Salicornia virginica), common tule (Schoenoplectus acutus), Olney's bulrush (Schoenoplectus americanus), cattails (Typha spp.), and vegetation diversity, and a negative relationship with vegetation species richness (Table 5b).

At the landscape level, Common Yellowthroats were most sensitive to variation in surrounding land use at the 2,000-m scale (Table 6). Common Yellowthroats had a higher probability of occurrence in areas with a higher proportion of agriculture within 2,000 m (Table 7b). A negative relationship occurred with patch perimeter/area ratio (Fig. 4). These variables predicted 38.5% of the variation in presence/absence. Using AIC minimization methods, several additional landscape variables were significantly associated

TABLE 7. REGRESSION COEFFICIENTS AND MODEL STATISTICS FOR LANDSCAPE REGRESSION MODELS.

(a). SONG SPARROW (*Melospiza melodia*)			
$r^2 = 0.188$[a]	B + SE	P	Partial r^2
Distance to nearest water edge (meters)	3.04e-4 + 9.12e-5	0.001	0.024
Distance to nearest non-marsh edge (meters)		NS	
Distance to nearest upland edge (meters)		NS	
Distance to nearest urban edge (meters)		NS	
Marsh patch size (hectares)		NS	
Log (marsh patch size, hectares)	0.128 + 0.0236	< 0.001	0.076
Distance to nearest marsh patch (meters)		NS	
Marsh patch perimeter/area ratio (meters/hectares)		NS	
Marsh patch fractal dimension		NS	
Tidal-marsh proportion within 500 m	-0.326 + 0.164	0.048	0.008
Non-tidal-marsh proportion within 500 m	-1.30 + 0.247	<0.001	0.058
Urban proportion within 500 m		NS	
Agriculture proportion within 500 m	-2.06 + 0.439	<0.001	0.046
Natural upland proportion within 500 m	0.755 + 0.322	0.019	0.012
Salt pond proportion within 500 m		NS	
Bay main effect		0.040	0.014
N	392		

[a] AIC minimization model contained no additional variables.

(b). COMMON YELLOWTHROAT (*Geothlypis trichas*)			
Pseudo $r^2 = 0.385$[a]	B + SE	P	Partial r^2
Distance to nearest water edge (meters)	(-)	NS	
Distance to nearest non-marsh edge (meters)		NS	
Distance to nearest upland edge (meters)	(+)	NS	
Distance to nearest urban edge (meters)		NS	
Marsh patch size (hectares)		NS	
Log (marsh patch size, hectares)	(-)	NS	
Distance to nearest marsh patch (meters)		NS	
Marsh patch perimeter/area ratio (meters/hectares)	-6.93e-3 + 0.307e-3	0.015	0.012
Marsh patch fractal dimension		NS	
Tidal-marsh proportion within 2000 m		NS	
Non-tidal-marsh proportion within 2,000 m		NS	
Urbanization proportion within 2,000 m		NS	
Agriculture proportion within 2,000 m	8.13 + 1.64	<0.001	0.073
Natural upland proportion within 2,000 m	(+)	NS	
Salt pond proportion within 2,000 m		NS	
Bay main effect		<0.001	0.333
N	392		

[a] AIC minimization model pseudo $r^2 = 0.401$, N = 392.
Note: NS = not significant. Signs in parentheses indicate direction of relationship for additional variables retained in an alternative inclusive final model developed using Akaike information criteria (AIC) minimization. For (b), (c), and (d), P values refer to likelihood ratio tests.

with Common Yellowthroat presence: distance to nearest upland edge and proportion of natural upland within 2,000 m (positive), distance to the nearest water edge and log marsh size (negative; Table 7b).

Combining the local habitat and landscape variables resulted in a model explaining 51.9% of the variation in probability of occurrence (Table 8). All variables remained significant in this final model, including two that were retained in the AIC minimization procedure for the local or landscape models: common tule and distance to nearest water edge. Additional variables retained here using AIC minimization were cover of rushes, pickleweed, Olney's bulrush, cattails, vegetation species richness, and vegetation species diversity. A significant interaction was found between bay and patch perimeter/area ratio; the association with perimeter/area ratio was negative in San Francisco and San Pablo bays, but positive in Suisun Bay (Table 9). A significant interaction occurred between bay and pepperweed cover, although it was not possible to test the slopes of all three bays, probably due to small sample size in San Francisco Bay (Table 9).

MARSH WREN

Comparing across bays, Marsh Wrens were detected at more points in Suisun Bay, followed by San Pablo and San Francisco bays; this

TABLE 7. CONTINUED.

(c). MARSH WREN (*Cistothorus palustris*)

Pseudo r^2 = 0.319[a]	B + SE	P	Partial r^2
Distance to nearest water edge (meters)	(-)	NS	
Distance to nearest non-marsh edge (meters)	(+)	NS	
Distance to nearest upland edge (meters)	1.61 e-3 + 2.96e-4	<0.001	0.068
Distance to nearest urban edge (meters)	(+)	NS	
Marsh patch size (hectares)		NS	
Log (marsh patch size, hectares)	(-)	NS	
Distance to nearest marsh patch (meters)		NS	
Marsh patch perimeter/area ratio (meters/hectare)	-8.71e-3 + 2.35e-3	<0.001	0.030
Fractal dimension		NS	
Percent tidal-marsh within 2,000 m		NS	
Percent non-tidal-marsh within 2,000 m	(+)	NS	
Percent urban within 2,000 m		NS	
Percent agriculture within 2,000 m	5.79 + 1.12	<0.001	0.057
Percent natural uplands within 2,000 m	(+)	NS	
Percent salt ponds within 2,000 m	-3.20 + 1.19	0.005	0.015
Bay main effect		<0.001	0.202
N	392		

[a] AIC minimization model pseudo r^2 = 0.349, N = 392.

(d). BLACK RAIL (*Laterallus jamaicensis*)

Pseudo r^2 = 0.126[a]	B + SE	P	Partial r^2
Distance to nearest water edge (meters)	-1.50 e-3 + 7.21 e-4	0.033	0.016
Distance to nearest non-marsh edge (meters)	(-)	NS	
Distance to nearest upland edge (meters)		NS	
Distance to nearest urban edge (meters)		NS	
Marsh patch size (hectares)		NS	
Log (marsh patch size, hectares)		NS	
Distance to nearest marsh patch (meters)	0.017 + 6.96 e-3	0.003	0.032
Marsh patch perimeter/area ratio (meters/hectare)		NS	
Marsh patch fractal dimension		NS	
Tidal-marsh proportion within 1,000 m	2.39 + 0.866	0.005	0.029
Non-tidal-marsh proportion within 1,000 m		NS	
Urbanization proportion within 1,000 m		NS	
Agriculture proportion within 1,000 m	10.8 + 3.54	<0.001	0.044
Natural upland proportion within 1,000 m	2.60 + 1.13	0.021	0.019
Salt pond proportion within 1,000 m		NS	
Bay main effect		NS	
N	204		

[a] AIC minimization model pseudo r^2 = 0.134.

Note: NS = not significant. Signs in parentheses indicate direction of relationship for additional variables retained in an alternative inclusive final model developed using Akaike information criteria (AIC) minimization. For (b), (c), and (d), P values refer to likelihood ratio tests.

difference was statistically significant ($F_{2,419}$ = 42.27, P < 0.001; Table 4). The local-habitat model (including bay) explained 40.5% of the variation in Marsh Wren probability of occurrence (42.8% using AIC minimization; Table 5c). Marsh Wren presence was positively associated with the percent marsh habitat within 50 m, as well as with coyote brush and alkali bulrush cover (Fig. 5); and negatively associated with saltgrass (*Distichlis spicata*) cover and distance to closest channel (Table 5c). Using AIC minimization methods, Marsh Wren abundance was also positively associated with total vegetation cover and cover of common tule, California bulrush, and Olney's bulrush (Table 5c).

Land-use composition variables within a 2,000-m radius were the best explanation of variation in Marsh Wren probability of occurrence (Table 6). Landscape variables (including bay) explained 31.9% of the variation in Marsh Wren presence (34.9% using AIC minimization; Table 7c). Probability of occurrence was negatively associated with the proportion of salt ponds within 2,000 m and with marsh perimeter/area ratio; and positively associated with the proportion of agriculture in the surrounding landscape and with the distance to the nearest upland edge (Table 7c). Using AIC minimization methods, Marsh Wren probability of occurrence was also positively associated

TABLE 8. REGRESSION COEFFICIENTS AND MODEL STATISTICS FOR FINAL COMBINED LOCAL AND LANDSCAPE REGRESSION MODELS.

(a). SONG SPARROW (*Melospiza melodia*)
Model statistics: $r^2 = 0.322$, $F_{(11,371)} = 16.03$, $P < 0.001$[a]

Independent variables	B + SE	P	Partial r^2
Coyote brush (*Baccharis pilularis*) cover	2.98 + 0.985	0.003	0.019
Gumplant (*Grindelia stricta*) cover	2.73 + 0.468	<0.001	0.066
Rushes (*Juncus* spp.) cover	-2.58 + 0.611	<0.001	0.035
Smooth cordgrass (*Spartina alterniflora*) cover	(-)		
Pond/pan cover	-0.0151 + 3.53e-3	<0.001	0.040
Distance to nearest water edge (meters)	2.64e-4 + 8.44e-5	0.002	0.020
Log (marsh patch size, hectares)	0.102 + 0.0171	<0.001	0.080
Non-tidal-marsh proportion within 500 m	-0.963 + 0.176	<0.001	0.058
Agriculture proportion within 500 m	-1.96 + 0.396	<0.001	0.048
Natural upland proportion within 500 m	0.717 + 0.315	0.023	0.012
Bay main effect		0.036	0.013

[a] AIC minimization model $r^2 = 0.328$.

(b). COMMON YELLOWTHROAT (*Geothlypis trichas*)
Model statistics: pseudo $r^2 = 0.519$, likelihood ratio $\chi^2 = 229.9$, $P < 0.001$, $N = 342$[a]

Independent variables	B + SE	P	Partial r^2
Rushes (*Juncus* spp.) cover	(+)		
Pepperweed (*Lepidium latifolium*) cover	16.1 + 4.69	<0.001	0.039
Pickleweed (*Salicornia virginica*) cover	(+)		
Common tule and California bulrush (*Schoenoplectus acutus-S. californicus*) cover	8.32 + 3.89	<0.001	0.026
Olney's bulrush (*Schoenoplectus americanus*) cover	(+)		
Alkali bulrush (*Bolboschoenus maritimus*) cover	4.44 + 1.19	<0.001	0.033
Cattail (*Typha* spp.) cover	(+)		
Vegetation species richness	(-)		
Vegetation species diversity	(+)		
Shrub cover	0.0935 + 0.0320	0.003	0.020
Distance to closest channel (meters)	-0.0192 + 9.00e-3	0.016	0.013
Distance to water (meters)	-1.81e-3 + 5.44e-4	<0.001	0.029
Marsh patch perimeter/area ratio (meters /hectare)	-7.85e-3 + 3.58e-3	0.019	0.012
Agriculture proportion within 2,000 m	10.4 + 2.42	<0.001	0.063
Bay main effect		<0.001	0.191

[a] AIC minimization model pseudo $r^2 = 0.548$, $N = 341$.
Note: Signs in parentheses indicate direction of relationship for additional variables retained in an alternative inclusive final model developed using AIC minimization. For (b), (c), and (d), P values refer to likelihood ratio tests.

with distance to nearest non-marsh edge and nearest urban edge, and the proportion of non-tidal marsh and natural uplands within 2,000 m; and negatively associated with distance to water, marsh-patch size, and distance to nearest marsh (Table 7c).

Combining local-habitat and landscape variables, the resulting model explained 50.2% of the variation in Marsh Wren probability of occurrence (Table 8). The proportion of marsh within the 50-m point-count radius and the proportion of agriculture in the surrounding 2,000 m were not significant in this final combined model. Using AIC minimization methods, cover of common tule, California bulrush, and Olney's bulrush, distance to water edge, and proportion salt ponds and non-tidal marsh in the surrounding 2,000 m were significant. In the final model a significant interaction was found between bay and coyote brush cover—the relationship was positive in San Pablo Bay and not significant in the other two bays (Table 9).

BLACK RAIL

Black Rails were detected at more points in San Pablo Bay than in Suisun Bay, although this difference was not statistically significant ($F_{1,180} = 0.89$; $P = 0.35$; Table 4). At the local scale, Black Rail presence was negatively associated with common tule, California bulrush, and saltgrass cover; and positively with the number of tidal channels <1-m wide (Table 5d). The local model (including bay) accounted for only 10.2% of the variance in probability of occurrence. Using AIC minimization methods, additional significant variables included a positive relationship with gumplant and total vegetation cover, and a negative relationship with pickleweed and

TABLE 8. CONTINUED.

(c). MARSH WREN (*Cistothorus palustris*)
Model statistics: pseudo r^2 = 0.502, likelihood ratio χ^2 = 230.72, P < 0.001, N = 343[a]

Independent Variables	B + SE	P	Partial r^2
Coyote Brush (*Baccharis pilularis*) cover	49.5 + 21.6	0.002	0.020
Saltgrass (*Distichlis spicata*) cover	-5.28 + 2.16	0.009	0.015
Common tule and California bulrush (*Schoenoplectus acutus- S. californicus*) cover	(+)		
Olney's bulrush (*Schoenoplectus americanus*) cover	(+)		
Alkali bulrush (*Bolboschoenus maritimus*) cover	25.4 + 4.87	<0.001	0.147
Distance to nearest channel (meters)	-0.0225 + 7.27e-3	<0.001	0.026
Distance to nearest water edge (meters)	(-)		
Distance to nearest upland edge (meters)	6.23e-4 + 2.63e-4	0.014	0.013
Distance to nearest non-marsh edge (meters)	3.70e-3 + 1.81e-3	0.034	0.010
Marsh patch perimeter/area ratio (meters/hectare)	-0.0121 + 3.18e-3	<0.001	0.040
Non-tidal-marsh proportion within 2,000 m	(+)		
Salt pond proportion within 2,000 m	(-)		
Bay main effect		<0.001	0.230

[a] AIC minimization model pseudo r^2 = 0.542, N = 343.

(d). BLACK RAIL (*Laterallus jamaicensis*)
Model statistics: pseudo r^2 = 0.180, likelihood ratio χ^2 = 44.34, P < 0.001, N = 180[a]

Independent variables	B + SE	P	Partial r^2
Total vegetation cover	0.0379 + 0.0188	0.034	0.018
Distance to nearest channel (meters)	(-)		
Distance to nearest marsh patch (kilometers)	-0.0176 + 6.96e-3	0.002	0.038
Tidal-marsh proportion within 1,000 m	2.63 + 0.956	0.005	0.032
Agriculture proportion within 1,000 m	11.7 + 3.66	<0.001	0.055
Natural upland proportion within 1,000 m	5.08 + 1.48	<0.001	0.055
Bay main effect		0.579	0.001

[a] AIC minimization model pseudo r^2 = 0.234, N = 163.

Note: Signs in parentheses indicate direction of relationship for additional variables retained in an alternative inclusive final model developed using AIC minimization. For (b), (c), and (d), P values refer to likelihood ratio tests.

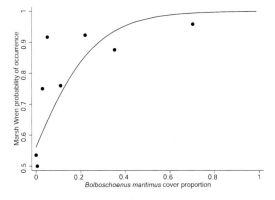

FIGURE 5. Marsh Wren (*Cistothorus palustris*) probability of occurrence versus alkali bulrush (*Bolboschoenus maritimus*) cover within 50 m. The logistic regression line represents the effect of alkali bulrush cover on Marsh Wren probability of occurrence without controlling for all other variables in the final multi-scale model. Points represent the mean probability of occurrence for each fifth percentile of alkali bulrush cover values (i.e., 5% of the observations contained in each category, with 64% equal to zero). These mean values are shown for illustration purposes only and were not used to calculate regression line.

California cordgrass (*Spartina foliosa*) and with distance to closest tidal channel; this model accounted for 16.8% of variance in probability of occurrence (Table 5d).

Land-use composition variables within a 1,000-m radius were the best explanation of variation in Black Rail probability of occurrence (Table 6). At the landscape level, Black Rail presence was positively associated with the proportion of tidal-marsh, agriculture and natural uplands within a 1,000-m radius; and negatively with the distance to the nearest marsh patch and distance to nearest water edge. The landscape model (including bay) predicted 12.6% of the variance in probability of occurrence among points (Table 7d). Using AIC minimization, distance to nearest non-marsh edge was also significantly negatively associated with Black Rail presence, and partial r^2 was slightly higher at 13.4% (Table 7d)

When considering local and landscape variables together, the only local-habitat variable that remained significant was total vegetation cover (which had a positive relationship to Black Rail presence; Table 8d). The landscape variables that remained significant were: distance to nearest

TABLE 9. VARIABLES WHOSE EFFECTS VARIED AMONG BAYS.

Species	Interaction terms Overall interaction P value	Bay-specific slopes		
		San Francisco	San Pablo	Suisun
Tidal-marsh Song Sparrow (*Melospiza melodia*)	Bay × (Non-tidal-marsh proportion within 500 m) P = 0.006	β = -1.46 +0.390 P < 0.001	β = 0.196 +0.412 P = 0.634	β = -1.08 +0.200 P < 0.001
	Bay × (patch size) P = 0.002	β = 0.184 +0.031 P < 0.001	β = 0.0866 +0.0228 P < 0.001	β = 0.0423 +0.0308 P = 0.170
Marsh Wren (*Cistothorus palustris*)	Bay × (coyote brush [*Baccharis pilularis*] cover) P < 0.001	NS P > 0.9	β = 61.8 +25.5 P = 0.015	NS P > 0.9
Salt Marsh Common Yellowthroat (*Geothlypis trichas*)	Bay × (perimeter / area ratio) P = 0.015	β = -0.0142 +0.0071 P = 0.038	β = -0.0351 +0.0145 P = 0.016	β = 0.0221 +0.0108 P = 0.041
	Bay × (pepperweed cover)[a] P = 0.011	[b]		
California Black Rail (*Laterallus jamaicensis*)	Bay × (vegetation cover) P = 0.013		NS P > 0.9	β = 0.139 +0.0641 P = 0.030

[a] We were not able to estimate the slopes separately for each bay.
[b] Black Rails were not surveyed in San Francisco Bay.
Note: We examined for interaction all variables in the final combined models shown in Table 8.

marsh patch (negative); and the three variables related to surrounding land use in the surrounding 1,000 m: tidal marsh, agriculture, and natural uplands, all of which were positively related to rail presence (Table 8d). The variables in the final combined model accounted for 18.0% of the variance in probability of occurrence among points. One additional variable, distance to closest channel, was retained using AIC minimization methods. A significant interaction occurred between bay and vegetation cover—the relationship between Black Rail presence and vegetation cover proportion was positive in Suisun Bay and not significant in San Pablo Bay (Table 9).

DISCUSSION

Our results demonstrated that each of the four tidal-marsh species we examined responded to both local-habitat features and to broader-scale characteristics of the habitat patch and the surrounding landscape. For each species, we were able to develop separate predictive models that accounted for substantial variation in the distribution or abundance of that species based solely on local habitat features or solely on patch and landscape characteristics. And in each case, the final combined model included both local-habitat and landscape variables. This suggests that the distribution and abundance of tidal-marsh birds is influenced by a range of ecological processes, operating at both small (local) and large (landscape) spatial scales. Clearly, these species are responding to local-habitat (mainly vegetation) characteristics, but vegetation alone may not indicate quality habitat.

Interestingly, landscape models were comparable in terms of predictive ability to local-habitat models in accounting for the local distribution of these species. This suggests that information based on remote sensing data (i.e., aerial photos or satellite imagery) can be used to develop useful broad-scale guidelines for conservation and management of the tidal-marsh bird community. Nevertheless, better predictive models can be developed by incorporating multi-scale data. These results are consistent with other studies in shrub (Bolger et al. 1997) and wetland (Naugle et al. 2001) habitats that have demonstrated the importance of multi-scale habitat-landscape models for predicting variation in bird distribution and abundance. In contrast, others have found that landscape models are either better than local-habitat models (Saab 1999) or worse than local-habitat models (Scott et al. 2003).

This study also demonstrated important differences among species, including the degree to which variation in abundance or probability of occurrence can be explained by the suite of local habitat and landscape variables that we examined. For Marsh Wrens and Common Yellowthroats, the explanatory power of our models was relatively high; for Black Rails, the explanatory power was fairly low; and for Song Sparrows, our results were intermediate. Marsh Wren and Common Yellowthroat presence were better predicted by local-habitat characteristics while for Song Sparrows and Black Rails landscape-level characteristics were better. Several potential explanations may account for these differences, including variations in species detectability, and the degree of habitat specialization. Black Rail presence has previously been shown to be related to variation in marsh area (Evens and Nur 2001), but here we have evidence of a weak relationship between Black Rail presence at individual points and a large suite of variables at several scales. It is likely that Black Rails may respond more strongly to vegetation or habitat characteristics that we did not quantify, or that their presence is primarily controlled by other ecosystem processes such as predation.

In contrast, the moderately low predictive ability of models for the Song Sparrow—the most abundant tidal-marsh bird species in the estuary being present at 97% of the points surveyed—may reflect the relative generalist nature of this species which is found in a wide range of wet and/or scrubby habitats across North America (Nice 1937, Marshall 1948a, Aldrich 1984, Hochachka et al. 1989, Arcese et al. 2002). The tidal-marsh subspecies, in particular, have the highest reported densities for the species (Johnston 1956b) and are well distributed throughout different parts of tidal marshes, including levees and other upland edges (Cogswell 2000). The San Francisco Bay subspecies (*M. m. pusillula*), has even been found to nest—with low success—in areas invaded by non-native smooth cordgrass (Guntenspergen and Nordby, *this volume*).

Common Yellowthroats and Marsh Wrens responded most strongly to vegetation characteristics, having somewhat more specialized habitat preferences within the marsh primarily related to vegetation structure and height for nesting (Foster 1977a, b; Leonard and Picman 1987, Rosenberg et al. 1991, Marshall and Dedrick 1994).

INTER-BAY DIFFERENCES

The highest relative abundance of Song Sparrow and presence of Black Rail was in San Pablo Bay, while the Marsh Wren and Salt Marsh Common Yellowthroat were present at more points in Suisun Bay. For each of the species, inter-bay differences in presence and abundance are likely due primarily to differences in local-habitat characteristics (determined ultimately by salinity, elevation, tidal influence, local seed sources, and disturbance regime), and surrounding land use. In general, San Pablo Bay and San Francisco Bay are higher in salinity than Suisun Bay, and consequently, the vegetation communities are different with more pickleweed-dominated saltmarsh in the former two bays. In the Suisun Bay and in the upper reaches of rivers draining into San Pablo Bay, more tall plant species adapted to brackish or fresh conditions occur; these areas also have a higher plant-species diversity (Josselyn 1983). Many of the brackish plant species more commonly found in Suisun Bay (e.g., bulrush and cattail species) are taller than the high-salinity species (e.g., pickleweed and saltgrass), provide more structure at greater heights, and provide preferred nesting habitat and cover for the Marsh Wren and Common Yellowthroat; whereas the more saline marshes of San Pablo Bay are apparently preferred by the Song Sparrow and Black Rail, both which nest regularly in pickleweed. Black Rails do not regularly nest in San Francisco Bay, likely due to the scarcity of high-marsh habitat (Trulio and Evens 2000).

Controlling for local-habitat and landscape conditions, however, Song Sparrows and Black Rails still had significantly different probabilities of occurrence across bays, suggesting that vegetation influences the abundance of these species but does not completely determine their regional distribution patterns. For both of these species landscape level characteristics were the strongest predictors of abundance or presence.

LOCAL HABITAT ASSOCIATIONS

Our local-habitat models provided more specific information on the regional-habitat associations of each species than has been previously reported. Even after controlling for bay and landscape setting, the tidal-marsh bird species examined in this study appeared to respond to species-specific vegetation composition as well as to general vegetation structure and habitat features. One implication of this result is that monitoring and research studies should collect both types of data at the local scale. A long-standing tradition in avian ecology is the obtaining of information on general vegetation structure, but researchers do not always collect information on species-specific vegetation composition, which can be just as important as structure to particular bird species (Wiens and Rotenberry 1981).

The results of our local-habitat models highlight the different habitat associations of each tidal-marsh species evaluated, and accordingly, different management needs. While the Song Sparrow and Marsh Wren both exhibited positive associations with coyote brush, an upland shrub often found in higher elevations on levees and at marsh edges, their similarities ended there. Song Sparrows were more abundant in areas with higher relative cover of the halophytic wetland shrub, gumplant (Fig. 3), which is known to be one of their preferred nesting substrates (Johnston 1956a, b; Nur et al. 1997; PRBO, unpubl. data), and less abundant in rushes which are short, brackish-marsh plants not typically used for nesting, as well as in smooth cordgrass, a non-native cordgrass that is sometimes used for nesting except where it occurs in monotypic stands (Guntenspergen and Nordby, *this volume*). Neither do they typically nest in coyote brush which is found primarily on levees and upland edges, but their positive association with this species was probably due to its value for song perches and cover. Collins and Resh (1985) also found a positive relationship between Song Sparrow density and coyote brush in Petaluma Marsh, an old high-elevation marsh, where coyote brush is common along the high banks of tidal channels. Although the relationships were not significant when controlling for variability in other local-habitat variables (primarily vegetation), Song Sparrows appeared to be most abundant near channels (where vegetation tends to be thickest and highest, especially in saline marshes) and in areas with more medium-width to narrow channels. Others have already demonstrated that Song Sparrow territories tend to be established along channels, sloughs, and mosquito ditches (Johnston 1956a, b; Collins and Resh 1985). Our results suggest, however, that this channel affinity is likely due to the higher availability of shrubs or other dense vegetation along channels, which is preferred for nesting.

Common Yellowthroats were also more likely to be present at points nearer to tidal channels with a greater cover of tall plants, including alkali bulrush, Olney's bulrush, rush and cattail species, and high overall shrub cover. Common Yellowthroats were also strongly associated with the non-native invasive pepperweed, a tall and dense plant found in brackish marshes, particularly in higher elevation areas and along channels. Pepperweed appears to be expanding throughout the region and is difficult to control. While its expansion may be positive for Common Yellowthroat distribution, more information on other effects of pepperweed on Common Yellowthroats and other bird species is needed. For example, the impact of pepperweed on the food web (particularly on invertebrate populations), and its relative utility as cover for nesting and refuge from predators are unknown.

Marsh Wrens also appeared to be highly associated with channels and with several sedge species, tall plants that are often used for nesting—alkali bulrush (*Bolboschoenus maritimus*; saline, primarily in San Pablo Bay; Fig. 5), common tule and California bulrush (*Schoenoplectus acutus* and *S. maritimus*; fresh-brackish, primarily in Suisun Bay), and Olney's bulrush (*S. americanus*; brackish, primarily in Suisun Bay). The Marsh Wren demonstrated a negative relationship with saltgrass, a short grass found in saltier high-marsh areas that are not likely to be used by the species for nesting or cover. These results are not surprising in that Marsh Wren nests are usually found at approximately 1 m above the ground in tidal marshes (PRBO, unpubl. data); thus they require tall vegetation for nesting.

Black Rails were not positively associated with any particular plant species (other than a weak relationship with gumplant only in the inclusive model), but they did exhibit negative associations with saltgrass, common tule and California bulrush. Saltgrass is used occasionally as a nesting substrate, particularly when mixed with pickleweed and/or alkali bulrush, but areas with large contiguous areas of saltgrass do not apparently make preferred Black Rail habitat. Common tule and California bulrush, unlike the other sedge species in local marshes, are found along and within channels; these species grow most commonly in low-elevation areas subject to regular tidal flooding and generally have little or no vegetation cover beneath them in which Black Rails can nest. The stems are thick, smooth, and rigid and are commonly used as nest substrate only by Marsh Wrens and Red-winged Blackbirds (*Agelaius phoeniceus*). Black Rails were also most likely to be present closer to channels, and in areas with more channels <1 m in width, which are likely to be third- and fourth-order channels found in upper-marsh areas. However, when controlling for landscape variables, the only local-habitat variables that remained significantly associated with Black Rail presence were overall vegetation cover (significant in San Pablo Bay only) and distance to channel; no individual plant species cover variables were significant.

LANDSCAPE ASSOCIATIONS

Significant landscape-level predictors of abundance also varied among species, although some relationships were common across several species. With respect to edge-proximity relationships, we observed some differentiation among species. Song Sparrows had higher abundances away from the water edge (usually open bay), while Marsh Wrens and Common Yellowthroats were more likely to be present away from the upland edge and closer to the water edge, even while controlling for vegetation variables. This may be due in part to the demonstrated vegetation preference of these species, with Song Sparrows preferring to nest in high marsh, in shrubs along marsh edges and channels, and Marsh Wrens and Common Yellowthroats preferring sedge species which are more tolerant of conditions along the bay edge, particularly in San Pablo Bay. However, the relationships between upland water-edge proximity and abundance/presence were similar to the relationships of these variables with various measures of Song Sparrow nest survivorship (PRBO, unpubl. data), suggesting that edge aversion may be related to species-specific predation pressures. For the Black Rail, there was an affinity for marsh edge and water edge, but these variables were not significant when controlling for vegetation cover.

With respect to patch configuration (size and shape), all species except the Black Rail exhibited a strong association with either log-transformed patch size (positive) or patch perimeter/area ratio (negative). The lack of a Black Rail response may be due to the fact that marsh patches as we defined them included non-tidal wetlands that may not be used by this species. This species was not detected in marshes of <8 ha, suggesting that there may at least be a threshold size below which Black Rails do not occur; but our sample size was too small to detect a significant difference at that level (only four marshes smaller than 8 ha were surveyed).

For all three passerine species, the relationship of abundance or presence with size and perimeter/area ratio were correlated (i.e., opposite relationships of similar magnitude), indicating that the negative associations with perimeter/area ratio may have been driven more by patch size than by patch shape. This was also borne out by the lack of importance of the fractal-dimension index, a scale-independent measure of fragmentation or patch shape (McGarigal and Marks 1995). A weak response to landscape pattern (i.e., patch shape), above and beyond landscape composition (marsh size) is consistent with the findings of other recent studies (Fahrig 1997, Harrison and Bruna 1999). Some researchers believe that landscape pattern becomes important only in landscapes with low proportions of suitable habitat (Andrén 1994) or for species with certain life-history traits (Hansen and Urban 1992).

Nonetheless, the relative importance of patch size and shape differed by species. For the Song Sparrow, only the effect of log-transformed patch size was significant (Fig. 2), and only in Suisun and San Francisco bays. This was the most significant of all variables examined for the Song Sparrow. While we found few marshes that did not contain Song Sparrows, their relative per-point abundance (a measure of relative density, rather than the total number of individuals within a given patch) was higher in large versus small patches, suggesting lower habitat quality in smaller patches, or reduced survivorship due to predation or other factors (Takekawa et al., chapter 11, *this volume*). If survival or reproductive success is reduced in small patches, and recolonization rare, then extirpation could occur over time. The relationship of Song Sparrow density to area was somewhat non-linear—the largest patches did not have the highest densities of Song Sparrows.

For the Common Yellowthroat and Marsh Wren, which did not occur in patches of <8 ha, probability of occurrence (at the survey-point level) increased with patch size but perimeter/area ratio was a stronger predictor of occurrence, suggesting that marsh fragmentation (resulting in a higher perimeter/area ratio) may have some detrimental effects on these species, perhaps by increasing their exposure to edge-associated predators or other negative upland-associated factors. Alternatively, vegetation composition and structure may differ between marsh edges and marsh interiors, due to differences in elevation and hydrology, which may in turn affect these species' distributions. For the Common Yellowthroat, the negative relationship with perimeter/area ratio was evident in San Francisco and San Pablo bays while in Suisun Bay the relationship was actually positive, indicating a probable difference in edge quality among bays.

We also observed an effect of marsh isolation (i.e., reduced probability of occurrence with increase in distance to nearest marsh patch) for all but the Song Sparrow. This is potentially due to the high affinity of these three species for wetland areas, especially the Black Rail, which, in the San Francisco Bay region, is found exclusively in tidal-marsh habitats (Evens et al. 1991). The Common Yellowthroat and Song Sparrow are more likely to use adjacent upland habitats such as ruderal scrub (Song Sparrow) and riparian woodland (Common Yellowthroat and Song Sparrow) during the non-breeding season and therefore are likely to have different barriers to movement and dispersal than the other species (Cogswell 2000, Terrill 2000); the barriers for Black Rails are probably more extensive than for Song Sparrows.

While all four species responded to surrounding land use, their strongest responses were at different spatial scales ranging from 500–2,000 m. The tidal-marsh Song Sparrow, a year-round tidal-marsh resident with a small territory size (Marshall 1948; Johnston 1956a, b), was most strongly influenced by more immediate landscape conditions (i.e., within 500 m, rather than 1,000 or 2,000 m), and the Black Rail, a secretive species also expected to be fairly sedentary in its habits, responded most strongly to conditions within a 1,000-m radius. The Marsh Wren and Common Yellowthroat were most responsive to land-use characteristics within a 2,000-m radius. The wider-scale sensitivity of the latter two species may be related to the fact that they are less philopatric and are quite mobile during the non-breeding season, with the Common Yellowthroat apparently moving to wetlands outside the San Francisco Bay during the winter (Grinnell and Miller 1944).

Only the Black Rail exhibited a positive relationship with the proportion of tidal-marsh habitat in the surrounding landscape; it is apparently the most tidal-marsh dependent of the four species. However, when controlling for other variables, the proportion of natural upland and agriculture were more important to Black Rails than overall marsh cover (see below). Landscape variables other than tidal marsh were also more important for the other species we examined. Song Sparrows actually exhibited a negative relationship with the proportion of marsh within 500 m.

Song Sparrows were positively associated with natural uplands and negatively associated with tidal and non-tidal marsh and agriculture in the surrounding landscape, reflecting the upland edge and shrub affiliation of this species. Marsh Wrens were negatively associated with the proportion of salt ponds in the surrounding 2,000 m and positively associated with agriculture and natural uplands in the surrounding landscape, also reflecting the use of uplands by this species.

Common Yellowthroats, Marsh Wrens, and Black Rails all exhibited positive associations with the proportion of agricultural land use in the surrounding area, controlling for other variables. Because none of these species actually occur in agricultural fields or pastures, and given that this effect was primarily driven by San Pablo Bay, where agricultural land use is most prevalent, it may actually represent the absence of urban development, or the potential co-occurrence of agricultural lands with less saline marsh conditions away from the bay edge (for Common Yellowthroats and Marsh Wrens). Alternatively, agricultural lands may actually contain suitable habitats such as riparian woodland (Common Yellowthroats) or freshwater wetland (used by both species). For the Black Rail, which was also positively associated with the proportion of surrounding natural uplands, agricultural lands (and natural uplands) may provide refugia from predation during high tides (when birds are forced out of the marsh onto higher elevations), known to be a period of significant mortality for this species (Evens and Page 1986).

CONSERVATION AND MANAGEMENT IMPLICATIONS

The specific local and landscape-level habitat associations quantified herein can provide land managers with the specific information needed to manage for or restore key habitat elements for specific bird species, e.g., gumplant for Song Sparrows, sedge species for Common Yellowthroats and Marsh Wrens, and numerous small channels for Black Rails and Common Yellowthroats.

The range of responses among species to local- and landscape-level habitat factors highlights the importance of preserving a heterogeneous mosaic of tidal-marsh habitat throughout the San Francisco Bay estuary, representing the entire salinity gradient and the resulting diversity of estuarine habitats. In addition, habitat diversity within a site, representing the full elevational and tidal inundation spectrum of a natural marsh, is equally important for providing the habitat elements needed by the full range of tidal-marsh-dependent species. Thus we suggest that large areas of contiguous tidal marsh and adjacent natural uplands be protected and restored, in order to preserve biological and physical heterogeneity at the ecosystem level.

Our results also suggest that landscape context is important for tidal-marsh birds. In particular, marshes surrounded by natural or agricultural uplands appear to be more valuable than those surrounded by urbanization. This finding should be considered in the evaluation of bayland sites for potential tidal-marsh restoration, as a potential predictor of restoration success.

ACKNOWLEDGMENTS

We would like to thank the many field assistants and biologists who collected data for this study, especially J. Wood, E. Brusati, J. Hammond, J. Caudill, S. Cashen, A. Ackerman, S. Webb, E. Strauss, G. Downard, L. Hug, R. Leong, W. Neville, T. Eggert, C. Millett, and S. Macias. G. Geupel, T. Gardali, S. Zack, and J. Evens contributed greatly to the initiation of the study. V. Toniolo helped calculate GIS-based spatial metrics. We would also like to thank J. E. Maldonado and two anonymous reviewers for suggested improvements to the manuscript and to J. Evens for insights related to Black Rails. This study was made possible by grants from the Mary Crocker Trust, the USDI Fish and Wildlife Service Coastal Program, the Bernard Osher Foundation, Calfed Bay/Delta Program, and the Gabilan Foundation. We thank the following agencies and individuals for permission to access their property and for facilitating our field studies: USDI Fish and Wildlife Service (Don Edwards San Francisco Bay National Wildlife Refuge, with special thanks to J. Albertson, C. Morris, and M. Kolar; and San Pablo Bay National Wildlife Refuge, with special thanks to G. Downard, L. Vicencio and B. Winton), California Department of Fish and Game, California Department of Parks and Recreation, East Bay Regional Park District, Hayward Regional Shoreline, the City of Vallejo, Sonoma Land Trust, Solano County Farmlands and Open Space, Peninsula Open Space Trust, the City of Palo Alto, Wickland Oil Martinez, and Marin County Parks. This work was conducted under a memorandum of understanding with California Department of Fish and Game. This is PRBO contribution number 1090.

THE CLAPPER RAIL AS AN INDICATOR SPECIES OF ESTUARINE-MARSH HEALTH

James M. Novak, Karen F. Gaines, James C. Cumbee, Jr., Gary L. Mills, Alejandro Rodriguez-Navarro, and Christopher S. Romanek

Abstract. Clapper Rails (*Rallus longirostris*) can potentially serve as an indicator species of estuarine-marsh health because of their strong site fidelity and predictable diet consisting predominantly of benthic organisms. These feeding habits increase the likelihood of individuals accumulating significant amounts of contaminants associated with coastal sediments. Moreover, since Clapper Rails are threatened in most of their western range, additional study of the effects of potential toxins on these birds is essential to conservation programs for this species. Here we present techniques (DNA strand breakage, eggshell structure, and human-consumption risk) that can be used to quantify detrimental effects to Clapper Rails exposed to multiple contaminants in disturbed ecosystems as well as humans who may eat them. Adult birds collected near a site contaminated with polychlorinated biphenyls (PCBs) and metals in Brunswick, Georgia had a high degree of strand breakage, while those collected from a nearby reference area had no strand breakage. Although, results showed that eggshell integrity was compromised in eggs from the contaminated sites, these results were more diffuse, re-emphasizing that multiple endpoints should be used in ecological assessments. This study also shows that techniques such as eggshell integrity on hatched eggs and DNA strand breakage in adults can be used as non-lethal mechanisms to monitor the population health of more threatened populations such as those in the western US. We also present results from human-based risk assessment for PCBs as a third toxicological endpoint, since these species are hunted and consumed by the public in the southeastern US. Using standard human-risk thresholds, we show a potential risk to hunters who consume Clapper Rails shot near the contaminated site from PCBs because of the additional lifetime cancer risk associated with that consumption.

Key Words: Clapper Rail, DNA strand breakage, eggshell integrity, indicator species, metals, polychlorinated biphenyl, *Rallus longirostris*.

EL RASCÓN PICUDO COMO ESPECIE INDICADORA DE LA SALUD DE MARISMAS ESTUARINOS.

Resumen. Los Rascones Picudos (*Rallus longirostris*) pueden servir potencialmente como una especie indicadora de la salud de marismas estuarinos, gracias a su fuerte fidelidad al sitio y a su predecible dieta que consiste predominantemente en organismos bentónicos. Estos hábitos alimenticios incrementan la posibilidad de individuos que acumulan cantidades significativas de contaminantes asociados con sedimentos costeros. Además, ya que los Rascones Picudos se encuentran en peligro en casi todo su rango oeste, estudios adicionales de los efectos de toxinas potenciales en estas aves es esencial para los programas de conservación para estas especies. Aquí presentamos técnicas (rompimiento de ADN, estructura de cáscara de huevo, y riesgo de consumo humano) que pueden ser utilizadas para cuantificar efectos detrimentales para los Rascones Picudos, expuestas a múltiples contaminantes en ecosistemas en disturbio, como también en humanos que los consumen. Las aves adultas colectadas cerca de un sitio contaminado con bifenil policlorinatado (BPC) y metales en Brunswick, Georgia tienen un alto grado de rompimiento de ADN, mientras que aquellos colectados de un área de referencia cercana no tenían rompimiento de ADN. A pesar de que los resultados muestran que la integridad de la cáscara de huevo estuvo comprometida en huevos del agua contaminada, estos resultados fueron más difusos, re-enfatizando que múltiples puntos finales deberían ser utilizados en valoraciones ecológicas. Este estudio también muestra que técnicas tales como integridad de cáscara de huevo en huevos eclosionados y rompimiento de ADN en adultos pueden ser utilizados como mecanismos no letales, para el monitoreo de la salud de la población de más poblaciones en peligro, tales como aquellas en el oeste de EU. También presentamos resultados de valoración del riesgo basado en el humano para BPC como un punto final toxicológico tercero, ya que estas especies son cazadas y consumidas por el público en el sureste de EU. Utilizando umbrales estándar de riesgo humano, mostramos un potencial riesgo para los cazadores que consumen Rascones Picudos matadas cerca de sitios contaminados por BPC, debido al riesgo de cáncer adicional de toda la vida, asociado con ese consumo.

Saltmarsh habitats along the Atlantic and Pacific coasts are biologically and economically valuable natural resource areas. These areas not only attract tourists for recreation, but the abundance of the marsh's seemingly unending resources have lured industries to capitalize on the easy access to the open ocean's busy shipping lanes. Consequently, it becomes increasingly important to protect these fragile ecosystems from the effects of pollution and other anthropogenic disturbances. Since it is not practical to monitor every potential response to environmental impacts, studies must choose appropriate endpoints. In estuarine systems, wildlife can be extremely useful as indicators of the overall health of associated marshlands, especially to address the consequences of ecotoxicological disturbances. In the case of environmental pollution in saltmarsh systems, disturbances can have effects at multiple spatial scales. For example, due to their geochemical properties and mode of introduction into the environment, pollutants can often be studied at the local scale (hectares), whereas others tend to spread to the landscape scale requiring a spatial extent of many square kilometers (Hooper et al 1991, Crimmins et al. 2002). Therefore, to address concerns that may be approached at multiple scales, the proper species must be utilized to indicate if there are deleterious effects. In such cases, birds, specifically rails (Rallidae), are excellent species to monitor since they utilize these systems at both the local and landscape level. Further, genotoxicological and reproductive endpoints can be used to quantify and better understand the long-term effects that these disturbances may have to an estuary.

The Clapper Rail (*Rallus longirostris*) is a secretive marsh bird found throughout coastal saltmarshes from the Gulf of Mexico to Rhode Island and along California's Pacific coastline. The rail's strong site fidelity (Zembal et al. 1989) and predictable diet (Terres 1991) makes it an ideal organism to study the movement and fate of contaminants in disturbed ecosystems. In addition, this species is an integral part of the saltmarsh ecosystem, feeds relatively high on the food chain, is abundant throughout the East Coast, and is a popular game species in the Southeast. Conversely, the Pacific coastal populations are threatened due to habitat destruction and pollution (Eddleman and Conway 1998) and are thus not as amenable to study and experimentation. Using the Clapper Rail as an indicator species not only provides a way of assessing ecosystem health, but information of the relative toxicant burdens can be used to inform the public about potential health risks in areas where they may fish or hunt. Further, since rails are hunted, consuming birds that have inhabited contaminated areas may also present a direct risk to humans.

In coastal Georgia, large expanses of saltmarsh have abundant populations of Clapper Rails throughout the year. In the coastal city of Brunswick, Georgia, with its proximity to major shipping lanes, these marshes are host to many industries making them susceptible to industrial contamination. For example, a chlor-alkali plant discharged as much as 1 kg of mercury (Hg) a day for a period of 6 yr ending in 1972 in this region (Gardner et al. 1978). This site still has elevated levels of Hg as well as the polychlorinated biphenyl (PCB) Aroclor 1268 and other contaminants as will be shown in this paper. The effects of these pollutants have been of concern for many years to the residents of Brunswick as well as to government agencies such as the Environmental Protection Agency (EPA) and USDI Fish and Wildlife Service. Therefore, using Clapper Rails as indicator species in this estuary can address toxicant issues in the Brunswick area, as well as similarly impacted East Coast populations, and can contribute to information needed for the management of endangered subspecies such as the Light-footed Clapper Rail populations in California estuaries where individuals cannot be studied as intensely as in this investigation (Lonzarich et al. 1992).

Contaminant loads for rails, their food items, and their habitat have been established for the endangered California populations but only one study has been conducted in Brunswick, Georgia (Gardner et al. 1978, Lonzarich et al. 1992). San Francisco Bay has had a history of contamination of PCB's since the 1950s. Eggs from the Light-footed Clapper Rail were found to have elevated levels of PCBs as well as selenium (Se) and Hg. However, the effects of these toxicants could not be pursued any further because of limitations of sampling methodology and the rails endangered status (Lonzarich et al. 1992). Therefore, little information exists concerning lethal and sublethal effects that may have occurred or are occurring due to toxicant exposure which could have serious implications for the recovery of this species. We present the results of a study performed in Brunswick that will provide an example of how this species was used to address toxicological issues at the local scale by looking at reproductive effects (eggshell integrity), and at the landscape level by looking at genotoxicological effects to the rails themselves (DNA strand breakage). Finally, we determine what the probability of humans developing cancer would be from consuming Clapper Rail flesh (additional lifetime cancer risk defined as the probability of contracting cancer over the individual's lifetime compared to the expected probability of contracting cancer if the individual had no contaminant exposure) from a marsh located next to a chemical plant that released PCBs and metals as well as other marshes located a few kilometers away from the chemical plant to provide an toxicological endpoint that addresses human risk.

ECOLOGICAL ENDPOINTS

DNA Strand Breakage

Contaminants can interact directly and indirectly with DNA to cause damage. One of the most obvious genotoxic interactions of contaminants with DNA is the induction of DNA strand breaks. DNA strand breaks are among the most easily detected and quantified types of DNA damage (Theodorakis et al. 1994, Sugg et al. 1995). A variety of metal species (including Cr [VI], Ni [II], Co [II], Fe [III], Cd [II], and Pb [II]) are known to induce DNA strand breaks (Hartwig 1995), therefore this technique is extremely useful as an endpoint to quantify damage from toxicant effects. Further, increased DNA strand breakage within living cells has been correlated with PAHs, PCBs, and heavy metals (Theodorakis et al. 1994, Sugg et al. 1995, Siu et al. 2003). In addition, contaminants such as Hg are known to interfere with DNA repair mechanisms (Snyder and Lachman 1989) and indeed Sugg et al. (1995) found that the synergistic effects of Hg and ^{137}Cs on strand breakage were greater than the effects of each contaminant alone. Thus, strand break assays represent a useful endpoint for assessing the consequences of exposure to a mixture of contaminants.

Eggshell Integrity

The eggshell protects the developing embryo against both mechanical impacts and bacterial invasions. Further, it controls the exchange of water and gases through the pores, is the main calcium reservoir for skeletal formation, and supplies some of the magnesium required during embryogenesis (Richards and Packard 1996, Nys et al. 1999). Therefore, it is of fundamental importance for an adequate development of the embryo to ensure the quality and integrity of the eggshell. DDTs and other organochlorides (e.g., PCBs) can affect enzyme activity involved in calcium transportation and consequently eggshell thickness (Cooke 1973, Baird 1995). Further, trace-metal contaminants can also influence the mineralization of eggshell. Specifically, they can reduce the availability of calcium in the diet, interfere with calcium metabolism, and can interfere with the mineralization process itself by affecting the precipitation rate, mineralogy, size, and morphology of crystals that make up the eggshell (Rodriguez-Navarro et al. 2002b). Therefore, exploring eggshell mineralization and thickness can help to better understand how these co-contaminants may affect eggshell integrity, which provides an ecological endpoint for reproductive effects.

Lifetime Human Cancer Risk

In the Southeast and especially in the Brunswick area, Clapper Rails are a popular game species with hunters often achieving their bag limits (K. Giovengo, pers comm.). This was evident to the authors during our collections because many local residents asked us if we would give them the rail carcasses when we came back to the boat landing. Interestingly, after informing them that these birds were collected near an area contaminated with PCBs, some individuals were still interested in consuming the birds. Therefore, since the birds are hunted and consumed by humans, using additional lifetime cancer risk to humans from consuming Clapper Rails contaminated with PCBs in this region is an extremely appropriate endpoint. Studies in humans provide supportive evidence for potential carcinogenic and non-carcinogenic effects of PCBs (Environmental Protection Agency 1996). Further, the EPA provides a framework to evaluate risk to humans who may consume PCBs in their food items. This quantification takes into account the amount of PCB's in the muscle tissue of the food items, the ingestion rate, exposure rate, the body weight, and the expected lifetime of the exposed individual (Environmental Protection Agency 1992).

METHODS

Study Area

This study was conducted in the estuarine marshes near Brunswick, Glynn County, Georgia. Clapper Rails were collected to compare ecotoxicological data from a saltmarsh contaminated with PCBs and metals near a contaminated high-priority Superfund Site—Linden Chemicals and Plastics (LCP)—to other similar saltmarsh locations located a few kilometers away from LCP that were not directly contaminated from point sources. The contaminated LCP site is a saltmarsh system located on the western shore of the Brunswick peninsula and is similar in vegetation structure to nearby saltmarsh systems. It has been classified as a Comprehensive Environmental Response, Compensation, and Liability Act (CERCLA; also known as Superfund Site) by the EPA, primarily due to contamination by Hg and PCBs (Aroclor 1268). Both contaminants are present in elevated levels in the sediments and resident fauna (fiddler crabs [*Uca* spp.]) of this marsh (J. M. Novak et al., unpubl. data). Further, this specific Aroclor has not been produced by any other company on the East Coast of the US. Therefore, it can be used as a marker indicating that animals with measurable levels of this contaminant accumulated it because of its release from the LCP site. The reference marsh areas, Troupe Creek, Mackay River, and Blythe Island are all located near Brunswick. These sites were chosen as representative of the surrounding areas—having a similar vegetation profile, tidal influence, tidal-creek diversity, and water chemistry (Gaines et al. 2003). The vegetation and habitat structure of these marshes are consistent with most other southeastern saltmarshes, consisting primarily of cordgrass (*Spartina* spp.) interspersed with small patches of rushes (*Juncus* spp.) and intersected by tidal creeks.

COLLECTION TECHNIQUES

Adult Clapper Rails (N = 30) were collected from November–December 1999 from three locations in the saltmarsh estuary in Brunswick: the LCP marsh (N = 10), the Mackay River (N = 8), and Troupe Creek (N = 12). Birds were collected in the field during the full-moon high tide, using a shotgun. Upon collection, blood was immediately taken from the bird by opening up the chest cavity and puncturing the heart and/or major blood vessels. Only three–five drops of blood were collected and stored in STE buffer (see strand breakage methods below) for DNA strand-breakage analyses and frozen in liquid nitrogen. The bird was then placed in a cooler and transported to the Savannah River Ecology Laboratory (SREL) immediately following the collection period. In the laboratory, birds were either immediately dissected for muscle and liver tissue or placed in a refrigerator and dissected the following day. These samples were stored in a standard scintillation vial and stored in a -20 C freezer until PCB and metal analysis (see methods below). All blood samples were immediately stored in an ultra-cold freezer. During the dissection process, birds were aged by bursa and plumage examination, sexed, and weighed.

Clapper Rail nest searches were performed from 15 March–June 2000 in the Blythe Island and the LCP marshes. If a nest was found with four or more eggs, the eggs were removed from the nest and brought back to SREL within 4 hr. If nests had <four eggs, they were revisited within a few days to collect a larger clutch. The width, length, and weight of each egg were measured and, on return to the laboratory, eggs were immediately put into an incubator. All eggs were incubated at 37.2 C at 87% relative humidity and rotated automatically every 12 hr. Eggshells from each clutch were saved for mineralization analyses. Eggs were monitored on a daily basis and detailed notes were taken when pipping was initiated. The number of eggs hatched, total incubation time, and pipping activity (including the number of eggs not hatched but pipped) were quantified for each clutch. After hatching was complete (determined by the chick being fully out of its shell for at least 12 hr) hatchlings were weighed, euthanized by cervical displacement, and then frozen for further analysis.

EGGSHELL MINERALIZATION

Eggshell mineralization and thickness were quantified as described in Rodriguez-Navarro et al. (2002a). In brief, the mineral composition of eggshell was determined using a powder X-ray diffractometer (Scintag X1). A diffractogram was collected from a sample of ground shell to be used as a reference pattern for crystals having a completely random orientation (I_o). The structure can be characterized by measuring the area of peaks, calculating the area of the peak divided by a ground shell reference (I/I_o) and plotting the ratios against the angle for the normal of each peak compared to a reference plane. The breadth of this distribution at half the maximum height (e.g., at $I/I_o = 0.5$) is called full width at half maximum (FWHM). FWHM is used as a gauge of crystal orientation in a composite structure. The smaller the number of peaks and the narrower the FWHM distribution, the higher is the degree of crystal orientation in the shell. It is preferable for the crystal orientation of the shell to be low. That is, as the crystal orientation increases, the eggshell integrity will decrease and the shell will become weak. These weaknesses may be offset by the thickness of the eggshell. Therefore, the thickness of the shell was also measured at four points separated by 90° at the egg waist using a micrometer.

DNA STRAND BREAKAGE

Basic strand-breakage protocols were modified from Theodorakis et al. (1994) with modifications as listed below. Red blood cells were collected from each Clapper Rail taken in the field. Blood samples were stored in STE buffer (100 mM NaCl, 100 mM Tris pH 8.0, 100 mM EDTA) until they were prepared for electrophoresis. Four separate plugs were made from each blood sample. This provided replication of individuals within gels (replications 1 and 2) and between gels (replications 3 and 4). Instead of using standard agarose gel electrophoresis, we used a pulsed-field agarose gel electrophoresis assay similar to the system described by Blocher et al. (1989) to measure double-strand breakage of DNA with the following modifications: (1) red blood cells were used as the source of intact nuclei, (2) a BioRad CHEF DR III system was used for pulsed-field agarose gel electrophoresis, (3) DNA was stained with Sybr Gold™ after electrophoresis, and (4) Samples were loaded into every other lane such that every sample was flanked by a negative control lane and three lanes were loaded with three different DNA size ladders to serve as positive controls. The specific run conditions were: run time of 16 hr, electrical potential of 2.5V/cm, pulse angle of 120°, angle change ramped from 40–120 sec over the run, 2.2 l of 0.5× TBE buffer cooled to 14 C and pumped at approximately 0.8 l/min. A commercial gel image analysis system (Eagle Eye II, Stratagene, La Jolla, CA) was used to capture images of the flourescence from the UV-illuminated DNA. Each gel was imaged using 17 different exposures for fine scale quantification of the amount of DNA damage.

ADDITIONAL LIFETIME HUMAN CANCER RISK

The additional lifetime cancer risk from ingesting food items contaminated with PCBs is determined using a tiered approach from existing information provided by the EPA (1996). Specifically, slope factors are derived from linear extrapolation of dose response studies. This slope factor is multiplied by lifetime average exposure levels to estimate the risk of cancer. These calculations are based on generalized studies and therefore are not specific to individual Aroclors (Environmental Protection Agency 1996). The specific calculations used were

$$LADD = \frac{C \times IR \times ED}{(BW \times LT)} \quad \text{Equation 1}$$

$$Risk = LADD \times Slope \quad \text{Equation 2}$$

where:
  $LADD$ = lifetime average daily dose
  C = concentration of PCBs in Clapper Rail flesh (µg/kilogram dry weight)
  IR = intake rate (gram/day)
  ED = Exposure duration (years)
  BW = body weight (kilogram)
  LT = lifetime (years)
  $Slope$ = USEPA derived slope factor appropriate for food chain exposure.

Calculations were based on a hunting season from September through December with a 15 bird/day bag limit. This essentially provides 24 hunting opportunities during the season, since birds are usually hunted over a 3-d period during the full-moon high tide (two tides per 24 hr period).

ANALYSIS OF POLYCHLORINATED BIPHENYLS AS AROCLOR 1268

PCBs were extracted from the tissue using ultrasonic extraction (EPA Method 3550B). Muscle tissue was used from adults but the whole hatchling was ground since they were too small to dissect individual tissues. Tissues were freeze dried and macerated prior to extraction. Dibromooctofluorobyphenyl and tetrachlorometa zylene added as internal surrogate standards. The extractions were performed by sonicating the tissues in 150 ml of acetone: hexane (1:1v/v) using a Tekmar sonic disruptor operated at 100% power in the pulsed mode with a 50% duty cycle for 3 min. The mixture was filtered and the extraction repeated twice with fresh solvent. The combined solvent extracts were dried with Na_2SO_4 solvent, exchanged, and concentrated. Lipids were removed by treatment with 1:1 sulfuric acid solution and the solution back-extracted into hexane. The aqueous phase was discarded and the procedure repeated until a clear hexane extract was obtained. The hexane extracts were concentrated to about 1 ml and then charged onto a pre-cleaned silica gel column to isolate the PCBs from other organic contaminants. The column was sequentially eluted with a series of organic solvents and the PCB fraction collected. The isolated fraction was then concentrated and analyzed using gas chromatography (GC) and gas chromatography-mass spectrometry (GC-MS).

PCB analyses were performed on a gas chromatograph equipped with an electron capture detector (ECD), splitless injection, electronic pressure control (EPC), and autoinjector. Separation of PCB congeners was achieved using a 30 m DB-5 (0.025 mm I.D., 0.25 mm film thickness) capillary chromatographic column (J & W Scientific, Folsom, CA). Samples were quantified as Aroclor 1268 using a five-point calibration curve derived from dilutions of certified standards. Six characteristic peaks were selected from the Aroclor mixtures. All selected congener peaks were at least 25% of the highest Aroclor component. A Hewlett Packard 5890 Series II gas chromatograph with splitless injection, EPC, and a 5972 mass spectrometer (GC-MS) was used to confirm GC-ECD identifications. All samples were analyzed by GC-MS using the selected ion monitoring (SIM) acquisition mode. Selected samples were also analyzed using full scan acquisition in a separate sample injection/analysis. All of the 12 congeners in the Aroclor 1268 mixture were determined in the GC-MS analysis. Selected ions in the SIM mode for different retention time windows were determined from the analysis of an Aroclor 1268 standard. Analysis of spectra obtained in the full-scan mode (mass 50–550) were performed by comparing the mass spectra with Aroclor 1268 standards as well as the National Institute of Standards and Technology (NIST) reference library.

Metal Analyses

Wet tissue and eggshell samples were digested with nitric acid and hydrogen peroxide using microwave digestion protocols. Approximately 25 mg of homogenized sample was placed in a Teflon microwave digestion vessel to which 5 ml of redistilled 70% HNO_3 was added. The vessel was capped and microwave digested using a variable powered program with increasing microwave power applied over 1 hr. After cooling, the vessels were uncapped and 1 ml of 30% H_2O_2 was added; the vessels were then recapped and subject to an identical microwave heating procedure. After the vessels had cooled the digest was brought to a final volume of 25 ml using volumetric flasks. Two duplicate samples, one blank and two standard reference materials (SRMs; DORM-2 and DOLT-2 NRC-CNRC, Ottawa, Ontario, Canada) were included per digestion set. Analysis data with <95% SRM recovery was the rejection criteria. No samples fell below this range. The digested tissue samples were analyzed for V, Cr, Mn, Fe, Co, Ni, Cu, Zn, As, Se, Mo, Cd, Sb, Pb, and Hg following the methodology outlined in EPA method 6020. Quality-control procedures were based on EPA SW-846.

Statistical Analyses

A chi-square test was used to determine if DNA strand breakage differed between the contaminated LCP site and the reference locations. Logistic regression (Hosmer and Lemeshow 2000) was used to determine which toxicants (Aroclor 1268 and metals) contributed to DNA strand breakage. All metal concentrations were log-transformed prior to analysis to meet assumptions of normality. A response variable of one was used for observations that had strand breakage and a response variable of zero was used for observations that had no strand breakage. A full model was fit initially and independent variables were removed one at a time based upon their beta values. At each step the corrected Akaike information criteria (AIC) was calculated and compared to the previous model. The model fitting was stopped when AIC was not smaller than the previous model. A randomization function was employed as the statistical validation procedure to evaluate the final logistic regression model's prediction strength (Manly 1998). The leave-one-out cross-validation procedure was used to produce the predicted binomial observation (0 vs. 1) by dropping the data of one observation from the dependant variable and reestimating the response from the tested model (Neter et al. 1990). The observation was then put back into the data set and the procedure was repeated until all observations were used. The model's validity was then judged by comparing the number of accurate predictions to the number of inaccurate predictions.

The relationship between eggshell thickness and FWHM as measures of eggshell structural integrity was explored using a simple correlation as well as a principal component analysis (PCA). Further, the relationship between the toxicants found in the eggshell was also quantified using a PCA. All metal concentrations were log-transformed prior to analysis to meet assumptions of normality. Each measure of integrity was independently tested to determine if they differed based on site using a t-test. A general linear model was then employed to determine if eggshell integrity was dependent upon toxicant load using the principal components from the respective PCA as the response and dependent variables. The first principal component for eggshell integrity was used as the response variable. Site (LCP vs. Blythe Island) was used as an additional dependent variable along with appropriate interaction terms within the model. All analyses were performed based on the results of the clutch rather than the individual egg. This was because individuals within a clutch could not be separated during incubation, which made it impossible to determine from which shell a hatchling hatched. Eggshells were analyzed for metals by grinding material from each egg of the clutch into one matrix, and the hatchling was used to measure PCB load since PCBs, due to their lipophillic nature, will accumulate in lipid biomass and not in the predominately inorganic egg shell matrix (Rassussen et al., 1990, Schwartzenbach et al., 2003). The FWHM and eggshell thickness were averaged to give one observation for each clutch. A full model was fit initially and independent variables were removed one at a time based upon their beta values. At each step the corrected AIC was calculated and compared to the previous model. The model fitting was stopped when AIC was not smaller than the previous model.

RESULTS

The initial replication (four replicates/individual) and multiple imaging, proved unnecessary for site-level comparisons. The distinction between broken and unbroken sample morphology is distinct and replicable (Fig. 1). In every case, if an individual had broken DNA, all four replicates exhibited a broken morphology. Likewise, for individuals with unbroken DNA, all four replicates exhibited no broken morphology.

All 10 birds from LCP (100%) exhibited broken DNA, one of eight birds (12.5%) from Mackay River had broken DNA and one of 12 birds (8.3%) from Troupe Creek exhibited broken DNA. When the birds from Mackay River and Troupe Creek were combined into a single reference sample, two of 20 birds (10%) exhibited broken DNA. Using this frequency to generate the expected values for LCP, a chi-square analysis resulted in a highly significant value ($G_1 = 46.05$, $P = 1.15 \times 10^{-11}$). Thus, at the population level, Clapper Rails from LCP exhibited a significantly higher frequency of double stranded DNA breaks compared to birds from Mackay River and Troupe Creek. The final logistic regression model showed that Hg had a model probability of 99% with a positive relationship between breakage and Hg concentration while the probability for Pb was 1% and had a negative relationship between breakage and Pb concentration (Table 1). No other metals showed significant relationships (Body burdens for all metals in adult clapper rails are listed in Appendix 1 and for PCB levels in adults and hatchlings in Appendix 2). The take one out cross validation procedure showed that the

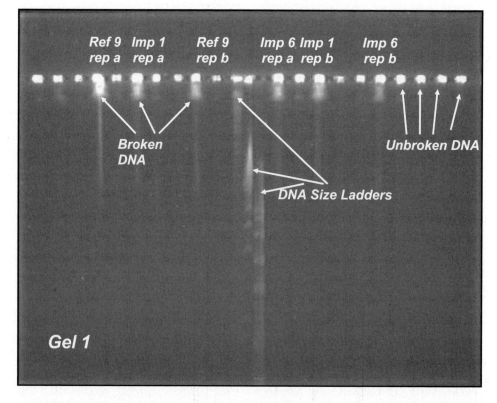

FIGURE 1. Example of broken, unbroken, and DNA size ladders used on a gel run to measure DNA strand breakage in adult Clapper Rails. This image shows that the distinction between broken and unbroken sample morphology is distinct and replicable. Ref refers to the reference site and Imp the impacted site. The numbers refer to individual birds collected from each site and rep a and rep b refer to replicate samples run on the same gel. Thus Imp 1, rep a refers to the first sample from the first bird collected from the impacted site.

TABLE 1. RESULTS OF THE LOGISTIC REGRESSION ANALYSIS OF DNA STRAND BREAKAGE AND CONTAMINANT CONCENTRATION.

Parameter	df	Estimate	SE	χ^2	P	POC (%)
Intercept	1	0.490	3.351	0.021	0.8836	
Hg	1	20.469	9.795	4.367	0.0366	99.086
Pb	1	-4.255	3.271	1.692	0.1933	0.914

Note: POC is the probability of change computed from the odds-ratio.

logistic model predicted 10 of the 12 broken samples correctly and 17 of the 18 unbroken samples correctly.

Thirty-four nests (21 from reference areas and 13 from LCP) comprising 146 eggs were used for eggshell integrity-toxicant analyses. Thickness and FWHM display a negative correlation (r = -0.733, P < 0.0001, N = 34). Further, there was no difference between the LCP and the reference locations for either measure. The PCA showed that the first component explained 87% of the variation for the two measures. The PCA for metals in the eggshell and PCB's in hatchlings (Appendix 2) indicated that 79.1% of the total variation was contained in the first four components (Table 2). The simple linear regression showed that PC2 (mostly explained by Mn, Zn, Hg, and PCB) was a significant variable as was the interaction of site and PC4 (mostly explained by Cu and Zn) (Table 4; Fig. 3). Site, PC1 and PC4 were left in the final model because of constraints imposed when using PC components and from significant interaction terms.

Additional lifetime human cancer risk from consuming PCB contaminated birds over a 30-yr period for a 70 kg adult using the maximum PCB value from birds collected from LCP and reference areas were both above the 1×10^{-6} risk (expectation for unexposed individuals) threshold used by the EPA to determine risk (Table 5). Specifically, the additional lifetime cancer risk for LCP was 1.41×10^{-3}, while the general Brunswick area was lower with the estimated risk being 1.37×10^{-4}.

DISCUSSION

Aroclor 1268, as well as other toxicants, are bioavailable in the Brunswick estuary. The DNA strand breakage study and the PCB additional lifetime cancer risk estimates showed that these anthropogenic insults are impacting the estuary at both the landscape and local scale. During the late fall and early winter, Clapper Rails have larger home ranges than during their breeding season (Meanley 1985). Further, although it is thought that the Brunswick population is non-migratory, this winter population could also have been mixed with other migratory populations from the north. Regardless, the group-level DNA strand breakage analyses showed that birds that were collected and assumed to reside in and around the LCP site had a higher percentage of individuals with broken DNA compared to those collected from other areas only a few kilometers away. One possibility is that birds collected from the LCP site were resident birds that may have spent large amounts of time in the areas, possibly even breeding at that site. Another possibility is that the birds may have only used that area for over wintering, which would imply that the contamination in that marsh might accumulate and show toxicity response very quickly.

The results from the logistic regression analysis indicate that Hg is primarily responsible for increased levels of strand breakage. However, it is also possible that breakage can be elevated due to the synergistic effects of a contaminant mixture that would not be readily detectable without much larger sample sizes. One of the birds collected from the reference sites that was scored as having broken DNA, had levels of Hg in the same range as birds from LCP and the other did not. This individual-level analysis helps explain one data point from our group-level analysis but also gives pause in ascribing too much weight to our result of Hg and breakage. While we can certainly state that Hg concentrations are influencing the levels of strand breakage we have less confidence in stating that

TABLE 2. PRINCIPAL COMPONENT ANALYSIS (PCA) OF THE METALS MEASURED IN THE EGGSHELLS FROM EGGS COLLECTED AT THE LPC AND REFERENCE SITES—EIGENVALUES FOR THE 11 PCS CALCULATED FOR THE METALS.

PC	Eigenvalue	POV	CPOV
1	3.36697651	**0.3061**	**0.3061**
2	2.68185851	**0.2438**	**0.5499**
3	1.39326853	**0.1267**	**0.6766**
4	1.25509985	**0.1141**	**0.7907**
5	0.72052911	0.0655	0.8562
6	0.67220944	0.0611	0.9173
7	0.34738532	0.0316	0.9488
8	0.21722518	0.0197	0.9686
9	0.15444438	0.0140	0.9826
10	0.12547491	0.0114	0.9940
11	0.06552826	0.0060	1.0000

Notes: The PCA was based upon the correlation matrix. Significant PCs are in bold. POV is the proportion of variance explained by each component, and CPOV is the cumulative proportion of variance explained by the component and all previous components.

TABLE 3. PRINCIPAL COMPONENT ANALYSIS (PCA) OF THE METALS MEASURED IN THE EGGSHELLS FROM EGGS COLLECTED AT THE LCP AND REFERENCE SITES — COMPONENT LOADINGS FOR THE FOUR SIGNIFICANT EIGENVECTORS.

Variable	PC1	PC2	PC3	PC4
Ni	0.157079	0.257248	**0.363281**	0.264882
Mg	-0.134219	0.314552	**0.544807**	0.257439
Al	**0.484737**	0.065041	0.203060	-0.113849
P	**-0.391599**	0.239709	0.218198	-0.252857
Mn	**0.352640**	**0.354443**	-0.052157	-0.249082
Fe	**0.483298**	0.117830	0.169759	-0.184244
Cu	-0.103615	0.217930	-0.045241	**0.676402**
Zn	-0.215307	**0.417834**	0.201922	**-0.342930**
Pb	**0.345458**	-0.128209	0.086392	0.322304
Hg	-0.069768	**0.470641**	**-0.423272**	0.062197
PCB	0.172501	**0.416722**	**-0.470519**	0.088362

Note: Loadings with an absolute value >0.33 are in bold.

TABLE 4. GENERAL LINEAR MODEL ANALYSIS TO DETERMINE EGGSHELL INTEGRITY DEPENDENCY UPON MEASURED PARAMETERS.

Effect	df	F	P
Site	1	0.70	0.411
PC1	1	3.74	0.065
PC2	1	9.87	0.004
PC4	1	0.84	0.369
Site × PC4	1	7.69	0.011

Notes: The response variable of the regression is the principal component of eggshell integrity variance. The dependent variables are site (contaminated vs. reference) and the principal components of the PCB levels from the hatchlings and the metal levels within the eggshells.

other contaminants in the mixture have little or no effect.

It is also unclear what affect increased levels of double-strand breakage may have on the rails on either an ecological or evolutionary time scale. This will be determined by the extent of the breakage, the specific tissue it occurs in and the efficiency and quality of the repair. In somatic cells, if the break is not repaired it will most likely lead to a loss of cell function and depending on the tissue, apoptosis (Rich et al. 2000). If it is repaired with error then the mutated cell may become cancerous (Kasprzak et al. 1999). It is unlikely, that levels of cell death could be high enough or rates of cancer production fast enough to influence survival rate for Clapper Rails in the wild. However, the increased energy demands of these processes may have measurable effects on both survival and reproduction under stressful conditions (Hoffmann and Parsons 1991). In gametic cells, repair mechanisms that result in mutations will increase the base mutation rate of progeny and thus has the potential to change the evolutionary trajectory of populations and species (Fox 1995). Recent research has implicated multiple mechanisms in vertebrates for the repair of double-strand breaks (Liang et al. 1998), some also being involved in translocation events (Kanaar et al. 1998) and generation of antibody variability (Karran 2000). Thus, the evolutionary effects of double-strand breaks may be more pervasive than expected.

All birds that were collected did have measurable PCB levels (Appendix 2), which can be judged by the calculations to estimate the additional lifetime cancer risk from consuming PCB contaminated flesh (Table 5). Measurable levels of Aroclor 1268 as well as metals have been found in the soil and rail food items in both the LCP and reference areas (J. M. Novak et al., unpubl. data), indicating that this toxicant is bioavailable at the landscape level. Therefore, it is likely that birds collected from these reference areas are picking up Aroclor 1268 from the respective areas. However, because these birds are likely to have winter home ranges that may encompass the reference areas as well as LCP, they may have accumulated the PCB from the LCP site itself. Home-range studies of resident rails are needed to address these questions.

TABLE 5. ADDITIONAL LIFETIME HUMAN CANCER RISK FROM CONSUMING PCB-CONTAMINATED CLAPPER RAILS.

Site	PCB concentration (μg/kg)	Ingestion rate (g/d)	Lifetime average daily dose (mg/kg -d)	Risk
LCP	1.76×10^{-02}	6.58	7.07×10^{-04}	1.41×10^{-03}
Reference	1.70×10^{-03}	6.58	6.85×10^{-05}	1.37×10^{-04}

Notes: From areas near the LCP site (N = 10) and reference locations (N = 17) in the Brunswick area. Calculations were based on a 30-yr exposure over a 70-yr lifetime for a 70-kg adult using a USEPA derived slope factor of 2 for the maximum PCB levels found from each site.

Since eggshell integrity should be indicative of the toxicity of contaminants at both the local and landscape level it can be a useful ecological endpoint. Clapper Rails in the southeastern US will start to set up breeding territories in early February and their home ranges will focus around those areas through the breeding season (Meanley 1985; J. M. Novak et al., unpubl. data). Therefore, the toxicants those female birds accumulate in the months before breeding should be representative of the area where they breed. However, toxicants that have been accumulated prior to the nesting season from other areas may still be persistent in the birds' organs and therefore depurated into the egg as well. The integrity of some of eggshells from the LCP site as well as the reference areas did show signs of structural problems. Although the nature of the matrix of contaminants is extremely complex, some interesting interpretations of the data can be made. For example, since the plot of the principal component 2 from the PCA showed that individuals with higher Mn, Zn, Hg, and Aroclor 1268 (PCB) had thinner eggshells that were less oriented and that the trends were the same from both the LCP and reference sites, may imply that ecosystem integrity has been compromised for the entire Brunswick estuary (Fig. 2). That is, the bioavailability of Aroclor 1268 at the Brunswick landscape level coupled with the possibility of birds from the reference areas using the contaminated site during the winter, contributed to the structural problems found within the eggshells. In summary, since the slopes of the lines did not differ between the sites, the effect of these contaminants appears to be at a scale greater than the individual sites themselves.

Another interesting finding is that principal component 4 has a significant interaction with site (Fig. 3). In this case, higher levels of Cu and lower levels of Zn are associated with stronger eggs at the reference site but weaker eggs at LCP. This most likely indicates localized effects at each site that mitigates the relationship between these toxicants and eggshell integrity. Since the contaminants represent a very complex mixture, it is unlikely these relationships can be further disentangled without using an experimental approach. It is difficult to speculate how these findings would have influenced egg survival in the wild. Although more oriented eggshells have less integrity, the greater thickness may compensate for this flaw. This portion of the study was inspired by the fact that when eggs were marked with a pencil, some tended to quickly shatter even with the lightest touch of a pencil, indicating structural problems with the eggs. Further study of the possible compensatory nature of eggshell

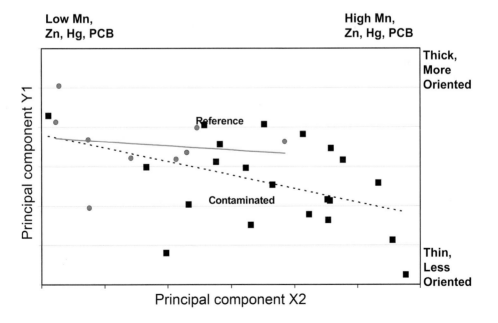

FIGURE 2. Plot of principal component Y1 (from the eggshell thickness and orientation PCA) vs. X2 (from eggshell and hatchling contaminant level PCA, Table 3) showing that for both the reference (circles) and impacted (squares) location, individuals with higher Mn, Zn, Hg, and Aroclor 1268 (PCB) had thinner eggshells that were less oriented.

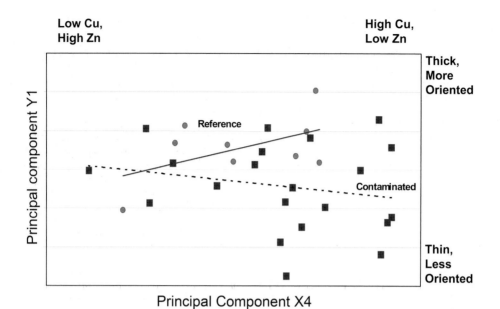

FIGURE 3. Plot of principal component Y1 (from the eggshell thickness and orientation PCA) vs. X4 (from eggshell and nestling contaminant level PCA, Table 3) showing that higher levels of Cu and lower levels of Zn are associated with stronger eggs at the reference site but weaker eggs at the impacted site.

integrity with crystal orientation must be pursued to strengthen the use of these measures as endpoints of survivability.

Estimating the additional lifetime cancer risk from consuming PCB contaminated meat is the most widely used and easy to understand measure for a human toxicological endpoint. Further, cancer studies comparing commercial and environmental mixtures, especially those found in the food chain, warn that food chain risks could be underestimated (Environmental Protection Agency 1992). In addition, PCBs have been shown to cause a variety of health effects to humans and other animals beyond carcinogenicity. Specifically, studies have revealed that PCBs can cause a variety of immune, reproductive, neurological, as well as endocrine effects (Environmental Protection Agency 1992). It is unclear how environmental processes alter the composition and subsequent toxicity of PCB mixtures and exposure to a myriad of contaminants can further complicate interpretations of PCB toxicity as shown in this study. Our data do show that Aroclor 1268 is bioavailable throughout the Brunswick estuary, and poses a potential health risk to individuals who may consistently ingest Clapper Rail flesh.

Since Clapper Rails are an integral part of the ecosystem structure of the Brunswick estuary, their use as indicator species can be very helpful in understanding how anthropogenic activities can affect estuarine systems. Based on two separate ecological endpoints, it appears that the contamination by metals and PCBs in the estuary is having detrimental effects on the resident Clapper Rail population. Further, based on the PCB levels alone, it appears that those who consistently hunt Clapper Rails in this region should consider hunting outside the immediate Brunswick area to avoid possible detrimental health effects. The information and techniques outlined in this paper can be used as a template for other regions of the US that have large rail populations, to use these species as ecological indicators. DNA strand breakage and eggshell integrity offer an additional value in that they are non-lethal techniques and therefore can be used to study threatened and endangered populations.

ACKNOWLEDGMENTS

We thank K. Hastie and G. Masson for logistical help. We would also like to thank S. Boring, P. Shull, and S. Murray for help in the field. Two anonymous reviewers provided insight that improved the quality of this manuscript. This project was funded through the USDI Fish and Wildlife Service and Financial Assistance Award DE-FC09-96SR18546 from the U.S. Department of Energy to the University of Georgia Research Foundation.

APPENDIX 1. CONTAMINANT BURDENS (PPM WET WEIGHT) FOUND IN ADULT CLAPPER RAIL MUSCLE.

LCP (N=10)	Al	Cr	Mn	Fe	Co	Ni	Cu	Zn	As	Se	Rb	Sr	Cd	Ba	Pb	Hg
Mean	8.17	0.36	0.46	62.53	0.02	0.19	3.16	12.42	0.28	0.32	2.03	0.32	0.01	0.09	0.11	1.40
Standard error	2.05	0.03	0.05	6.06	0.01	0.04	0.31	1.67	0.03	0.06	0.05	0.06	0.00	0.03	0.01	0.21
Median	7.00	0.34	0.43	54.97	0.01	0.15	3.13	10.89	0.29	0.38	2.05	0.27	0.01	0.05	0.12	1.16
Standard deviation	6.48	0.10	0.16	19.16	0.02	0.11	0.97	5.28	0.10	0.18	0.15	0.18	0.01	0.11	0.04	0.67
Minimum	0.00	0.25	0.26	47.32	0.01	0.09	1.80	8.48	0.11	0.00	1.79	0.15	0.00	0.03	0.05	0.61
Maximum	22.43	0.60	0.79	108.69	0.07	0.44	5.62	26.62	0.42	0.50	2.24	0.75	0.02	0.39	0.16	2.52
Reference (N = 20)	Al	Cr	Mn	Fe	Co	Ni	Cu	Zn	As	Se	Rb	Sr	Cd	Ba	Pb	Hg
Mean	6.48	0.44	0.45	71.72	0.06	0.91	3.81	12.05	0.52	0.38	2.23	0.26	0.01	0.05	0.19	0.44
Standard error	0.96	0.03	0.02	3.90	0.02	0.26	0.29	0.62	0.06	0.04	0.08	0.02	0.00	0.01	0.03	0.03
Median	7.32	0.42	0.45	67.62	0.02	0.24	3.44	11.67	0.45	0.38	2.28	0.24	0.01	0.04	0.16	0.41
Standard deviation	4.30	0.13	0.07	17.46	0.08	1.16	1.28	2.79	0.28	0.16	0.36	0.09	0.01	0.03	0.14	0.11
Minimum	0.00	0.26	0.29	44.38	0.01	0.08	2.15	8.77	0.20	0.00	1.68	0.13	0.00	0.02	0.04	0.27
Maximum	12.87	0.73	0.56	105.62	0.31	3.98	6.10	21.33	1.23	0.66	2.87	0.49	0.03	0.12	0.54	0.72

APPENDIX 2. SUMMARY STATISTICS OF THE AMOUNT OF AROCLOR 1268 (PPM DRY WEIGHT) FOUND IN ADULT CLAPPER RAIL MUSCLE COLLECTED FROM THE LCP MARSH AND REFERENCE AREAS IN BRUNSWICK, GEORGIA, DURING THE 1999 HUNTING SEASON AND IN CLAPPER RAIL HATCHLINGS IMMEDIATELY AFTER HATCHING THAT WERE COLLECTED AS EGGS FROM THE LCP MARSH AND THE BLYTHE ISLAND REFERENCE AREA IN BRUNSWICK, GEORGIA, DURING THE 2000 NESTING SEASON. HATCHLING CONCENTRATIONS WERE AVERAGED FOR EACH NEST TO PROVIDE COMPARISONS FOR EGGSHELL ANALYSES.

	Adults		Hatchlings	
Statistic	LCP	Reference	LCP	BLYTHE
Mean	6.7619	0.4308	147.3442	31.8828
Standard error	2.0016	0.1243	19.1795	6.3084
Median	3.9930	0.3000	81.7666	25.7578
Standard deviation	6.0048	0.5124	144.8020	26.7641
Minimum	1.3200	0.0030	13.9373	9.2856
Maximum	17.5600	1.7020	659.6327	114.1756
N	9	17	57	18

A UNIFIED STRATEGY FOR MONITORING CHANGES IN ABUNDANCE OF BIRDS ASSOCIATED WITH NORTH AMERICAN TIDAL MARSHES

COURTNEY J. CONWAY AND SAM DROEGE

Abstract. An effective approach to species conservation involves efforts to prevent species from becoming threatened with extinction before they become listed as endangered. Standardized monitoring efforts provide the data necessary to estimate population trajectories of many species so that management agencies can identify declining species before they reach the point of endangerment. Species that occur in tidal saltmarshes in North America are under sampled by existing broad-scale monitoring programs. We summarize existing local and regional survey efforts for saltmarsh birds and propose a standardized continental protocol for assessing the status and population trends of birds that breed in saltmarshes in North America. The objective of this proposed survey effort is to create a series of interconnected monitoring efforts that will provide information on the status and the changes in status of terrestrial birds living in saltmarsh systems of North America. We describe detailed field protocols for standardized surveys of saltmarsh birds across North America. We recommend morning point-count surveys with an initial 5-min passive period followed by a period of call broadcast. Surveyors record all individual birds detected (regardless of distance) for all species that are associated with saltmarshes and estimate the distance to each individual bird detected. We provide recommendations for standardizing distance between adjacent survey points, how repeat detections across points are recorded, daily and seasonal timing of surveys, timing of surveys relative to tidal cycles, number of replicate surveys per year, and focal species for this standardized survey effort. Recommended survey protocols include methods that allow estimation of various components of detection probability so that stronger inferences can be made based on trends in count data. We explain why the various survey recommendations are made so that potential participants understand the rationale for various aspects of the survey protocols. We also provide sample data forms and an example of how to fill out a data form. These protocols build upon the *Standardized North American Marsh-Bird Monitoring Protocols* by encouraging those interested in saltmarsh passerines (and other saltmarsh birds) to conduct surveys using a standardized protocol similar to that being used for secretive marsh birds. Standardization of this sort will allow data from surveys focusing on saltmarsh passerines to be easily pooled with data from surveys focusing on secretive marsh birds. Implementing these standardized surveys in saltmarshes across North America will help document regional and continental patterns in distribution and abundance of all birds associated with tidal marshes.

Key Words: monitoring, saltmarsh birds, saltmarsh endemics, tidal marsh.

UNA ESTRATEGIA UNIFICADA PARA MONITOREAR CAMBIOS EN LA ABUNDANCIA DE AVES ASOCIADAS CON LAS MARISMAS DE MAREA DE NORTE AMÉRICA

Resumen. Un enfoque efectivo para la conservación de especies incluye esfuerzos para prevenir que las especies se conviertan en peligro de extinción antes de que se enlisten como amenazadas. Esfuerzos de monitoreo estandarizados proveen de datos necesarios para estimar trayectorias de las poblaciones de muchas especies, es por ello que las agencias de manejo pueden identificar especies en declive antes de que alcancen el punto de amenazadas. Las especies que se presentan en marismas saladas en Norte América se encuentran sub muestreadas por programas existentes de monitoreo de amplia escala. Resumimos esfuerzos de muestreos locales y regionales existentes para aves de marismas saladas y proponemos un protocolo continental estandarizado para la valoración del estatus y tendencias de población de aves que se reproducen en marismas saladas en Norte América. El objetivo de este muestreo propuesto es crear una serie de esfuerzos interconectados de monitoreo, que proveerán información sobre el estatus y los cambios en el estatus de aves terrestres que viven en sistemas de marisma salada en Norte América. Describimos protocolos de campo detallados para muestreos de aves de marisma salada a través de Norte América. Recomendamos muestreos de conteo-punto matutinos con un periodo pasivo inicial de 5-minutos, seguido de un periodo de llamado por emisión. Los investigadores grabaron cada ave detectada (a pesar de la distancia) para todas las especies que se encontraban asociadas con marismas saladas y estimaron la distancia para cada individuo de ave detectada. Proveemos recomendaciones para distancia estandarizada entre puntos de muestreo adyacentes, cómo las repeticiones a través de los puntos son grabadas, el ritmo diario y de estación de los muestreos, la sincronía de muestreos relacionados a ciclos de marea, numero de muestreos replicados por año, y especies focales para este esfuerzo de muestro estandarizado. Los protocolos de muestreo recomendado incluyen métodos que permitan la estimación de varios componentes de probabilidad de detección, para que inferencias más fuertes puedan realizarse basándose en tendencias de datos contados. Explicamos por qué las recomendaciones de muestreo se hacen, para que los potenciales participantes entiendan el motivo de varios aspectos de los protocolos de muestreo. También proveemos formatos para datos de muestreo y un ejemplo de como llenar un formato para datos. Estos protocolos construidos basados en los *Protocolos de Monitoreo de Aves de Marisma Estandarizados de Norte América* fomentarán el interés en aquellos interesados en colorines de marisma salada (y otras aves de marisma salada) para conducir muestreos utilizando un protocolo estandarizado similar a aquel utilizado para aves de marisma sigilosas. Una estandarización de este tipo permitiría que los datos de los estudios enfocados en colorines de marisma salada fueran fácilmente de reunir con datos de estudios enfocados en aves sigilosas de marisma salada. Implementar estos estudios estandarizados en marismas saladas a través de Norteamérica ayudaría a documentar patrones regionales y continentales en la distribución y abundancia de todas las aves asociadas con marismas de marea.

Conserving endemic species diversity and preventing extinction and local extirpation are goals of many land-management agencies and non-profit organizations in North America. In the US, the Endangered Species Act (ESA) of 1973 protects species that are at the greatest risk of extinction. However, we cannot afford to wait until species are listed under the ESA to initiate recovery efforts. The average wait time between listing and approval of a recovery plan is currently unacceptably long (Tear et al. 1995) and additional species are listed as endangered faster than they can be recovered. Moreover, recovery efforts for listed species typically involve high costs and low probability of success (Tear et al. 1995). Population monitoring is critical to effective species conservation because monitoring allows us to identify problems before populations are threatened with extinction (Goldsmith 1991, Hagan et al. 1992). Indeed, early detection of declining populations allows more effective and less-costly recovery efforts (Green and Hirons 1991). Hence, a more effective and efficient approach to species conservation is to prevent species from becoming endangered in the first place (Miller 1996). This approach requires identifying declining species before they become endangered.

Standardized monitoring efforts provide the data necessary for more scientifically credible listing and de-listing decisions (Gerber et al. 1999). Accurate estimates of population trajectory can save management agencies money and reduce contentious interactions with industry and the general public (Gerber et al. 1999). Large-scale monitoring efforts such as the North American Breeding Bird Survey (BBS) have been useful at identifying declining species before they reach the point of endangerment. The BBS has been useful in helping target management efforts towards several species of terrestrial birds that were declining throughout their range. But the BBS does have limitations—limited success estimating population trends for species or subspecies with restrictive distributions and/or those that have very narrow habitat requirements. Hence, we need to develop standardized monitoring efforts that focus on species or vegetative communities that are not sampled effectively by existing broad-scale monitoring efforts. A good example of an ecosystem that is under sampled by existing broad-scale monitoring programs and needs focused monitoring efforts is tidal saltmarshes in North America.

Tidal-marsh ecosystems in North America are unique in that they support numerous species and subspecies of endemic birds (Greenberg and Maldonado, *this volume*). However, while the number of hectares of saltmarshes in the US has declined by 30–40% (Horwitz 1978), we lack information on the status of saltmarsh birds because the BBS does not adequately sample birds in marshes (Bystrak 1981, Robbins et al. 1986, Gibbs and Melvin 1993, Sauer et al. 2000). The presence of taxa endemic to tidal marshes presents scientists and land managers with the responsibility of ensuring their persistence. Ensuring population viability of these unique species needs immediate attention due to anthropogenic treats to these environments. Indeed, a large number of bird species associated with tidal marshes are considered species of conservation concern, rare, threatened, endangered, or have already gone extinct (Pashley et al. 2000).

Many hectares of tidal marsh in North America have been altered or eliminated as a result of land reclamation, ditching, pesticide application, and other public-works activities. Relatively few studies have focused on saltmarshes despite the fact that these systems are often on publicly owned or protected land. The result is that one of the earth's most unique ecosystems has been allowed to deteriorate and the species associated with these systems have been comparatively unstudied. We need to increase our understanding of saltmarshes and the species they support because rising sea levels and increased mosquito control efforts pose immediate threats to many saltmarsh systems in North America.

Numerous local or regional avian monitoring efforts already exist in North American saltmarshes (Table 1). Most of the coordinated regional monitoring efforts in saltmarshes are restricted to vocal surveys (Erwin et al. 2002). However, other monitoring activities can provide additional information not possible with vocal surveys alone. For example, collecting capture-recapture or mark-resighting data is useful to estimate local population size (and annual survival). Monitoring demographic parameters associated with reproduction (e.g., nesting success and annual fecundity) can provide insight into proximate causes of population change and is useful for long-term studies tracking change over time at specific locations.

Point-count surveys where observers count the number of birds seen or heard during a fixed-time interval are commonly used to estimate population trends across a broad geographic area. Moreover, point-count surveys can be designed so that observers differentiate nest-departure calls (Greenberg 2003) from other vocalizations. Recording the number of nest-departure calls allows surveyors to provide an index of reproductive activity that could be compared across locations or over time. Ideally, a comprehensive monitoring program targeting saltmarsh birds would include point-count surveys to estimate population trends at broad geographic scales as well as nest monitoring and capture-recapture methods for estimating demographic parameters at specific locations. Studies comparing demographic parameters among sites undergoing different management treatments would be particularly helpful for incorporating the needs of saltmarsh birds into future management plans. Conducting long-term demographic studies in marshes that also are sampled as part of a broad-scale vocal survey effort has many benefits (i.e., provides a correlation between survey data and demographic parameters).

The purpose of this chapter is to outline standardized methods for assessing the status of birds that breed in saltmarshes. The objective of this proposed survey effort was to create a series of monitoring efforts that will provide information on the status and the changes in status of terrestrial birds living in saltmarsh systems of North America. We have information on current status of bird populations within only a few of the tidal systems in North America, and we lack appropriate data to estimate population trends (Shriver et al. 2004) or to compare avian abundance among tidal wetlands with any sort of confidence. In contrast, we have over 30 yr of count data from the BBS for assessing population trends for several hundred species of land birds. This document aims to encourage a monitoring effort that will help correct that discrepancy by establishing standardized surveys within tidal-marsh systems throughout North America.

The information contained here builds upon the protocols in Conway (2005) by encouraging those interested in saltmarsh passerines (and other saltmarsh birds) to conduct surveys using a standardized protocol similar to that being used for secretive marsh birds (i.e., rails, moorhens, gallinules, and bitterns). The standardized protocols in Conway (2005) focus on secretive marsh birds and over 100 organizations and biologists throughout North America are already conducting surveys following this protocol (Conway and Timmermans 2005, Conway and Nadeau 2006). However, most participants only record secretive marsh birds during their surveys. This document outlines standardized survey methods that focus on saltmarsh passerines such that these data can be collected simultaneously with surveys focusing on secretive marsh birds. The document also describes a standardized survey protocol for those only interested in surveying saltmarsh passerines. Standardization of this sort will allow data from surveys focusing on saltmarsh passerines to be easily pooled with data from surveys focusing on secretive marsh birds. Implementing these standardized surveys in saltmarshes across North America will help document regional and continental patterns in distribution and abundance of all birds associated with tidal marshes.

In addition to this protocol's broad-scale use to estimate population trends, we recommend that it also be used to inventory poorly known species or subspecies that breed in saltmarshes. Examples include the various subspecies of Savannah Sparrows (*Passerculus sandwichensis*) in coastal California and northwestern Mexico (Wheelwright and Rising 1993), the Coastal Plain Swamp Sparrow (*Melospiza geogiana nigrescens*) in the northeastern US (Greenberg and Droege 1990), the three subspecies of Song Sparrows (*Melospiza melodia*) that occur in tidal saltmarshes in San Francisco Bay, California (Marshall 1948a, b; Arcese et al. 2002), and the Eastern Black Rail (*Laterallus jamaicensis jamaicensis*), and California Black Rail (*L. j. coturniculus*; Eddleman et al. 1994, Conway et al. 2004). Many of the species targeted here have very patchy breeding distributions. The patchy distribution of these species needs to be taken into account when developing a sampling frame to implement these survey protocols.

The methods outlined here may still not be sufficient for some species of saltmarsh birds. For example, Saltmarsh Sharp-tailed Sparrows (*Ammodramus caudacutus*) and Black Rails rarely

TABLE 1. DESCRIPTION OF EXISTING AVIAN SURVEY EFFORTS WITHIN TIDAL SALTMARSHES IN NORTH AMERICA.

Name of survey effort	Target species	Surveyors record other species	Lead agency	Year effort began	Frequency	Location	N points or transects	Technique	Contact
Light-footed Clapper Rail Survey	Light-footed Clapper Rail	No	USFWS	1979	Annual	Southwestern CA, northern Mexico	n/a	Territory mapping	RZembal@ocwd.com.
California Clapper Rail Survey	California Clapper Rail	Other rails	USFWS	1972	Annual or biannual	San Francisco Bay, CA	50–75	Winter high tide airboat survey; breeding call-broadcast survey	Joy_Albertson@r1.fws.gov.
San Francisco Estuary Wetlands Regional Monitoring Program	All birds	Yes	PRBO Conservation Science	1996	Annual	San Francisco Bay, CA	>1,000	Point counts; territory mapping; demographic monitoring	mherzog@prbo.org.
Belding's Savannah Sparrow Survey	Belding's Savannah Sparrow	No	USFWS	1986	Every 5 yr	Southern CA	n/a	Territory mapping	RZembal@ocwd.com.
North American marsh bird Survey[a]	Rails, bitterns, grebes	Some do	USGS	1999	Annual	all North America	~4,200[b]	Point counts with call broadcast	cconway@usgs.gov.
Coastal Plain Swamp Sparrow Survey	Coastal Plain Swamp Sparrow	No	Smithsonian	2000	Annual	Chesapeake and, Delaware Bays, MD, DE, NJ	141	Roadside point counts	GreenbergR@si.edu.
Waterbird Monitoring Program at Cape Cod National Seashore	Secretive marsh birds	No	NPS	1999	Every 3–5 yr	Cape Cod National Seashore, MA	42	Point counts with call broadcast	Robert_Cook@nps.gov.
New England Survey of Saltmarsh Birds	All saltmarsh birds	Yes	Massachusetts Audubon, Maine Dept. Inland Fisheries and Wildlife	1997	One time survey	ME, NH, MA, RI, CT	911	Point counts	gshriver@udel.edu

TABLE 1. CONTINUED.

Name of survey effort	Target species	Surveyors record other species	Lead agency	Year effort began	Frequency	Location	N points or transects	Technique	Contact
Gulf of Mexico Winter Survey of Saltmarsh Birds	Saltmarsh passerines	Yes	Mississippi Department of Marine Resources	2003	Annual	MS	17	Line transects during winter	msw103@ra.msstate.edu.
Galilee Bird Sanctuary	All birds	Yes	University of Rhode Island	1993	Annual or biannual	Southern RI	31	Point counts	ppaton@URI.edu.
Region 5 National Wildlife Refuge Surveys of Saltmarsh Birds	All birds (emphasis waterbirds)	Some do	USGS, USFWS	2000	Annual	DE to ME	30–40	Spring, fall winter point counts, walking transects	rme5g@c.ms.mail.virginia.edu
North American Breeding Bird Survey	All birds	Yes	USGS	1966	Annual	All North America	~300	Roadside point counts	kpardieck@usgs.gov.

[a] Incorporates the survey strategy described in this document.
[b] Includes all points including those in freshwater marshes.

vocalize. For Black Rails, we recommend use of call-broadcast surveys to increase vocalization probability. The methods for such broadcasts are discussed in Conway (2005). For Saltmarsh Sharp-tailed Sparrows, a second phase of more intense monitoring methods may need to be added in locations where these hard-to-detect species breed. For example, line-transect surveys that radiate out from each survey point could be used at a subset of marshes whereby observers record the number of birds detected while walking the line transects.

SURVEY AREA AND DEFINITION OF ANALYSIS UNITS

This document is meant to provide guidance to those wishing to conduct surveys for diurnal passerine birds within any tidal marsh in North America from Mexico north through Canada. These protocols are intended to be useful for monitoring birds in marshes dominated by shrubs, emergent wetland plants, and grasses, but not mangrove wetlands. As with all survey efforts, one must define the size of the smallest unit of land that will be analyzed for population changes. The size of that land unit, along with the statistical issues of precision, accuracy, and the analytical model used to calculate change will dictate how many samples the monitoring program will need to meet program objectives. We envision that the smallest analysis unit for this monitoring effort is formed from ecological units of saltmarshes, sometimes bounded by state and provincial boundaries. The list includes natural groupings of saltmarshes based on location and natural history. The following list of potential regions for monitoring birds associated with saltmarshes includes all the major tidal systems on the continent (this list does not imply priority or rank): southeastern Alaska and British Columbia; Strait of Georgia-Puget Sound; coastal Washington to Northern California; San Francisco Bay (with Suisun, San Pablo Bay, and south and central San Francisco Bay subregions); southern California; Baja, and Gulf of California (including Sonora and Sinaloa coastal plains plus Nayarit Marismas Nacionales); Pacific Coast from Jalisco to Chiapas; Gulf of Mexico coast from Rio Bravo (Grande) to Rio Tonala; Tabasco and Campeche Wetlands; Yucatan Peninsula coastal wetlands (including Cozumel); coastal Texas and Louisiana; Mississippi Delta; coastal Mississippi and Alabama; Gulf Coast of Florida; Atlantic Coast of Florida; coastal Georgia; coastal South Carolina; coastal North Carolina and Virginia north to the Chesapeake Bay; western shore of the Chesapeake Bay; eastern shore of the Chesapeake Bay; coastal Virginia, Maryland, and Delaware north of the Chesapeake Bay; Delaware portion of the Delaware Bay; New Jersey portion of the Delaware Bay; coastal New Jersey and Long Island; Long Island Sound; Rhode Island east to Cape Cod's south shore including Martha's Vineyard and Nantucket; outer Cape Cod, Cape Cod Bay, and north to the Gulf of Maine; coastal Nova Scotia; Bay of Fundy; Gulf of Saint Lawrence excluding Newfoundland; and Newfoundland.

Sub-sampling within any of these units can provide detailed information at smaller scales such as individual states, counties, and refuges within each saltmarsh system. Our purpose here is to recommend a sampling methodology so that data can be shared and compared among saltmarsh systems in different parts of the continent. If biologists use different approaches to survey marsh birds within each saltmarsh system, then estimates of parameters such as relative abundance are not comparable among areas. Moreover, standardization of survey methods improves efficiency of data sharing and data management. For rare species that are of regional or national conservation concern, we may ultimately need to combine all available survey data (regardless of the survey methods used) to generate a trend estimate. We need careful planning and standardization to insure that all available survey data can be pooled to yield regional or range-wide estimates of population trends. Conforming to a standard sampling protocol may require compromises, but participants benefit by allowing them to put their results into a regional perspective and having the data they collect add to our understanding of marsh-bird dynamics at regional and continental scales.

MONITORING APPROACH

Point-count surveys have been the most common method used to monitor land birds in North America. For marsh birds, some efforts have incorporated call-broadcast, distance estimates, and fixed-radius circular plots into the basic technique of counting birds from a single point (Conway and Gibbs 2005). Line-transect surveys and plot-based searches, i.e., spot mapping, are alternative methods of monitoring marsh birds, but point-count surveys provide the most efficient way of monitoring population trends of marsh birds across a large geographic area and allow survey data to be pooled with data collected for secretive marsh birds (Conway 2005).

Participants at the October 2003 workshop on tidal-marsh vertebrates agreed that the methods outlined here should constitute the minimum information collected by everyone working on marsh birds in tidal systems. Individual collaborators may decide or agree to collect additional information pertinent to each area or each set of study objectives, but participants felt that these core variables were sufficient to meet the goal of creating statistically-informative indices relevant to determining the status of tidal-marsh birds:

1. Conduct initial 5-min passive point-count surveys at each survey point followed by a period of call-broadcast.
2. Record all individuals detected (irregardless of distance) for all species that are associated with saltmarshes (Appendix 1).
3. Record each individual bird detected on a separate data line and record whether each bird was heard and/or seen (and whether each was flying over).
4. Estimate the distance to each bird detected.
5. Include a column for repeats, so that observers can denote an individual bird detected at a point that is thought to be one that was already counted at a previous point.
6. Count only birds heard or seen in the tidal marshes (or flying over the marsh) even though upland areas may be within the counting radii.
7. Count only from dawn to 3 hr after dawn. Surveys conducted within the first 2 hr after dawn are optimal because detection probability of many species tends to decline after that, but detection remains relatively high for many species for 3 hr after dawn.
8. Use 400 m between adjacent points. If a participant wants adjacent points to be closer than 400 m due to local reasons, we recommend they use increments of 400 m (i.e., 200 m). Distance between adjacent points must be ≥200 m if a participant wants to calculate density estimates based on number of birds detected within a 100-m radius of each point.
9. Begin surveys after the bulk of spring migration for resident marsh birds has occurred (typically sometime between early March and mid-June depending on latitude) and should be completed prior to the date when detection probability of target species declines (typically sometime between May and early July depending on latitude and species of interest). In general, surveys should be conducted when calling frequency is highest for focal species. For many tidal-marsh systems this is a survey window of approximately 5 wk. Potential participants are encouraged to contact one of the authors for information on optimal survey timing in their region.
10. Surveys should occur during the first week following a high spring tide because many saltmarsh passerines are forced to renest and detection probability is high following these high tides.
11. Immediately following the 5-minute passive survey, broadcast calls of secretive marsh birds to elicit vocalizations of rails, bitterns, and other secretive marsh birds. See Conway (2005) for explanation of format for call-broadcast.
12. For secretive marsh birds, record whether or not each individual bird was detected during each 1-min interval during both the passive and call-broadcast periods. See Conway (2005) for list of secretive marsh birds. For saltmarsh passerines and other marsh birds, participants should only record detection data within the 1-min intervals if doing so is logistically feasible in their study area. Recording non-marsh species should be avoided as it takes time away from estimating distance for the focal species.

The data produced from these surveys will provide analysts with several options for calculating abundance indices, trend estimates, and detection probability based on the raw counts. An example of a completed data sheet for these survey efforts is attached (Appendix 2).

Because the variability in counts of birds is usually greater among points than within points, surveying more points is sometimes a better strategy for estimating population change than conducting repeated surveys at a smaller number of points (Link et al. 1994). However, other benefits are associated with conducting replicate surveys at each point. Conducting replicate surveys per year at each point expands the possible number of analyses that can be performed on the count data. Replicate surveys reduce the variance of the counts, permitting a more precise measurement of any changes to the index. Replicate surveys are especially useful during the first few years of a monitoring effort so analysts can learn more about the factors affecting these counts and to provide a basis for estimating the sample size needed to detect changes in abundance for target species. Once several years of data are collected in various tidal marshes across North America, analysts can determine the value of replicate surveys for monitoring and make appropriate adjustments to the standardized protocol. Having repeated counts also allows analysts to estimate the number of points that should have detected the species out of the collection of points that never once recorded the species (MacKenzie et al. 2002). Moreover, recent analyses indicate that repeated counts at points can be used to create another estimate of the average abundance of birds across a set of points (Kéry et al. 2005).

Participants should conduct three surveys annually during the presumed peak breeding season for marsh birds in their area. Each of the three replicate surveys should be conducted during a 10-d window, and each of the 10-d windows should be separated by 5 d. Seasonal timing of these three replicate survey windows will vary regionally depending on migration and breeding chronology of the primary marsh birds breeding in an area.

Participants should focus on bird species that breed in association with saltmarsh vegetation (Appendix 1). Individuals of these species flying over the marsh and individuals along the marsh-upland edge will also be counted. We also encourage participants to use methods similar to those outlined here to conduct winter surveys for saltmarsh passerines. Our knowledge of distribution, habitat use, and population trends during winter is poor for most saltmarsh passerines. Some examples of possible response variables that the resultant survey data would produce include:

1. An index of abundance based on the total number of birds detected (regardless of distance) along a survey route or within a marshland.
2. An index of breeding density based on the numbers of birds detected within a certain radius (i.e., 50 or 100 m) of each point.
3. An estimate of breeding density based on distance sampling to correct for the fact that detection probability typically declines with distance from the surveyor.
4. An estimate of breeding density that incorporates both distance sampling and capture-recapture models (based on data from the five 1-min intervals) to account for detection probability being less than 100%.

Additional indices and methods for accounting for variation in detection probability are possible if all (or a subset) of points are surveyed three (or more) times per year. Replicate surveys at a point can provide estimates of site occupancy and estimates of the probability of missing a species at a point where it is indeed present (MacKenzie et al. 2002). Replicate surveys at a point also provide a method of calculating the percent area occupied by each species. For these reasons, we recommend that participants conduct three replicate surveys per year at each point (but those who are only able to conduct one or two replicate surveys per year are still encouraged to participate and follow these survey methods).

Several factors are known to affect detection probability of birds in tidal marshes. Some of these factors can be measured and accounted for during the data-analysis stage either by eliminating survey data that do not meet minimal conditions or adding the factor as a covariate in the analyses. Below is a list of necessary information that needs to be collected at each point.

ANCILLARY INFORMATION AT EACH POINT

In addition to using standardizing methods for conducting bird surveys in marshlands, we recommend that surveyors collect ancillary information (e.g., salinity of water, moon phase, tide stage, water depth, vegetation measurements, and current or ongoing management actions) at each survey point. These ancillary data may help document patterns of association between bird populations and geographic locations, habitats, and management actions. Such patterns may help generate hypotheses regarding possible causes of population change. Required ancillary information: (1). date, (2). name of marsh or study site, (3). full name of surveyor, (4). survey number (whether current survey is the first, second, or third at that point this year), (5). unique station number identifying the location of the point count, (6). latitude and longitude to four decimal places using a GPS receiver, (7). start time, (8). wind speed (Beaufort Code), (9). ambient temperature, (10). percent cloud cover, (11). amount of precipitation during past 24 hrs, (12). days since last full moon, (13). time of last high tide, (14). salinity of water, (15). an estimate of distance to each bird detected, (16). type(s) of call given, (17). water depth of the marsh at the time of the survey, and (18). characterization of plant-species composition and land-use types within a 50-m radius of each survey point. Information on plant species and land-use should be recorded annually if possible, but at least once every 5 yr. See Conway (2005) for more details on recording plant composition and land-use data at each survey point.

Rationale for Ancillary Information

Weather variables

Wind speed, ambient temperature, percent cloud cover, and amount of precipitation during the past 24 hr are factors that can influence vocalization probability of marsh birds (Conway and Gibbs 2001, 2005) and the ability of observers to hear marsh-bird calls. Hence, recording these parameters can help explain some variation across years in number of marsh birds counted.

Moon phase

Amount of moon light can potentially affect detection probability of some marsh birds. For example, the number of Black Rails detected on surveys in California was positively correlated with amount of moon light the preceding night (Spear et al. 1999). Relatively few studies have examined the influence of moon phase on detection probability of saltmarsh birds, so recording the number of days since last full moon in a broad-scale monitoring effort will provide guidance for revised protocols and future survey efforts.

Tide stage

Stage of the tidal cycle can potentially affect detection probability and habitat occupancy of some marsh birds. For example, the number of Black Rails detected on surveys in California was negatively correlated with tide height (Spear et al. 1999). Tide stage can also affect access to some saltmarshes. Relatively few studies have examined the influence of tide stage on detection probability of saltmarsh birds, so including this parameter in a broad-scale monitoring effort will provide guidance for revised protocols and future survey efforts. Until more information is available on the effects of tide stage, surveys in tidal marshes should always be conducted at a similar tidal stage at each point for each replicate survey both within and across years. The tidal stage within which to conduct local marsh-bird surveys should be based on when highest numbers of marsh birds are likely to be detected in your area; optimal tidal stage for surveys may vary among regions. Many saltmarsh passerines are forced to renest during the peak spring high tide, and detection probability for these species is highest during the week after a high spring tide (Shriver 2002). Clapper Rail (*Rallus longirostris*) surveys have been conducted during high tide since 1972 at San Francisco Bay National Wildlife Refuge, but high tide was a period of reduced vocalization probability for Clapper Rails in southern California (Zembal and Massey 1987) and for Black Rails in northern California (Spear et al. 1999). Current guidelines for conducting Clapper Rail surveys in San Francisco Bay suggest that surveys should not be conducted during high tides or during full moon periods and should be conducted when tidal sloughs are no more than bank full (M. Herzog, Point Reyes Bird Observatory, unpubl. data). As a general guideline, surveys in tidal marshes should not be conducted on mornings or evenings when high or low tide falls within the morning (or evening) survey time period. We need additional research designed to quantify the effects of tide stage on detection probability for all species of saltmarsh birds. Conway and Gibbs (2001) provide a review of previous studies that have examined the effects of environmental factors on detection probability of secretive marsh birds.

Salinity of water

Salinity varies spatially both within and among marshes and can also vary over time. Participants are encouraged to record the salinity content of the water directly in front of each point on each survey. Salinity level may affect a site's use by certain species. Such information is relatively easy to collect and can be used as a covariate to control for variation in models estimating population change. Handheld salinity meters are available for <$30.

Distance to each bird

Surveyors should estimate the distance to every bird detected at each point with no maximum limit or upper threshold. Recording the distance to a calling bird that is not visible will often require the surveyor to provide a rough estimate of distance based on the volume of the call. Obviously these distance estimates will not always be accurate, but with a large pooled sample size the pooled data set can be used to produce a distance-detection function for each species which will allow the estimation of detection probability using distance-sampling methodology. We realize that distance estimation is difficult and accuracy of any one distance estimate is suspect. That is acceptable. Surveyors should just try to ensure that they are not always underestimating or always overestimating distance to birds. Participants should note in the comments column of the data sheet their perceived accuracy of their distance estimates. Having observers put each bird into distance categories (rather than estimate distance) may make them feel a little better, but the potential for bias is still the same (analyses will require that we make the distance variable continuous and use the mid-point of each category). Estimating whether a bird is 80–100 or 100–120 m away is just as problematic as estimating actual distance to each bird. We can always convert distance estimates to distance categories after the fact if observers estimate distance. Ultimately, some analysts may use the count data while ignoring the distance data and others can use the distance data for what it is worth. See Conway and Nadeau (2006) for more information on rationale for estimating distance to each bird.

Call type

Including the types of calls given by each bird detected allows analysts to account for variation among observers in their ability to identify different calls and account for the fact that the probability of detection differs among different call types. Controlling for call type may improve our ability to estimate population trends across time by accounting for variation in observers' ability to identify species' calls. Each focal species of secretive marsh bird has two–five common calls. Some of these calls are loud, raucous, easy to learn, and unique (easy to hear at a great distance and difficult to confuse with other calls). Others are soft or easy to confuse with other species' calls. Hence, the observer detection probability for a particular species likely differs depending on the type of call given. Recording the call(s) given by each bird allows observers to estimate population trends of a particular species in several ways: (1) using all detections regardless of call given, (2) restricting the analysis to include only birds that gave the most common call for that species, or (3) restricting the analysis to include only birds that gave the most distinguishable call for that species. Data on calls given by each species can also help account for the potential bias associated with long-term surveys if the timing of the breeding season changes over time. Many marsh birds have particular calls (e.g., the Virginia Rail's [*Rallus limicola*] *ticket* and the Clapper Rail's *kek*) that are only given during the pairing and early mating season. The proportion of these calls relative to calls given by mated pairs (e.g., the Virginia Rail's *grunt* and the Clapper Rail's *clatter*) can provide a basis for testing whether the timing of the breeding season has changed over time and whether or not surveys were conducted during the same stage of the breeding cycle in different locations.

These data can also be used to refine the seasonal survey windows in the continental protocol so that surveys are conducted during the same stage of the breeding cycle in each region of North America (to the extent possible).

Water depth

Water depth affects the suitability of a marshland for many species of marsh birds and water depth can change over time in response to both natural and anthropogenic processes. Recording water depth during each survey can help explain some of the variation in the number of birds counted each year. Water depth is known to affect abundance of marsh birds and water depth in marshlands often varies greatly across years and even across replicate surveys within a year. Recording water depth will allow analysts to use this important parameter as a covariate in models used to estimate population change. To do so, place a permanent device for recording water depth within each marsh at which surveys are conducted and record water depth before or after each survey (i.e., before the first survey point or after the final survey point on each morning that surveys are conducted).

Plant species composition and land use

Participants should include information on the management actions (spraying, burning, draw downs, or other management activities that might affect bird abundance) that have recently occurred in the 100-m radius surrounding each point. Participants should also document plant composition surrounding each survey point. Plant composition within a tidal marsh naturally changes over time. The rate of such changes may increase due to predicted increases in sea levels. Changes in plant composition within tidal marshes may also be exacerbated by man-made hydrological changes resulting from such actions as manipulation of sediment deposition, changes in nutrient inputs, changes in farming practices in the surrounding landscape, and manipulation of the way water enters and exits a marsh. Characterizing the changes in plant composition surrounding each survey point will allow analysts to determine whether changes in bird abundance are correlated with changes in plant composition. Similarly, recording the date of management actions (spraying, burning, draw downs, or other management activities that might affect bird abundance) that have recently occurred in the 100-m radius surrounding each point will allow analysts to determine whether certain management actions adversely affect marsh-bird populations.

A recommended, but optional, component in each survey area involves the use of multiple observers at a subset of surveys (Conway and Nadeau 2006). The double-observer technique (Nichols et al. 2000) is a very useful way of detecting differences in observer detection probability (i.e., observer bias) among surveyors. However, it does have the drawback of requiring that two observers be present at a point. Moreover, the method only corrects for biases associated with differences caused by observer bias. Because many people travel in marshes in pairs there will be times when no additional person-hours would be required to conduct double-observer surveys. Double-observer surveys are also a very useful method of determining whether newly trained surveyors are ready to conduct surveys independently. Comparing survey results after a survey is complete provides a useful means of giving surveyors feedback on particular species or groups of species for which they need more practice. Double-observer surveys do not need to be conducted at every point and participants may want to conduct these surveys at a subset of points each year to have estimates of observer bias and to identify individuals who have poor hearing or low detection abilities.

SAMPLING METHODOLOGY

Conducting surveys in tidal marshes can present some logistical difficulties. Many tidal marshes are in remote locations, terrain can be treacherous, access is often limited, and changing tides can pose challenges for coordinating safe entry and departure routes. Consequently, conducting surveys at a system of point-count stations placed randomly or systematically throughout a large tidal marsh would be logistically difficult in many systems. Hence, workshop participants explored alternative approaches for locating survey stations within a tidal marsh. Participants discussed five alternative sampling methodologies: (1) random or systematic selection of points, (2) roadside access points, (3) water access points, (4) points within interior marsh, and (5) points placed at special locations.

Locating points via some form of random or systematic approach is ideal. Spatial variation in marsh-bird abundance is typically high within a marshland; birds are often clumped within particular areas. Points can be stratified to account for difficulty of access, patterns of marsh vegetation, hydrology, or perceived importance of particular areas within the marshes in a region (e.g., marshlands on national wildlife refuges). Using a systematic grid placed over a map of the marshland to locate sampling points is a good way to ensure that a marshland is adequately sampled. Tide stage affects behavior of saltmarsh birds and needs to be considered when choosing locations of survey points.

Roadside access points can be used effectively in situations where roads come in close contact with marshlands. Examples include bridge crossings, roads through marshlands, boat-access points, and impoundment roads. Conducting point-count surveys at roadside access points has numerous logistical benefits. These areas are usually easily accessible, safe, dry, and appealing to potential surveyors. However, using roadside-access points to survey tidal-marsh birds causes large sections of interior marshland to go un-sampled and prevents analysts from making inferences to the entire marshland. One compromise would be to include some roadside- and water-access points and some interior-marsh points. Survey points along roadsides and waterways should be established at 400-m intervals along all roads and waterways within the marshland. If all of the points cannot be surveyed, the participant should subdivide the marshland into sectors such that each sector has an equal number of potential survey points. The participant should then randomly select which of the sectors will be sampled and all suitable points in that sector should be sampled. Because the location of suitable marsh vegetation can change over time, participants may need to add additional survey points (but never eliminate points) in future years to ensure that all suitable areas within the sector are sampled. If the marsh vegetation surrounding a pre-existing survey point is no longer present (and hence the area is no longer suitable for any marsh birds), surveyors should record the point on the data form and note that the survey was not conducted because of insufficient habitat. One difficulty with water-access points is that marshes can sometimes overtake small channels or open water areas, making it difficult for surveyors to access these points in future years.

Any location within a marsh that is not within 400 m of a road or an accessible waterway (a somewhat arbitrary distance beyond which many birds cannot be heard from a point) is considered un-sampled marsh interior. These areas need to be defined and then sampling locations can be regularly spaced throughout as a way to supplement or complement roadside and water access points. The spacing and number of points will be determined by the sample size requirements for the region and the ease by which those points can be sampled.

Participants may also want to survey special places, either because they are known to be important areas for target species or because they are of interest for special management or research efforts. Departures from regular spacing or surveying special places outside of a defined sampling frame would either need to have a statistical justification (such as a stratification scheme) or the additional points treated separately during analysis. For example, it is completely appropriate to put in a point at a spot simply because that location is known to have high numbers of birds. Indeed, you might have some high counts or discover rare species there, but that point would have to be treated separately in analyses.

An investigator or group of investigators may employ any combination of the five sampling approaches discussed above, but the results from those surveys must always be tempered by an explicit reminder of the limits to the inferences which can be made using each of these approaches. Moreover, participants need to record explicitly how each survey point was identified and to which of the five sampling approaches that point contributes. This information needs to be in the database and will be very important to analysts who will need to know the scope of inference possible from the data collected at each site.

Reviewing the consequences of using any of these five different sampling strategies using geographic information system (GIS) overlays is recommended. The portions of marshlands that would go un-sampled using any of the above combinations of sampling strategies and the relative costs in terms of number of points and access time will be more apparent. This approach would allow sampling alternatives to be scrutinized prior to the start of sampling.

NUMBERS OF SAMPLING POINTS

Determining optimal or sufficient sample sizes for a region requires someone to estimate the temporal variability of the proposed counts, choose a period of years over which estimates of change are desired, define the minimum amount of change in population size that is thought to be important to detect, define the minimum levels of statistical precision needed to detect those changes, and choose an analytical approach to measuring change that permits sample sizes to be estimated using some form of power analysis. Hence, an estimate of the number of sampling points needed to estimate trends in saltmarsh birds is not currently possible, but will be available once various individuals collect data following this protocol.

The ability to yield range-wide trend estimates largely depends on the sampling frame used to locate survey points, the number of points surveyed that detect more than one individual of a particular species, and variation in detection probability of that species. This manuscript summarizes a standardized survey method, and does not address or make recommendations regarding the number of survey points at which this protocol will or should be used.

CONDUCTING SURVEYS FOR SALTMARSH PASSERINES ONLY

A standardized marsh-bird monitoring protocol that targets rails, bitterns, and other secretive marsh birds (Conway 2005) is already developed and being used by hundreds of biologists in a variety of federal, state, and nongovernmental organizations across North America (Conway and Nadeau 2006). This marsh-bird monitoring effort includes the use of call broadcast to increase detection probability for certain species that are otherwise difficult to detect. Individuals currently participating in this program have the option of recording all marsh birds, including those not on their broadcast sequence (e.g., saltmarsh passerines). Hence, individuals wanting information on saltmarsh passerines are encouraged to include a call-broadcast portion following the initial 5-min passive point count so that their data will be compatible with other marsh-bird surveys in their region. However, some organizations or biologists may not want to include call-broadcast for certain reasons. These individuals are to follow the survey methods outlined here for the first 5-min passive point-count survey. Doing so will allow their data to be comparable to the initial 5-min of data from other marsh-bird survey efforts. Substantial benefits come from having all individuals conducting surveys within both fresh and saltwater marshes in North America using as similar methods as possible. Organizations interested in potentially conducting avian surveys within any marshland system in North America are encouraged to contact the authors of this paper to discuss standardization of survey methods and the extent to which they can or cannot follow the protocols outlined in this document.

PERSONNEL AND TRAINING

All observers should have the ability to identify all common calls of marsh-bird species in their area. Observers should listen to recorded calls of the species common in their area and also practice call identification at marshes (outside the intended survey area if necessary) where the common species in their region are frequently heard calling. All observers should take and pass a self-administered vocalization identification exam each year prior to conducting surveys. All observers should also be trained to estimate distance to calling marsh birds, and to identify the common species of emergent plants on their area. Methods for training observers to estimate distance include: (1) broadcast calls in the marsh at an known distance and have observers estimate distance, (2) choose a piece of vegetation in the marsh where the bird is thought to be calling from and use a range finder to determine distance, and (3) have an observer estimate the distance to a bird that is calling with regularity and is near a road or marsh edge, have a second observer walk along the road or edge until adjacent to that calling bird, and then measure this distance by pacing or use of a GPS receiver. Surveyors should use some combination of these three methods prior to the survey season to practice estimating distances to calling birds. Double-observer surveys are very useful in this regard. After a survey is complete, the two observers can discuss not only what species they heard, but how far each person estimated the distance to each bird. Periodic double-observer surveys not only produce estimates of observer bias but also allow participants to determine whether one person is constantly underestimating or overestimating distance to calling birds. All surveyors should also have a hearing test at a qualified hearing or medical clinic before, during, or immediately after the survey season each year. These data can be included as a covariate and will help control for observer bias in trend analyses. New participants should do at least one trial run before their first data-collection window begins because it takes time to get used to the data sheet and recording the data appropriately.

EQUIPMENT AND MATERIALS

If possible, fixed survey points should be permanently marked with inconspicuous markers and numbered. Portable GPS receivers should be used to mark survey points onto aerial maps. GPS coordinates of each permanent survey point should be recorded and saved for reference in future years. A compact disc (CD) with calls of secretive marsh birds in your area should be obtained from the first author of this document (C. Conway) and new CDs should be requested if quality declines. CD players and amplified speakers should be good quality and batteries should be changed or re-charged frequently (before sound quality declines). Participants should routinely ask themselves if the quality of the broadcast sound is high. Observers should always carry replacement batteries on all surveys. A sound level meter with ±5 dB precision (Radio Shack model #33-2050 for $34.99; or EXTECH sound-level meter, $99 from Forestry Suppliers, Inc.) should be used to standardize broadcast volume. A small boat or canoe may be useful for surveying larger wetland habitats adjacent to open water, reducing travel time between survey points. When using a boat, use the same boat and engine on each survey each year to control for possible effects of engine noise on detection probability. If a different boat or different engine is used, make a note of the change in the comments column on the datasheet (Appendix 2). A spare CD player should be carried in case the primary unit fails to operate. A prototype field data form for use on vocal surveys is attached to this document (Appendix 2). The number of columns on the datasheet will vary among survey areas depending on the number of bird species included in the call-broadcast segment of your survey so participants will have to tailor the datasheet below to suite their own broadcast sequence. Contact the first author for an electronic copy of the data form.

ACKNOWLEDGMENTS

This document is based on many discussions with researchers, managers, and biometricians on the need and the optimal methods for improved monitoring of all birds associated with marshlands. In particular, the participants of the marsh-bird monitoring workshop held in 1998 (Ribic et al. 1999), the participants of the tidal-marsh vertebrate workshop held in October 2003, and the encouragement of J. Taylor from Region 5 of the USDI Fish and Wildlife Service, laid the groundwork and created the impetus for holding a workshop on creating a system of tidal-marsh-bird surveys. C. Hunter, J. Bart, R. Greenberg, and C. Marti provided helpful comments that improved the manuscript.

APPENDIX 1. SALTMARSH BIRD SPECIES (AND BIRD BANDING LABORATORY ALPHA CODES) OF PARTICULAR IMPORTANCE FOR SURVEYS. SEE CONWAY (2005) FOR COMPLETE LIST OF MARSH BIRDS (THOSE ASSOCIATED WITH BOTH FRESHWATER AND SALTWATER MARSHES).

AMBI	American Bittern	*Botaurus lentiginosus*
LEBI	Least Bittern	*Ixobrychus exilis*
GRHE	Green Heron	*Butorides virescens*
GBHE	Great Blue Heron	*Ardea herodias*
TRHE	Tricolored Heron	*Egretta tricolor*
LBHE	Little Blue Heron	*Egretta caerulea*
SNEG	Snowy Egret	*Egretta thula*
GREG	Great Egret	*Ardea alba*
CAEG	Cattle Egret	*Bubulcus ibis*
YCNH	Yellow-crowned Night Heron	*Nyctanassa violacea*
BCNH	Black-crowned Night Heron	*Nycticorax nycticorax*
GLIB	Glossy Ibis	*Plegadis falcinellus*
WFIB	White-faced Ibis	*Plegadis chihi*
WHIB	White Ibis	*Eudocimus albus*
NOHA	Northern Harrier	*Circus cyaneus*
OSPR	Osprey	*Pandion haliaetus*
BLRA	Black Rail	*Laterallus jamaicensis*
YERA	Yellow Rail	*Coturnicops noveboracensis*
SORA	Sora	*Porzana carolina*
VIRA	Virginia Rail	*Rallus limicola*
CLRA	Clapper Rail	*Rallus longirostris*
BNST	Black-necked Stilt	*Himantopus mexicanus*
WILL	Willet	*Catoptrophorus semipalmatus*
WISN	Wilson's Snipe	*Gallinago delicata*
FOTE	Forster's Tern	*Sterna forsteri*
BEKI	Belted Kingfisher	*Ceryle alcyon*
SEWR	Sedge Wren	*Cistothorus platensis*
MAWR	Marsh Wren	*Cistothorus palustris*
COYE	Common Yellowthroat	*Geothlypis trichas*
SSTS	Saltmarsh Sharp-tailed Sparrow	*Ammodramus caudacutus*
NSTS	Nelson's Sharp-tailed Sparrow	*Ammodramus nelsoni*
SESP	Seaside Sparrow	*Ammodramus maritimus*
SOSP	Song Sparrow	*Melospiza melodia*
SWSP	Swamp Sparrow	*Melospiza georgiana*
SAVS	Savannah Sparrow	*Passerculus sandwichensis*
RWBL	Red-winged Blackbird	*Agelaius phoeniceus*
BTGR	Boat-tailed Grackle	*Quiscalus major*
GTGR	Great-tailed Grackle	*Quiscalus mexicanus*

APPENDIX 2. Example of completed data form for marsh bird surveys in North American tidal marshes.

Marsh Bird Survey Data Sheet

Date (eg 10-May-04): 20 April 2006
Name of marsh or route : Hidden Shores Marsh
Observer(s) (list all)*: Chris Nadeau, Bob Blabla
Survey replicate # : 2
Water depth (cm): 12
Salinity content (ppt): 20

	Before	After
Temperature (°F):	10	21
Wind speed (mph):	0	1
Cloud cover (%):	15	60
Precipitation:	none	none
Days since full moon:	5	
Time of last high tide:	0513	

*list all observers in order of their contribution to the data collected
put an "S" in the appropriate column if the bird was seen, a "1" if the bird was heard, and "1S" if both heard and seen

Station#	Start Time (military)	Background noise	Species	Before	Pass 0-1	Pass 1-2	Pass 2-3	Pass 3-4	Pass 4-5	BLRA 5-6	LEBI 6-7	VIRA 7-8	CLRA 8-9	After	# Detected	Call Type(s)	Direction	Distance (m)	Detected at a Previous Point	Comments
HSM1	0610	0	BLRA		1	1				1						grr	↙	95	N	
			RWBL		1	1	1	1	1							song	↓	20	N	
			SOSP		1		1									song	↑	45	N	
			BLRA			1				1	1	1	1	1		kkkerr	↓	110	N	
			SOSP			1	s									song	↙	80	N	
			VIRA				1s	1				1				ticket, grunt	↙	30	N	
			SWSP				1									NDC	↙	12	N	

APPENDIX 2. CONTINUED.

HSM2	0621	1	CLRA	1							p-cltr	◐	40	N
			CLRA	1							p-cltr	◐	45	N
			VIRA			1					grunt	◐	100	Y
			CLRA					1			Throaty hoo	◐	10	N
HSM3														Not surveyed; no habitat
HSM4	0650	1	COMO	1	1	1		1			wipeout	◐	150	N
			MAWR	1	1	1s					song	◐	10	N
			MAWR	1s	1s	1					song	◐	40	N
			MAWR	1	1						call	◐	65	N
			MAWR		1				1		song	◐	120	N
			SORA								perweep	◐	210	N
			RWBL				16					◯		too many to list by line
HSM5	0705	2										◯		No birds detected
												◯		
												◯		

Call Types: BLRA: *kkkerr, grr, churt* CLRA: *cltr, kburr, kek, khurrah* LEBI: *coo, kak, ert* VIRA: *grunt, ticket, kicker*

If the call type is not one of the above listed types, describe the call in the comments column. See Conway (2005) for complete list of call types.

Background noise: 0 *no noise* 1 *faint noise* 2 *moderate noise (probably can't hear some birds beyond 100m)*
3 *loud noise (probably can't hear some birds beyond 50m)* 4 *intense noise (probably can't hear some birds beyond 25m)*

AN AGENDA FOR RESEARCH ON THE ECOLOGY, EVOLUTION, AND CONSERVATION OF TIDAL-MARSH VERTEBRATES

THE SYMPOSIUM CONTRIBUTORS

In this volume, we have taken a major step toward a more holistic view of the ecology, evolution, and conservation of tidal-marsh vertebrates. We provide strong evidence that numerous global issues of environmental and conservation concern face tidal-marsh biota, and that tidal marshes are a model system used in studying fundamental issues of biogeography and evolutionary biology. Future investigations into the ecology of tidal-marsh vertebrates will require a more comprehensive and comparative approach than has been employed. For beginning investigators, in particular, who are interested in doing pioneering work based on comparison and synthesis of processes among regions, the tidal-marsh system offers considerable promise. In an effort to further catalyze this research, we offer the following menu of activities, themes, and questions that provide a framework for progress in the study of tidal-marsh vertebrates:

1. A uniform global inventory of the distribution of tidal marshes, categorized by salinity and vegetation type. Such an inventory needs to be made available on a web site for tidal-marsh researchers throughout the world.
2. An increase of research on the Quaternary (Pleistocene, Holocene, and very recent) history of tidal marshes, focusing on their extent, distribution, and floral composition through time. Every effort should be made to apply the research broadly and with an explicitly geographic comparative component. Results of such a comprehensive historical survey could be presented along with the current tidal-marsh distribution, in a web atlas that is periodically revised and updated.
3. Standardized inventory and monitoring data for tidal-marsh vertebrates. This effort can begin by application of the tidal-marsh monitoring protocols for birds developed in Conway and Droege (*this volume*), and expanded to small mammals and reptiles and conducted in representative tidal marshes along all coastlines.
4. More comparative work focused on tidal-marsh taxa (vertebrate, invertebrate, and plant) living along different coastlines. Most current work, as seen in this volume, has been concentrated in North America. Globally comparative research would provide better tests for functional and adaptive hypotheses that were originally developed from studies focused on individual marsh systems.
5. More work integrating the role of physiological, trophic, life-history, and social factors in shaping adaptations to tidal marsh environments. Along with this, we need to develop models for the factors that drive and inhibit divergence of tidal-marsh populations from their inland source and sister populations.
6. More fine-scale genetic and morphological studies from less well known coastlines to determine if the differentiation described for North American taxa is mirrored elsewhere.
7. Continued inventory work on basic distributional data on which vertebrate species occur in tidal marshes and how dependent they are on this habitat. Such data may exist already in the published or un-reviewed literature, but they need to be compiled into usable and accessible formats. In cases where published data are lacking, faunal inventories should be initiated.
8. More work on tidal-marsh trophic-resource relationships throughout the world and how these compare to freshwater systems. This would include more information on the diet of terrestrial vertebrates, quantitative monitoring of terrestrial marsh arthropods, marine invertebrates, seeds, fruits, and other edible vegetation, and experimental analysis of the relationship between vertebrates and their food base.
9. Empirical monitoring and modeling approaches to determine how the distribution of different species will respond to regional habitat loss and changes in the floristic composition and vegetative structure of tidal marshes resulting from sea-level rise. We have seen how the reproductive success of tidal-marsh birds (and probably mammals and reptiles as well) is delicately balanced

between avoiding predation and flood loss. We need to develop predictive models for how changes in mean sea level and changes in the frequency of severe storms might influence flooding regimes, and how this will effect the survival and reproductive output of tidal-marsh vertebrates. Sea-level rise and changes in the hydrology of coastal estuaries also portend changes in salinity; how this might affect less specialized saltmarsh species would be an interesting question of applied physiological ecology.

10. Research focused on the facility with which tidal-marsh species adapt to radical changes in dominant vegetation caused by the advent of invasive species. Tidal marshes appear to be particularly prone to invasions by dominant species of plants. Given the low diversity of the tidal-marsh flora, such invasions have a major impact on the structure and function of marsh ecosystems. Similar studies should be conducted also on the impact of invasive fauna (such as rats, mice, nutrias, and opossums). Finally, more work needs to assess the impact of invasive mollusks and other non-native invertebrates on tidal-marsh systems.

11. More research on the effect of pollution and toxic chemicals on vertebrates in the marsh systems, especially those that are threatened and endangered. Tidal marshes are particularly vulnerable to the inputs of pollutants, nutrients, and wastes from agricultural and industrial development applied within the watersheds that support them, as well as chemical spills and contaminants introduced from the coastal marine waters. Furthermore, tidal marshes are often the recipient of broad-spectrum pesticides for mosquito control applied repeatedly throughout the season, a practice that may be exacerbated by the perception that the habitat is a source of vectors for emerging diseases, such as West Nile virus.

12. Greater integration with efforts to maintain populations of threatened and endangered species associated with non-tidal-marsh habitats in the same estuarine systems.

In spite of the aforementioned holes in our scientific understanding of vertebrates in tidal-marsh ecosystems, coastal-wetland ecosystems have hosted considerable research efforts. One of the problems we must overcome is the tendency for the balkanization of past research not only by discipline, but also by focal taxa and coastline and estuary of interest. We therefore recommend the formation of an international congress focused on biological conservation and ecological research on tidal-marsh wildlife, with the explicit goal of increasing the interchange of information between researchers on different continents and coastlines. We would encourage the program to be broadly interdisciplinary, covering areas that we as ornithologists and wildlife biologists need to understand in order to make sense of the ecology, evolution, and conservation of tidal-marsh vertebrates. We believe that, coupled with a renewed research agenda, we can go forward towards catalyzing a global approach to tidal-marsh biodiversity.

Finally, researchers of tidal-marsh systems are often motivated by their desire to positively influence the conservation of this fascinating ecosystem. Regular meetings of scientists and policy makers could advance conservation of tidal marshes from the different coastal areas to share and evaluate approaches that are working to conserve and restore coastal wetlands. At this time, tidal-marsh conservation is not the priority focus of any one organization approaching the issue from a global perspective. The development of such an organization or network focused on tidal marshes from a global perspective would be a major step forward in meeting this goal of drawing attention to the heretofore under recognized tidal-marsh global resources. We cannot afford to do otherwise.

LITERATURE CITED

ABERLE, B. L. 1993. The biology and control of introduced *Spartina* (cordgrass) worldwide and recommendations for its control in Washington. M.S. thesis, The Evergreen State College, Olympia, WA.

ABLE, K.W., S. M. HGAN, AND S. A. BROWN. 2003. Mechanisms of marsh habitat alteration due to *Phragmites*: response of young-of-the-year mummichog (*Fundulus heteroclitus*) to treatment for *Phragmites* control. Estuaries 26:484–494.

ACCURSO, L. M. 1992. Distribution and abundance of wintering waterfowl on San Francisco Bay 1988–1990. M.S. thesis, Humboldt State University, Arcata, CA.

ACKERMAN, J. T., J. Y. TAKEKAWA, K. L. KRUSE, D. L. ORTHMEYER, J. L. YEE, C. R. ELY, D. H. WARD, K. S. BOLLINGER, AND D. M. MULCAHY. 2004. Using radiotelemetry to monitor cardiac response of free-living Tule Greater White-fronted Geese (*Anser albifrons elgasi*) to human disturbance. Wilson Bulletin 116:146–151.

ADAM, P. 1990. Saltmarsh ecology. Cambridge University Press, Cambridge, UK.

ADAMS J. M., AND H. FAURE (EDITORS). 1997. Review and atlas of palaeovegetation: preliminary land ecosystem maps of the world since the last glacial maximum. Oak Ridge National Laboratory, TN. <http://www.esd.ornl.gov/projects/qen/adams1.html> (2 January 2006).

ALBERTSON, J. D. 1995. Ecology of the California Clapper Rail in south San Francisco Bay. M.S. thesis, San Francisco State University, San Francisco, CA.

ALBERTSON, J. D., AND J. G. EVENS. 2000. California Clapper Rail. Pp. 332–341 *in* P. R. Olofson (editor). Baylands ecosystem species and community profiles: life histories and environmental requirements of key plants, fish and wildlife. San Francisco Bay Regional Water Quality Control Board, Oakland, CA.

ALDRICH, J. W. 1984. Ecogeographical variation in size and proportions of Song Sparrows (*Melospiza melodia*). Ornithological Monographs 35:1–134.

ALEXANDER, C. E., M. A. BROUTMAN, AND D. W. FIELD. 1986. An inventory of coastal wetlands of the USA. U.S. Department of Commerce, National Oceanic and Atmospheric Administration, Washington, DC.

ALLANO, L., P. BONNET, P. CONSTANT, AND M. C. EYBERT. 1994. Habitat structure and population density in the Bluethroat *Luscinia svecica* Nannetum Mayqud. Revue d' Ecologie La Tierre et la Vie 49: 21–33.

ALLEN, A. A. 1914. The Red-winged Blackbird: a study in the ecology of a cat-tail marsh. Proceedings of the Linnaean Society of New York 24–25:43–128.

ALLISON, S. K. 1992. The influence of rainfall variability on the species composition of a northern California salt marsh plant assemblage. Vegetatio 101:145–160.

ALLISON, S. K. 1995. Recovery from small-scale anthropogenic disturbances by northern California salt marsh plant assemblages. Ecological Applications 5:693–702.

ALPERS, C., AND M. HUNERLACH. 2000. Mercury contamination from historic gold mining in California. USGS Fact Sheet FS-061-00. Department of the Interior, U.S. Geological Survey <http://ca.water.usgs.gov/mercury/fs06100.html> (6 June 2006).

ÁLVAREZ-CASTAÑEDA, S. T., AND J. L. PATTON. 1999. Mamíferos del Noroeste de México. Centro de Investigaciones Biologicas del Noroeste, S. C. Baja California, México.

AMBUEL, B., AND S. A. TEMPLE. 1983. Area-dependent changes in the bird communities and vegetation of southern Wisconsin forests. Ecology 64: 1057–1068.

AMERICAN GEOPHYSICAL UNION. 2004. Salt marsh geomorphology: physical and ecological effects on landform. Chapman Conference Program. Halifax, NS, Canada.

AMERICAN ORNITHOLOGISTS' UNION. 1931. Check-list of North American birds, 4th ed. Lancaster, PA.

AMERICAN ORNITHOLOGISTS' UNION. 1957. Check-list of North American birds, 5th ed. Lord Baltimore Press, Baltimore, MD.

AMERICAN ORNITHOLOGISTS' UNION. 1973. Thirty-second supplement to the American Ornithologists' Union check-list of North American birds. Auk 90:411–419.

AMERICAN ORNITHOLOGISTS' UNION. 1983. Check-list of North American birds, 6th ed. Allen Press, Lawrence, KS.

AMERICAN ORNITHOLOGISTS' UNION. 1995. Fortieth supplement to the American Ornithologists' Union Checklist of North American Birds. Auk 112:819–830.

AMERICAN ORNITHOLOGISTS' UNION. 1998. Check-list of North American birds, 7th edition. American Ornithologists' Union, Washington, DC.

AMMON, E. M. 1995. Lincoln's Sparrow (*Melospiza lincolnii*). *In* A. Poole, and F. Gill (editors). The birds of North America, No. 191. The Academy of Natural Sciences, Philadelphia, PA and The American Ornithologists' Union, Washington, DC.

AMOS, C. L., AND B. F. N. LONG. 1987. The sedimentary character of the Minas Basin, Bay of Fundy. Pp. 123–152 *in* S. B. McCann (editor). The coastline of Canada. Geological Survey of Canada Paper 80-10. Minister of Supply and Services. Ottawa, ON, Canada.

AMOS, C. L., AND K. T. TEE. 1989. Suspended sediment transport processes in Cumberland Basin. Journal of Geophysical Research 94:14407–14417.

ANDERSON, P. K. 1961. Variation in populations of brown snakes genus *Storeria* bordering on the Gulf of Mexico. American Midland Naturalist 66: 235–249.

ANDRÉN, H. 1994. Effects of habitat fragmentation on birds and mammals in landscapes with different proportions of suitable habitat: a review. Oikos 71:355–366.

ANDRESEN, H., J. P. BAKKER, M. BRONGERS, B. HEYDEMANN, AND U. IRMLER. 1990. Long-term changes of salt marsh communities by cattle grazing. Vegetatio 89:137–148.

ANDREWS, A. 1990. Fragmentation of habitat by roads and utility corridors: a review. Australian Zoologist 2:130–141.

ANDREWS, H. F. 1980. Nest-related behavior of the Clapper Rail (*Rallus longirostris*). Ph.D. dissertation, Rutgers University, Newark, NJ.

ARBOGAST, B. S., S. V. EDWARDS, J. WAKELEY, P. BEERLI, AND J. B. SLOWINSKI. 2002. Estimating divergence times from molecular data on phylogenetic and population genetic timescales. Annual Review of Ecology and Systematics 33:707–740.

ARBOGAST, B. S., AND G. J. KENAGY. 2001. Comparative phylogeography as an integrative approach to historical biogeography. Journal of Biogeography 28:819–825.

ARCESE, P., J. N. M. SMITH, W. M. HOCHACHKA, C. M. ROGERS, AND D. LUDWIG. 1992. Stability, regulation, and the determination of abundance in an insular Song Sparrow population. Ecology 73:805–822.

ARCESE, P., A. K. SOGGE, A. B. MARR, AND M. A. PATTEN. 2002. Song Sparrow (*Melospiza melodia*). *In* A. Poole, and F. Gill (editors). The birds of North America, No. 704. The Academy of Natural Sciences, Philadelphia, PA and The American Ornithologists' Union, Washington, DC.

ASHMOLE, N. P. 1968. Body size, prey size, and ecological segregation in five sympatric tropical terns (Aves: Laridae). Systematic Zoology 17:292–304.

ASHTON, R. E., AND P. S. ASHTON. 1991. Handbook of reptiles and amphibians of Florida. Part two: lizards, turtles, and crocodilians. Revised 2nd edition. Windward Publication, Miami. FL.

ATWATER, B. F. 1979. Ancient processes at the site of southern San Francisco Bay: movement of the crust and changes in sea level. Pp. 31–45 *in* T. J. Conomos (editor). San Francisco Bay: the urbanized estuary. Pacific Division, American Association for the Advancement of Science, San Francisco, CA.

ATWATER, B. F., AND D. F. BELKNAP. 1980. Tidal wetland deposits of the Sacramento-San Joaquin Delta, California. Pp. 89–103 *in* M. E. Field (editor). Quaternary depositional environments of the Pacific coast,. Society of Economic Paleontologists and Mineralogists, Pacific Section, Los Angeles, CA.

ATWATER, B. F., S. G. CONARD, J. N. DOWDEN, C. W. HEDEL, R. L. MACDONALD, AND W. SAVAGE. 1979. History, landforms, and vegetation of the estuary's tidal marshes. Pp. 347–380 *in* T. J. Conomos (editor). San Francisco Bay: the urbanized estuary. Pacific Division/American Association for the Advancement of Science, San Francisco, CA.

ATWATER, B. F., C. W. HEDEL, AND E. J. HELLEY. 1977. Late quaternary depositional history, Holocene sea-level changes, and vertical crustal movement, southern San Francisco Bay, California. U.S. Geological Survey Professional Paper 1014, Menlo Park, CA.

ATWATER, B. F., AND E. HEMPHILL-HALEY. 1997. Recurrence intervals for great earthquakes of the past 3,500 years at northeastern Willapa Bay, Washington. U.S. Geological Survey Professional Paper 1576, Seattle, WA.

AUGER, P. J. 1989. Sex ratio and nesting behavior in a population of *Malaclemys terrapin* displaying temperature-dependent sex-determination. Ph.D. dissertation, Tufts University, Medford, MA.

AUSTIN-SMITH, P. J. 1998. Tantramar dykeland status report. Canadian Wildlife Service, Sackville, NB, Canada.

AUSTIN-SMITH, P., D. KIDSTON, A. MOSCOSO, AND M. RADER. 2000. Habitat delineation and verification of the Cole Harbour saltmarsh. Unpublished report Cole Harbour Joint Project. Cole Harbour, NS, Canada.

AVISE, J. C. 2000. Phylogeography: the history and formation of species. Harvard University Press, Cambridge, MA.

AVISE, J. C., AND R. M. BALL, JR. 1990. Principles of genealogical concordance in species concepts and biological taxonomy. Pp. 47–61 *in* D. Futuyma, and J. Antonovics (editors). Oxford Surveys in Evolutionary Biology. Oxford University Press, New York, NY.

AVISE, J. C., AND R. A. LANSMAN. 1983. Polymorphism of mitochondrial DNA in populations of higher animals. Pp. 147–164 *in* M. Nei, and R. K. Koehn (editors). Evolution of genes and proteins. Sinauer Associates, Inc., Sunderland, MA.

AVISE, J. C., AND W. S. NELSON. 1989. Molecular genetic relationships of the extinct Dusky Seaside Sparrow. Science 243:646–648.

AVISE, J. C., J. C. PATTON, AND C. F. AQUADRO. 1980. Evolutionary genetics of birds II. Conservative protein evolution in North American sparrows and relatives. Systematic Zoology 29:323–334.

AVISE, J. C., AND D. WALKER. 1998. Pleistocene phylogeographic effects on avian populations and the speciation process. Proceedings of the Royal Society of London Series B Biological Sciences 265:457–463.

AVISE, J. C., AND R. M. ZINK. 1988. Molecular genetic divergence between avian sibling species: King and Clapper Rails, Long-billed and Short-billed Dowitchers, Boat-tailed and Great-tailed Grackles, and Tufted and Black-crested Titmice. Auk 105:516–528.

AYRES, D. R., P. BAYE, AND D. R. STRONG. 2003. *Spartina foliosa* (Poaceae) — common species on the road to rarity? Madroño 50:209–213.

AYRES, D. R., D. GARCIA-ROSSI, H. G. DAVIS, AND D. R. STRONG. 1999. Extent and degree of hybridization between exotic (*Spartina alterniflora*) and native (*S. foliosa*) cordgrass (Poaceae) in California, USA determined by random amplified polymorphic DNA (RAPDs). Molecular Ecology 8: 1179–1186.

AYRES, D. R., D. L. SMITH, K. ZAREMBA, S. KLOHR, AND D. R. STRONG. 2004. Spread of exotic cordgrasses and hybrids (*Spartina* sp.) in the tidal marshes of San Francisco Bay, CA, USA. Biological Invasions 6:221–231.

BAGLEY, M. J., AND J. B. GELLER. 2000. Microsatellite DNA analysis of native and invading populations of European green crabs. Pp. 241–242 *in* J. Pederson (editor) Marine bioinvasions:

proceedings of the first national conference. MIT Sea Grant College Program, Cambridge, MA.

BAIRD, C. 1995. Environmental chemistry. W.H. Freeman, New York, NY.

BAJA CALIFORNIA WETLAND INVENTORY. 2004. <http://proesteros.cicese.mx/investigacion/inv_hum/cont/intro.htm> (20 January 2006),

BAKER, J. L. 1973. Preliminary studies of the Dusky Seaside Sparrow on the St. Johns National Wildlife Refuge. Proceedings of the Annual Conference of Southeastern Association of Game and Fish Commissioners 27:207–214.

BAKER, R. H. 1940. Effects of burning and grazing on rodent populations. Journal of Mammalogy 21:223.

BAKER, W. S., F. E. HAYES, AND E. W. LATHROP. 1992. Avian use of vernal pools at the Santa Rosa Plateau Preserve, Santa Ana Mountains, California. Southwestern Naturalist 37:392–403.

BAKKER, K. K., D. E. NAUGLE, AND K. F. HIGGINS. 2002. Incorporating landscape attributes into models for migratory grassland bird conservation. Conservation Biology 16:1638–1646.

BALABAN, E. 1988. Cultural and genetic variation in the Swamp Sparrow (*Melospiza georgiana*). I. Song variation, genetic variation and their relationship. Behaviour 105:250–291.

BALL, R. M., JR., AND J. C. AVISE. 1992. Mitochondrial DNA phylogeographic differentiation among avian populations and the evolutionary significance of subspecies. Auk 109:626–636.

BALL, R. M., JR., S. FREEMAN, F. C. JAMES, E. BERMINGHAM, AND J. C. AVISE. 1988. Phylogeographic population structure of Red-winged Blackbirds assessed by mitochondrial DNA. Proceedings of the National Academy of Sciences 85:1558–1562.

BALLARD, J. W. O., AND M. C. WHITLOCK. 2004. The incomplete natural history of mitochondria. Molecular Ecology 13:729–744.

BANGS, O. 1898. The land mammmals of peninsular Florida and the coast region of Georgia. Proceedings of the Boston Society of Natural History 27:1–6.

BARD, E., B. HAMELIN, M. ARNOLD, L. MONTAGGIONI, G. CABIOCH, G. FAURE, AND F. ROUGERIE. 1996. Deglacial sea level record from Tahiti corals and the timing of global meltwater discharge. Nature 32: 241–244.

BARDWELL, E., C. W. BENKMAN, AND W. R. GOULD. 2001. Adaptive geographic variation in Western Scrub-Jays. Ecology 82:2617–2627.

BARKAY, T., M. GILLMAN, AND R. R. TURNER. 1997. Effects of dissolved organic carbon and salinity on bioavailability of mercury. Applied Environmental Microbiology 63:4267–4271.

BARRIE, J. V., AND K. W. CONWAY. 2002. Rapid sea-level change and coastal evolution on the Pacific margin of Canada. Sedimentary Geology 150: 171–183.

BARROWCLOUGH, G. F. 1983. Biochemical studies of microevolutionary processes. Pp. 223–261 *in* A. H. Brush, and G. A. J. Clark (editors). Perspectives in ornithology. Cambridge University Press, New York, NY.

BARROWCLOUGH, G. F., R. J. GUTIERREZ, AND J. G. GROTH. 1999. Phylogeography of spotted owl (*Strix occidentalis*) populations based on mitochondrial DNA sequences: gene flow, genetic structure, and a novel biogeographic pattern. Evolution 53:919–931.

BART, D., AND J. M. HARTMAN. 2003. The role of large rhizome dispersal and low salinity windows in the establishment of common reed, *Phragmites australis*, in salt marsh marshes: new links to human activities. Estuaries 2:436–443.

BARTHOLOMEW, G. A., AND T. J. CADE. 1963. The water economy of land birds. Auk 80:504–539.

BARTLETT, D. S., G. J. WHITING, AND J. M. HARTMAN. 1990. Use of vegetation indices to estimate intercepted solar radiation and net carbon dioxide exchange of a grass canopy. Remote Sensing of Environment 30:115–128.

BASHAM, M. P., AND L. R. MEWALDT. 1987. Salt water tolerance and the distribution of south San Francisco Bay Song Sparrows. Condor 89:697–709.

BASKIN, Y. 1994. California's ephemeral vernal pools may be a good model for speciation. BioScience 44:384–388.

BASSINOT, F. C., L. D. LABEYRIE, E. VINCENT, X. QUIDELLEUR, N. J. SHACKLETON, AND Y. LANCELOT. 1994. The astronomical theory of climate and the age of the Brunhes-Matuyama magnetic reversal. Earth and Planetary Science Letters 126:91–108.

BATT, B. D. J. (EDITOR). 1998. The Greater Snow Goose: report of the arctic goose habitat working group. Arctic Goose Joint Venture Special Publication. USDI Fish and Wildlife Service, Washington DC and Canadian Wildlife Service, Ottawa, ON, Canada.

BATZER, D., AND V. RESH. 1992. Wetland management strategies that enhance waterfowl habitats can also control mosquitoes. American Mosquito Control Association Journal 8:117–125.

BATZLI, G. O. 1986. Nutritional ecology of the California vole: effects of food quality on reproduction. Ecology 67:406–412.

BAYE, P. R., P. M. FABER, AND B. GREWELL. 2000.Tidal marsh plants of the San Francisco estuary. Pp. 9–33 *in* P. R. Olofson (editor). Baylands ecosystem species and community profiles: life histories and environmental requirements of key plants, fish and wildlife. San Francisco Bay Regional Water Quality Control Board, Oakland, CA.

BEAL, F. E. L. 1907. Birds of California in relation to the fruit industry. USDA Biological Survey Bulletin No. 30., U.S. Government Printing Office, Washington, DC.

BEECHER, W. J. 1951. Adaptations for food-getting in the American blackbirds. Auk 68:411–440.

BEECHER, W. J. 1955. Late Pleistocene isolation in salt-marsh sparrows. Ecology 36:23–28.

BEECHER, W. J. 1962. The bio-mechanics of the bird skull. Bulletin of the Chicago Academy of Sciences 11:10–33.

BEER, J. R., AND D. TIBBITS. 1950. Nesting behavior of the Red-winged Blackbird. Flicker 22:61–77.

BEIER, P., AND R. F. NOSS. 1998. Do habitat corridors provide connectivity? Conservation Biology 12: 1241–1252.

BELANGER, L., AND J. BEDARD. 1994. Role of ice scouring and goose grubbing in marsh plant dynamics. Journal of Ecology 82:437–445.

BELETSKY, L. D. 1996. The Red-winged Blackbird: the biology of a strongly polygynous songbird. Academic Press, San Diego, CA.

BELETSKY, L. D., AND G. H. ORIANS. 1990. Male parental care in a population of Red-winged Blackbirds, 1983-1988. Canadian Journal of Zoology 68: 606-609.

BELETSKY, L. D., AND G. H. ORIANS. 1991. Effects of breeding experience and familiarity on site fidelity in female Red-winged Blackbirds. Ecology 72: 787-796.

BELETSKY, L. D., AND G. H. ORIANS. 1996. Red-winged Blackbirds: decision-making and reproductive success. University of Chicago Press, Chicago, IL.

BELL, C., W. ACEVEDO, AND J. T. BUCHANAN. 1995. Pp. 723-734 in Dynamic mapping of urban regions: growth of the San Francisco Sacramento region. Proceedings, Urban and Regional Information Systems Association, San Antonio, TX. <http://landcover.usgs.gov/urban/umap/pubs/urisa_cb.asp> (16 January 2006).

BELL, D. M., M. J. HAMILTON, C. W. EDWARDS, L. E. WIGGINS, R. M. MARTINEZ, R. E. STRAUSS, R. D. BRADLEY, AND R. J. BAKER. 2001. Patterns of karyotypic megaevolution in *Reithrodontomys*: evidence from a cytochrome-b phylogenetic hypothesis. Journal of Mammalogy 82:81-91.

BELL, S. S., M. S. FONSECA, AND L. B. MOTTEN. 1997. Linking restoration and landscape ecology. Restoration Ecology 5:318-323.

BELLROSE, F. C. 1976. Ducks, geese, and swans of North America. Stackpole Books, Harrisburg, PA.

BELSLEY, D. A. 1980. Regression diagnostics: identifying influential data and sources of collinearity. Wiley, New York, NY.

BENDER, D. J., T. A. CONTRERAS, AND L. FAHRIG. 1998. Habitat loss and population decline: a meta-analysis of the patch size effect. Ecology 79:517-533.

BENKMAN, C. W. 1993. Adaptation to single resources and the evolution of crossbill (*Loxia*) diversity. Ecological Monographs 63:305-325.

BENOIT, L. K. 1997. Impact of the spread of *Phragmites* on populations of tidal marsh birds in Connecticut. M.A. thesis, Connecticut College, New London, CT.

BENOIT, L. K., AND R. A. ASKINS. 1999. Impact of the spread of *Phragmites* on the distribution of birds in Connecticut tidal marshes. Wetlands 19: 194-208.

BENOIT, L. K., AND R. A. ASKINS. 2002. Relationship between habitat area and the distribution of tidal marsh birds. Wilson Bulletin 114:314-323.

BENT, A. C. 1962a. Life histories of North American shorebirds. Part 1. Dover Publications, New York, NY.

BENT, A. C. 1962b. Life histories of North American shorebirds: part 2. Dover Publications, New York, NY.

BENT, A. C. 1963a. Life histories of North American gulls and terns. Dover Publications, New York, NY.

BENT, A. C. 1963b. Life histories of North American marsh birds. Dover Publications, New York, NY.

BERGIN, T. M., L. B. BEST, K. E. FREEMARK, AND K. J. KOEHLER. 2000. Effects of landscape structure on nest predation in roadsides of a midwestern agroecosystem: a multiscale analysis. Landscape Ecology 15:131-143.

BERTNESS, M. D. 1999. The ecology of Atlantic shorelines. Sinauer Associates, Sunderland, MA.

BERTNESS, M. D., AND A. M. ELLISON. 1987. Determinants of pattern in a New England marsh plant community. Ecological Monographs 57:129-147.

BERTNESS, M. D., P. J. EWANCHUK, AND B. R. SILLIMAN. 2002. Anthropogenic modification of New England salt marshes. Proceedings National Academy of Sciences 99:1395-1398.

BERTNESS, M. D., AND S. C. PENNINGS. 2000. Spatial variation in process and pattern in saltmarsh plant communities in eastern North America. Pp. 39-57 in M. P. Weinstein, and D. A. Kreeger (editors). Concepts and controversies in tidal marsh ecology. Kluwer Academic Publishers, Dordecht, The Netherlands.

BERTNESS, M. D., B. S. SILLIMAN, AND R. JEFFRIES. 2004. Saltmarshes under siege. American Scientist 92: 54-61.

BEUCHAT, C. A. 1990. Body size, medullary thickness, and urine concentrating ability in mammals. American Journal of Physiology 258:298-308.

BEUCHAT, C. A., M. R. PREEST, AND E. J. BRAUN. 1999. Glomerular and medullary architecture in the kidney of Anna's Hummingbird. Journal of Morphology 240:95-100.

BHATTACHARYYA, A., AND V. CHAUDHARY. 1997. Is it possible to demarcate floristically the Pleistocene/Holocene transition in India? Palaeobotanist 46: 186-190.

BIAS, M. A., AND M. L. MORRISON. 1999. Movements and home range of salt marsh harvest mice. Southwestern Naturalist 44:348-353.

BICKHAM, J. W., T. LAMB, P. MINX, AND J. C. PATTON. 1996. Molecular systematics of the genus *Clemmys* and the intergeneric relationships of emydid turtles. Herpetologica 52:89-97.

BISHOP, J. M. 1983. Incidental capture of diamondback terrapin by crab pots. Estuaries 6:426-430.

BISHOP, R., R. ANDREWS, AND R. BRIDGES. 1979. Marsh management and its relationship to vegetation, waterfowl, and muskrats. Proceedings of the Iowa Academy of Sciences 86:50-56.

BJÄRVALL, A., AND S. ULSTRÖM. 1986. The Mammals of Britain and Europe. Croom Helm, London, UK.

BLAKLEY, N. R. 1976. Successive polygyny in upland nesting Red-winged Blackbirds. Condor 78: 129-133.

BLAUSTEIN, A. R. 1980. Behavioral aspects of competition in a three species rodent guild of coastal southern California. Behavioural Ecology and Sociobiology 6:247-255.

BLEAKNEY, J. S. 2004. Sods, soils and spades: the Acadians at Grand Pre and their dykeland legacy. McGill-Queen's University Press. Montreal, QC, Canada.

BLEAKNEY, J. S., AND K. B. MEYER. 1979. Observations on saltmarsh pools, Minas Basin, Nova Scotia, 1965-1977. Proceedings of the Nova Scotia Institute of Science 29:353-371.

BLOCHER, D., M. EINSPENNER, AND J. ZAJACKOWSKI. 1989. CHEF electrophoresis: a sensitive technique for the determination of DNA double-strand breaks.

International Journal of Radiation Biology 56: 437–448.

Bó, M. S., J. P. Isacch, A. I. Malizia, and M. M. Martínez. 2002. Lista comentada de los Mamíferos de la Reserva de Biósfera Mar Chiquita, Provincia de Buenos Aires, Argentina. Mastozoología Neotropical 9:5–11.

Boag, P. T., and P. R. Grant. 1981. Intense natural selection in a population of Darwin's finches (Geospizinae) in the Galápagos, Ecuador. Science 214:82–85.

Boarman, W. I., and K. H. Berry. 1995. Common Ravens in the southwestern United States, 1968–92. Pp. 73–75 in E. L. LaRoe, G. S. Farris, and C. E. Puckett (editors). Our living resources: a report to the nation on the distribution, abundance, and health of U.S. plants, animals, and ecosystems. USDI National Biological Service, Washington, DC.

Boesch, D., D. Levin, D. Nummedal, and K. Bowles. 1983. Subsidence in coastal Louisiana: causes, rates, and effects in wetlands. USDI Fish and Wildlife Service, FWS/OBS-83/26. Washington, DC.

Bolger, D. T., A. C. Alberts, and M. E. Soulé. 1991. Occurrence patterns of bird species in habitat fragments: sampling, extinction, and nested species subsets. American Naturalist 137:155–166.

Bolger, D. T., T. A. Scott, and J. T. Rotenberry. 1997. Breeding bird abundance in an urbanizing landscape in coastal southern California. Conservation Biology 11:406–421.

Bongiorno, S. F. 1970. Nest-site selection by adult Laughing Gulls (*Larus atricilla*). Animal Behaviour 18:434–444.

Botti, F., D. Warenycia, and D. Becker. 1986. Utilization by salt marsh harvest mice *Reithrodontomys raviventris halicoetes* of a non-pickleweed marsh. California Fish and Game 72:62–64.

Boulenger, G. A. 1989. Catalogue of the chelonians, ryncocephalians, and crocodiles in the British Museum (Natural History). The British Museum, London, UK.

Boumans, R. M., and J. W. Day, Jr. 1993. High precision measurements of sediment elevation in shallow coastal areas using a sedimentation-erosion table. Estuaries 16:375–380.

Bowman, R. I. 1961. Morphological differentiation and adaptation in the Galapagos Finches. University of California Press, Berkeley, CA.

Boyer, K. B., and J. B. Zedler. 1998. Effects of nitrogen additions on the vertical structure of a constructed cordgrass marsh. Ecological Applications 8: 692–705.

Bradley, R. A. 1973. A population census of the Belding's Savannah Sparrow (*Passerculus sandwichensis beldingi*). Western Bird Bander 48:40–43.

Bradley, R. A. 1994. Cultural change and geographic variation in the songs of the Belding's Savannah Sparrow (*Passerculus sandwichensis beldingi*). Bulletin of the Southern California Academy of Sciences 93:91–109.

Bradley, R. S. 1985. Quaternary paleoclimatology: methods of paleoclimatic reconstruction. Unwin Hyman, Winchester, MA.

Brandl, R., A. Kristin, and B. Leisler. 1994. Dietary niche breadth in a local community of passerine birds: an analysis using phylogenetic contrasts. Oecologia 98:109–116.

Bratton, J. F., S. M. Colman, R. E. Thieler, and R. R. Seal. 2003. Birth of the modern Chesapeake Bay estuary between 7.4 and 8.2 ka and implications for global sea-level rise. Geo-Marine Letters 22: 188–197.

Brawley, A. H. 1994. Birds of the Connecticut River estuary: relating patterns of use to environmental conditions. Report to the Nature Conservancy—Connecticut Chapter, Conservation Biology Research Program, Hartford, CT.

Brawley, A. J., R. S. Warren, and R. A. Askins. 1998. Bird use of restoration and reference marshes within the Barn Island Wildlife Management Area, Stonington, Connecticut, USA. Environmental Management 22:625–633.

Brawn, J. D., S. K. Robinson, and F. R. Thompson, III. 2001. The role of disturbance in the ecology and conservation of birds. Annual Review of Ecology and Systematics 32:251–276.

Brazil, M. 1991. The birds of Japan. Smithsonian Institution Press, Washington, DC.

Breaux, A. M. 2000. Non-native predators: Norway rat and roof rat. Pp. 249–250 in P. R. Olofson (editor). Baylands ecosystem species and community profiles: life histories and environmental requirements of key plants, fish and wildlife. San Francisco Bay Area Wetlands Ecosystem Goals Project, and San Francisco Bay Regional Water Quality Control Board, Oakland, CA.

Bredy, J. P., and R. G. Botzler. 1989. The effects of six environmental variables on *Pasteurella multocida* populations in water. Journal of Wildlife Diseases 25: 232–239.

Brinson, M., R. Christian, and L. K. Blum. 1995. Multiple states in the sea-level induced transition from terrestrial forest to estuary. Estuaries 18:648–659.

Bronikowski, A. M., and S. J. Arnold. 2001. Cytochrome b phylogeny does not match subspecific classification in the western terrestrial garter snake, *Thamnophis elega*ns. Copeia 2001:508–513.

Brown, A. F., and P. W. Atkinson. 1996. Habitat associations of coastal wintering passerines. Bird Study 43:188–200.

Brown, M., and J. J. Dinsmore. 1986. Implications of marsh size and isolation for marsh bird management. Journal of Wildlife Management 50: 382–397.

Brush, T., R. Lent, T. Hruby, B. Harrington, R. Marshall, and W. Montgomery. 1986. Habitat use by salt marsh birds and response to open marsh water management. Colonial Waterbirds 9:189–195.

Buchanan, J. B. 2003. *Spartina* invasion of Pacific coast estuaries in the United States: implications for shorebird conservation. Wader Study Group Bulletin 100:47–49.

Buckley, F. G., and P. A. Buckley. 1982. Microenvironmental determinants of survival in saltmarsh-nesting Common Terns. Colonial Waterbirds 5:39–48.

BURDICK, D. M., M. DIONNE, R. M. BOUMAN, AND F. T. SHORT. 1997. Ecological responses to tidal restoration of two northern New England salt marshes. Wetlands Ecology and Management 4:129–144.

BURGER, J. 1977. Nesting behavior of Herring Gulls: invasion into *Spartina* salt marsh areas of New Jersey. Condor 79:162–169.

BURGER, J. 1979. Nest repair behavior in birds nesting in salt marshes. Journal of Comparative and Physiological Psychology 93:189–199.

BURGER, J. 1981. The effect of human activity on birds at a coastal bay. Biological Conservation 21:231–241.

BURGER, J. 1982. The role of reproductive success in colony-site selection and abandonment in Black Skimmers (*Rynchops niger*). Auk 99:109–115.

BURGER, J. 1993. Shorebird squeeze. Natural History 102:8–14.

BURGER, J. 1998. Effects of motorboats and personal watercraft on flight behavior over a colony of Common Terns. Condor 100:528–534.

BURGER, J., AND M. G. GOCHFELD. 1991a. The Common Tern: its breeding biology and social behavior. Columbia University Press, New York, NY.

BURGER, J., AND M. GOCHFELD. 1991b. Human activity influence and diurnal and nocturnal foraging of Sanderlings (*Calidris alba*). Condor 93:259–265.

BURGER, J., AND F. LESSER. 1978. Selection of colony sites and nest sites by Common Terns *Sterna hirundo* in Ocean County, New Jersey. Ibis 120:433–449.

BURGER, J., I. C. T. NISBET, C. SAFINA, AND M. GOCHFELD. 1996. Temporal patterns in reproductive success in the endangered Roseate Tern *Sterna dougallii* nesting on Long Island, New York, and Bird Island, Massachusetts. Auk 113:131–142.

BURGER, J., AND J. K. SHISLER. 1978a. The effects of ditching a salt marsh on colony and nest site selection by Herring Gulls (*Larus argentatus*). American Midland Naturalist 100:544–563.

BURGER, J., AND J. SHISLER. 1978b. Nest-site selection of Willets in a New Jersey saltmarsh. Wilson Bulletin. 90:599–607.

BURGER, J., AND J. SHISLER. 1980. Colony and nest site selection in Laughing Gulls in response to tidal flooding. Condor 82:251–258.

BURGER, J., J. K. SHISLER, AND F. R. LESSER. 1982. Avian utilization on six salt marshes in New Jersey. Biological Conservation 23:187–212.

BURGMAN, M. A., S. FERSON, AND H. R. AKOAKAYA. 1993. Risk assessment in conservation biology. Chapman and Hall. London, UK.

BURNELL, K. L. 1996. Genetic and cultural evolution in an endangered songbird, the Belding's Savannah Sparrow. Ph.D. dissertation, University of California, Santa Cruz, CA.

BURNHAM, K. P., AND D. R. ANDERSON. 2002. Model selection and inference: a practical information-theoretic approach, 2nd edition. Springer Verlag, New York, NY.

BYERS, J. E. 2000. Competition between two estuarine snails: implications for invasions of exoctic species. Ecology 81:1225–1239.

BYRNE, R., B. L. INGRAM, S. STARRATT, F. MALAMUD-ROAM, J. N. COLLINS, AND M. E. CONRAD, 2001. Carbon-isotope, diatom and pollen evidence for late Holocene salinity change in a brackish marsh in the San Francisco estuary. Quaternary Research 55:66–76.

BYSTRAK, D. 1981. The North American breeding bird survey. Studies in Avian Biology 6:34–41.

CADE, T. J., AND G. A. BARTHOLOMEW. 1959. Seawater and salt utilization by Savannah Sparrows. Physiological Zoology 32:230–238.

CADE, T. J., AND L. GREENWALD. 1966. Nasal salt secretion in falconiform birds. Condor 68:338–350.

CADWALLADR, D. A., AND J. V. MORLEY. 1973. Sheep grazing preferences on a saltings pasture and their significance for Widgeon (*Anas penelope* L.) conservation. British Grasslands Society 28:235–242.

CAHOON, D. R., J. W. DAY, JR., AND D. REED. 1999. The influence of surface and shallow subsurface soil processes on wetland elevation: a synthesis. Current Topics in Wetland Biogeochemistry 3:72–88.

CAHOON, D. R., J. C. LYNCH, AND R. M. KNAUS. 1996. Improved cryogenic coring device for sampling wetland soils. Journal of Sedimentary Research 66:1025–1027.

CAHOON, D. R., J. C. LYNCH, P. HENSEL, R. BOUMANS, B. C. PEREZ, B. SEGURA, AND J. W. DAY, JR. 2002. High-precision measurements of wetland sediment elevation: I. recent improvements to the sedimentation-erosion table. Journal of Sedimentary Research 72:730–733.

CAHOON, D. R., AND R. E. TURNER. 1989. Accretion and canal impacts in a rapidly subsiding wetland: II. feldspar marker horizon technique. Estuaries 12:260–268.

CALFLORA. 2003. Calflora species profile for *Spartina foliosa* <http://www.calflora.org> (6 June 2006).

CALIFORNIA ENERGY COMMISSION. 2003. California's oil refineries <http://www.energy.ca.gov/oil/refineries.html> (6 June 2006).

CALLAWAY, J. C., AND M. N. JOSSELYN. 1992. The introduction and spread of smooth cordgrass (*Spartina alterniflora*) in south San Francisco Bay. Estuaries 15:218–226.

CAM, E., J. E. HINES, J.-Y. MONNAT, J. D. NICHOLS, AND E. DANCHIN. 1998. Are adult nonbreeders prudent parents? The Kittiwake model. Ecology 79:2917–2930.

CAPEHART, A. A., AND C. T. HACKNEY. 1989. The potential role of roots and rhizomes in structuring salt-marsh benthic communities. Estuaries 12:119–122.

CARLTON, J. T. 1979. History, biogeography, and ecology of the introduced marine and estuarine invertebrates of the Pacific coast of North America. Ph.D. dissertation, University of California, Davis, CA.

CARR, A. 1952. Handbook of turtles: the turtles of the United States, Canada, and Baja California. Cornell University Press, Ithaca, NY.

CARROLL, C. R. 1992. Ecological management of sensitive areas. Pp. 237–256 *in* P. Fiedler, and S. Jain (editors). Conservation biology: the theory and practice of nature conservation preservation and management. Chapman and Hall, New York, NY.

CARSON, R., AND G. S. SPICER. 2003. A phylogenetic analysis of the emberizid sparrows based on three mitochondrial genes. Molecular Phylogenetics and Evolution 29:43–57.

CASE, N. A., AND O. H. HEWITT. 1963. Nesting and productivity of the Red-winged Blackbird in relation to habitat. Living Bird 2:7–20.

CASOTTI, G., AND E. J. BRAUN. 2000. Renal anatomy in sparrows from different environments. Journal of Morphology 243:283–291.

CASTELLE, A. J., C. CONOLLY, M. EMERS, E. D. METZ, S. MEYER, M. WITTER, S. MAUERMAN, T. ERICKSON, AND S. S. COOKE. 1992. Wetland buffers: use and effectiveness. Adolfson Associates, Inc., Shorelands and Coastal Zone Management Program, Publication No. 92-10. Washington Department of Ecology, Olympia, WA

CATCHPOLE, E. A., Y. FAN, B. J. T. MORGAN, T. H. CLUTTON-BROCK, AND T. COULSON. 2004. Sexual dimorphism, survival and dispersal in red deer. Journal of Agricultural, Biological, and Environmental Statistics 9:1–26.

CENTER FOR DISEASE CONTROL. 1998. Preventing emerging infectious disease: a strategy for the 21st Century. Centers for Disease Control and Prevention, U.S. Public Health Service, Department of Health and Human Services. Atlanta, GA.

CENTER FOR DISEASE CONTROL. 2001. Epidemic/epizootic West Nile virus in the United States: revised guidelines for surveillance, prevention, and control. Centers for Disease Control and Prevention, National Center for Infectious Diseases, Division of Vector-Borne Infectious Diseases. Fort Collins, CO.

CHABRECK, R. A. 1960. Coastal marsh impoundments for ducks in Louisiana. Proceedings of the Southeastern Association of Game and Fish Commissioners 14:24–29.

CHABRECK, R. A. 1968. The relation of cattle and cattle grazing to marsh wildlife and plants in Louisiana. Proceedings of the Annual Conference of Southeastern Association of Game and Fish Commissioners 22:55–58.

CHABRECK, R. A. 1981 Effect of burn date on regrowth rate of *Scirpus olneyi* and *Spartina patens*. Proceedings of the Annual Conference of the Southeastern Association of Fish and Wildlife Agencies 35:201–210.

CHABRECK, R. A. 1988. Coastal marshes: ecology and wildlife management. University of Minnesota Press, Minneapolis, MN.

CHABRECK, R. A., R. K. YANCEY, AND L. McNEASE. 1974. Duck usage of management units in the Louisiana coastal marsh. Proceedings of the Annual Meeting of the Southeastern Fish and Wildlife Association 28:507–516.

CHAGUÉ-GOFF, C., T. S. HAMILTON, AND D. B. SCOTT. 2001. Geochemical evidence for the recent changes in a saltmarsh, Chezzetcook Inlet, Nova Scotia, Canada. Proceedings of the Nova Scotia Institute of Science 41:149–159.

CHALFOUN, A. D., F. R. THOMPSON II, AND M. J. RATNASWAMY. 2002. Nest predators and fragmentation: a review and meta-analysis. Conservation Biology 16:306–318.

CHAMBERS, R. M. 1997. Porewater chemistry associated with *Phragmites* and *Spartina* in a Connecticut tidal marsh. Wetlands 17:360–367.

CHAMBERS, R. M., L. M. MEYERSON, AND K. SALTONSTALL. 1999. Expansion of *Phragmites australis* into tidal wetlands of North America. Aquatic Botany 64:261–273.

CHAMBERS, R. M., D. T. OSGOOD, D. J. BART, AND F. MONTALTO. 2003. *Phragmites australis* invasion and expansion in tidal wetlands: interactions among salinity, sulfide, and hydrology. Estuaries 28:398–406.

CHAN, Y., AND P. ARCESE. 2002. Subspecific differentiation and conservation of Song Sparrows (*Melospiza melodia*) in the San Francisco Bay region inferred by microsatellite loci analysis. Auk 119:641–657.

CHAN, Y., AND P. ARCESE. 2003. Morphological and microsatellite differentiation at a microgeographic scale. Journal of Evolutionary Biology 16:939–947.

CHAPMAN, J. A., AND G. A. FELDHAMMER (EDITORS). 1982. Wild mammals of North America: biology, management, and economics. Johns Hopkins University Press, Baltimore, MD.

CHAPMAN, V. J. 1974. Salt marshes and salt deserts of the world. J. Cramer, 2nd ed. Lehre, Austria.

CHAPMAN, V. J., (EDITOR). 1977. Wet coastal ecosystems. Ecosystems of the world. Vol. 1. Elsevier Scientific, Amsterdam, The Netherlands.

CHAPPELL, J., AND H. POLACH. 1991. Postglacial sea-level rise a coral record at Huon Peninsula, Papua-New Guinea. Nature 349:147–149.

CHAPPLE, D. G. 2003. Ecology, life history and behavior of the Australian genus *Egernia* with comments on the evolution of sociality in lizards. Herpetological Monographs 17:145–180.

CHEN, J. H., H. A. CURRAN, B. WHITE, AND G. J. WASSERBURG. 1991. Precise chronology of the last interglacial period: 234U-230Th data from fossil coral reefs in the Bahamas. Geological Society of America Bulletin 103:82–97.

CHMURA, G. L. 1997. Salt meadow hay (*Spartina patens*). Pp. 107–115 *in* M. Burt, (editor). Habitat identification of critical species in the Quoddy region of the Gulf of Maine. Report to the Gulf of Maine Council on the Marine Environment, Augusta, ME.

CHMURA, G. L., P. CHASE, AND J. BERCOVITCH. 1997. Climatic controls of the middle marsh zone in the Bay of Fundy. Estuaries 20:689–699.

CHMURA, G. L., AND G. A. HUNG. 2004. Controls on saltmarsh accretion: a test in saltmarshes of eastern Canada. Estuaries 27:70–81.

CLAGUE, J. J., AND T. S. JAMES. 2002. History and isostatic effects of the last ice sheet in southern British Columbia. Quaternary Science Reviews 21:71–87.

CLARK, D. R. JR., K. S. FOERSTER, C. M. MARN, AND R. L. HOTHEM. 1992. Uptake of environmental contaminants by small mammals in pickleweed habitats at San Francisco Bay, California. Archives of Environmental Contamination and Toxicology 22:389–396.

CLARK, P. U., A. M McCABE, A. C. MIX, AND A. J. WEAVER. 2004. Rapid rise of sea level 19,000 years ago and its global implications. Science 304:1141–1144.

CLARKE, J., B. A. HARRINGTON, T. HRUBY, AND F. E. WASSERMAN. 1984. The effect of ditching for mosquito control on salt marsh use by birds in Rowley, Massachusetts. Journal of Field Ornithology 55:160–180.

CLAY, W. M. 1938. A synopsis of North American water snakes. Copeia 1938:173–182.

CLIBURN, J. W. 1960. The phylogeny and zoogeography of North American *Natrix*. Ph.D. dissertation, University of Alabama, Tuscaloosa, AL.

CLOERN, J. E. 1982. Does the benthos control phytoplankton biomass in south San Francisco Bay (California USA)? Marine Ecology-Progress Series 9:191–202.

CODY, M. 1968. Ecological aspects of reproduction. Pp. 461–512 *in* D. S. Farner, and J. R. King (editors). Avian biology. Vol. 1. Academic Press, New York, NY.

CODY, M. L. 1985. Habitat selection in birds. Academic Press, Orlando, FL.

COGSWELL, H. L. 2000. Song Sparrow. Pp. 374–385 *in* M. Olufson (editor). Baylands ecosystem species and community profiles: life histories and environmental requirements of key plants, fish, and wildlife. San Francisco Bay Area Wetlands Ecosystem Goals Project. San Francisco Bay Regional Water Quality Control Board, Oakland, CA.

COHEN, A. N. 1997. The invasion of the estuaries. Pp. 6–9 *in* K. Patten (editor). Proceedings second international *Spartina* conference. Washington State University/Cooperative Extension, Long Beach, WA.

COHEN, A. N., AND J. T. CARLTON. 1995. Nonindigenous aquatic species in a United States estuary: a case study of the biological invasions of the San Francisco Bay and delta. Report for the USDI Fish and Wildlife Service, Washington, DC.

COHEN, A. N., AND J. T. CARLTON. 1997. Transoceanic transport mechanisms: the introduction of the Chinese mitten crab *Eriocheir sinensis* to California. Pacific Science 51:1–11

COHEN, A. N., AND J. T. CARLTON. 1998. Accelerating invasion rate in a highly invaded estuary. Science 279:555–558.

COHEN, A. N., J. T. CARLTON, AND M. C. FOUNTAIN. 1995. Introduction, dispersal and potential impacts of the green crab *Carcinus maenas* in San Francisco Bay. Marine Biology 122:225–237.

COHEN, A. N., AND B. FOSTER. 2000. The regulation of biological pollution: preventing exotic species invasions from ballast water discharged into California coastal waters. Golden Gate University Law Review 30:787–883.

COLLETTE, L. 1995. New Brunswick dykeland-strategy for protection, maintenance and development. Pp. 42–47 *in* J. Bain, and M. Evans (editors). Dykelands Enhancement Workshop. Canadian Wildlife Service, Sackville, NB, Canada.

COLLIER, R. E. L., M. R. LEEDER, M. TROUT, G. FERENTINOS, E. LYBERIS, AND G. PAPATHEODOROU. 2000. High sediment yields and cool, wet winters: test of last glacial paleoclimates in the northern Mediterranean. Geology 28:999–1002.

COLLINS, J. N. 2002. Invasion of San Francisco Bay by smooth cordgrass, *Spartina alterniflora*: a forecast of geomorphic effects on the intertidal zone. Unpublished report of San Francisco Estuary Institute, Oakland, CA.

COLLINS, J. N., AND V. H. RESH. 1985. Utilization of natural and man-made habitats by the salt marsh Song Sparrow, *Melospiza melodia samuelis* (Baird). California Fish and Game 71: 40–52.

COLLINS, J. N., AND V. H. RESH. 1989. Guidelines for the ecological control of mosquitoes in non-tidal wetlands of the San Francisco Bay area. California Mosquito and Vector Control Association, Elk Grove, CA.

COLLINS, P. W., AND S. B. GEORGE. 1990. Systematics and taxonomy of island and mainland populations of western harvest mice (*Reithrodontomys megalotis*) in southern California. Natural History Museum of Los Angeles County. Contributions in Science 420:1–26.

COLLOTY, B. M., J. B. ADAMS, AND G. C. BATE. 2000. The botanical importance of the estuaries in former Ciskei/Transkei. Water Resource Commission report 812/1/00. Water Research Commission, Pretoria, South Africa.

COLMAN, S. M., AND R. B. MIXON. 1988. The record of major Quaternary sea-level changes in a large coastal plain estuary, Chesapeake Bay, eastern United States. Palaeogeography, Palaeoclimatology, and Palaeoecology 68:99–116.

COLWELL, R. 2004. Using climate to predict infectious disease outbreaks: a review. Report for the Department of Communicable Diseases Surveillance and Response, the Department of Protection of the Human Environment, and the Roll Back Malaria Department. World Health Organization. Publication No. WHO/SDE/OEH/04.01. Geneva, Switzerland.

COMBS, S. M., AND R. G. BOTZLER. 1991. Correlations of daily activity with avian cholera mortality among wildfowl. Journal of Wildlife Diseases 27: 543–550.

CONANT, R. 1963. Evidence of the specific status of the water snake, *Natrix fasciata*. American Museum Novitates 2212:1–38.

CONANT, R. 1969. A review of the water snakes of the genus *Natrix* in Mexico. Bulletin of the American Museum of Natural History 142:1–140.

CONANT, R. 1975. A field guide to the reptiles and amphibians of eastern and central North America. Houghton Mifflin Co., Boston, MA.

CONANT, R., AND J. T. COLLINS. 1991. A field guide to reptiles and amphibians: eastern and central North America. Houghton Mifflin, Boston, MA.

CONANT, R., J. T. COLLINS, I. H. CONANT, T. R. JOHNSON, AND S. L. COLLINS. 1998. A field guide to reptiles and amphibians of eastern and central North America. Houghton Mifflin Co., New York, NY.

CONANT, R., AND J. D. LAZELL, JR. 1973. The Carolina salt marsh snake: a distinct form of *Natrix sipedon*. Brevoria 400:1–13.

CONGDON, J. D., A. E. DUNHAM, AND R. C. VAN LOBEN SELS. 1993. Delayed sexual maturity and demographics of Blanding's turtles (*Emydoidea blandingii*): implications for conservation and management of long-lived organisms. Conservation Biology 7: 826–833.

CONNER, K. 2002. Assessment of community response to EHJV enhancement of Saint John River floodplain wetlands. Wetlands and Coastal Habitat

Program-New Brunswick Department of Natural Resources and Energy, Fredericton, NB, Canada.

CONROY, C. J., AND J. A. COOK. 2000. Phylogeography of a post-glacial colonizer: *Microtus longicaudus* (Rodentia: Muridae). Molecular Ecology 9:165-175.

CONTRA COSTA MOSQUITO AND VECTOR CONTROL DISTRICT. 1997. Initial study and mitigated negative declaration: the integrated vector management program of the Contra Costa Mosquito and Vector Control District. Concord, CA.

CONWAY, C. J. 2005. Standardized North American marsh bird monitoring protocols. Wildlife Research Report No. 2005-04, U.S. Geological Survey, Arizona Cooperative Fish and Wildlife Research Unit, Tucson, AZ.

CONWAY, C. J., AND J. P. GIBBS. 2001. Factors influencing detection probability and the benefits of call-broadcast surveys for monitoring marsh birds. Final Report, U.S Geological Survey, Patuxent Wildlife Research Center, Laurel, MD.

CONWAY, C. J., AND J. P. GIBBS. 2005. Effectiveness of call-broadcast surveys for monitoring marsh birds. Auk 122:26-35.

CONWAY, C. J., AND C. P. NADEAU. 2006. Development and field-testing of survey methods for a continental marsh bird monitoring program in North America. Wildlife Research Report No. 2005-11. U.S. Geological Survey, Arizona Cooperative Fish and Wildlife Research Unit, Tucson, AZ.

CONWAY, C. J., C. SULZMAN, AND B. A. RAULSTON. 2004. Factors affecting detection probability of California Black Rails. Journal of Wildlife Management 68:360-370.

CONWAY, C. J., AND S. T. A. TIMMERMANS. 2005. Progress toward developing field protocols for a North American marsh bird monitoring program. Pages 997-1005 *in* C. J. Ralph, and T. D. Rich (editors). Bird conservation implementation and integration in the Americas: Proceedings of the Third International Partners in Flight Conference. Volume 2. USDA Forest Service General Technical Report PSW-GTR-191. USDA Forest Service, Pacific Southwest Research Station, Albany, CA.

COOKE, A. S. 1973. Shell thinning in avian eggs by environmental pollutants. Environmental Pollution 4:85-152.

CORDELL, J. R., AND S. M. MORRISON. 1996. The invasive Asian copepod *Pseudodiaptomus inopinus* in Orgeon, Washington, and British Columbia estuaries. Estuaries 19:629-638.

CORY, C. B., AND C. E. HELLMAYR. 1927. The catalogue of the birds of the Americas. Part V. Field Museum of Natural History publication No. 242. Chicago, IL.

COSTA, C. S. B., J. C. MARANGONI, AND A. M. G. AZEVEDO. 2003. Plant zonation in irregularly flooded salt marshes: relative importance of stress tolerance and biological interactions. Journal of Ecology 91:951-965.

COTTAM, C. 1938. The coordination of mosquito control with wildlife conservation. Proceedings of the New Jersey Mosquito Extermination Association 25:130-137.

COULOMBE, H. N. 1970. The role of succulent halophytes in the water balance of salt marsh rodents. Oecologia 4:223-247.

COWAN, F. B. M. 1969. Gross and microscopic anatomy of the orbital glands of *Malaclemys* and other emydine turtles. Canadian Journal of Zoology 47:723-729.

COWAN, F. B. M. 1971. The ultrastructure of the lachrymal "salt" gland and the Harderian gland in the euryhaline *Malaclemys* and some closely related stenohaline emydines. Canadian Journal of Zoology 49:691-697.

COWIE, I. D., AND P. A. WARNER. 1993. Alien plant species invasive in Kakadu National Park, tropical northern Australia. Biological Conservation 63:127-135.

COYER, J. A., A. F. PETERS, W. T. STAM, AND J. L. OLSEN. 2003. Post-ice age recolonization and differentiation of *Fucus serratus* L. (Phaeophyceae; Fucaceae) populations in northern Europe. Molecular Ecology 12:1817.

CRAIG, R. J., AND K. G. BEAL. 1992. The influence of habitat variables on marsh bird communities of the Connecticut River Estuary. Wilson Bulletin 104:295-311.

CRAMP, S. (EDITOR). 1988. Handbook of the birds of Europe, the Middle East, and North Africa. The birds of the western Palearctic. Vol. V. tyrant flycatchers to thrushes. Oxford University Press, Oxford, UK.

CRAMP, S. (EDITOR). 1992. Handbook of the birds of Europe, the Middle East, and North Africa. The birds of the western Palearctic. Vol. VI. warblers. Oxford University Press, Oxford, UK.

CRAMP, S., AND C. M. PERRINS (EDITORS). 1993. Handbook of the birds of Europe, the Middle East, and North Africa: The birds of the western Palearctic. Vol. VII. flycatchers to shrikes. Oxford University Press, Oxford, UK.

CRAMP, S., AND C. M. PERRINS (EDITORS). 1994. Handbook of the birds of Europe, the Middle East, and North Africa. The birds of the western Palearctic. Vol. VIII. crows to finches. Oxford University Press, Oxford, UK.

CRANDALL, K. A., O. R. P. BININDA-EMONDS, G. MACE, AND R. K. WAYNE. 2000. Considering evolutionary processes in conservation biology. Trends in Ecology and Evolution 15:290-295.

CRIMMINS, B. S., P. DOELLING-BROWN, D. P. KELSO, AND G. D. FOSTER. 2002 Bioaccumulation of PCBs in aquatic biota from a tidal freshwater marsh ecosystem. Archives of Environmental Contamination and Toxicology 42:296-404.

CRONK, Q. C. B., AND J. L. FULLER. 1995. Plant invaders. Chapman and Hall, London, UK.

CROOKS, J. A. 1998. Habitat alteration and community-level effects of an exotic mussel, *Musculista senhousia*. Marine Ecology Progress Series 162:137-152.

CROUSE, D. T., L. B. CROWDER, AND H. CASWELL. 1987. A stage-based population model for loggerhead sea turtles and implications for conservation. Ecology 68:1412-1423.

CUBASCH, U., AND G. A. MEEHL. 2001. Projections of future climate change. Pp. 525-582 *in* J. T.

Houghton, Y. Ding, D. J. Griggs, M. Noguer, P. J. van der Linden, X. Dai, K. Maskell, and C. A. Johnson (editors). Climate change 2001: the scientific basis. Cambridge University Press, Cambridge, UK.

Cuneo, K. 1987. San Francisco Bay salt marsh vegetation geography and ecology: a baseline for use in impact assessment and restoration planning. Ph.D. dissertation, University of California. Berkeley, CA.

Curnutt, J. L., A. L. Mayer, T. M. Brooks, L. Manne, O. Bass, Jr., D. M. Fleming, M. P. Nott, and S. L. Pimm. 1998. Population dynamics of the endangered Cape Sable Seaside Sparrow. Animal Conservation 1:11–21.

Cutter, G. A., and M. L. C. San Diego-McGlone. 1990. Temporal variability of selenium fluxes in San Francisco Bay. Science of the Total Environment 97/98:235–250.

D'Antonio, C. M., and P. M. Vitousek. 1992. Biological invasions by exotic grasses, the grass/fire cycle and global change. Annual Review of Ecology and Systematics 23:63–87.

Dachang, Z. 1996. Vegetation of coastal China. Beijing. Ocean Press, Beijing, China. (in Chinese)

Daehler, C. C. 1998. Variation in self-fertility and the reproductive advantage of self-fertility for an invading plant (Spartina alterniflora). Evolutionary Ecology 12:553–568

Daehler, C. C., and D. R. Strong. 1996. Status, prediction and prevention of introduced cordgrass Spartina spp. invasions in Pacific estuaries, USA. Biological Conservation 78:51–58.

Daehler, C. C., and D. R. Strong. 1997. Hybridization between introduced smooth cordgrass (Spartina alterniflora; Poaceae) and native California cordgrass (S. foliosa) in San Francisco Bay, California, USA. American Journal of Botany 84:607–611.

Dahlgren, R. B., and C. E. Korschgen. 1992. Human disturbances of waterfowl: an annotated bibliography. USDI Fish and Wildlife Service, Resource Publication 188. Washington, DC.

Daiber, F. C. 1982. Animals of the tidal marsh. Van Nostrand Reinhold, New York, NY.

Daiber, F. C. 1986. Conservation of tidal marshes. Van Nostrand Reinhold, New York, NY.

Daiber, F. 1987. A brief history of tidal marsh mosquito control, Pp. 233–252 in W. R. Whitman, and W. H. Meredith (editors). Waterfowl and wetlands symposium: proceedings of a symposium on waterfowl and wetlands management in the coastal zone of the Atlantic Flyway. Delaware Coastal Management Program, Delaware Department of Natural Resources and Environmental Control, Dover, DE.

Dale, P. E. R., and K. Hulsman. 1990. A critical review of salt marsh management methods for mosquito control. Reviews in Aquatic Sciences 3:281–311.

Dale, P. E. R., P. T. Dale, K. Hulsman, and B. H. Kay. 1993. Runnelling to control saltmarsh mosquitoes: long-term efficacy and environmental impacts. Journal American Mosquito Association 9:174–181.

Dame, R. D., M. Alber, D. Allen, M. Mallin, C. Montague, A. Lewitus, A. Chalmers, R. Gardner, C. Gilman, B. Kjerfve, J. Pinckney, and N. Smith. 2000. Estuaries of the south Atlantic coast of North America: their geographical signature. Estuaries 23:793–819.

Daniels, R. C., T. W. White, and K. K. Chapman. 1993. Sea-level rise: destruction of threatened and endangered species habitat in South Carolina. Environmental Management 17:373–385.

Davenport, J., and E. A. Macedo. 1990. Behavioral osmotic control in the euryhaline diamondback terrapin Malaclemys terrapin: responses to low salinity and rainfall. Journal of Zoology (London) 220:487–496.

Davenport, J., and T. M. Wong. 1986. Observations on the water economy of the estuarine turtles Batagur baska (Gray) and Callagur borneoensis (Schlegel and Muller). Comparative Biochemistry and Physiology 84A:703–707.

Davenport, J., T. M. Wong, and J. East. 1992. Feeding and digestion in the omnivorous estuarine turtle Batagur baska (Gray). Herpetological Journal 2: 133–139.

Davidson, N. C., and P. I. Rothwell. 1993. Disturbance to waterfowl on estuaries: the conservation and coastal management implications of current knowledge. Wader Study Group Bulletin 68: 98–105.

Davis, D. S., and S. Browne (editors). 1996a. The natural history of Nova Scotia—topics and habitats. Government of Nova Scotia and Nimbus Publishing, Halifax, NS, Canada.

Davis, D. S., and S. Browne (editors). 1996b. The natural history of Nova Scotia—theme regions. Government of Nova Scotia and Nimbus Publishing, Halifax, NS, Canada.

Davis, F. W., P. A. Stine, D. M. Stoms, M. I. Borchert, and A. D. Hollander. 1995. Gap analysis of the actual vegetation of California 1. The southwestern region. Madroño 42:40–78.

Davis, J. A., M. D. May, B. K. Greenfield, R. Fairey, C. Roberts, G. Ichikawa, M. S. Stoelting, J. S. Becker, and R. S. Tjeerdema. 2002. Contaminant concentrations in sportfish from San Francisco Bay, 1997. Marine Pollution Bulletin 44:1117–1129.

Davis, S. D., J. B. Williams, W. J. Adams, and S. L. Brown. 1984. The effect of egg temperature on attentiveness in the Belding's Savannah Sparrow. Auk 101:556–566.

Day, R. H., R. K. Holz, and J. W. Day, Jr. 1990. An inventory of wetland impoundments in the coastal zone of Louisiana, USA: historical trends. Environmental Management 14:229–240.

De Groot, D. S. 1927. The California Clapper Rail: its nesting habits, enemies and habitat. Condor 29: 259–270.

Department of Water Resources. 1993–1999. County land use survey data. California Department of Water Resources, Division of Planning and Local Assistance, Sacramento, CA. <http://www.landwateruse.water.ca.gov/basicdata/landuse/landusesurvey.cfm> (2 May 2006).

DeRagon, W. R. 1988. Breeding ecology of Seaside and Sharp-tailed sparrows in Rhode Island salt marshes. M.S. thesis, University of Rhode Island, Kingston, RI.

Desplanque, C., and D. J. Mossman. 2000. Fundamentals of Fundy tides. Pp. 178–203 *in* T. Chopin, and P. G. Wells (editors). Opportunities and challenges for protecting, restoring and enhancing coastal habitats in the Bay of Fundy. 4th Bay of Fundy Science Workshop. Environment Canada-Atlantic Region, Saint John, NB, Canada.

Desplanque, C., and D. J. Mossman. 2004. Tides and their seminal impact on the geology, geography, history, and socio-economics of the Bay of Fundy, eastern Canada. Atlantic Geology 40: 1–130.

DeSzalay, F. A., and V. H. Resh. 1997. Responses of wetland invertebrates and plants important in waterfowl diets to burning and mowing of emergent vegetation. Wetlands 17:149–156.

Dettinger, M. D., and D. R. Cayan. 2003. Interseasonal covariability of Sierra Nevada streamflow and San Francisco Bay salinity. Journal of Hydrology 277:164–181.

Dettinger, M., W. Bennett, D. Cayan, J. Florsheim, M. Hughes, B. L. Ingram, N. Knowles, F. Malamud-Roam, D. Peterson, K. Redmond, and L. Smith. 2003. Climate science issues and needs of the Calfed Bay-Delta Program. American Meteorological Society, 83rd Annual Meeting, Impacts of Water Variability Symposium. Long Beach, CA. <http://ams.confex.com/ams/pdfpapers/52444.pdf> (6 June 2006).

Dias, R. A., and G. N. Maurío. 1998. Lista preliminar de avifauna da extrimidante sudoeste de Mangueira e arredores e Rio Grande do Sul. Ornitologicas 86: 10–11.

Differnbaugh, N. S., and L. C. Sloan. 2004. Mid-Holocene orbital forcing of regional-scale climate: a case study of western North America using a high-resolution RCM. Journal of Climate 17: 2927–2937.

Dijkema, K. S. 1990. Salt and brackish marshes around the Baltic Sea and adjacent parts of the North Sea: their development and management. Biological Conservation 51:191–209.

DiQuinzio, D. A. 1999. Population and breeding ecology of the promiscuous Saltmarsh Sharp-tailed Sparrow in Rhode Island salt marshes. M.S. thesis, University of Rhode Island, Kingston, RI.

DiQuinzio, D. A., P. W. C. Paton, and W. R. Eddleman. 2001. Site fidelity, philopatry, and survival of promiscuous Saltmarsh Sharp-tailed Sparrows in Rhode Island. Auk 118:888–899.

DiQuinzio, D. A., P. W. C. Paton, and W. R. Eddleman. 2002. Nesting ecology of Saltmarsh Sharp-tailed Sparrows in a tidally restricted salt marsh. Wetlands 22:179–185.

Dixon, J. 1908. A new harvest mouse from the salt marshes of San Francisco Bay, California. Proceedings of the Biological Society of Washington 21:197–198.

Doak, D., P. Kareiva, and B. Klepetka. 1994. Modeling population viability for the desert tortoise in the western Mohave Desert. Ecological Applications 4:446–460.

Dolbeer, R. A. 1976. Reproductive rate and temporal spacing of nesting of Red-winged Blackbirds in upland habitat. Auk 93:343–355.

Donnelly, J. P., and M. D. Bertness. 2001. Rapid shoreward encroachment of salt marsh cordgrass in response to accelerated sea-level rise. Proceedings of the National Academy of Sciences 98:14218–14223.

Douglas, B. 1991. Global sea level rise. Journal of Geophysical Research 96:6981–6992.

Drake, J. A., H. A. Mooney, F. Dicastri, R. H. Groves, F. J. Kruger, M. Rejmanek, and M. Williamson (editors). 1989. Biological invasions: a global perspective. Scope 37. John Wiley and Sons, Chichester, UK.

Dramstad, W. E., J. D. Olson, and R. T. T. Forman. 1996. Landscape ecology principles in landscape architecture and land-use planning. Island Press, Washington, DC.

Dreyer, G. D., and W. A. Niering. 1995. Tidal marshes of Long Island Sound: ecology, history, and restoration. Bulletin No. 34. Connecticut College Arboretum, New London, CT.

Dunham, A. E., B. W. Grant, and K. L. Overall. 1989. The interface between biophysical ecology and population ecology of terrestrial vertebrate ectotherms. Physiological Zoology 62:335–355.

Dunson, W. A. 1980. The relationship of sodium and water balance to survival in sea water of estuarine and freshwater races of the snakes *Nerodia fasciata*, *N. sipedon*, and *N. valida*. Copeia 1980:268–280.

Dunson, W. A. 1981. Behavioral osmoregulation by the key mud turtle (*Kinosternon b. baurii*). Journal of Herpetology 15:163–173.

Dunson, W. A. 1985. Effect of water salinity and food salt content on growth and sodium efflux of hatchling diamondback terrapins (*Malaclemys*). Physiological Zoology 58:736–747.

Dunson, W. A. 1986. Estuarine populations of the snapping turtle (*Chelydra*) as a model for the evolution of marine adaptation in reptiles. Copeia 1986:741–756.

Dunson, W. A., and F. J. Mazzotti. 1989. Salinity as a limiting factor in the distribution of reptiles in Florida Bay: a theory for the estuarine origin of marine snakes and turtles. Bulletin of Marine Science 44:229–244.

Dunson, W. A., and J. Travis. 1994. Patterns in the evolution and physiological specialization in salt-marsh animals. Estuaries 17:102–110.

Dunton, K. H., B. Hardegree, and T. E. Whitledge. 2001. Response of estuarine marsh vegetation to interannual variations in precipitation. Estuaries 24:851–861.

Durso, S. L. 1996. The biology and control of mosquitoes in California. Mosquito and Vector Control Association of California, Elk Grove, CA.

Dyer, M. I., J. Pinowski, and B. Pinowska. 1977. Population dynamics. Pp. 53–105 *in* J. Pinowski, and S. C. Kendeigh (editors). Granivorous birds in ecosystems. Cambridge University Press, Cambridge, UK.

Earle, J. C., and K. A. Kershaw. 1989. Vegetation patterns in James Bay (Canada) coastal marshes: III. Salinity and elevation as factors influencing plant zonations. Canadian Journal of Botany 67: 2967–2974.

Eckert, S. A., and L. Sarti, 1997. Distant fisheries implicated in the loss of the world's largest

leatherback nesting population. Marine Turtle Newsletter 78:2–7.
EDDLEMAN, W. R., AND C. J. CONWAY. 1998. Clapper Rail (*Rallus longirostris*). *In* A. Poole, and F. Gill (editors). The birds of North America, No. 340. The Academy of Natural Sciences, Philadelphia, PA and The American Ornithologists' Union, Washington, DC.
EDDLEMAN, W. R., R. E. FLORES, AND M. L. LEGARE. 1994. Black Rail (*Laterallus jamaicensis*). *In* A. Poole, and F. Gill (editors). The birds of North America, No. 123. The Academy of Natural Sciences, Philadelphia, PA and The American Ornithologists' Union, Washington, DC.
EDWARDS, R. L., J. W. BECK, G. S. BURR, D. J. DONAHUE, J. M. A. CHAPPELL, A. L. BLOOM, E. R. M. DRUFFEL, AND F. W. TAYLOR. 1993. A large drop in atmospheric C-14/C-12 and reduced melting in the Younger Dryas, documented with TH-230 ages of corals. Science 260:962–968.
EDWARDS, S. V., AND P. BEERLI. 2000. Perspective: gene divergence, population divergence, and the variance in coalescence time in phylogeographic studies. Evolution 54:1839–1854.
EHRLICH, P. R., D. S. DOBKIN, AND D. WHEYE. 1988. The birder's handbook: a field guide to the natural history of North American birds. Simon and Schuster Inc., New York, NY.
EISENBERG, J. F., AND K. H. REDFORD. 1999. Mammals of the Neotropics: the central Neotropics. University of Chicago Press, Chicago, IL.
EL-BEGEARMI, M. M., M. L. SUNDE, AND H. E. GANTHER. 1977. A mutual protective effect of mercury and selenium in Japanese quail. Poultry Science 56:313–322.
ELBRØND, V. S., V. DANTZER, T. M. MAYHEW, AND E. SKADHAUGE. 1993. Dietary and aldosterone effects on the morphology and electrophysiology of the chicken coprodeum. Pp. 217–226 *in* P. J. Sharp (editor). Avian endocrinology. Society for Endocrinology, Bristol, UK.
ELEUTERIUS, L. N. 1990. Tidal marsh plants. Pelican Publishing Company, Gretna, LA.
ELKIE, P., R. REMPEL, AND A. CARR. 1999. Patch analyst user's manual. TM-002, Ontario Ministry of Natural Resources, Northwest Science and Technology, Thunder Bay, ON, Canada.
ELLIS, H. K., III. 1980. Ecology and breeding biology of the Swamp Sparrow in a southern Rhode Island peatland. M. S. thesis, University of Rhode Island, Kingston, RI.
EMERY, K. O., AND D. G. AUBREY. 1991. Sea levels, land levels, and tide gauges. Springer-Verlag, New York, NY.
EMERY, N., P. EWANCHUCK, AND M. D. BERTNESS. 2001. Nutrients, mechanisms of competition, and the zonation of plants across salt marsh landscapes. Ecology 82:2471–2484.
EMMETT, R., R. LLANSO, J. NEWTON, R. THOM, M. HORNBERGER, C. MORGAN, C. LEVINGS, A. COPPING, AND P. FISHMAN. 2000. Geographic signatures of North American west coast estuaries. Estuaries 23:765–792.
EMMONS, L. H. 1990. Neotropical rainforest mammals: a field guide. University of Chicago, Chicago, IL.

ENS, B., J. D. GOSS-CUSTARD, AND T. P. WEBER. 1995. Effects of climate change on bird migration strategies along the east Atlantic flyway. Report No. 410 100 075. Dutch National Research Program on Global Air Pollution and Climate Change, Texel, The Netherlands.
ENVIRONMENTAL PROTECTION AGENCY. 1992. Guidelines for exposure assessment. Federal Register 57:22888–22938.
ENVIRONMENTAL PROTECTION AGENCY. 1996. PCBs: cancer dose-response assessment and application to environmental mixtures. National Center for Environmental Assessment. Office of Research and Development. Environmental Protection Agency/600/P-96/001F. Washington, DC.
ENVIRONMENTAL SYSTEMS RESEARCH INSTITUTE. 1999. Spatial analyst 1 extension for ArcView 3.x. Environmental Systems Research Institute, Redlands, CA.
ENVIRONMENTAL SYSTEMS RESEARCH INSTITUTE. 2000. ArcView 3.2a. Environmental Systems Research Institute, Redlands, CA.
ENVIRONMENTAL SYSTEMS RESEARCH INSTITUTE. 2002. ArcGIS 8.2. Environmental Systems Research Institute, Redlands, CA.
EPSTEIN, M. B., AND R. L. JOYNER. 1988. Waterbird use of brackish wetlands managed for waterfowl. Proceedings of the Southeastern Association of Fish and Wildlife Agencies 42:476–490.
ERNST, C. H., AND R. W. BARBOUR. 1989. Turtles of the world. Smithsonian Institution Press, Washington, DC.
ERNST, C. H., J. E. LOVICH, AND R. W. BARBOUR. 1994. Turtles of the United States and Canada. Smithsonian Institution Press, Washington, DC.
ERRINGTON, P. 1961. Muskrats and marsh management. Stackpole Company, Harrisburg, PA.
ERSKINE, A. J. 1992. Atlas of breeding birds of the Maritime Provinces. Nimbus Publishing and the Nova Scotia Museum, Halifax, NS, Canada.
ERWIN, R. M. 1980. Breeding habitat use by colonially nesting waterbirds in two mid-Atlantic U.S. regions under different regimes of human disturbance. Biological Conservation 18:39–51.
ERWIN, R. M., C. J. CONWAY, AND S. W. HADDEN. 2002. Species occurrence of marsh birds at Cape Cod National Seashore, Massachusetts. Northeastern Naturalist 9:1–12.
ERWIN, R. M., D. DAWSON, D. STOTTS, L. MCALLISTER, AND P. GEISSLER. 1991. Open marsh water management in the mid-Atlantic region: aerial surveys of waterbird use. Wetlands 11:209–227.
ERWIN, R. M., J. S. HATFIELD, M. A. HOWE, AND S. S. KLUGMAN. 1994. Waterbird use of saltmarsh ponds created for open marsh water management. Journal of Wildlife Management 58:516–524.
ERWIN, R. M., J. S. HATFIELD, AND T. J. WILMERS. 1995. The value and vulnerability of small estuarine islands for conserving metapopulations of breeding waterbirds. Biological Conservation 71:187–191.
ERWIN, R. M., D. JENKINS, AND D. H. ALLEN. 2003. Created versus natural coastal islands: Atlantic waterbird populations, habitat choices and management implications. Estuaries 26:949–955.
ERWIN, R. M., B. R. TRUITT, AND J. E. JIMENEZ. 2001. Ground-nesting waterbirds and mammalian

carnivores in the Virginia barrier island region: running out of options. Journal of Coastal Research 17:292-296.

ESSELINK, P., W. ZIJLSTRA, K. DIJKEMA, AND R. VAN DIGGELEN. 2000. The effects of decreased management on plant-species distribution patterns in a saltmarsh nature reserve in the Wadden Sea. Biological Conservation 93:61-76.

ESTUARY RESTORATION ACT. 2000. Pub. L. 106-457, title I, Nov. 7, 2000. 114 Stat. 1958 (33 U.S.C. 2901 et seq.)

EVENS, J. G., AND N. NUR. 2002. California Black Rails in the San Francisco Bay region: spatial and temporal variation in distribution and abundance. Bird Populations 6:1-12.

EVENS, J., AND G. W. PAGE. 1986. Predation on Black Rails during high tides in salt marshes. Condor 88:107-109.

EVENS, J., G. W. PAGE, S. A. LAYMON, AND R. W. STALLCUP. 1991. Distribution, relative abundance, and status of the California Black Rail in western North America. Condor 93:952-966.

EYLER, T. B., R. M. ERWIN, D. B. STOTTS, AND J. S. HATFIELD. 1999. Aspects of hatching success and chick survival in Gull-billed Terns in coastal Virginia. Waterbirds 22:54-59.

FABER, P. M. 1996. Common wetland plants of coastal California. Pickleweed Press, Mill Valley, CA.

FABER, P. M. 2000. *Spartina densiflora*. Pp. 301-302 *in* C. C. Bossard, J. M. Randall, and M. C. Hoshovsky (editors). Invasive plants of California's wildlands. University of California Press, Berkeley, CA.

FAHRIG, L. 1997. Relative effects of habitat loss and fragmentation on population extinction. Journal of Wildlife Management 61:603-610.

FAIRBAIRN, S. E., AND J. J. DINSMORE. 2001. Local and landscape-level influences on wetland bird communities of the prairie pothole region of Iowa, USA. Wetlands 21:41-47.

FAIRBANKS, R. G., 1989. A 17,000-year glacio-eustatic sea level record; influence of glacial melting rates on the Younger Dryas event and deep-ocean circulation. Nature 342:637-642.

FAIRBANKS, R. G. 1992. Holocene marine coastal evolution of the United States. Pp. 9-20 *in* Special Publication No. 48. Society of Economic and Petroleum Mineralogy, Quaternary coasts of the United States: marine and lacustrine systems, Tulsa, OK.

FAIREY, R., K. TABERSKI, S. LAMERDIN, E. JOHNSON, R. P. CLARK, J. W. DOWNING, J. NEWMAN, AND M. PETREAS. 1997. Organochlorines and other environmental contaminants in muscle tissue of sportfish collected from San Francisco Bay. Marine Pollution Bulletin 34:1058-1071.

FEINBERG, J. A. 2003. Nest predation and ecology of terrapins, *Malaclemys terrapin terrapin*, at the Jamaica Bay Wildlife Refuge. Pp. 5-12 *in* C. W. Swarth, W. M. Roosenberg, and E. Kiviat (editors). Conservation and ecology of turtles of the mid-Atlantic region; a symposium. Bibliomania, Salt Lake City, UT.

FELL, P. E., S. P. WEISSBACH, D. A. JONES, M. A. FALLON, J. A. ZEPPIERI, E. K. FAISON, K. A. LENNON, K. J. NEWBERRY, AND L. K. REDDINGTON. 1998. Does invasion of oligohaline tidal marshes by reed grass, *Phragmites australis* (Cav.) Trin. Ex Steud. Affect the availability of prey resources for the mummichog, *Fundulus hetroclitus* L. Journal of Experimental Marine Biology and Ecology 222:59-77.

FENGER, J., E. BUCH, AND P. R. JAKOBSEN. 2001. Technical and political aspects of sea level rise in Denmark. World Resource Review 13:540-554.

FENSOME, R. A., AND G. L. WILLIAMS (EDITORS). 2001. The last billion years: a geological history of the Maritime Provinces of Canada. Atlantic Geoscience Society, Halifax, NS, Canada.

FERRIGNO, F., AND D. JOBBINS. 1968. Open marsh water management. Proceedings of the New Jersey Mosquito Extermination Association 55:104-115.

FIALA, K. L., AND J. D. CONGDON. 1983. Energetic consequences of sexual size dimorphism in nestling Red-winged Blackbirds. Ecology 64:642-647.

FIELD, D. W., A. J. REYER, P. V. GENOVESE, AND B. D. SHEARER. 1991. Coastal wetlands of the United States. National Oceanographic and Atmospheric Administration and USDI Fish and Wildlife Service, Washington, DC.

FINDLEY, J. S. 1955. Speciation of the wandering shrew. University of Kansas Publications, Museum of Natural History 9:1-68.

FINKELSTEIN, K., AND C. S. HARDAWAY. 1988. Late Holocene sedimentation and erosion of estuarine fringing marshes, York River, Virginia (USA). Journal of Coastal Research 4:447-456.

FINLAYSON, C. M., G. E. HOLLIS, AND T. J. DAVIS (EDITORS). 1992. Managing Mediterranean wetlands and their birds. Special Publication No. 20, International Waterfowl and Wetlands Research Bureau, Slimbridge, UK.

FISLER, G. F. 1962, Ingestion of sea water by *Peromyscus maniculatus*. Journal of Mammalogy. 43:416-417.

FISLER, G. F. 1963. Effects of salt water on food and water consumption and weight of harvest mice. Ecology 44:604-606.

FISLER, G. F. 1965. Adaptation and speciation in harvest mice of the marshes of San Francisco Bay. University of California Publications in Zoology 77:1-108.

FLATHER, C. H., AND J. R. SAUER. 1996. Using landscape ecology to test hypotheses about large-scale abundance patterns in migratory birds. Ecology 77:28-35.

FLEISCHER, R. C., G. FULLER, AND D. B. LEDIG. 1995. Genetic structure of endangered Clapper Rail (*Rallus longirostris*) populations in southern California. Conservation Biology 9:1234-1243.

FLEISCHER, R. C., AND C. E. MCINTOSH. 2001. Molecular systematics and biogeography of the Hawaiian avifauna. Studies in Avian Biology 22:51-60.

FLETCHER, C. H., H. J. HARLEY, AND J. C. KRAFT. 1990. Holocene evolution of an estuarine coast and tidal wetlands. Geological Society of America Bulletin 102:283-297.

FLETCHER, C. H., III, J. E. VAN PELT, G. S. BRUSH, AND J. SHERMAN. 1993. Tidal wetland record of Holocene sea-level movements and climate history. Palaeogeography Palaeoclimatology Palaeoecology 102:177-213.

FLORES, C., AND D. BOUNDS. 2001. First year response of prescribed fire on wetland vegetation in Dorchester County, MD. Wetland Journal 13:15–23.

FOERSTER, K. S., AND J. E. TAKEKAWA. 1991. San Francisco Bay National Wildlife Refuge predator management plan and final environmental assessment. USDI Fish and Wildlife Service, San Francisco Bay National Wildlife Refuge, Newark, CA.

FOERSTER, K. S., J. E. TAKEKAWA, AND J. D. ALBERTSON. 1990. Breeding density, nesting habitat and predators of the California Clapper Rail. USDI Fish and Wildlife Service, San Francisco Bay National Wildlife Refuge, Newark, CA.

FOIN, T. C., AND J. L. BRENCHLEY-JACKSON. 1991. Simulation model evaluation of potential recovery of endangered Light-footed Clapper Rail populations. Biological Conservation 58:123–148.

FOOTE, L. 1996. Coastal wetlands—questions of management. International Waterfowl Symposium 7:149–158.

FORMAN, R. T. T. 1995. Land mosaics, the ecology of landscapes and regions. Cambridge University Press, Cambridge, UK.

FORSTMEIER, W., B. LEISLER, AND B. KEMPENAERS. 2001. Bill morphology reflects female independence from male parental help. Proceedings of the Royal Society of London, Series B: Biological Sciences 268:1583–1588.

FORSTNER, M. R. J., G. PARKS, L. MILLER, B. HERBERT, K. HALBROOK, AND B. K. MEALEY. 2000. Genetic variability and geographic structure of *Malaclemys terrapin* populations in Texas and south Florida. Final report to the Species at Risk program of the U.S. Geological Survey, Reston, VA.

FOSTER, M. L. 1977a. A breeding season study of the Salt Marsh Yellowthroat (*Geothlypis trichas sinuosa*) of the San Francisco Bay area, California. M.A. thesis, San Jose State University, San Jose, CA.

FOSTER, M. L. 1977b. Status of the Salt Marsh Yellowthroat (*Geothlypis trichas sinuosa*) in the San Francisco Bay area, California 1975–1976. California Department of Fish and Game, Sacramento, CA.

FOSTER, M. S. 1974. A model to explain molt-breeding overlap and clutch size in some tropical birds. Evolution 28:182–190.

FOX, G. A. 1995. Tinkering with the tinkerer: pollution versus evolution. Environmental Health Perspectives 103:93–100.

FRANKHAM, R., J. D. BALLOU, AND D. A. BRISCOE. 2002. Introduction to conservation genetics. Cambridge University Press, Cambridge, UK.

FRASER, D. J., AND L. BERNATCHEZ. 2001. Adaptive evolutionary conservation: towards a unified concept for defining conservation units. Molecular Ecology 10:2741–2752.

FREDERICK, P. 1987. Chronic tidally-induced nest failure in a colony of White Ibises. Condor 89:413–419.

FRENKEL, R. E. 1987. Introduction and spread of cordgrass (*Spartina*) into the Pacific Northwest. Northwest Environmental Journal 3:152–154.

FREY, R. W., AND P. B. BASAN. 1985. Coastal salt marshes. *In* R. A. Davis (editor) Coastal sedimentary environments. Springer-Verlag, New York, NY.

FRIEND, M. 1987. Avian cholera. Pp. 69–82 *in* M. Friend (editor). Field guide to wildlife diseases. USDI Fish and Wildlife Service, Washington DC.

FRIEND, M. 1992. Environmental influences on major waterfowl diseases. Transactions of the North American Wildlife and Natural Resources Conference 57:517–525.

FROST, D. R. 2004. Amphibian species of the World: an online reference. Version 3.0. American Museum of Natural History, New York, NY. <http://research.amnh.org/herpetology/amphibia/index.html> (17 January 2006),

FRY, A. J., AND R. M. ZINK. 1998. Geographic analysis of nucleotide diversity and Song Sparrow (Aves: Emberizidae) population history. Molecular Ecology 7:1303–1313.

FUNDERBURG, J. B., JR., AND T. L. QUAY. 1983. Distributional evolution of the Seaside Sparrow. Pp. 19–27 *in* T. L. Quay, J. B. Funderburg, Jr., D. S. Lee, E. Potter, and C. S. Robbins (editors). The Seaside Sparrow, its biology and management. North Carolina Biological Survey, Raleigh, NC.

FUNK, D. J., AND K. E. OMLAND. 2003. Species-level paraphyly and polyphyly: frequency, causes, and consequences with insights from animal mitochondrial DNA. Annual Review of Ecology and Systematics 34:397–423.

FURBISH, C., AND M. ALBANO. 1994. Selective herbivory and plant community structure in a mid-Atlantic salt marsh. Ecology 75:1015–1022.

GABREY, S. W., AND A. D. AFTON. 2000. Effects of winter marsh burning on abundance and nesting activity of Louisiana Seaside Sparrows in the Gulf Coast Chenier Plain. Wilson Bulletin 112:365–372.

GABREY, S. W., AND A. D. AFTON. 2001. Plant community composition and biomass in Gulf Coast Chenier Plain marshes: responses to winter burning and structural marsh management. Environmental Management 27:281–293.

GABREY, S. W., AND A. D. AFTON. 2004. Composition of breeding bird communities in Gulf Coast Chenier Plain marshes: effects of winter burning. Southeastern Naturalist 3:173–185.

GABREY, S. W., A. D. AFTON, AND B. C. WILSON. 1999. Effects of winter burning and structural marsh management on vegetation and winter bird abundance in the Gulf Coast Chenier Plain, USA. Wetlands 19:594–606.

GABREY, S. W., A. D. AFTON, AND B. C. WILSON. 2001. Effects of structural marsh management and winter burning on plant and bird communities during summer in the Gulf Coast Chenier Plain. Wildlife Society Bulletin 29:218–231.

GABREY, S. W., B. C. WILSON, AND A. D. AFTON. 2002. Success of artificial bird nests in burned Gulf Coast Chenier Plain marshes. Southwestern Naturalist 47:532–538.

GAFFNEY, E. S., AND P. A. MEYLAN. 1988. A phylogeny of turtles. Pp. 157–219 *in* M. J. Benton (editor). The phylogeny and classification of the tetrapods, Vol. 1. Clarendon Press, Oxford, UK.

GAHAGAN AND BRYANT ASSOCIATES, BECHTEL CORPORATION, ENTRIX, AND PHILIP WILLIAM AND ASSOCIATES, LTD. 1994. A review of the physical and biological performance of tidal marshes constructed with dredged materials in San Francisco Bay, California. Report to the U.S. Army Corps of Engineers, San Francisco District, San Francisco, CA.

GAINES, K. F., J. C. CUMBEE, JR., AND W. L. STEPHENS, JR. 2003. Nest characteristics of the Clapper Rail from coastal Georgia. Journal of Field Ornithology 74: 152–156.

GALBRAITH, H., R. JONES, R. PARK, J. CLOUGH, S. HERROD-JULIUS, B. HARRINGTON, AND G. PAGE. 2002. Global climate change and sea level rise: potential losses of intertidal habitat for shorebirds. Waterbirds 25: 173–183.

GANONG, W. F. 1903. The vegetation of the Bay of Fundy salt and diked marshes: an ecological study. Botanical Gazette 36:161–186.

GARDNER, W. S., D. R. KENDALL, R. R. ODOM, H. L. WINDOM, AND J. A. STEPHENS. 1978. The distribution of methyl mercury in a contaminated salt marsh ecosystem. Environmental Pollution 15:243–251.

GAUL, R. W., JR. 1996. An investigation of the genetic and ecological status of the Carolina salt marsh snake, *Nerodia sipedon williamengelsi*. M.S. thesis, East Carolina University, Greenville, NC.

GAUVIN, J. M. 1979. Etude de la vegetation des marais sales du Parc National de Kouchibouguac, N.-B. M.Sc. thesis, Universite de Moncton, Moncton, NB, Canada.

GAVIN, T. A., R. A. HOWARD, AND B. MAY. 1991. Allozyme variation among breeding populations of Red-winged Blackbirds: the California conundrum. Auk 108:602–611.

GEISSEL, W., H. SHELLHAMMER, AND H. T. HARVEY. 1988. The ecology of the salt-marsh harvest mouse (*Reithrodontomys raviventris*) in a diked salt marsh. Journal of Mammalogy 69:696–703.

GERBER, L. R., D. P. DEMASTER, AND P. M. KAREIVA. 1999. Gray whales and the value of monitoring data in implementing the U.S. Endangered Species Act. Conservation Biology 13:1215–1219.

GETZ, L. L. 1966. Salt tolerance of salt marsh meadow voles. Journal of Mammalogy 47:201–207.

GIBBONS, J. W., J. E. LOVICH, A. D. TUCKER, N. N. FITZSIMMONS, AND J. L. GREENE. 2001. Demography and ecological factors affecting conservation and management of the diamondback terrapin (*Malaclemys terrapin*). Chelonian Conservation and Biology 4:66–74.

GIBBS, J. P., AND S. M. MELVIN. 1993. Call-response surveys for monitoring breeding waterbirds. Journal of Wildlife Management 7:27–34.

GILBERT, C. 1981. Le comportement social du Pinson a ueue Aigue. M.S. thesis, L'Universite Laval, Laval, Quebec, ON, Canada.

GILL, R. E., JR. 1973. The breeding birds of the south San Francisco Bay estuary. M.A. thesis, San Jose State University, San Jose, CA.

GILLES-BAILLEN, M. 1970. Urea and osmoregulation in the diamondback terrapin *Malaclemys centrata centrata* (Latreille). Journal of Experimental Biology 52:691–697.

GILLES-BAILLIEN, M. 1973. Hibernation and osmoregulation in the diamondback terrapin *Malaclemys centrata centrata* (Latreille). Journal of Experimental Biology 52:45–51.

GILMOUR, C. C., G. S. RIEDEL, M. C. EDERINGTON, J. T. BELL, J. M. BENOIT, G. A. GILL, AND M. C. STORDAL. 1998. Methylmercury concentrations and production rates across a trophic gradient in the northern Everglades. Biogeochemistry 40: 327–345.

GIORGI, F., B. HEWITSON, J. CHRISTIANSEN, M. HULME, H. VON STORCH, P. WHETTON, R. JONES, L. MEARNS, AND C. FU. 2001. Regional climate information-evaluation and projections. Pp. 583–638 *in* J. T. Houghton, Y. Ding, D. J. Griggs, M. Noguer, P. J. van der Linden, X. Dai, K. Maskell, and C. A. Johnson (editors). The scientific basis. Contributions of Working Group I to the third assessment report of the Intergovernmental Panel on Climate Change. Cambridge University Press, Cambridge, UK, and New York, NY.

GIVENS, L. S. 1962. Use of fire on southeastern wildlife refuges. Proceedings of the Tall Timbers Fire Ecology Conference 1:121–126.

GLOOSCHENKO, V., I. P. MARTINI, AND K. CLARKE-WHISTLER. 1988. Saltmarshes of Canada. Pp. 348–376 *in* National Wetlands Working Group. Wetlands of Canada. Sustainable Development Branch, Environment Canada, and Polyscience Publications Inc., Ottawa, ON, Canada.

GLUE, D. E. 1971. Salt marsh reclamation stages and their associated bird life. Bird Study 18:187–198.

GOALS PROJECT. 1999. Baylands ecosystem habitat goals. A report of habitat recommendations prepared by the San Francisco Bay Area Wetlands Ecosystems Goals Project. Joint publication of the U.S. Environmental Protection Agency, San Francisco, California, and San Francisco Bay Regional Water Quality Control Board, Oakland, CA.

GOALS PROJECT. 2000. Baylands ecosystem species and community profiles: life histories and environmental requirements of key plants, fish and wildlife. Prepared by the San Francisco Bay Area Wetlands Ecosystem Goals Project. U.S. Environmental Protection Agency, San Francisco, California, and San Francisco Bay Regional Water Quality Control Board, Oakland, CA.

GODDARD, L. B., A. E. ROTH, W. K. REISEN, AND T. W. SCOTT. 2002. Vector competence of California mosquitoes for West Nile virus. Emerging Infectious Diseases 8:1385–1391.

GOLDSMITH, B. 1991. Monitoring for conservation and ecology. Chapman and Hall, London, UK.

GOLDSTEIN, D. L., AND S. D. BRADSHAW. 1998. Regulation of water and sodium balance in the field by Australian honeyeaters (Aves: Meliphagidae). Physiological Zoology 71:214–225.

GOLDSTEIN, D. L., AND E. J. BRAUN. 1986. Proportions of mammalian-type and reptilian-type nephrons in the kidneys of two passerine birds. Journal of Morphology 187:173–180.

GOLDSTEIN, D. L., AND E. J. BRAUN. 1989. Structure and concentrating ability in the avian kidney. American Journal of Physiology 256:R501–R509.

GOLDSTEIN, D. L., AND E. SKADHAUGE. 1999. Renal and extra-renal osmoregulation. Pp. 265–297 *in* G. C. Whittow (editor). Sturkie's avian physiology. Academic Press, San Deigo, CA.

GOLDSTEIN, D. L., J. B. WILLIAMS, AND E. J. BRAUN. 1990. Osmoregulation in the field by Salt-marsh

Savannah Sparrows *Passerculus sandwichensis beldingi*. Physiological Zoology 63:669–682.

GOMAN, M. 1996. A history of Holocene environmental change in the San Francisco estuary. Ph.D. dissertation, University of California. Berkeley, CA.

GOMAN, M. 2001. Statistical analysis of modern seed assemblages from the San Francisco Bay: applications for the reconstruction of paleo-salinity and paleo-tidal inundation. Journal of Paleolimnology 24:393–409.

GOMAN, M., AND E. WELLS. 2000. Trends in river flow affecting the northeastern reach of the San Francisco Bay estuary over the past 7,000 years. Quaternary Research 54:206–217.

GORDON, D. C., JR., AND P. J. CRANFORD. 1994. Export of organic matter from macrotidal saltmarshes in the upper Bay of Fundy, Canada. Pp. 257–264 *in* W. J. Mitsch (editor). Global wetlands: Old World and New. Elsevier, New York, NY.

GORDON, D. C., JR., P. J. CRANFORD, AND C. DESPLANQUE. 1985. Observations on the ecological importance of saltmarshes in the Cumberland Basin, a macrotidal estuary in the Bay of Fundy (Canada). Estuarine Coastal and Shelf Science 20:205–228.

GORDON, D. C., JR., AND C. DESPLANQUE. 1983. Dynamics and environmental effects of ice in the Cumberland Basin of the Bay of Fundy (Canada). Canadian Journal of Fisheries and Aquatic Sciences 40:1331–1342.

GORDON, D. H., B. T. GRAY, AND R. M. KAMINSKI. 1998. Dabbling duck-habitat associations during winter in coastal South Carolina. Journal of Wildlife Management 62:569–580.

GORDON, D. H., B. T. GRAY, R. D. PERRY, M. B. PREVOST, T. H. STRANGE, AND R. K. WILLIAMS. 1989. South Atlantic coastal wetlands. Pp. 57–92 *in* L. M. Smith, R. L. Pederson, and R. M. Kaminski (editors). Habitat management for migrating and wintering waterfowl in North America. Texas Tech University Press, Lubbock, TX.

GOSS-CUSTARD, J. D., AND M. E. MOSER. 1988. Rates of change in the numbers of Dunlin, *Calidris alpina*, wintering in British estuaries in relation to the spread of *Spartina anglica*. Journal of Applied Ecology 25:95–109.

GRAY, A. J., AND R. J. MOGG. 2001. Climate impacts on pioneer saltmarsh plants. Climate Research 18:105–112.

GREEN, M. M. 1932. An unrecognized shrew from New Jersey. University of California Publications in Zoology 38:387–389.

GREEN, M. T., P. E. LOWTHER, S. L. JONES, S. K. DAVIS, AND B. C. DALE. 2002. Baird's Sparrow (*Ammodramus bairdii*). *In* A. Poole, and F. Gill (editors). The birds of North America, No. 638. The Academy of Natural Sciences, Philadelphia, PA and The American Ornithologists' Union, Washington, DC.

GREEN, R. E., AND G. J. M. HIRONS. 1991. The relevance of population studies to the conservation of threatened birds. Pp. 594–633 *in* C. M. Perrins, J.-D. Lebreton, and G. J. M. Hirons (editors). Bird population studies: relevance to conservation and management. Oxford University Press, London, UK.

GREENBERG, C. H., S. H. CROWNOVER, AND D. R. GORDON. 1997. Roadside soil: a corridor for invasion of xeric scrub by nonindigenous plants. Natural Areas Journal 17:99–100.

GREENBERG, R. 1981. Dissimilar bill shapes in New-World tropical vs. temperate forest foliage gleaning birds. Oecologia 49:143–147.

GREENBERG, R. 2003. On the use of nest departure calls for surveying Swamp Sparrows. Journal of Field Ornithology 74:12–16.

GREENBERG, R., P. J. CORDERO, S. DROEGE, AND R. C. FLEISCHER. 1998. Morphological adaptation with no mitochondrial DNA differentiation in the Coastal Plain Swamp Sparrow. Auk 115:706–712.

GREENBERG, R., AND S. DROEGE. 1990. Adapations to tidal marshes in breeding populations of the Swamp Sparrow. Condor 92:393–404.

GREENBERG, R., P. P. MARRA, AND M. J. WOOLLER. In press. Stable isotope (C, N, H) analyses locate the unknown winter range of the coastal plain Swamp Sparrow (*Melospiza georgiana nigrescens*). Auk.

GREENHAUGH, M. 1971. The breeding bird community of Lancashire salt marshes. Bird Study 18:199–212.

GREENLAW, J. S. 1983. Microgeographic distribution of breeding Seaside Sparrows on New York salt marshes. Pp. 99–114 *in* T. L. Quay, J. B. Funderburg, D. S. Lee, E. F. Potter, and C. S. Robbins (editors). The Seaside Sparrow: its biology and management. North Carolina Biological Survey, Raleigh, NC.

GREENLAW, J. S. 1989. On mating systems in passerine birds of American marshlands. Pp. 2597–2612 *in* H. Ouellet (editor). ACTA XIX Congressus Internationalis Ornithologici, Vol. II. University of Ottawa Press, Ottawa, ON, Canada.

GREENLAW, J. S. 1993. Behavioral and morphological diversification in Sharp-tailed Sparrows (*Ammodramus caudacutus*) of the Atlantic Coast. Auk 110:286–303.

GREENLAW, J. S., AND W. POST. 1985. Evolution of monogamy in Seaside Sparrows, *Ammodramus maritimus*: tests of hypotheses. Animal Behaviour 33:373–383.

GREENLAW, J. S., AND J. D. RISING. 1994. Sharp-tailed Sparrow (*Ammodramus caudacutus*). *In* A. Poole, and F. Gill (editors). The birds of North America, No. 112. Academy of Natural Sciences, Philadelphia, PA and The American Ornithologists' Union, Washington, DC.

GRENIER, J. L. 2004. Ecology, behavior, and trophic adaptations of the salt marsh Song Sparrow *Melospiza melodia samuelis*: the importance of the tidal influence gradient. Ph.D. dissertation, University of California, Berkeley, CA.

GRENIER, J. L., AND R. GREENBERG. 2005. A biogeographic pattern in sparrow bill morphology: parallel adaptation to tidal marshes. Evolution 59:1588–1595.

GRIFFITH, B., J. M. SCOTT, J. CARPENTER, AND C. REED. 1989. Translocation as a species conservation tool: status and strategy. Science 245:477–480.

GRINNELL, J. 1909. Three new Song Sparrows from California. University of California Publications in Zoology 5:265–269.

GRINNELL, J. 1913. Note on the palustrine faunas of west-central California. University of California Publications in Zoology 10:191–194.

GRINNELL, J. 1933. Review of the recent mammal fauna of California. University of California Publications in Zoology 40:71-234.

GRINNELL, J., AND A. H. MILLER. 1944. The distribution of the birds of California. Pacific Coast Avifauna 27. Cooper Ornithological Society, Berkeley, CA.

GROSHOLZ, E. 2002. Ecological and evolutionary consequences of coastal invasions. Trends in Ecology and Evolution 17:22-27

GROSJEAN, M., J. F. N. VAN LEEUWEN, W. O. VAN DER KNAAP, M. A. GEYH, B. AMMANN, W. TANNER, B. MESSERLI, L. A. NUNEZ, B. L. VALERO-GARCES, AND H. VEIT. 2001. A 22,000 ^{14}C year B.P. sediment and pollen record of climate change from Laguna Miscanti 23␣S, Northern Chile. Global and Planetary Change 28:35-51.

GROSSINGER, R., J. ALEXANDER, A. N. COHEN, AND J. N. COLLINS. 1998. Introduced tidal marsh plants in the San Francisco estuary: regional distribution and priorities for control. San Francisco Estuary Institute, Richmond, CA.

GROSSINGER, R., J. ALEXANDER, A. N. COHEN, AND J. N. COLLINS. 1998. Introduced tidal marsh plants in the San Francisco Estuary. San Francisco Estuary Institute, Richmond, CA.

GUSTAFSON, E. J., AND G. R. PARKER. 1994. Using an index of habitat patch proximity for landscape design. Landscape and Urban Planning 29:117-130.

GUZY, M. J., AND G. RITCHISON. 1999. Common Yellowthroat (*Geothlypis trichas*). *In* A. Poole, and F. Gill (editors). The birds of North America, No. 448. The Academy of Natural Sciences, Philadelphia, PA and The American Ornithologists' Union, Washington, DC.

HACKNEY, C. T., AND A. A. DE LA CRUZ. 1981. Effects of fire on brackish marsh communities: management implications. Wetlands 1:75-86.

HAGAN, J. M., T. L. LLOYD-EVANS, J. L. ATWOOD, AND D. S. WOOD. 1992. Long-term changes in migratory landbirds in the north-eastern United States: evidence from migration capture data. Pp. 115-130 *in* J. M. Hagan, and D. W. Johnson (editors). Ecology and conservation of neotropical migrant landbirds. Smithsonian Institution Press, Washington, DC.

HAIG, S. M., D. W. MEHLMAN, AND L. W. ORING. 1998. Avian movements and wetland connectivity in landscape conservation. Conservation Biology 12:749-758.

HALL, E. R. 1981. The mammals of North America. 2nd. ed. John Wiley and Sons, Inc., New York, NY.

HALL, E. R., AND K. R. KELSON. 1959. The mammals of North America. Ronald Press, New York, NY.

HAMPTON, S., R. G. FORD, H. R. CARTER, C. ABRAHAM, AND D. HUMPLE. 2003. Chronic oiling and seabird mortality from the sunken vessel S.S. Jacob Luckenbach in central California. Marine Ornithology 31:35-41.

HANSEN, A. J., AND D. L. URBAN. 1992. Avian response to landscape pattern—the role of species life histories. Landscape Ecology 7:163-180.

HANSEN, G. L. 1979. Territorial and foraging behaviour of the Eastern Willet *Catoptrophorus semipalmatus semipalmatus* (Gmelin). M.S. thesis, Acadia University, Wolfville, NS, Canada.

HANSON, A. 2004. Breeding bird use of saltmarsh habitat in the Maritime Provinces. Technical Report Series No. 414. Canadian Wildlife Service-Atlantic Region. Sackville, NB, Canada.

HANSON, A. R., D. BERUBE, D. L. FORBES, S. O'CARROLL, J. OLLERHEAD, AND L. OLSEN. 2005. Impacts of sea-level rise and residential development on salt marsh area in southeastern New Brunswick 1944-2001. Proceedings 12th Canadian Coastal Conference, Dartmouth, NS, Canada.

HANSON, A., AND L. CALKINS. 1996. Wetlands of the Maritime provinces: revised documentation for the wetlands inventory. Technical Report Series No. 267. Canadian Wildlife Service-Atlantic Region, Sackville, NB, Canada.

HARRIS, V. R. 1953. Ecological relationships of meadow voles and rice rats in tidal marshes. Journal of Mammalogy 34:479-487.

HARRIS, V. T., AND F. WEBERT. 1962. Nutria feeding activity and its effect on marsh vegetation in southwestern Louisiana. Special Scientific Report 64, USDI Fish and Wildlife Service, Washington, DC.

HARRISON, S., AND E. BRUNA. 1999. Habitat fragmentation and large-scale conservation: what do we know for sure? Ecography 22:225-232.

HART, K. M. 1999. Declines in diamondbacks: terrapin population modeling and implications for management. M.E.M. thesis, Duke University, NC

HARTWIG, A. 1995. Current aspects in metal genotoxicity. BioMetals 8:3-11.

HARVEY, T. E. 1988. Breeding biology of the California Clapper Rail in south San Francisco Bay. Transactions of the Western Section of the Wildlife Society 24:98-104.

HARVEY, T. E., K. J. MILLER, R. L. HOTHEM, M. J. RAUZON, G. W. PAGE, AND R. A. KECK. 1992. Status and trends report on wildlife of the San Francisco Estuary. USDI Fish and Wildlife Service report for the San Francisco Estuary Project. U.S. Environmental Protection Agency, San Francisco, CA and San Francisco Bay Regional Water Quality Control Board, Oakland, CA.

HATVANY, M. G. 2001. "Wedded to the marshes": salt-marshes and socio-economic differentiation in early Prince Edward Island. Acadiensis 30:40-55.

HAY, O. P. 1908. The fossil turtles of North America. Carnegie Institution of Washington Publication 75:1-568.

HAYDEN, B., R. DUESER, J. T. CALLAHAN, AND H. H. SHUGART. 1991. Long-term research at the Virginia Coast Reserve. BioScience 41:310-318.

HAYS, J. D., IMBRIE, J., AND N. J. SHACKLETON. 1977. Variations in the Earth's orbit: pacemaker of the ice ages? Science 198:529-530.

HAYS, W. S. T., AND W. Z. LIDICKER, JR. 2000. Winter aggregations, Dehnel effect, and habitat relations in the Suisun shrew *Sorex ornatus sinuosus*. Acta Theriologica 45:433-442.

HAZELDEN, J., AND L. A. BOORMAN. 2001. Soils and 'managed retreat' in south east England. Soil Use and Management 17:150-154.

HEATWOLE, H., AND J. TAYLOR. 1987. Ecology of reptiles. Beatty and Sons Pty. Ltd., Sidney, New South Wales, Australia

HEINZ, G. H. 1996. Selenium in birds. Pp. 447–458 *in* W. N. Beyer, G. H. Heinz, and A. W. Redmon-Norwood (editors). Environmental contaminants in wildlife: interpreting tissue concentrations. CRC Lewis Publishers, New York, NY.

HEINZ, G. H., AND D. J. HOFFMAN. 1998. Methylmercury chloride and selenomethionine interactions on health and reproduction in Mallards. Environmental Toxicology and Chemistry 17:139–145.

HELLMAYER, C. E. 1932. The catalogue of birds of the Americas. Part IX. Publication No. 381, Field Museum of Natural History, Chicago, IL.

HELLMAYER, C. E. 1938. The catalogue of birds of the Americas. Part IX. Publication No. 430, Field Museum of Natural History, Chicago, IL.

HELLMAYR, C. E., AND B. CONOVER. 1942. The catalogue of birds of the Americas. Part I. Publication No 514, Field Museum of Natural History, Chicago, IL.

HEPPELL, S. S. 1998. Application of life-history theory and population model analysis to turtle conservation. Copeia 1998:367–375.

HEPPELL, S. S., H. CASWELL, AND L. B. CROWDER. 2000. Life histories and elasticity patterns: perturbation analysis for species with minimal demographic data. Ecology 81:654–665.

HEPPELL, S. S., AND L. B. CROWDER. 1998. Prognostic evaluation of enhancement efforts using population models and life history analysis. Bulletin of Marine Lacince 62:495–507.

HEPPELL, S. S., L. B. CROWDER, AND T. R. MENZEL. 1999. Life table analysis of long-lived marine species with implications for conservation and management. American Fisheries Society Symposium 23. Bethesda, MD.

HERKERT, J. R., P. D. VICKERY, AND D. E. KROODSMA. 2002. Henslow's Sparrow (*Ammodramus henslowii*). *In* A. Poole, and F. Gill (editors). The birds of North America, No. 672. The Academy of Natural Sciences, Philadelphia, PA and The American Ornithologists' Union, Washington, DC.

HIGGINS, P. J. (EDITOR). 1999. Handbook of Australian, New Zealand, and Antarctic birds. Vol. 4. parrots to dollar birds. Oxford University Press, Oxford, UK.

HIGGINS, P. J., J. M. PETER, AND W. K. STEELE (EDITORS). 2001. Handbook of Australian, New Zealand, and Antarctic birds. Vol. 5. tyrant flycatchers to chats. Oxford University Press, Oxford, UK.

HILDEBRAND, S. F. 1929. Review of experiments on artificial culture of diamond-back terrapin. Bulletin of the U.S. Bureau of Fisheries 45:25–70.

HILDEBRAND, S. F. 1932. Growth of diamondback terrapins, size attained, sex ratio, and longevity. Zoologica 9:551–563.

HILDEBRAND, S. F., AND C. HATSEL. 1926. Diamond-back terrapin culture at Beaufort, N.C. U.S. Bureau of Fisheries, Economic Circular 60:1–20.

HILDEBRAND, S. F., AND H. F. PRYTHERCH. 1947. Diamondback terrapin culture. USDI Fish and Wildlife Service Leaflet 216:1–5.

HILL, C. E., AND W. POST. 2005. Extra-pair paternity in Seaside Sparrows. Journal of Field Ornithology 76:119–126.

HOBSON, K., P. PERRINE, E. B. ROBERTS, M. L. FOSTER, AND P. WOODIN. 1986. A breeding season survey of Salt Marsh Yellowthroats (*Geothlypis trichas sinuosa*) in the San Francisco Bay Region. Report of the San Francisco Bay Bird Observatory to USDI Fish and Wildlife Service, Sacramento, CA.

HOCHACHKA, W. M., J. N. M. SMITH, AND P. ARCESE. 1989. Song Sparrow. Pp. 135–152 *in* I. Newton (editor). Lifetime reproduction in birds. Academic Press, New York, NY.

HOCKEY, P. A. R., AND J. K. TURPIE. 1999. Estuarine birds in South Africa. Pp. 235–268 *in* B. Allanson, and D. Baird (editors). Estuaries of South Africa. Cambridge University Press, Cambridge, UK.

HODGMAN, T. P., W. G. SHRIVER, AND P. D. VICKERY. 2002. Redefining range overlap between the Sharp-tailed Sparrows of coastal New England. Wilson Bulletin 114:38–43.

HOFFMAN, D. J., C. P. RICE, AND T. J. KUBIAK. 1996. PCBs and dioxins in birds. Pp. 165–207 *in* W. N. Beyer, G. H. Heinz, and A. W. Redmon-Norwood (editors). Environmental contaminants in wildlife, interpreting tissue concentrations. CRC Lewis Publishers, New York, NY.

HOFFMANN, A. A., AND P. A. PARSONS. 1991. Evolutionary genetics and environmental stress. Oxford University Press, Oxford, UK.

HOFFPAUIR, C. M. 1961. Methods of measuring and determining the effects of marsh fires. M.S. thesis, Louisiana State University, Baton Rouge, LA.

HOFFPAUIR, C. M. 1968. Burning for coastal marsh management. Pp. 134–139 *in* J. Newsom (editor). Proceedings of the marsh and estuary management symposium. Louisiana State University, Baton Rouge, LA.

HOLCOMB, L. C. 1966. Red-winged Blackbird nestling development. Wilson Bulletin 78:283–288.

HOLCOMB, L. C., AND G. TWIEST. 1970. Growth rates and sex ratios of Red-winged Blackbird nestlings. Wilson Bulletin 82:294–303.

HOLLAND, R. F., AND S. K. JAIN. 1977. Vernal pools. Pp. 515–533 *in* M. G. Barbour, and J. Major (editors). Terrestrial vegetation of California. John Wiley and Sons, New York, NY.

HOLLAND, R. F., AND S. K. JAIN. 1984. Spatial and temporal variation in plant species diversity of vernal pools. Pp. 198–209 *in* S. Jain, and P. Moyle (editors). Vernal pools and intermittent streams. Publication number 28, University of California Institute of Ecology, Davis, CA.

HOLT, E. R., AND R. BUCHSBAUM. 2000. Bird use of *Phragmites australis* in coastal marshes of northern Massachusetts. Proceedings of the First National Conference on Marine Bioinvasions. Massachusetts Institute of Technology, Cambridge, MA.

HOOD, C. S., L. W. ROBBINS, R. J. BAKER, AND H. S. SHELLHAMMER. 1984. Chromosomal studies and evolutionary relationships of an endangered species, *Reithrodontomys raviventris*. Journal of Mammalogy 65:655–667.

HOOD, W. G. 2004. Indirect environmental effects of dikes on estuarine tidal channels: thinking outside of the dike for habitat restoration and monitoring. Estuaries 27:273–282.

HOOPER, E. T. 1952. A systematic revision of the harvest mice (genus *Reithrodontomys*) of Latin America. Miscellaneous Publications of the Museum of Zoology, University of Michigan 77:1–255.

HOOPER, S., C. PETTIGREW, AND G. SAYLER. 1991. Ecological fate, effects, and prospects for the elimination of environmental polychlorinated biphenyls (PCBs). Environmental Toxicology and Chemistry 9:655–667.

HORWITZ, E. L. 1978. Our nation's wetlands. Council on Environmental Quality, Washington, DC.

HOSMER, D. W., AND S. LEMESHOW. 1989. Applied logistic regression. John Wiley and Sons, New York, NY.

HOSMER, D. W., AND S. LEMESHOW. 2000. Applied logistic regression, 2nd edition. John Wiley and Sons, Inc, New York, NY.

HOTKER, H. 1992. Reclaimed Norstrand Bay—consequences for wetland birds and management. Netherlands Journal of Sea Research 20:257–260.

HOWE, M. A. 1982. Social organization in a nesting population of Eastern Willets (*Catoptrophorus semipalmatus*). Auk 99:88–102.

HOWE, R. W., G. J. DAVIS, AND V. MOSCA. 1991. The demographic significance of 'sink' populations. Biological Conservation 57:239–255.

HOYLE, M. E., AND J. W. GIBBONS. 2000. Use of a marked population of diamondback terrapins (*Malaclemys terrapin*) to determine the impacts of recreational crab pots. Chelonian Conservation and Biology 3:735–736.

HRABAR, H., K. DE NAGY, AND M. PERRIN. 2002. The effect of bill structure on seed selection by granivorous birds. African Zoology 37:67–80.

HURD, L. E., G. W. SMEDES, AND T. A. DEAN. 1979. An ecological study of a natural population of diamondback terrapins (*Malaclemys t. terrapin*) in a Delaware saltmarsh. Estuaries 2:28–33.

INGLES, L. G. 1967. The Mammals of the Pacific states. Stanford University Press, Stanford, CA.

INGRAM, B. L., M. E. CONRAD, AND J. C. INGLE. 1996. Stable isotope record of late Holocene salinity and river discharge in San Francisco Bay, California. Earth and Planetary Science Letters 141:237–247.

INTERAGENCY ECOLOGICAL PROGRAM. 2003. Interagency ecological program environmental monitoring program review and recommendations: Final Report. CALFED Bay-Delta Authority, Sacramento, CA.

INTERGOVERNMENTAL PANEL ON CLIMATE CHANGE. 2001. Summary for policymakers: climate change 2001. Impacts, adaptation, and vulnerability. Intergovernmental Panel on Climate Change, Geneva, Switzerland.

ISACCH, J. P., C. S. COSTA, B., RODRÍGUEZ-GALLEGO, L. CONDE, D. ESCAPA, M. GAGLIARDINI, D. S., AND O. O. IRIBARNE. 2006. Distribution of saltmarsh plant communites associated with environmental factors along a latitudinal gradient on the SW Atlantic coast. Journal of Biogeography 33:888–900.

ISACCH, J. P., S. HOLZ, L. RICCI, AND M. M. MARTINEZ. 2004. Post-fire vegetation change and bird use of a salt marsh in coastal Argentina. Wetlands 24:235–243

ISSAR, A. S. 2003. Climate changes during the Holocene and their impact on hydrological systems. International Hydrology Series. Cambridge University Press, Cambridge, UK.

IVERSON, J. B. 1991. Patterns of survivorship in turtles (order Testudines). Canadian Journal of Zoology 69:385–391.

JACKSON, H. H. 1928. A taxonomic review of the American long-tailed shrews. North American Fauna 51:1–238.

JACKSON, J. A. 1983. Adaptive response of nesting Clapper Rails to unusually high water. Wilson Bulletin 95:308–309.

JACOBSEN, H. A., G. L. JACOBSEN, JR., AND J. T. KELLEY. 1987. Distribution and abundance of tidal marshes along the coast of Maine. Estuaries 10:126–131.

JAFFE, B., R. SMITH, AND L. ZINK-TORRESAN. 1998. Sedimentation and bathymetric change in San Pablo Bay 1856–1983. U.S. Geological Survey Open File Report 98-759. <http://geopubs.wr.usgs.gov/open-file/of98-759> (6 June 2006).

JAMES, F. C. 1983. Environmental component of morphological differentiation in birds. Science 221:184–186.

JAMES-PIRRI, M. J., R. M. ERWIN, D. J. PROSSER, AND J. TAYLOR. 2004. Monitoring salt marsh responses to open marsh water management. Environmental Restoration 22:55–56.

JANES, D. N. 1997. Osmoregulation by Adelie Penguin chicks on the Antarctic peninsula. Auk 114:488–495.

JAQUES, D. 2002. Shorebird status and effects of *Spartina alterniflora* at Willapa National Wildlife Refuge. Report to USDI Fish and Wildlife Service, Willapa National Wildlife Refuge, Cathlemet, WA.

JASSBY, A. D., AND J. E. CLOERN. 2000. Organic matter sources and rehabilitation of the Sacramento-San Joaquin Delta (California, USA). Aquatic Conservation 10:323–352.

JASSBY, A. D., J. E. CLOERN, AND B. E. COLE. 2002. Annual primary production: patterns and mechanisms of change in a nutrient-rich tidal ecosystem. Limnology and Oceanography 47:698–712.

JASSBY, A. D., W. J. KIMMERER, S. G. MONISMITH, C. ARMOR, J. E. CLOERN, T. M. POWELL, J. R. SCHUBEL, AND T. J. VENDLINSKI. 1995. Isohaline position as a habitat indicator for estuarine populations. Ecological Applications 5:272–289

JEFFRIES, R. L., AND R. F. ROCKWELL. 2002. Foraging geese, vegetation loss, and soil degradation in an arctic salt marsh. Applied Vegetation Science 5:7–16.

JOHNSON, D. H., AND L. D. IGL. 2001. Area requirements of grassland birds: a regional perspective. Auk 118:24–34.

JOHNSON, O. W. 1974. Relative thickness of the renal medulla in birds. Journal of Morphology 142:277–284.

JOHNSON, O. W., AND J. N. MUGAAS. 1972. Quantitative and organizational features of the avian renal medulla. Condor 72:288–292.

JOHNSON, O. W., AND R. D. OHMART. 1973. Some features of water economy and kidney microstructure in the Large-billed Savannah Sparrow (*Passerculus sandwichensis rostratus*). Physiological Zoology 46:276–284.

JOHNSTON, R. F. 1954. Variation in breeding season and clutch size in Song Sparrows of the Pacific coast. Condor 56:268–273

JOHNSTON, R. F. 1955. Influence of winter tides on two populations of saltmarsh Song Sparrow. Condor 57:308–309.

JOHNSTON, R. F. 1956a. Population structure in salt marsh Song Sparrows. Part I. Environment and annual cycle. Condor 58:24–44.

JOHNSTON, R. F. 1956b. Population structure in salt marsh Song Sparrows. Part II. Density, age structure, and maintenance Condor 58:254–272.

JOHNSTON, R. F. 1957. Adaptation of salt marsh mammals to high tides. Journal of Mammology 38:529–531.

JOHNSTON, R. F., AND R. L. RUDD. 1957. Breeding of the salt marsh shrew. Journal of Mammalogy 38:157–163.

JOSSELYN, M. 1983. The ecology of San Francisco Bay tidal marshes: a community profile. USDI Fish and Wildlife Service, Biological Services Program. FWS/OBS-83/82. Washington, DC.

JOSSELYN, M., M. MARTINDALE, AND J. DUFFIELD. 1989. Public access and wetlands: Impacts of recreational use. Technical Report 9. Romberg Tiburon Centers, Center for Environmental Studies, San Francisco State University, Tiburon, CA.

JUNGIUS, H., AND U. HIRSCH. 1979. Herzfrequenzanderungen bei brutvogeln in Galapagos als Folge von Storungen durch Besucher. Journal fur Ornithologie 120:299–310.

KAISER, M., AND E. FRITZELL. 1984. Effects of river recreationists on Green-backed Heron behavior. Journal of Wildlife Management 48:561–567.

KALE, H. W., II. 1965. Ecology and bioenergetics of the Long-billed Marsh Wren, *Telmatodytes palustris griseus* (Brewster) in Georgia salt marshes. Publication Nuttall Ornithological Club. No. 5. Cambridge, MA.

KANAAR, R., J. H. J. HOEIJMAKERS, AND D. C. VAN GENT. 1998. Molecular mechanisms of DNA double-strand break repair. Trends in Cell Biology 8:483–489.

KANTRUD, H. A., AND R. E. STEWART. 1984. Ecological distribution and crude density of breeding birds in prairie wetlands. Journal of Wildlife Management 48:426–437.

KARRAN, P. 2000. DNA double strand break repair in mammalian cells. Current Opinion in Genetics and Development 10:144–150.

KASPRZAK, K. S., W. BAL, D. W. PORTER AND K. BIALKOWSKI. 1999. Studies on oxidative mechanisms of metal-induced carcinogenesis. Pp. 193–208 *in* M. Dizdaroglu, and A. E. Karakaya (editors). Advances in DNA damage and repair: oxygen radical effects, cellular protection, and biological consequences. Kluwer Academic Publishers, New York, NY.

KASTLER, J., AND P. L. WIBERG. 1996. Sedimentation and boundary changes of Virginia salt marshes. Estuarine, Coastal and Shelf Science 42:683–700.

KAYE, C. A., AND G. W. STUCKEY. 1973. Nodal tidal cycle of 18.6 yrs. Geology 1:141–145.

KEARNEY, M. S. 1996. Sea-level change during the last thousand years in Chesapeake Bay. Journal of Coastal Research 12:977–983.

KEARNEY, M. S., J. C. STEVENSON, AND L G. WARD. 1994. Spatial and temporal changes in marsh vertical accretion rates at Monie Bay: implications for sea level rise. Journal of Coastal Research 10:1010–1020.

KEER, G., AND J. B. ZEDLER. 2002. Saltmarsh canopy architecture differs with the number and composition of species. Ecological Applications 12:456–473.

KEIPER, R. 1985. The Assateague ponies. Tidewater Publishers, Centerville, MD.

KELDSEN, T. J. 1997. Potential impacts of climate change on California Clapper Rail habitat of south San Francisco Bay. M.S. thesis, Colorado State University. Fort Collins, CO.

KELLER, L. F., AND D. M. WALLER. 2002. Inbreeding effects in wild populations. Trends in Ecology and Evolution 17:230–241.

KELLEY, J. T., D. F. BELKNAP, G. L. JACOBSON, JR., AND H. A. JACOBSON. 1988. The morphology and origin of saltmarshes along the glaciated coastline of Maine, USA. Journal of Coastal Research 4:649–666.

KELLY, C. A., J. W. M. RUDD, R. A. BODALY, N. T. ROULET, V. L. ST. LOUIS, A. HEYES, T. R. MOORE, R. ARAVENA, B. DYCK, R. HARRISS, S. SCHIFF, B. WARNER, AND G. EDWARDS. 1997. Increases in fluxes of greenhouse gases and methyl mercury following flooding of an experimental reservoir, Environmental Science and Technology 31:1334–1344,

KÉRY, M., J. A. ROYLE, AND H. SCHMID. 2005. Modeling avian abundance from replicated counts using binomial mixture models. Ecological Applications 15:1450–1461.

KINLER, N. W., G. LINSCOMBE, AND P. R. RAMSEY. 1987. Nutria. Pp. 327–342 *in* M. Novak, J. Baker, M. Obbard, and B. Malloch (editors). Wild furbearer management and conservation in North America. Ministry of Natural Resources, Ottawa, ON, Canada.

KIRBY, R. E., S. J. LEWIS, AND T. N. SEXXON. 1988. Fire in North American wetland ecosystems and fire-wildlife relations: an annotated bibiography. Biological Report 88, USDI Fish and Wildlife Service, Washington, DC.

KJELMYR, J. E., G. W. PAGE, W. D. SHUFORD, AND L. E. STENZEL. 1991. Shorebird numbers in wetlands of the Pacific flyway: a summary of spring, fall, and winter counts in 1988, 1989, and 1990. Point Reyes Bird Observatory, Stinson Beach, CA.

KLEIN, M. 1993. Waterbird behavioral response to human disturbance. Wildlife Society Bulletin 21:31–39.

KLICKA, J., AND R. M. ZINK. 1997. The importance of recent ice ages in speciation: a failed paradigm. Science 277:1666–1669.

KNIGHT, R. L, AND D. N. COLE. 1991. Effects of recreational activity on wildlife in wildlands. Transactions of the North American Wildlife and Natural Resources Conference 56:238–247.

KNOPF, F. L., AND J. A. SEDGWICK. 1987. Latent population responses of summer birds to a catastrophic, climatological event. Condor 89:869–873.

KNOWLTON, S. M. 1971. Geomorphological history of tidal marshes, Eastern Shore, Virginia 1852–1966. M.S. thesis, University of Virginia, Charlottesville, VA.

KOCKELMAN, W. J., T. J. CONOMOS, AND A. E. LEVITON. 1982. San Francisco Bay: use and protection. Sixty-first annual meeting of the American Association for the Advancement of Science, Pacific Division, Davis, CA.

Komarek, E. V., Sr. 1969. Fire and animal behavior. Proceedings of the Annual Tall Timbers Fire Ecology Conference 9:160–207.

Komarek, E. V., Sr. 1984. Remarks on fire ecology and fire management particularly pertinent to coastal regions. Pp. 4–10 in M. K. Foley, and S. P. Bratton (editors). Barrier islands: critical fire management problems: proceedings of a workshop. Canaveral National Seashore, Titusville, FL.

Koneff, M. D., and J. A. Royle. 2004. Modeling wetland change along the United States Atlantic Coast. Ecological Modeling 177:41–59.

Koolhaas, A., A. Dekinga, and T. Piersma. 1993. Disturbance of foraging Knots by aircraft in the Dutch Wadden Sea in August–October 1992. Wader Study Group Bulletin 68:20–22.

Koppelman, L. E., P. K. Weyl, M. Grant Gross, and D. S. Davies. 1976. The urban sea: Long Island Sound. Praeger Publishers, New York, NY.

Kotliar, N. B., and J. A. Wiens. 1990. Multiple scales of patchiness and patch structure: a hierarchical framework for the study of heterogeneity. Oikos 59:253–260.

Kozicky, E. L., and F. V. Schmidt. 1949. Nesting habits of the Clapper Rail in New Jersey. Auk 66:355–364.

Krakauer, T. H. 1970. The ecological and physiological control of water loss in snakes. Ph.D. dissertation, University of Florida, Gainesville, FL.

Kramer, V. L., J. N. Collins, and C. Beesley. 1995. Reduction of *Aedes dorsalis* by enhancing tidal action in a northern California marsh. Journal of the American Mosquito Control Association 11:389–395.

Krebs, C. 1989. Ecological methodology. Harper and Row, New York, NY.

Krone, R. B. 1996. Recent sedimentation in the San Francisco Bay system. Pp. 63–67 in J. T. Hollibaugh (editor) San Francisco Bay: the ecosystem. Further investigations into the natural history of the San Francisco Bay and delta with reference to the influence of man. American Association for the Advancement of Sciences, Pacific Division, San Francisco, CA.

Kroodsma, D. E., and J. Verner. 1997. Marsh Wren (*Cistothorus palustris*). In A. Poole, and F. Gill (editors). The birds of North America, No. 308. The Academy of Natural Sciences, Philadelphia, PA and The American Ornithologists' Union, Washington, DC.

Kruchek, B. L. 2004. Use of tidal marshland upland habitats by the marsh rice rat, (*Oryzomys palustris*). Journal of Mammology 85:569–576.

Kruse, K. L., J. R. Lovvorn, J. Y. Takekawa, and J. Mackay. 2003. Long-term productivity of Canvasbacks in a snowpack-driven desert marsh. Auk 120:107–119.

Kuhn-Campbell, M. 1979. A tale of two dykes — the story of Cole Harbour. Lancelot Press, Hantsport, NS, Canada.

Kutzbach, J., R. Gallimore, S. Harrison, P. Behling, R. Selin, and F. Laarif. 1998. Climate and biome simulations for the past 21,000 years. Quaternary Science Reviews 17:473–506.

Lack, D. 1947. The significance of clutch-size. Ibis 89:302–352.

Lafferty, K. D., C. C. Swift, and R. F Ambrose. 1999. Extirpation and recolonization in a metapopulation of an endangered fish, the tidewater goby. Conservation Biology 13:1447–1453.

Lamb, T., and J. C. Avise. 1992. Molecular and population aspects of mitochondrial DNA variability in the diamondback terrapin, *Malaclemys terrapin*. Journal of Heredity 83:262–269.

Lamb, T., and M. F. Osentoski. 1997. On the paraphyly of *Malaclemys*: a molecular genetic assessment. Journal of Herpetology 31:258–265.

Langis, R., M. Zalejko, and J. B. Zedler. 1991. Nitrogen assessment in a constructed and a natural salt marsh of San Diego Bay. Ecological Applications 1:40–51.

Larison, B., S. A. Laymon, P. L. Williams, and T. B. Smith. 2001. Avian responses to restoration: nest-site selection and reproductive success in Song Sparrows. Auk 118:432–442.

Larssen, T. 1976. Composition and density of the bird fauna in Swedish shore meadows. Ornis Scandinavica 7:1–12.

Lathrop, R. G., L. Windham, and P. Montesano. 2003. Does *Phragmites* expansion alter the structure and function of marsh landscapes? Patterns and processes revisited. Estuaries 26:423–435.

Laudenslayer, W. F., Jr., W. E. Grenfell, Jr., and D. C. Zeiner. 1991. A check-list of the amphibians, reptiles, birds, and mammals of California. California Fish and Game 77:109–141.

Laverty G., and R. F. Wideman, Jr. 1989. Sodium excretion rates and renal responses to acute salt loading in the European Starling. Journal of Comparative Physiology 159B:401–408.

Lawson, R. A., P. Meier, G. Frank, and P. E. Moler. 1991. Allozyme variation and systematics of the *Nerodia fasciata-Nerodia clarkii* complex of water snakes (Serpentes: Colubridae). Copeia 1991:638–659.

Lazell, J. D., Jr. 1979. Diamondback terrapins at Sandy Neck aquasphere. New England Aquarium 13:28–31.

Lazell, J. D., Jr., and P. J. Auger. 1981. Predation on diamondback terrapin (*Malaclemys terrapin*) eggs by dunegrass (*Ammophila breviligulata*). Copeia 1981:723–724.

Lebreton, J. D., K. P. Burnham, J. Clobert, and D. R. Anderson. 1992. Modeling survival and testing biological hypotheses using marked animals: a unified approach with case studies. Ecological Monographs 62:67–118.

Leck, M. A. 1989. Wetland seed banks. Pp. 283–306 in M. A. Leck, V. T. Parker, and R. L. Simpson (editors). Ecology of soil seed banks. Academic Press, San Diego, CA.

Lefeuvre, J. C., and R. F. Dame. 1994. Comparative studies of salt marsh processes in the New and Old Worlds: an introduction. Pp. 139–153 in W. J. Mitsch (editor). Global wetlands: Old World and New. Elsevier, Amsterdam, The Netherlands.

Legare, M. L., D. B. McNair, W. C. Conway, and S. A. Legare. 2000. Swamp Sparrow winter site fidelity records in Florida. Florida Field Naturalist 28:73–74.

LEISLER, B., H. WINKLER, AND M. WINK. 2002. Evolution of breeding systems in Acrocephaline warblers. Auk 119:379–390.

LEONARD, M. L., AND J. PICMAN. 1987. Nesting mortality and habitat selection by Marsh Wrens. Auk 104:491–495.

LEVIN, P., J. ELLIS, R. PETRIK, AND M. HAY. 2002. Indirect effects of feral horses on estuarine communities. Conservation Biology 16:1364–1371.

LEVINE, J., S. BREWER, AND M. D. BERTNESS. 1998. Nutrient availability and the zonation of marsh plant communities. Journal of Ecology 86:285–292.

LEWIS, J. C., AND K. L. SALEE. 1999. Introduction and range expansion of nonnative red foxes (*Vulpes vulpes*) in California. American Midland Naturalist 142:372–381.

LIANG, F., M. HAN, P. J. ROMANIENKO, AND M. JASIN. 1998. Homology-directed repair is a major double-strand break repair pathway in mammalian cells. Proceedings of the National Academy of Sciences, USA 95:5172–5177.

LIDICKER, W. Z., JR. 2000. California vole *Microtus californicus*. Pp. 229–231 in P. R. Olofson (editor). Baylands ecosystem species and community profiles: life histories and environmental requirements of key plants, fish and wildlife. San Francisco Bay Area Wetlands Ecosystem Goals Project, San Francisco Bay Regional Water Control Board, Oakland, CA.

LINDAU, C. W., AND L. R. HOSSNER. 1981. Substrate characterization of an experimental marsh and three natural marshes. Soil Science Society of America Journal 45:1171–1176.

LINK, W. A., R. J. BARKER, J. R. SAUER, AND S. DROEGE. 1994. Within-site variability in surveys of wildlife populations. Ecology 75:1097–1108.

LINVILLE, R. G., S. N. LUOMA, L. CUTTER, AND G. CUTTER. 2002. Increased selenium threat as a result of the invasion of the exotic bivalve *Potamacorbula amurensis* in the San Francisco Bay-Delta. Aquatic Toxicology 57:51–64.

LINZEY, D. W. 1998. The mammals of Virginia. The McDonald and Woodward Publishing Company, Blacksburg, VA.

LOCKWOOD, J. L., K. H. FENN, J. L. CURNUTT, D. ROSENTHAL, K. L. BALENT, AND A. L. MAYER. 1997. Life history of the endangered Cape Sable Seaside Sparrow. Wilson Bulletin 109:720–731.

LOMOLINO, M. V., 2000. A species-based theory of insular biozoography. Global Ecology and Biogeography 9:39–58.

LONSDALE, W. M., AND A. M. LANE. 1994. Tourist vehicles as vectors of weed seeds in Kakadu National Park, Northern Australia. Biological Conservation 69:277–283.

LONZARICH, D. G., T. E. HARVEY, AND J. E. TAKEKAWA. 1992. Trace element and organochlorine concentrations in California Clapper Rails (*Rallus longirostris obsoletus*) eggs. Archives of Environmental Contamination and Toxicology 23:147–153.

LOPEZ-LABORDE, J. 1997. Geomorphological and geological setting of the Rio de la Plata. Pp. 1–16 in P. G. Wells, and G. R. Daborn (editors). The Rio de la Plata: An environmental overview. Dalhousie University, Halifax, NS, Canada.

LOVICH, J. E., AND J. W. GIBBONS. 1990. Age at maturity influences adult sex ratio in the turtle *Malaclemys terrapin*. Oikos 59:126–134.

LOWERY, G. H., JR. 1974. The mammals of Louisiana and its adjacent waters. Louisiana State University Press, Baton Rouge, LA.

LOWTHER, P. E. 1996. Le Conte's Sparrow (*Ammodramus leconteii*). In A. Poole, and F. Gill (editors). The birds of North America, No. 224. The Academy of Natural Sciences, Philadelphia, PA and The American Ornithologists' Union, Washington, DC.

LOWTHER, P. E., H. D. DOUGLAS III, AND C. L. GRATTO-TREVOR. 2001.Willet (*Cataptrophorus semipalmatus*). In A. Poole, and F. Gill (editors). The birds of North America, No. 579. The Academy of Natural Sciences, Philadelphia, PA and The American Ornithologists' Union, Washington, DC.

LOYN, R. H., B. A. LANE, C. CHANDLER, AND G. W. CARR. 1986. Ecology of Orange-bellied Parrots (*Neophema chrysogaster*) at their main remnant wintering site. Emu 86:195–206.

LUOMA, S. N., AND T. S. PRESSER. 2000. Forecasting selenium discharges to the San Francisco Bay-Delta Estuary: ecological effects of a proposed San Luis Drain extension. U.S. Geological Survey Open File Report 00-416, Water Resources Division, National Research Program, Menlo Park, CA.

LYNCH, J. J. 1941. The place of burning in management of the Gulf Coast wildlife refuges. Journal of Wildlife Management 5:454–457.

LYNCH, J. J., T. O'NEIL, D. LAY, AND A. EINARSEN. 1947. Management significance of damage by geese and muskrats to Gulf Coast marshes. Journal of Wildlife Management 11:50–76.

MACARTHUR, R. H., AND E. O. WILSON. 1963. An equilibrium theory of insular biogeography. Evolution 17:373–387.

MACARTHUR, R. H., AND E. O. WILSON. 1967. The theory of island biogeography. Monographs in population biology. Princeton University Press, Princeton, NJ.

MACDONALD, I. A. W., L. L. LOOPE, M. B. USHER, AND O. HAMANN. Wildlife conservation and the invasion of nature preserves by introduced species: A global perspective. 1989. Pp. 215–255 in J. A. Drake, H. A. Mooney, F. diCastri, R. H. Groves, F. J. Kruger, M. Rejmanek, and M. Williamson (editors). Biological invasions: a global perspective. Scope 37. John Wiley and Sons, Chichester, UK.

MACDONALD, K. B. 1977. Coastal salt marsh. Pp. 263–294 in M. G. Barbour, and J. Major (editors). Terrestrial vegetation of California. John Wiley and Sons, New York, NY.

MACDONALD, K. B. 1990. Marine ecological characterization, bay history and physical environment. South San Diego Bay enhancement plan, Vol. 1, Resources Atlas. San Diego Unified Port District, San Diego, CA.

MACK, R. N., D. SIMBERLOFF, W. M. LONSDALE, H. EVANS, M. CLOUT, AND F. A. BAZAZZ. 2000. Biotic invasions: causes, epidemiology, global consequences, and control. Ecological Applications 10:689–710.

MACKAY, K. 2000. Suisun shrew. Pp. 233–236 in P. R. Olofson (editor). Baylands ecosystem species and community profiles: life histories and environmental requirements of key plants, fish and wildlife. San Francisco Bay Regional Water Quality Control Board, Oakland, CA.

MacKenzie, D. I., J. D. Nichols, G. B. Lachman, S. Droege, J. A. Royle, and C. A. Langtimm. 2002. Estimating site occupancy rates when detection probabilities are less than one. Ecology 83: 2248–2255.

MacKinnon, K., and D. B. Scott. 1984. An evaluation of saltmarshes in Atlantic Canada. Centre for Marine Geology, Dalhousie University, Halifax, NS, Canada.

MacMillen, R. E. 1964. Water economy and salt balance in the western harvest mouse. Physiological Zoology 37:45–56.

Maffei, W. A. 2000. Invertebrates. Pp. 154–192 in P. R. Olufson (editor). Baylands ecosystem species and community profiles: life histories and environmental requirements of key plants, fish, and wildlife. San Francisco Bay Area Wetlands Ecosystem Goals Project and San Francisco Bay Regional Water Quality Control Board, Oakland, CA.

Mahall, B. E., and R. B. Park. 1976. The ecotone between *Spartina foliosa* trin. and *Salicornia virginica* l. in salt marshes of northern San Francisco Bay. Journal of Ecology 64:421–433.

Mahoney, S. A., and J. R. Jehl. 1985. Adaptations of migratory shorebirds to highly saline and alkaline lakes: Wilson's Phalarope and American Avocet. Condor 87:520–527.

Maillet, J. 2000. An examination of issues and threats to coastal habitats along the Northumberland Strait in New Brunswick. Fish and Wildlife Branch, New Brunswick Division of Natural Resources, Fredericton, NB, Canada.

Malamud-Roam, F. 2002. A late Holocene history of vegetation change in San Francisco estuary marshes using stable carbon isotopes and pollen analysis. Ph.D. dissertation, University of California. Berkeley, CA.

Malamud-Roam, F., and B. L. Ingram, 2001. Carbon isotopic compositions of plants and sediments of tide marshes in the San Francisco estuary. Journal of Coastal Research 17:17–19.

Malamud-Roam, F., and B. L. Ingram, 2004. Late Holocene ^{13}C and pollen records of paleosalinity from tidal marshes in the San Francisco Bay estuary, California. Journal of Quaternary Research 62: 134–145.

Malamud-Roam, F., L. Ingram, M. Hughes, and J. Florsheim. 2006. Holocene paleoclimate records from a large California estuarine system and its watershed region: linking watershed climate and bay conditions. Quaternary Science Reviews 25: 1570–1598.

Malamud-Roam, K. 2000. Tidal regimes and tidal marsh hydroperiod in the San Francisco estuary: theory and implications for ecological restoration. Ph.D. dissertation, University of California. Berkeley, CA.

Maldonado, J. E., C. Vilà, and F. Hertel. 2004. Discordant patterns of morphological variation in genetically divergent populations of ornate shrews (*Sorex ornatus*). Journal of Mammalogy 85:886–896.

Maldonado, J. E., C. Vilà, and R. K. Wayne. 2001. Tripartite genetic subdivisions in the ornate shrew (*Sorex ornatus*). Molecular Ecology 10: 127–147.

Mangold, R. E. 1974. Clapper Rail studies. Final Report. USDI Fish and Wildlife Service, and State of New Jersey Division of Fish, Game, and Shellfisheries, Trenton, NJ.

Manly, B. F. J. 1998. Randomization, bootstrap and Monte Carlo methods in biology, 2nd edition. Chapman & Hall, London, UK.

Mann, T. M. 1995. Population surveys for diamondback terrapins (*Malaclemys terrapin*) and gulf salt marsh snakes (*Nerodia clarkii clarkii*) in Mississippi. Technical Report No. 37, Mississippi Museum of Natural Science, Jackson, MS.

Marshall, J. T. 1948a. Ecologic races of Song Sparrows in the San Francisco Bay region. Part I. habitat and abundance. Condor 50:193–215.

Marshall, J. T. 1948b. Ecologic races of Song Sparrows in the San Francisco Bay region. Part II. geographic variation. Condor 50:233–256.

Marshall, J. T., and K. G. Dedrick. 1994. Endemic Song Sparrows and Yellowthroats of San Francisco Bay. Studies in Avian Biology 15:316–327.

Marshall, R. M., and S. E. Reinert. 1990. Breeding ecology of Seaside Sparrows in a Massachusetts salt marsh. Wilson Bulletin 102:501–513.

Martin, A. C., H. S. Zim, and A. L. Nelson. 1961. American wildlife and plants: a guide to wildlife food habits. Dover Publications, New York, NY.

Martin, T. E. 1993. Nest predation among vegetation layers and habitat types: revising the dogma. American Naturalist 141:897–913.

Martin, T. E. 1995. Avian life history evolution in relation to nest sites, nest predation, and food. Ecological Monographs 65:101–127.

Martin, T. E., P. R. Martin, C. R. Olson, B. J. Heidinger, and J. J. Fontaine. 2000. Parental care and clutch sizes in North and South American birds. Science 287:2454–2460.

Martinez, M. M., M. S. Bó, and J. P. Isacch. 1997. Hábitat y abundancia de *Porzana spiloptera* y *Coturnicops notata* en Mar Chiquita (Prov. de Buenos Aires). Hornero 14:74–277.

Marvin-DiPasquale, M. C., J. L. Agee, R. M. Bouse, and B. E. Jaffe. 2003. Microbial cycling of mercury in contaminated pelagic and wetland sediments of San Pablo Bay, California. Environmental Geology 43:260–267.

Mason, O. K., and J. W. Jordan. 2001. Minimal late Holocene sea level rise in the Chukchi Sea: Arctic insensitivity to global change? Global and Planetary Change 32:13–23.

Massey, B. W. 1979. The Belding's Savannah Sparrow. U.S. Army Corps of Engineers, DACW09-78-C-0008, Los Angeles, CA.

Massey, B. W., and E. Palacios. 1994. Avifauna of the wetlands of Baja California, Mexico: current status. Studies in Avian Biology 15:45–57.

Massey, B. W., and R. Zembal. 1987. Vocalizations of the Light-footed Clapper Rail. Journal of Field Ornithology 58:32–40.

Massey, B. W., R. Zembal, and P. D. Jorgensen. 1984. Nesting habitat of the Light-footed Clapper

Rail in southern California. Journal of Field Ornithology 55:67–80.

MATHEW, J. 1994. The status, distribution and habitat of the Slender-billed Thornbill *Acanthiza irdalei* in South Australia. South Australian Ornithologist 32:1–19.

MATHIESON, A. C., C. A. PENNIMAN, AND L. G. HARRIS. 1991. Northwest Atlantic rocky shore ecology. Pp. 109–191 *in* A. C. Mathieson, and P. H. Nienhuis (editors). Ecosystems of the World: intertidal and littoral ecosystems. Elsevier, Amsterdam, The Netherlands.

MATTA, J. F., AND C. L. CLOUSE. 1972. The effect of periodic burning on marshland insect populations. Virginia Journal of Science 23:113.

MAY, M. D. 1999. Vegetation and salinity changes over the last 2000 years at two islands in the northern San Francisco Estuary, California. M.A. thesis, University of California, Berkeley, CA.

MAYFIELD, H. F. 1961. Nesting success calculated from exposure. Wilson Bulletin 73:255–261.

MAYFIELD, H. F. 1975. Suggestions for calculating nest success. Wilson Bulletin 87:456–466.

MAYR, E. 1942. Systematics and the origin of species. Harper and Brothers, New York, NY.

MAYR, E. 1963. Animal species and evolution. Harvard University Press, Cambridge, MA.

MCATEE, J. W., C. J. SCIFRES, AND D. L. DRAWE. 1979. Improvement of Gulf Coast cordgrass range with burning or shredding. Journal of Range Management 32:372–375.

MCCABE, A. M., AND P. U. CLARK. 1998. Ice-sheet variability around the North Atlantic Ocean during the last deglaciation. Nature 392:373–377.

MCDONALD, M. V. 1986. Scott's Seaside Sparrows. Ph.D. dissertation, University of Florida, Gainesville, FL.

MCDONALD, M. V., AND R. GREENBERG. 1991. Nest departure calls in New World songbirds. Condor 93:365–374.

MCGARIGAL, K., AND B. J. MARKS. 1995. FRAGSTATS: spatial pattern analysis program for quantifying landscape structure. USDA Forest Service General Technical Report PNW-351. USDA Forest Service, Pacific Northwest Research Station, Corvallis, OR.

MCNAB, B. K. 1991. The energy expenditure of shrews. Special publication, Museum of Southwestern Biology, University of New Mexico 1:75–91.

MCNAIR, D. B. 1987. Egg data slips—are they useful for information on egg-laying dates and clutch size? Condor 89:369–376.

MEANLEY, B. 1985. The marsh hen: a natural history of the Clapper Rail of the Atlantic coast salt marsh. Tidewater Publishers, Centreville, MD.

MEANLEY, B. 1992. King Rail (*Rallus elegans*). *In* A. Poole, and F. Gill (editors). The birds of North America, No. 3. The Academy of Natural Sciences, Philadelphia, PA and The American Ornithologists' Union, Washington, DC.

MEANLY, B., AND D. K. WEATHERBEE. 1962. Ecological notes on mixed populations of King Rails and Clapper Rails in Delaware marshes. Auk 79:453–457.

MENARD, C., P. DUNCAN, G. FLEURANCE, J.-Y. GEORGES, AND M. LILA. 2002. Comparative foraging and nutrition of horses and cattle in European wetlands. Journal of Applied Ecology 39:120–133.

MEREDITH, W., AND D. SAVEIKIS. 1987. Effects of open marsh water management (OMWM) on bird populations of a Delaware tidal marsh, and OMWM's use in waterbird habitat restoration and enhancement. Pp. 298–321 *in* W. R. Whitman, and W. H. Meredith (editors). Waterfowl and wetlands symposium: proceedings of a symposium on waterfowl and wetlands management in the coastal zone of the Atlantic Flyway. Delaware Coastal Management Program, Delaware Department of Natural Resources and Environmental Control, Dover, DE.

MEREDITH, W., D. SAVEIKIS, AND C. STACHECKI. 1985. Guidelines for open marsh water management in Delaware's salt marshes—objectives, system designs, and installation procedures. Wetlands 5:119–133.

MERILÄ, J., AND B. C. SHELDON. 2001. Avian quantitative genetics. Current Ornithology 16:179–255.

MERINO, M., A. VILA, AND A. SERRET. 1993. Relevamiento Ecológico de la Bahía de Samborombón, Buenos Aires. Boletín Técnico No. 16. Fundación Vida Silvestre, Buenos Aires, Argentina.

MEYERSON, L. A., K. SALTONSTALL, L. WINDHAM, E. KIVIAT, AND S. FINDLAY. 2000. A comparison of *Phragmites australis* in freshwater and brackish water marsh environments in North America. Wetlands Ecology and Management 8:89–103.

MICHELI, F., AND C. H. PETERSON. 1999. Estuarine vegetated habitats as corridors for predator movements. Conservation Biology 13:869–881.

MILEWSKI, I., J. HARVEY, AND S. CALHOUN. 2001. Shifting sands: state of the coast in northern and eastern New Brunswick. Conservation Council of New Brunswick, Fredericton, NB, Canada.

MILLER, D. L., F. E. SMEINS, AND J. W. WEBB. 1996. Mid-Texas coastal marsh change (1939–1991) as influenced by Lesser Snow Goose herbivory. Journal of Coastal Research 12:462–476.

MILLER, F. W. 1928. A new white-tailed deer from Louisiana. Journal of Mammalogy 9:57–59.

MILLER, G. 1996. Ecosystem management: improving the Endangered Species Act. Ecological Applications 6:715–717.

MILLIGAN, D. C. 1987. Maritime dykelands. Nova Scotia Department of Government Services-Publishing Division, Halifax, NS, Canada.

MILOVICH, J. A., C. LASTA, D. A. GAGLIARDINI, AND B. GUILLAUMONT. 1992. Pp. 869–882 *in* Initial study on the structure of the salt marsh in the Samborombón Bay coastal area. Transactions of the First Thematic Conference on Remote Sensing for Marine and Coastal Environments. New Orleans, LA.

MITCHELL, M. S., R. A. LANCIA, AND J. A. GERWIN. 2001. Using landscape-level data to predict the distribution of birds on a managed forest: effects of scale. Ecological Applications 11:1692–1708.

MITCHELL-JONES, A. J., G. AMORI, W. BOGDANOWICZ, B. KRYSTUFEK, P. J. H. REIJNDERS, F. SPITZENBERGER, M. STUBBE, J. B. M THISSEN, V. VOHRALIK, AND J. ZIMA. 1999. The atlas of European mammals. Academic Press. London, UK.

Mitro, M. 2004. Demography and viability analyses of a diamondback terrapin population. Canadian Journal of Zoology 81:716–726.

Mitsch, W. J., and J. G. Gosselink. 2000. Wetlands. John Wiley, New York, NY.

Mizell, K. L. 1999. Effects of fire and grazing on Yellow Rail habitat in a Texas coastal marsh. Ph.D. dissertation, Texas A&M University, College Station, TX.

Moffitt, J. 1941. Notes on the food of the California Clapper Rail. Condor 43:270–273.

Møller, H. S. 1975. Danish salt marsh communities of breeding birds in relation to different types of management. Ornis Scandinavica 6:125–133.

Montagna, W. 1940. The Acadian Sharp-tailed Sparrows of Popham Beach, Maine. Wilson Bulletin 52:191–197.

Montagna, W. 1942. The Sharp-tailed Sparrows of the Atlantic Coast. Wilson Bulletin 54:107–120.

Montevecchi, W. A. 1975. Behavioral and ecological factors influencing the reproductive success of a tidal salt marsh colony of Laughing Gulls (*Larus atricilla*). Ph.D. dissertation, Rutgers University, Newark, NJ.

Montevecchi, W. A. 1978. Nest site selection and its survival value among Laughing Gulls. Behavioral Ecology and Sociobiology 4:143–161.

Montevecchi, W. A., and J. Burger. 1975. Aspects of the reproductive biology of the northern diamondback terrapin *Malaclemys terrapin terrapin*. American Midland Naturalist 94:166–178.

Mooney, H. A., and J. A. Drake (editors). 1986. Ecology of biological invasions of North America and Hawaii. Springer-Verlag, New York, NY.

Moore, W. S. 1995. Inferring phylogenies from mtDNA variation: mitochondrial-gene trees versus nuclear-gene trees. Evolution 49:718–726.

Moorman, T. E., and P. N. Gray. 1994. Mottled Duck (*Anas fulvigula*). *In* A. Poole, and F. Gill (editors). The birds of North America, No. 81. National Academy of Science, Philadelphia, PA and The American Ornithologists' Union, Washington, DC.

Morantz, D. L. 1976. Productivity and export from a marsh with a 15 m tidal range, and the effect of impoundment of selected areas. M.S. thesis, Dalhousie University, Halifax, NS, Canada.

Mori, Y., N. S. Sodhi, S. Kawanishi, and S. Yamagishi. 2001. The effect of human disturbance and flock composition on the flight distances of waterfowl species. Journal of Ethology 19:115–119.

Moritz, C. 1994. Defining 'evolutionary significant units' for conservation. Trends in Ecology and Evolution 9:373–375.

Moritz, C., S. Lavery, and R. Slade. 1995. Using allele frequency and phylogeny to define units for conservation and management. American Fisheries Society Symposium 17:249–262.

Morris, J. T., P. V. Sundareshwar, C. T. Nietch, B. Kjerfve, and D. R. Cahoon. 2002. Responses of coastal wetlands to rising sea level. Ecology 83:2869–2877.

Morton, E. S. 1975. Ecological sources of selection on avian sounds. American Naturalist 109:17–34.

Morton, J. M. 1995. Management of human disturbance and its effects on waterfowl. Pp. F56–F86 *in* W. R. Whitman, T. Strange, L. Widjeskog, R. Whittemore, P. Kehoe, and L. Roberts (editors). Waterfowl habitat restoration, enhancement and management in the Atlantic Flyway, 3rd edition. Environmental Management Communications, Atlantic Flyway Technical Section and Delaware Division of Fish and Wildlife, Dover, DE.

Mowbray, T. B. 1997. Swamp Sparrow (*Melospiza georgiana*). *In* A. Poole, and F. Gill (editors). The birds of North America, No. 279. The Academy of Natural Sciences, Philadelphia, PA and The American Ornithologists' Union, Washington, DC.

Muldal, A. M., J. D. Moffatt, and R. J. Robertson. 1986. Parental care of nestlings by male Red-winged Blackbirds. Behavioral Ecology and Sociobiology 19:105–114.

Mundy, N. I., N. S. Badcock, K. Scribner, K. Janssen, and N. J. Nadeau. 2004. Conserved genetic basis of a quantitative plumage trait involved in mate choice. Science 303:1870–1873.

Murray, B. G., Jr. 1969. A comparative study of the LeConte's and Sharp-tailed sparrows. Auk 86:199–231.

Murray, B. G., Jr. 1971. The ecological consequences of interspecific territorial behavior in birds. Ecology 52:414–423.

Murray, B. G., Jr. 1981. The origins of adaptive interspecific territorialism. Biological Review 56:1–22.

Myers, J. P., and R. T. Lester. 1992. Double jeopardy for migrating animals: multiple hits and resource asynchrony. Pp. 193–200 *in* R. L. Peters, and T. E. Lovejoy (editors). Global warming and biological diversity. Yale University Press, New Haven, CT.

Myers, P. A. 1988. The systematics of *Nerodia clarkii* and *N. fasciata*. M.S. thesis, Lousiana State University, Baton Rouge, LA.

National Academy of Sciences. 1987. Responding to changes in sea level: engineering implications. National Academy Press, Washington, DC.

National Ocean Service. 1974a. Bathymetric map, National Ocean Service 1307N-11B Vicinity Point Sur to Point Reyes, California. 1:250,000. U.S. Department of Commerce, Washington DC.

National Ocean Service. 1974b. Bathymetric Map, National Ocean Service 1307N-18. Point Reyes to Tolo Bank, California. 1:250,000. U.S. Department of Commerce, Washington DC.

National Oceanic and Atmospheric Administration. 2004a. Tide tables: West Coast of North and South America. International Marine Press. Camden, ME.

National Oceanic and Atmospheric Administration. 2004b. Tide tables: East Coast of North and South America. International Marine Press. Camden, ME.

National Park Service. 2002. AlaskaPak Functions Pack Extension for ArcView 3.2. Alaska Support Office, National Park Service, Anchorage, AK.

National Wildlife Health Center. 2003a. West Nile virus. <http://www.nwhc.usgs.gov/disease_information/west_nile_virus/index.jsp> (6 June 2006).

National Wildlife Health Center. 2003b. Avian cholera. <http://www.nwhc.usgs.gov/disease_information/avian_cholera/index.jsp> (6 June 2006).

NAUGLE, D. E., K. F. HIGGINS, S. M. NUSSER, AND W. C. JOHNSON. 1999. Scale-dependent habitat use in three species of prairie wetland birds. Landscape Ecology 14:267–276.

NAUGLE, D. E., R. R. JOHNSON, M. E. ESTEY, AND K. F. HIGGINS. 2001. A landscape approach to conserving wetland bird habitat in the prairie pothole region of eastern South Dakota. Wetlands 21:1–17.

NEILL, W. T. 1958. The occurrence of amphibians and reptiles in saltwater areas and a bibliography. Bulletin of Marine Science of the Gulf and Carribean 8:1–95

NELSON, K., R. J. BAKER, H. S. SHELLHAMMER, AND R. K. CHESSER. 1984. Test of alternative hypotheses concerning the origin of Reithrodontomys raviventris: genetic analysis. Journal of Mammalogy 65: 668–673.

NELSON, W. S., T. DEAN, AND J. C. AVISE. 2000. Matrilineal history of the endangered Cape Sable Seaside Sparrow inferred from mitochondrial DNA polymorphism. Molecular Ecology 9: 809–813.

NEREM, R. S., T. M. VAN DAM, AND M. S. SCHENEWERK. 1998. Chesapeake Bay subsidence monitored as wetlands loss continues. EOS Transactions, American Geophysical Union 79:149–157.

NERO, R. W. 1984. Redwings. Smithsonian Institution Press, Washington, DC.

NETER, J., W. WASSERMAN, AND M. H. KUTNER. 1990. Applied linear statistical models: regression, analysis of variance, and experimental designs. 3rd edition. Richard D. Irwin, Homewood, IL.

NEWMAN, J. R. 1976. Population dynamics of the wandering shrew Sorex vagrans. Wasmann Journal of Biology 34:235–250.

NICE, M. M. 1937. Studies in the life history of the Song Sparrow. I. A population study of the Song Sparrow. Transactions of the. Linnean Society of New York 4:1–247.

NICHOLLS, R., AND S. LEATHERMAN. 1996. Adapting to sea-level rise: relative sea-level trends to 2100 for the United States. Coastal Management 24:301–324.

NICHOLS, F. H., J. E. CLOERN, S. N. LUOMA, AND D. H. PETERSON. 1986. The modification of an estuary. Science 231:567–573.

NICHOLS, F. H., J. K. THOMPSON, AND L. E. SCHEMEL. 1990. Remarkable invasion of San Francisco Bay (California USA) by the Asian clam Potamocorbula Amurensis: displacement of a former community. Marine Ecology-Progress Series 66:95–102.

NICHOLS, J. D., J. E. HINES, J. R. SAUER, F. W. FALLON, J. E. FALLON, AND P. J. HEGLUND. 2000. A double-observer approach for estimating detection probability and abundance from point counts. Auk 117:393–408.

NIERING, W. A., AND R. S. WARREN. 1980. Vegetation patterns and processes in New England salt marshes. BioScience 30:301–307.

NIKITINA, D. L., J. E. PIZZUTO, R. A. SCHWIMMER, AND K. W. RAMSEY. 2000. An updated Holocene sea-level curve for the Delaware coast. Marine Geology 171:7–20.

NISBET, I. C. T., V. APANIUS, AND M. S. FRIAR. 2002. Breeding performance of very old Common Terns. Journal of Field Ornithology 73:117–124.

NIXON, S. M. 1982. The ecology of New England high salt marshes: a community profile. Office of Biological Services FWS/OBS-81-55. USDI Fish and Wildlife Service, Washington, DC.

NOAA. 2003. A biogeographic assessment off north/central California to support the Joint Management Plan Review for Cordell Bank, Gulf of the Farallones, and Monterey Bay National Marine Sanctuaries: phase I—marine fishes, birds, and mammals. NOAA, National Centers for Coastal Ocean Science (NCCOS) Biogeography Team, Silver Spring, MD.

NOAA. 2004a. Tide tables: west coast of North and South America. International Marine Press, Camden, ME.

NOAA. 2004b. Tide tables: east coast of North and South America. International Marine Press, Camden, ME.

NOAA. 2004c. Tide tables: Europe and the west coast of Africa. International Marine Press, Camden, ME.

NOAA. 2004d Tide tables: central and western Pacific Ocean and Indian Ocean. International Marine Press, Camden, ME.

NOCERA, J. J., G. J. PARSONS, G. R. MILTON, AND A. H. FREDEEN. 2005. Compatibility of delayed cutting regime with bird breeding and hay nutritional quality. Agriculture, Ecosystems and Environment 107:245–253.

NOL, E., AND H. BLOKPOEL. 1983. Incubation period of Ring-billed Gulls and the egg immersion technique. Wilson Bulletin 95:283–286.

NORRIS, K., T. COOK, B. O'DOWD, AND C. DURBIN. 1997. The density of Redshank (Tringa tetanus) breeding on the salt marshes of the wash in relation to habitat and its grazing management. Journal of Applied Ecology 34:995–1013.

NOWAK, R. M. 1999. Walker's mammals of the world. The Johns Hopkins University Press, Baltimore, MD.

NUR, N., S. ZACK, J. EVENS, AND T. GARDALI. 1997. Tidal marsh birds of the San Francisco Bay region: status, distribution, and conservation of five category 2 taxa. Final Report to U.S. Geological Survey-Biological Resources Division, Point Reyes Bird Observatory, Stinson Beach, CA.

NUSSER, J. A., R. M. GOTO, D. B. LEDIG, R. C. FLEISCHER, AND M. M. MILLER. 1996. RAPD analysis reveals low genetic variability in the endangered Light-footed Clapper Rail. Molecular Ecology 5: 463–472.

NYMAN, J. A., AND R. H. CHABRECK. 1995. Fire in coastal marshes: history and recent concerns. Proceedings of the Annual Tall Timbers Fire Ecology Conference 19:134–141.

NYS, Y., M. T. HINCKE, J. L. ARIAS, J. M. GARCIA-RUIZ, AND S. E. SOLOMAN. 1999. Avian eggshell mineralization. Poultry Avian Biological Review 10: 143–166.

O'BRIEN, S. J., AND E. MAYR. 1991. Bureaucratic mischief: recognizing endangered species and subspecies. Science 251:1187–1188.

ODUM, H. T. 1953. Factors controlling marine invasion into Florida fresh waters. Bulletin of Marine Sciences of the Gulf and Caribbean 3:134–156

Odum, W. E. 1988. Comparative ecology of tidal freshwater and salt marshes. Annual Review of Ecology and Systematics 19:147–176.

Odum, W. E., E. P. Odum, and H. T. Odum. 1995. Nature's pulsing paradigm. Estuaries 18:547–555.

Office of Spill Prevention and Response. 2003. SS Jacob Luckenbach oil removal project completed. California Office of Spill Prevention and Response 10:1–28.

Ogan, C. V., and R. M. Jurek. 1997. Biology and ecology of feral, free-roaming, and stray cats. Pp. 87–91 in J. E. Harris, and C. V. Ogan (editors). Mesocarnivores of northern California: biology, management, and survey techniques, workshop manual. Humboldt State University and The Wildlife Society, California North Coast Chapter, Arcata, CA.

Ohmart, R. D. 1972. Physiological and ecological observations concerning the salt-secreting nasal glands of the Roadrunner. Comparative Biochemical Physiology 43A:311–316.

Olsen, L., J. Ollerhead, and A. Hanson. 2005. Relationships between halophytic vascular plant species' zonation and elevation in salt marshes of the Bay of Fundy and Northumberland Strait, New Brunswick, Canada. Proceedings 12th Canadian Coastal Conference, Dartmouth, NS, Canada.

Olson, S. L. 1997. Towards a less imperfect understanding of the systematics and the biogeography of the Clapper and King rail complex (*Rallus longirostris* and *R. elegans*). Pp. 93–111 in R. W. Dickerman (editor). The era of Allan R. Phillips: a festschrift. Horizon Communications, Albuquerque, NM.

O'Neil, T. 1949. The muskrat in the Louisiana coastal marshes. Louisiana Wildlife and Fisheries Commission, New Orleans, LA.

Oney, J. 1954. Clapper Rail survey and investigative study. Final Report Federal Aid Project W-9-R. Georgia Game and Fish Commission, Atlanta, GA.

Orians, G. 1969. On the evolution of mating systems in birds and mammals. American Naturalist 103:589–603.

Orians, G. H., and G. M. Christman. 1968. A comparative study of the behavior of Red-winged, Tricolored, and Yellow-headed blackbirds. University of California Publications in Zoology 84, University of California, Berkeley, CA.

Orians, G. H., and M. F. Willson. 1964. Interspecific territories in birds. Ecology 45:736–745.

Orson, R. A. 1999. A paleoecological assessment of *Phragmites australis* in New England tidal marshes: changes in plant community structure during the last few millennia. Biological Invasions 1:149–158.

Osgood, D. T., D. J. Yozzo, R. M. Chambers, D. Jacobson, T. Hoffman, and J. Wnek. 2003. Tidal marsh hydrology and habitat utilization by resident nekton in *Phragmites* and non-phragmites marshes. Estuaries 26:522–533.

Owen, J. G., and R. S. Hoffmann. 1983. *Sorex ornatus*. Mammalian Species 212:1–5.

Padgett-Flohr, G. E., and L. Isakson. 2003. A random sampling of salt marsh harvest mice in a muted tidal marsh. Journal of Wildlife Management 67:646–653.

Page, G. W., L. E. Stenzel, and J. E. Kjelmyr. 1999. Overview of shorebird abundance and distribution in wetlands of the Pacific coast of the contiguous United States. Condor 101:461–471.

Palmer, A. J. M. 1979. Some observations on the relationships between high tide levels and coastal resources in the Cumberland Basin. Proceedings of the Nova Scotia Institute of Science 29:347–352.

Palmer, R. S. (editor). 1976. Handbook of North American birds: Vol. 2, waterfowl. Yale University Press, New Haven, CT.

Pan, J. J., and J. S. Price. 2002. Fitness and evolution in clonal plants: the impact of clonal growth. Evolutionary Ecology 15:583–600.

Papadopoulus, Y. A. 1995. The agricultural value of dykelands. Pp. 26–28 in J. Bain, and M. Evans (editors). Dykelands enhancement workshop. Canadian Wildlife Service, Sackville, NB, Canada.

Parsons, K. C. 2003. Reproductive success of wading birds using *Phragmites* marsh and upland nesting habitats. Estuaries 26:596–601.

Pashley, D. N., C. J. Beardmore, J. A. Fitzpatrick, R. P. Ford, W. C. Hunter, M. S. Morrison, and K. V. Rosenberg. 2000. Partners in flight: conservation of the land birds of the United States. American Bird Conservancy, The Plains, VA.

Patrick, W. H., and R. D. DeLaune. 1990. Subsidence, accretion, and sea-level rise in South San Francisco Bay marshes. Limnology and Oceanography 35:1389–1395.

Patterson, C. B. 1991. Relative parental investment in the Red-winged Blackbird. Journal of Field Ornithology 62:1–18.

Peabody, F. E., and J. M. Savage. 1958. Evolution of a coast range corridor in California and its effect on the origin and dispersal of living amphibians and reptiles. Pp. 159–186 in C. L. Hubbs (editor). Zoogeography. Publication 51. American Association for the Advancement of Science, Washington, DC.

Pearson, O. P. 1951. Additions to the fauna of Santa Cruz Island, California, with a description of a new subspecies of *Reithrodontomys megalotis*. Journal of Mammalogy 32:366–368.

Peltier, W. R. 1994. Ice age paleotopography. Science 265:195–201.

Peltier, W. R. 1996. Global sea level rise and glacial isostatic adjustment: an analysis of data from the east coast of North America. Geophysical Research Letters 23:717–720.

Peltier, W. R., I. Shennan, R. Drummond, and B. Horton. 2002. On the postglacial isostatic adjustment of the British Isles and the shallow viscoelastic structure of the Earth. Geophysical Journal International 148:443–475.

Pendleton, E. C., and J. C. Stevenson. 1983. Part V. Burning and production. Pp. 57–71 in Investigations of marsh losses at Blackwater Refuge. Reference No. 83-154-H.P.E.L. Horn Point Environmental Laboratories, University of Maryland, Cambridge, MD.

Pereira, W. E., T. L. Wade, F. D. Hostettler, and F. Parchaso. 1999. Accumulation of butyltins in sediments and lipid tissues of the Asian clam, *Potamocorbula amurensis*, near Mare Island Naval

Shipyard, San Francisco Bay. Marine Pollution Bulletin 38:1005–1010.

PERKINS, C. J. 1968. Controlled burning in the management of muskrats and waterfowl in Louisiana coastal marshes. Proceedings of the Annual Tall Timbers Fire Ecology Conference 8:269–280.

PERRY, M. C., AND A. DELLER. 1996. Review of factors affecting the distribution and abundance of waterfowl in shallow-water habitats of the Chesapeake Bay. Estuaries 19:272–278.

PETERS, J. L. 1942. The Canadian forms of the Sharp-tailed Sparrow *Ammospiza caudacata.* Annals of the Carnegie Museum 29:201–210.

PETERS, R. L., AND T. E. LOVEJOY (EDITORS). 1992. Global warming and biological diversity. Yale University Press, New Haven, CT.

PETERSON, D., D. CAYAN, J. DILEO, M. NOBLE, AND M. DETTINGER. 1995. The role of climate in estuarine variability. American Scientist 83:58–67.

PETERSON, L. P., G. W. TANNER, AND W. M. KITCHENS. 1995. A comparison of passerine foraging habits in two tidal marshes of different salinity. Wetlands 15:315–323.

PETHICK, J. S. 1992. Saltmarsh geomorphology. Pp. 41–62 *in* J. R. L Allen, and K. Pye (editors). Saltmarshes—morphodynamics, conservation and engineering significance. Cambridge University Press, Cambridge, UK.

PETTUS, D. 1958. Water relationships in *Natrix sipedon*. Copeia 1958:207–211.

PETTUS, D. 1963. Salinity and subspeciation in *Natrix sipedon*. Copeia 1963:499–504.

PFISTER, C., B. A. HARRINGTON, AND M. LAVINE. 1992. The impact of human disturbance on shorebirds at a migration staging area. Biological Conservation 60:115–126.

PHILIPPART, C. J. M. 1994. Interactions between *Arenicola marina* and *Zostera noltii* on a tidal flat in the Wadden Sea. Marine Ecology Progress Series 111:251–257.

PICMAN, J. 1977. Destruction of eggs by the Long-billed Marsh Wren (*Telmatodytes palustris palustris*). Canadian Journal of Zoology 55:1914–1920.

PICMAN, J. 1980. Impact of Marsh Wrens on reproductive strategy of Red-winged Blackbirds. Canadian Journal of Zoology 58:337–350.

PICMAN, J. 1981. The adaptive value of polygyny in marsh-nesting Red-winged Blackbirds: renesting, territory tenacity, and mate fidelity of females. Canadian Journal of Zoology 59:2284–2296.

PICMAN, J. 1984. Experimental study on the role of intra- and inter-specific competition in the role of nest-destroying behavior in Marsh Wrens. Canadian Journal of Zoology 62:2353–2356.

PICMAN, J., AND A. ISABELLE. 1995. Sources of nesting mortality and correlates of nesting success in Yellow-headed Blackbirds. Auk 112:183–191.

POAG, C. W. 1973. Late Quaternary sea levels in the Gulf of Mexico. Gulf Coast Association of Geology Society Transactions 23:394–400.

POMERANTZ, G. A., D. J. DECKER, G. R. GOFF, AND K. G. PURDY. 1988. Assessing impact of recreation on wildlife: a classification scheme. Wildlife Society Bulletin 16:58–62.

POOLE, A. (EDITOR). 2006. The birds of North America online. Cornell Laboratory of Ornithology, Ithaca, NY <http://bna.birds.cornell.edu/BNA> (1 March 2006).

POPPE, L. J., AND C. POLLONI (EDITORS). 1998. Long Island Sound environmental studies. U.S. Geological Survey Open-File Report 98-502. U.S. Geological Survey, Woods Hole, MA.

POST, W. 1974. Functional analysis of space-related behavior in the Seaside Sparrow. Ecology 55:564–575.

POST, W. 1981. The influence of rice rats *Oryzomys palustris* on the habitat use of the Seaside Sparrow *Ammospiza maritima*. Behavioural Ecology and Sociobiology 9:35–40.

POST, W. 1992. Dominance and mating success in male Boat-tailed Grackles. Animal Behaviour 44:917–929.

POST, W. 1998. The status of Nelson's and Saltmarsh Sharp-tailed sparrows on Waccassa Bay, Levy County, Florida. Florida Field Naturalist 26:1–6.

POST, W., AND F. ENDERS. 1970. Notes on a salt marsh Virginia Rail population. Kingbird 20:61–67.

POST, W., AND J. S. GREENLAW. 1975. Seaside Sparrow displays: their function in social organization and habitat. Auk 92:461–492.

POST, W., AND J. S. GREENLAW. 1982. Comparative costs of promiscuity and monogamy: a test of reproductive effort theory. Behavioral Ecology and Sociobiology 10:101–107.

POST, W. J., AND J. S. GREENLAW. 1994. The Seaside Sparrow (*Ammodramus maritimus*). *In* A. Poole, and F. Gill (editors). The birds of North America, No. 194. The Academy of Natural Sciences, Philadelphia, PA and The American Ornithologists' Union, Washington, DC.

POST, W., J. S. GREENLAW, T. L. MERRIAM, AND L. A. WOOD. 1983. Comparative ecology of northern and southern populations of the Seaside Sparrow. Pp. 123–136 *in* T. L. Quay, J. B. Funderburg, Jr., D. S. Lee, E. F. Potter, and C. S. Robbins (editors). The Seaside Sparrow, its biology and management. Occasional Papers of the North Carolina Biological Survey, Raleigh, NC.

POST, W., J. P. POSTON, AND G. T. BANCROFT. 1996. Boat-tailed Grackle (*Quiscalus major*). *In* A. Poole, and F. Gill (editors). The birds of North America, No. 207. The Academy of Natural Sciences, Philadelphia, PA and The American Ornithologists' Union, Washington, DC.

POST, W. M. 1974. Functional analysis of space-related behavior in the Seaside Sparrow. Ecology 55:564–575.

POULIN, B., G. LEFEBVRE, AND A. MAUCHAMP. 2002. Habitat requirements of passerines and reed-bed management in southern France. Biological Conservation 107:315–325.

POULSON, T. L. 1965. Countercurrent multiplication in avian kidneys. Science 148:389–391.

POULSON, T. L. 1969. Salt and water balance in Seaside and Sharp-tailed sparrows. Auk 86:473–489.

POULSON, T. L., AND G. A. BARTHOLOMEW. 1962. Salt balance in the Savannah Sparrow. Physiological Zoology 35:109–119.

Powell, A. N. 1993. Nesting habitat of Belding's Savannah Sparrows in coastal saltmarshes. Wetlands 13:129-133.

Powell, A. N., and C. L. Collier. 1998. Reproductive success of Belding's Savannah Sparrows in a highly fragmented landscape. Auk 115: 508-513.

Pribil, S., and J. Picman. 1997. The importance of using the proper methodology and spatial scale in the study of habitat selection by birds. Canadian Journal of Zoology 75:1835-1844.

Price, J. S., and M. K. Woo. 1988. Studies of a subarctic coastal marsh: I. Hydrology. Journal of Hydrology (Amsterdam) 103:275-292.

Pritchard, P. C. H. 1979. Encyclopedia of turtles. TFH Publications, Inc., Hong Kong, China.

Proche, S., and D. J. Marshall. 2001. Global distribution patterns of non-halacarid marine intertidal mites: implications for their origins in marine habitats. Journal of Biogeography 28:47-58.

Provost, M. W. 1969. Ecological control of salt marsh mosquitoes with side benefits to birds. Tall Timbers Conference on Ecological Animal Control by Habitat Management 1:193-206.

Pugh, D. T. 1987. Tides, surges, and mean sea-level. Wiley, New York, NY.

Pulliam, H. R. 1988. Sources, sinks, and population regulation. American Naturalist 132:653-661.

Pyle, P. 1997. Identification guide to North American Birds. Part I. Columbidae to Paloceidae. Slate Creek Press, Bolinas, CA.

Pysek, P., K. Prach, M. Rejmanek, and M. Wade (editors). 1995. Plant invasions. SPB Publishing, Amsterdam, The Netherlands.

Quinn, T. M. 2000. Shallow water science and ocean drilling challenges. EOS, Transactions of American Geophysical Union 81:397.

Raichel, D. L., K. W. Able, and J. M. Hartman. 2003. The influence of *Phragmites* (common reed) on the distribution, abundance, and potential prey of a resident marsh fish in the Hackensack Meadowlands, New Jersey. Estuaries 26: 511-521.

Ralph, C. J., G. R. Geupel, P. Pyle, T. E. Martin, and D. F. DeSante. 1993. Handbook of field methods for monitoring landbirds. USDA Forest Service General Technical Report PSW-GTR-144. USDA Forest Service, Pacific Southwest Research Station, Albany, CA.

Ralph, C. J., S. Droege, and J. R. Sauer. 1995. Managing and monitoring birds using point counts: standards and applications. Pp. 161-168 *in* C. J. Ralph, J. R. Sauer, and S. Droege (editors). Monitoring bird populations by point counts. USDA Forest Service General Technical Report PSW-GTR-149. USDA Forest Service Pacific, Southwest Research Station. Albany, CA.

Rassmussen, J. B., D. J. Rowan, D. R. S. Lean, and J. H. Carey. 1990. Food chain structure in Ontario lakes determines PCB levels in lake trout (*Salvelinus namaycush*) and other pelagic fish. Canadian Journal of Fisheries Aquatic Sciences 47:2030-2038.

Ratti, J. T., and E. O. Garton. 1996. Research and experimental design. Pp. 1-23 *in* T. A. Bookhout (editor). Research and management techniques for wildlife and habitats. Fifth edition, revised. The Wildlife Society, Bethesda, MD.

Redford, K., and J. Eisenberg. 1992. Mammals of the Neotropics, the Southern Cone. University of Chicago Press, Chicago, IL.

Reed, A., and A. D. Smith. 1972. Man and waterfowl in tidal shorelines of eastern Canada. Pp. 151-155 *in* Proceedings of coastal zone conference. Dartmouth, NS, Canada.

Reed, J. M., L. S. Mills, J. B. Dunning, Jr., E. S. Menges, K. S. McKelvey, R. Frye, S. R. Beissinger, M. C. Anstett, and P. Miller. 2002. Emerging issues in population viability analysis. Conservation Biology 16:7-19.

Reid, W. V., and M. C. Trexler. 1992. Responding to potential impacts of climate change on U.S. coastal diversity. Coastal Management 20:117-142.

Reinert, S. E. 1979. Breeding ecology of the Swamp Sparrow (*Melospiza georgiana*) in a southern Rhode Island peatland. M.S. thesis, University of Rhode Island, Kingston, RI.

Reinert, S. E., F. C. Golet, and W. R. DeRagon. 1981. Avian use of ditched and unditched salt marshes in southeastern New England: a preliminary report. Transactions of the Northeastern Mosquito Control Association 27:1-23.

Reinert, S. E., and M. J. Mello. 1995. Avian community structure and habitat use in a southern New England estuary. Wetlands 15:9-19.

Reinson, R. C. 1980. Variations in tidal inlet morphology and stability, northeast New Brunswick. Pp. 23-39 *in* S. B. McCann (editor). The coastline of Canada. Geological Survey of Canada, Paper 80-10. Minister of Supply and Services Canada, Ottawa, ON, Canada.

Reisen, W. K., V. L. Kramer, and L. S. Mian. 1995. Interagency guidelines for the surveillance and control of selected vector borne pathogens in California. Mosquito and Vector Control Association of California, Elk Grove, CA.

Ribic, C. A., S. Lewis, S. Melvin, J. Bart, and B. Peterjohn. 1999. Proceedings of the marshbird monitoring workshop. USDI Fish and Wildlife Service Region 3 Administrative Report, Fort Snelling, MN.

Rice, W. R., and E. E. Hostert. 1993. Laboratory experiments on speciation: what have we learned in 40 years? Evolution 47:1637-1653.

Rich, T., R. L. Allen, and A. H. Wyllie. 2000. Defying death after DNA damage. Nature 407:777-783.

Richards, M. P., and M. J. Packard. 1996. Mineral metabolism in avian embryos. Poultry Avian Biological Review 7:143-161.

Ricklefs, R. E. 1980. Geographical variation in clutch size among passerine birds: Ashmole's hypothesis. Auk 97:38-49.

Ricklefs, R. E., and E. Bermingham. 2002. The concept of the taxon cycle in biogeography. Global Ecology and Biogeography Letters 11:353-361.

Ricklefs, R. E., and G. Bloom. 1977. Components of avian breeding productivity. Auk 94:86-96.

Ridgely, R., and G. Tudor. 1989. The birds of South America. Vol. I. The oscine passerines. University of Texas Press, Austin, TX.

RIDGELY, R., AND G. TUDOR. 1994. The birds of South America. Vol. II. The suboscine passerines. University of Texas Press, Austin, TX.

RIDGWAY, R. 1899. New species of American birds.-iii. Fringillidae (continued). Auk 16:35–37.

RIDGWAY, R., AND H. FRIEDMANN. 1901. The birds of North and Middle America. Government Printing Office, Washington, DC.

RIITTERS, K. H., R. V. O'NEILL, AND K. B. JONES. 1997. Assessing habitat suitability at multiple scales: a landscape-level approach. Biological Conservation 81:191–202.

RIPLEY, S. D. 1977. Rails of the world. David R. Godine Publisher, Boston, MA.

RISING, J. D. 1989. Sexual dimorphism in *Passerculus sandwichensis* as it is. Evolution 43:1121–1123.

RISING, J. D. 1996. The sparrows of the United States and Canada. Academic Press, San Diego, CA.

RISING, J. D. 2001. Geographic variation in size and shape of Savannah Sparrows (*Passerculus sandwichensis*). Studies in Avian Biology 23.

RISING, J. D., AND J. C. AVISE. 1993. Application of genealogical-concordance principles to the taxonomy and evolutionary history of the Sharp-tailed Sparrow (*Ammodramus caudacutus*). Auk 110:844–856.

ROBBINS, C. S., D. BYSTRAK, AND P. H. GEISSLER. 1986. The breeding bird survey: its first fifteen years, 1965–1979. USDI Fish and Wildlife Service Resource Publication 157. Washington, DC.

ROBBINS, C. S., D. K. DAWSON, AND B. A. DOWELL. 1989. Habitat area requirements of breeding forest birds of the Middle Atlantic States. Wildlife Monographs 103:1–34.

ROBBINS, J. A., C. W. HOLMES, R. HALLEY, M. BOTHNER, E. SHINN, J. GRANEY, G. KEELER, M. TENBRINK, K. A. ORLANDINI, AND D. RUDNICK. 2000. Time-averaged fluxes of lead and fallout radionuclides to sediment in Florida Bay. Journal of Geophysical Research 105:805–821.

ROBERTS, B. A., AND A. ROBERTSON. 1986. Saltmarshes of Atlantic Canada: their ecology and distribution. Canadian Journal of Botany 64:455–467.

ROBERTS, L. A. 1989. Productivity and flux of organic material from two Maritime high saltmarshes. M.S. thesis, Acadia University, Wolfville, NS, Canada.

ROBERTS, L. A. 1993. Report on the status of saltmarsh habitat in New Brunswick. Wetlands and Coastal Habitat Program-New Brunswick Department of Natural Resources and Energy, Fredericton, NB, Canada.

ROBINSON, G. D., AND W. A. DUNSON. 1975. Water and sodium balance in the estuarine diamondback turtle (*Malaclemys*). Journal of Comparative Physiology A 105:129–152.

ROBINSON, G. D., AND W. A. DUNSON. 1976. Water and sodium balance in the estuarine diamondback terrapin (*Malaclemys*). Journal of Comparative Physiology 105:129–152.

ROCKE, T. E., AND M. FRIEND. 1999. Avian botulism. Pp. 271–286 *in* M. Friend, and J. C. Franson (editors). Field manual of wildlife diseases. U.S. Geological Survey, Biological Resources Discipline, National Wildlife Health Center, Madison, WI.

RODGERS, J. A., JR., AND H. T. SMITH. 1997. Buffer zone distances to protect foraging and loafing waterbirds from human disturbance in Florida. Wildlife Society Bulletin 25:39–145.

RODRIGUEZ-NAVARRO, A., K. F. GAINES, C. S. ROMANEK, AND G. R. MASSON. 2002a. Mineralization of Clapper Rail eggshell from a contaminated salt marsh system. Archives of Environmental Contamination and Toxicology 43:449–460.

RODRIGUEZ-NAVARRO, A., O. KALIN, Y. NYS, AND J. M. GARCIA-RUIZ. 2002b. Influence of the microstructure and crystallographic texture on the fracture strength of hen's eggshells. British Poultry Science 43:395–403.

RODRIGUEZ-ROBLES, J. A., D. F. DENARDO, AND R. E. STAUB. 1999. Phylogeography of the California mountain kingsnake, *Lampropeltis zonata* (Colubridae). Molecular Ecology 8:1923–1935.

RODRIGUEZ-ROBLES, J. A., G. R. STEWART, AND T. J. PAPENFUSS. 2001. Mitochondrial DNA-based phylogeography of North American rubber boas, *Charina bottae* (Serpentes: Boidae). Molecular Phylogenetics and Evolution 18:227–237.

ROLAND, A. E. 1982. Geological background and physiography of Nova Scotia. The Nova Scotian Institute of Science, Halifax, NS, Canada.

ROMAN, C. T., N. JAWORSKI, F. T. SHORT, S. FINDLAY, AND R. S. WARREN. 2000. Estuaries of the Northeastern United States: habitat and land use signatures. Estuaries 23:743–764.

ROMAN, C. T., W. A. NIERING, AND R. S. WARREN. 1984. Salt marsh vegetation change in response to tidal restriction. Environmental Management 8:141–150.

ROMAN, C. T., J. A. PECK, J. R. ALLEN, J. W. KING, AND P. G. APPLEBY. 1997. Accretion of a New England (U.S.A.) salt marsh in response to inlet migration, storms, and sea-level rise. Estuarine, Coastal, and Shelf Science 45:717–727.

ROMPRE, G., A. PAGE, AND F. SHAFFER. 1998. Status report on Nelson's Sharp-tailed Sparrow *Ammodramus nelsoni* in Canada. Environnement Canada, Service Canadien de la Faune, Chateauguay, QC, Canada.

ROOSENBURG, W. M. 1990. The diamondback terrapin: population dynamics, habitat requirements, and opportunities for conservation. Chesapeake Research Consortium Publication Number 137:227–234.

ROOSENBURG, W. M. 2003. The impact of crab pot fisheries on terrapin (*Malaclemys terrapin*) populations: where we are and where do we need to go? Pp. 23–30 *in* C. W. Swarth, W. M. Roosenberg, and E. Kiviat (editors). Conservation and ecology of turtles of the mid-Atlantic region; a symposium. Bibliomania, Salt Lake City, UT.

ROOSENBURG, W. M., W. CRESKO, M. MODESITTE, AND M. B. ROBBINS. 1997. Diamondback terrapin (*Malaclemys terrapin*) mortality in crab pots. Conservation Biology 11:1166–1172.

ROOSENBURG, W. M., AND A. E. DUNHAM. 1997. Clutch and egg size variation in the diamondback terrapin, *Malaclemys terrapin*. Copeia 1997:290–297.

ROOSENBURG W. M., K. L. HALEY, AND S. MCGUIRE. 1999. Habitat selection and movements of diamondback terrapins, *Malaclemys terrapin*, in a Maryland

estuary. Chelonian Conservation and Biology 3: 425–429.
Root, K. V. 1998. Evaluating the effects of habitat quality, connectivity, and catastrophes on a threatened species. Ecological Applications 8: 854–865.
Rooth, J. E., J. C. Stevenson, and J. C. Cornwell. 2003. Increased sedimentation rates following invasion by *Phragmites australis*: The role of litter. Estuaries 26:475–483.
Rosenberg, K. V., R. D. Ohmart, W. C. Hunter, and B. W. Anderson. 1991. Birds of the lower Colorado River valley. University of Arizona Press, Tucson, AZ.
Rosenberg, N. A., and M. Nordborg. 2002. Genealogical trees, coalescent theory and the analysis of genetic polymorphisms. Nature Reviews Genetics 3:380–390.
Rosza, R. 1995. Human impacts on tidal wetlands: history and regulations. Pp. 42–50 *in* G. D. Dreyer, and W. A. Niering (editors). Tidal marshes of Long Island Sound: ecology, history and restoration. Bulletin No. 34. Connecticut College Arboretum, New London, CT.
Rotenberry, J. T., and J. A. Wiens. 1980. Habitat structure, patchiness, and avian communities in North American steppe vegetation: a multivariate analysis. Ecology 61:1228–1250.
Rotenberry, J. T., R. J. Cooper, J. M. Wunderle, and K. G. Smith. 1995. When and how are populations limited? The roles of insect outbreaks, fire, and other natural perturbations. Pp. 165–171 *in* T. Martin, and D. Finch (editors). Ecology and management of neotropical migratory birds, a synthesis and review of critical issues. Oxford University Press, New York, NY.
Rottenborn, S. 1996. The use of coastal agricultural fields in Virginia as foraging habitat by shorebirds. Wilson Bulletin 108:783–796.
Rudd, R. L. 1953. Notes on maintenance and behavior of shrews in captivity. Journal of Mammalogy 34: 118–120.
Rudd, R. L. 1955. Population variation and hybridization in some California shrews. Systematic Zoology 4:21–34.
Ruddiman, W. F. 2001. Earth's climate past and future. W.H. Freeman and Co., New York, NY.
Ryan, H., H. Gibbons, J. W. Hendley II, and P. H. Stauffer. 1999. El Niño sea level rise wreaks havoc in California's San Francisco Bay region. U.S. Geological Survey Fact Sheet 175-99. U.S. Geological Survey, Menlo Park, CA.
Ryder, O. A. 1986. Species conservation and systematics: the dilemma of subspecies. Trends in Ecology and Evolution 1:9–10.
Saab, V. 1999. Importance of spatial scale to habitat use by breeding birds in riparian forests: a hierarchical analysis. Ecological Applications 9: 135–151.
Sabat, P. 2000. Birds in marine and saline environments: living in dry habitats. Revista Chilena de Historia Natural 73:243–252.
Sabat, P., and C. Martínez del Rio. 2002. Inter- and intraspecific variation in the use of marine food resources by three *Cinclodes* (Furnariidae, Aves) species: carbon isotopes and osmoregulatory physiology. Zoology 105:247–256.
Sæther, B. E., and Ø. Bakke. 2000. Avian life history variation and contribution of demographic traits to the population growth rate. Ecology 81:642–653.
Saintilan, N., and R. J. Williams. 1999. Mangrove transgression into saltmarsh environments in south-east Australia. Global Ecology and Biogeography 8:117–124.
Saltonstall, K. 2002. Cryptic invasions by a non-native genotype of the common reed, *Phragmites australis*, into North America. Proceedings of the National Academy of Sciences 99:2445–2449.
Saltonstall, K. 2003. Genetic variation among North American populations of *Phragmites australis*: implications for management. Estuaries 26:444–451.
San Francisco Bay Regional Water Quality Control Board. 1992. Mass emissions reduction strategy for selenium. Oakland, CA.
San Francisco Estuary Institute. 2000. San Francisco Bay area ecoatlas v. 1.50b4. San Francisco Estuary Institute, Richmond, CA.
San Francisco Estuary Institute. 2003. Pulse of the estuary: monitoring and managing contamination in the San Francisco Estuary. San Francisco Estuary Institute, Oakland, CA.
San Francisco Estuary Project. 1991. Status and trends report on wetlands and related habitats in the San Francisco Estuary. San Francisco Estuary Project, U.S. Environmental Protection Agency, Cooperative Agreement #815406-01-0. Oakland, CA.
San Francisco Estuary Project. 1992. State of the estuary: a report on conditions and problems in the San Francisco/Sacramento-San Joaquin delta estuary. Prepared by the Association of Bay Area Governments. Oakland, CA.
SAS. 2000. SAS/STAT software: changes and enhancements, Release 8.1. SAS Institute Inc., Cary, NC.
Sauer, J. R., J. E. Hines, and J. Fallon. 2004. The North American breeding bird survey, results and analysis 1966–2003. Version 2004.1. USGS Patuxent Wildlife Research Center, Laurel, MD.
Sauer, J. R., J. E. Hines, I. Thomas, J. Fallon, and G. Gough. 2000. The North American breeding bird survey, results and analysis 1966–1999. Version 98.1. U.S. Geological Survey, Patuxent Wildlife Research Center, Laurel, MD.
Saunders, D. A., R. J. Hobbs, and C. R. Margules. 1991. Biological consequences of ecosystem fragmentation: a review. Conservation Biology 5:8–32.
Sayce, K. 1988. Introduced cordgrass, *Spartina alterniflora* Loisel in saltmarshes and tidelands of Willapa Bay, Washington. Report to USDI Fish and Wildlife Service, Willapa National Wildlife Refuge, Cathlemet, WA.
Scheffer, T. H. 1945. The introduction of *Spartina alterniflora* to Washington with oyster culture. Leaflets of Western Botany 4:163–164.
Schlemon, R. J., and E. L. Begg. 1973. Late Quaternary evolution of the Sacramento-San Joaquin Delta, California. Pp. 259–266 *in* Proceedings of the Ninth Congress of the International Union for Quaternary Research. International Union for Quaternary Research, Christchurch, NZ.

SCHLUTER, D., AND J. N. M. SMITH. 1986. Natural selection on beak and body size in the Song Sparrow. Evolution 40:221–231.

SCHMALZER, P. A., AND C. R. HINKLE. 1993. Effects of fire on nutrient concentrations and standing crops in biomass of *J. roemerianus* and *Spartina bakeri* marshes. Castanea 58:90–114.

SCHMIDT-NIELSEN, K., AND Y.-T. KIM. 1964. The effect of salt intake on the size and function of the salt glands in ducks. Auk 81:160–172.

SCHNEIDER, M. F. 2001. Habitat loss, fragmentation and predator impact: spatial implications for prey conservation. Journal of Applied Ecology 38:720–735.

SCHOELLHAMER, D. H., G. G. SHELLENBARGER, N. K. GANJU, J. A. DAVIS, AND L. J. MCKEE. 2003. Sediment dynamics drive contaminant dynamics. Pp. 21–26 *in* Pulse of the estuary: monitoring and managing contamination in the San Francisco Estuary. San Francisco Estuary Institute, Oakland, CA.

SCHULZE-HAGEN, K., B. LEISTER, H. M. SCHÄFER, AND V. SCHMIDT. 1999. The breeding system of the Aquatic Warbler *Acrocephalus paludicola*—a review of new results. Vogelwelt 120:87–96.

SCHWARTZENBACH, R. P., P. M. GSCHEND, AND D. M. IMBODEN. 2003. Partitioning to living media—bioaccumulation and baseline. Chapter 10 *in* Environmental organic chemistry. John Wiley and Sons, Hoboken, NJ.

SCHWARZBACH, S., AND T. ADELSBACH. 2002. Assessment of ecological and human health impacts of mercury in the bay-selta watershed, Subtask 3B: field assessment of avian mercury exposure in the bay-delta ecosystem. Final Report to the CALFED Bay-Delta Mercury Project. CALFED Bay-Delta Authority, Sacramento, CA.

SCHWARZBACH, S. E., J. D. ALBERTSON, AND C. M. THOMAS. 2006. Effects of predation, flooding, and contamination on reproductive success of California Clapper Rails (*Rallus longirostris obsoletus*) in San Francisco Bay. Auk 123:45–60.

SCHWARZBACH, S. E., J. D. HENDERSON, C. M. THOMAS, AND J. D. ALBERTSON. 2001. Organochlorine concentrations and eggshell thickness in failed eggs of the California Clapper Rail from South San Francisco Bay. Condor 103:620–624.

SCOLLON, D. B. 1993. Spatial analysis of the tidal marsh habitat of the Suisun Song Sparrow. M.A. thesis, San Francisco State University, San Francisco, CA.

SCOTT, D. B. 1980. Morphological changes in an estuary: a historical and stratigraphical comparison. Pp. 199–205 *in* S. B. McCann (editor). The coastline of Canada. Geological Survey of Canada, Paper 80-10. Minister of Supply and Services Canada, Ottawa, ON, Canada.

SCOTT, D. M., R. E. LEMON, AND J. A. DARLEY. 1987. Relaying interval after nest failure in Gray Catbirds and Northern Cardinals. Wilson Bulletin 99:708–712.

SCOTT, M. L., S. K. SKAGEN, AND M. F. MERIGLIANO. 2003. Relating geomorphic change and grazing to avian communities in riparian forest. Conservation Biology 17:284–296.

SEARCY, W. A., AND K. YASUKAWA. 1989. Alternative models of territorial polygyny in birds. American Naturalist 134:323–343.

SEARCY, W. A., AND K. YASUKAWA. 1995. Polygyny and sexual selection in Red-winged Blackbirds. Princeton University Press, Princeton, NJ.

SEARCY, W. A., K. YASUKAWA, AND S. LANYON. 1999. Evolution of polygyny in the ancestors of Red-winged Blackbirds. Auk 116:5–19.

SEIGEL, R. A. 1980. Predation by raccoons on diamondback terrapins, *Malaclemys terrapin tequesta*. Journal of Herpetology 14:87–89.

SEIGEL, R. A. 1993. Apparent long-term decline in the diamondback terrapin populations at the Kennedy Space Center, Florida. Herpetological Review 24:102–103.

SEIGEL, R. A. 1994. Parameters of two populations of diamondback terrapins (*Malaclemys terrapin*) on the Atlantic coast of Florida. Pp. 77–87 *in* R. A. Seigel, L. E. Hunt, J. L. Knight, L. Malaret, and N. L. Zuschlag (editors). Vertebrate ecology and systematics—a tribute to Henry S. Fitch. Museum of Natural History, Special Publication 10, University of Kansas, Lawrence, KS.

SEIGEL, R. A., AND J. W. GIBBONS. 1995. Workshop on the ecology, status, and management of the diamondback terrapin (*Malaclemys terrapin*), final results and recommendations. Chelonian Conservation and Biology 1:240–243.

SEIGEL, S. W., AND P. A. M. BACHAND. 2002. Feasibility analysis of south bay salt pond restoration, San Francisco estuary, California. Wetlands and Water Resources, San Rafael, CA.

SELISKAR, D. M., AND J. L. GALLAGHER 1983. The ecology of tidal marshes of the Pacific Northwest coast: a community profile. Biological Services FWS/OBS-82-32. USDI Fish and Wildlife Service, Washington, DC.

SEUTIN, G., AND J. P. SIMON. 1988. Protein and enzyme uniformity in a new, isolated population of the Sharp-tailed Sparrow. Biochemical Systematics and Ecology 16:233–236.

SHACKLETON, N. J., AND N. D. OPDYKE. 1976. Oxygen isotope and palaeomagnetic stratigraphy of Pacific core V. 28-238: late Pliocene to latest Pleistocene. *In* R. M. Cline, and J. D. Hays (editors). Investigations of late Quaternary paleoceanography and paleoclimatology. Geological Society of America Memoirs 145: 449–64.

SHAFER, C. L. 1990. Nature reserve: island theory and conservation practice. Smithsonian Institution Press, Washington, DC.

SHAFFER, J. A., C. M. GOLDADE, M. F. DINKINS, D. H. JOHNSON, L. D. IGL, AND B. R. EULISS. 2003. Brown-headed Cowbirds in grasslands: their habitats, hosts, and response to management. Prairie Naturalist 35:145–186.

SHARP, H. F. 1967. Food ecology of the rice rat, *Oryzomys palustris* (Harlan) in a Georgian salt marsh. Journal of Mammology 48:557–563.

SHAW, J., R. B. TAYLOR, D. L. FORBES, M.-H. RUZ, AND S. SOLOMON. 1994. Sensitivity of the Canadian coast to sea-level rise. P. 2,377 *in* P. G. Wells, and P. J. Ricketts (editors). Coastal zone Canada '94, cooperation in the coastal zone. Coastal Zone Canada Association, Bedford Institute of Oceanography, Dartmouth, NS, Canada.

SHELLHAMMER, H. S. 1967. Cytotaxonomic studies of the harvest mice of the San Francisco Bay region. Journal of Mammalogy 48:549–556.

SHELLHAMMER, H. S. 2000. Salt marsh wandering shrew. Pp. 131–233 in P. R. Olofson (editor). Baylands ecosystem species and community profiles: life histories and environmental requirements of key plants, fish and wildlife. San Francisco Bay Regional Water Quality Control Board, Oakland, CA.

SHELLHAMMER, H. S., R. JACKSON, W. DAVILLA, A. M. GILROY, H. T. HARVEY, AND L. SIMONS. 1982. Habitat preferences of salt marsh harvest mice (*Reithrodontomys raviventris*). Wasmann Journal of Biology 40:102–114.

SHIELDS, G. F., AND A. C. WILSON. 1987. Calibration of mitochondrial DNA evolution in geese. Journal of Molecular Evolution 24:212–217.

SHISLER, J., AND F. FERRIGNO. 1987. The impacts of water management for mosquito control on waterfowl populations in New Jersey. Pp. 268–282 in W. R. Whitman, and W. H. Meredith (editors). Waterfowl and wetlands symposium: proceedings of a symposium on waterfowl and wetlands management in the coastal zone of the Atlantic Flyway. Delaware Coastal Management Program, Delaware Department of Natural Resources and Environmental Control, Dover, DE.

SHISLER, J., AND T. SCHULZE. 1976. Some aspects of open marsh water management procedures on Clapper Rail production. Proceedings of the Northeast Fish and Wildlife Conference 33:101–104.

SHRIVER, W. G. 2002. The conservation ecology of salt marsh birds in New England. Ph.D. dissertation, State University of New York, Syracuse, NY.

SHRIVER, W. G., AND J. P. GIBBS. 2004. Projected effects of sea-level rise on the population viability of Seaside Sparrows (*Ammodramus maritimus*). Pp. 397–401 in H. R. Akçakaya, M. A. Burgman, O. Kindvall, C. C. Wood, P. Sjögren-Gulve, J. S. Hatfield, and M. A. McCarthy (editors). Species conservation and management: case studies. Oxford University Press, Oxford, UK.

SHRIVER, W. G., T. P. HODGMAN, J. P. GIBBS, AND P. D. VICKERY. 2004. Landscape context influences saltmarsh bird diversity and area requirements in New England. Biological Conservation 119: 545–553.

SHUQING A. 2003. Ecological engineering of wetlands. Chemical Industry Press, Beijing, China. (in Chinese).

SIBLEY, C. G. 1955. The responses of salt-marsh birds to extremely high tides. Condor 57:241–242.

SIBLEY, C. G., AND B. L. MONROE, JR. 1990. Distribution and taxonomy of birds of the world. Yale University Press, New Haven, CT.

SIBLEY, D. 2000. The Sibley guide to birds. Knopf, New York, NY.

SILLIMAN, B. R., AND M. D. BERTNESS. 2004. Shoreline development drives invasion of *Phragmites australis* and the loss of plant diversity on New England salt marshes. Conservation Biology 18:1424–1434.

SIMENSTAD, C. A., M. DETHIER, C. LEVINGS, AND D. HAY. 1997. The land-margin interface of coastal temperate rain forest ecosystems: shaping the nature of coastal interactions. Pp. 149–187 in P. Schoonmaker, B. von Hagen, and E. Wolf (editors). The rain forests of home: profile of a North American bioregion. Ecotrust/Interain Pacific, Portland, Oregon, USA and Island Press, Covelo, CA.

SIU W. H. L., C. L. H. HUNG, H. L. WONG, B. J. RICHARDSON, AND P. K. S. LAM. 2003. Exposure and time dependent DNA strand breakage in hepatopancreas of green-lipped mussels (*Perna viridis*) exposed to Aroclor 1254, and mixtures of B[*a*]P and Aroclor 1254. Marine Pollution Bulletin 46: 1285–1293.

SKULASON, S., AND T. B. SMITH. 1995. Resource polymorphisms in vertebrates. Trends in Ecology and Evolution 10:366–370.

SKUTCH, A. F. 1949. Do tropical birds rear as many young as they can nourish? Ibis 91:430–455.

SMITH C. T., R. J. NELSON, C. C. WOOD, AND B. F. KOOP. 2001. Glacial biogeography of North American coho salmon (*Oncorhynchus kisutch*). Molecular Ecology 10:2775–2785.

SMITH, J. I. 1993. Environmental influence on the ontogeny, allometry, and behavior of the Song Sparrow (*Melospiza melodia*). Ph.D. dissertation, University of California, Berkeley, CA.

SMITH, J. N. M., M. J. TAITT, C. M. ROGERS, P. ARCESE, L. F. KELLER, A. L. E. V. CASSIDY, AND W. M. HOCHACHKA. 1996. A metapopulation approach to the population biology of the Song Sparrow *Melospiza melodia*. Ibis 138:120–128.

SMITH, J. N. M., AND R. ZACH. 1979. Heritability of some morphological characters in a Song Sparrow population. Evolution 33:460–467.

SMITH, L. M., J. A. KADLEC, AND P. V. FONNESBECK. 1984. Effects of prescribed burning on nutritive quality of marsh plants in Utah. Journal of Wildlife Management. 48:285–288.

SMITH, T. B., AND R. K. WAYNE. 1996. Molecular genetic approaches in conservation. Oxford University Press, New York, NY.

SMITH, T. B., R. K. WAYNE, D. J. GIRMAN, AND M. W. BRUFORD. 1997. A role for ecotones in generating rainforest biodiversity. Science 276:1855–1857.

SMITH, T. J., III, AND W. E. ODUM. 1981. The effects of grazing by Snow Geese on coastal salt marshes. Ecology 62:98–106.

SNYDER, R. D., AND P. J. LACHMANN. 1989. Thiol involvement in the inhibition of DNA repair by metals in mammalian cells. Molecular Toxicology 2: 117–128.

SOBCZAK, W. V., J. E. CLOERN, A. D. JASSBY, AND A. B. MULLER-SOLGER. 2002. Bioavailability of organic matter in a highly disturbed estuary: the role of detrital and algal resources. Proceedings of the National Academy of Sciences of the United States of America 99:8101–8105.

SOULÉ, M. E., D. T. BOLGER, A. C. ALBERTS, J. WRIGHT, M. SORICE, AND S. HILL. 1988. Reconstructed dynamics of rapid extinctions of chaparral requiring birds in urban habitat islands. Conservation Biology 2: 75–92.

SPAANS, H. 1994. The breeding birds of the Volkrech-Zioneer during the first five years after embankment. Limosa 67:15–26.

SPEAKMAN, J. 1997. Doubly labeled water—theory and practice. Kluwer Academic Publishers, New York, NY.

SPEAR, L. B., S. B. TERRILL, C. LENIHAN, AND P. DELEVORYAS. 1999. Effects of temporal and environmental factors on the probability of detecting California Black Rails. Journal of Field Ornithology 70:465–480.

STARRATT, S. W. 2004. Diatoms as indicators of late Holocene fresh water flow variation in the San Francisco Bay estuary, central California, U.S.A. Pp. 371–397 in M. Poulin (editor). Seventeenth International Diatom Symposium. Biopress Ltd., Bristol, UK.

STATACORP. 2003. Intercooled stata 8.0 for Windows, College Station, TX.

STATSOFT, INC.. 2003. STATISTICA. Version 6. <www.statsoft.com> (3 March 2006).

STEBBINS, R. C. 1954. Amphibians and reptiles of western North America. McGraw Hill, New York, NY.

STEERE, J. T., AND N. SCHAEFER. 2001. Restoring the estuary: an implementation strategy for the San Francisco Bay Joint Venture. San Francisco Bay Joint Venture. Oakland, CA.

STEPHENS, S. E., D. N. KOONS, J. J. ROTELLA, AND D. W. WILLEY. 2004. Effects of habitat fragmentation on avian nesting success: a review of the evidence at multiple spatial scales. Biological Conservation 115:101–110.

STEVENSON, J. C., J. ROOTH, M. S. KEARNEY, AND K. SUNDBERG. 2001. The health and long term stability of natural and restored marshes in Chesapeake Bay. Pp. 709–735 in M. P. Weinstein, and D. A. Kreeger (editors). Concepts and controversies in tidal marsh ecology. Kluwer Academic Press, Dordecht, The Netherlands.

STEVENSON, J. O., AND L. H. MEITZEN. 1946. Behavior and food habits of Sennett's White-tailed Hawk in Texas. Wilson Bulletin 58:198–205.

STEWART, R. E. 1951. Clapper Rail populations of the middle Atlantic states. Transactions of the North American Wildlife Conference 16:421–430.

STODDART, D. R., D. J. REED, AND J. R. FRENCH. 1989. Understanding salt-marsh accretion, Scolt Head Island, Norfolk, England. Estuaries 12:228–236.

STOREY, A. E. 1978. Adaptations in Common Terns and Forster's Terns for nesting in the salt marsh. Ph.D. dissertation, Rutgers University, Newark, NJ.

STOREY, A. E. 1987. Adaptations for marsh nesting in Common and Forster's terns. Canadian Journal of Zoology 65:1417–1420.

STOTZ, D. E., J. W. FITZPATRICK, T. A. PARKER III, AND D. K. MOKOVITS. 1996. Neotropical birds: ecology and conservation. University of Chicago Press, Chicago, IL.

STOUT, J. P. 1984. The ecology of irregularly flooded salt marshes of the northeastern Gulf of Mexico: a community profile. USDI Fish and Wildlife Service, Biological Report 85. Washington, DC.

STRALBERG, D., V. TONIOLO, G. PAGE, AND L. STENZEL. 2004. Potential Impacts of non-native Spartina spread on shorebird populations in south San Francisco Bay. February 2004 report to California Coastal Conservancy (contract #02-212). Point Reyes Bird Observatory Conservation Science, Stinson Beach, CA.

STRALBERG, D., N. WARNOCK, N. NUR, H. SPAUTZ, AND G. W. PAGE. 2005. Building a habitat conversion model for San Francisco Bay wetlands: a multi-species approach for integrating GIS and field data. Pp. 997–129 in C. J. Ralph, and T. D. Rich (editors). Bird conservation implementation and integration in the Americas: proceedings of the third international Partners in Flight conference. Volume 2. USDA Forest Service General Technical Report PSW-GTR-191. USDA Forest Service, Pacific Southwest Research Station, Albany, CA.

STREHL, C. E., AND J. WHITE. 1986. Effects of superabundant food on breeding success and behavior of the Red-winged Blackbird. Oecologia 70:178–186.

STUMPF, R. P., AND J. W. HAINES. 1998. Variations in tidal level in the Gulf of Mexico and implications for tidal wetlands. Estuarine, Coastal and Shelf Science 46:165–173.

SUGG, D. W., R. K. CHESSER, J. A. BROOKS, AND B. T. GRASMAN. 1995. The association of DNA damage to concentrations of mercury and radiocesium in largemouth bass. Environmental Toxicology and Chemistry 14:661–668.

SWIFT, B. L. 1989. Avian breeding habitats in Hudson River tidal marshes. Final Report to the Hudson River Foundation for Science and Environmental Research, Inc. New York State Department of Environmental Conservation, Albany, NY.

SYKES, G., 1937. The Colorado River delta. American Geographical Society Special Publication 19, New York, NY.

TABER, W. 1968a. *Passerculus sandwichensis beldingi* (Ridgway) Belding's Savannah Sparrow. Pp. 714–717 in A. C. Bent (editor). Life histories of North American cardinals, grosbeaks, buntings, towhees, finches, sparrows, and allies. Dover Publications, Inc., New York, NY.

TABER, W. 1968b. *Passerculus sandwichensis rostratus* (Cassin) Large-billed Savannah Sparrow. Pp. 722–724 in A. C. Bent (editor). Life histories of North American cardinals, grosbeaks, buntings, towhees, finches, sparrows, and allies. Dover Publications, Inc., New York, NY.

TAFT, A. C. 1944. Diamond-back terrapin introduced into California. California Fish and Game 30: 101–102.

TAILLANDIER, J. 1993. The breeding of the Fan-tailed Warbler (*Cisticola juncidis*) in salt marsh meadows. (Guerande Loire-Atlantique, western France). Alauda 61:39–51.

TAKEKAWA, J. 1993. The California Clapper Rail: turning the tide? Tideline 13:1–3, 11.

TAKEKAWA, J. Y., C. T. LU, AND R. T. PRATT. 2001. Avian communities in baylands and artificial salt evaporation ponds of the San Francisco Bay estuary. Hydrobiologia 466:317–328.

TAKEKAWA, J. Y., G. W. PAGE, J. M. ALEXANDER, AND D. R. BECKER. 2000. Waterfowl and shorebirds of the San Francisco Bay estuary. Pp. 309–316 in P. R. Olofson (editor). Baylands ecosystem species and community profiles: life histories and environmental requirements of key plants, fish and wildlife. San Francisco Bay Regional Water Quality Control Board, Oakland, CA.

TARR, C. L., AND R. C. FLEISCHER. 1993. Mitochondrial-DNA variation and evolutionary relationships in the amakihi complex. Auk 110:825–831.

TAYLOR, D. L. 1983. Fire management and the Cape Sable Sparrow. Pp. 147–152 in T. L. Quay, J. B. Funderburg, Jr., D. S. Lee, E. F. Potter, and C. S. Robbins (editors). The Seaside Sparrow: its biology and management. Occasional Papers of the North Carolina Biological Survey 1983-5. Raleigh, NC.

TAYLOR, R. 1998. Rails: a guide to the rails, crakes, gallinules and coots of the world. Princeton University Press, New Haven, CT.

TEAL, J. M. 1986. The ecology of regularly flooded salt marshes of New England: a community profile. USDI Fish and Wildlife Service, Biological Report 85 (7.4). Washington, DC.

TEAR, T. H., J. M. SCOTT, P. H. HAYWARD, AND B. GRIFFITH. 1995. Recovery plans and the Endangered Species Act: are criticisms supported by data? Conservation Biology 9:182–195.

TEMMERMAN, S., G. GOVERS, P. MEIRE, AND S. WARTEL. 2003. Modeling long-term tidal marsh growth under changing tidal conditions and suspended sediment concentrations, Scheldt estuary, Belgium. Marine Geology 193:151–169.

TERRES, J. K. 1991. The Audubon Society encyclopedia of North American birds. Wings Books, Avenel, NJ.

TERRILL, S. 2000. Salt Marsh Common Yellowthroat Geothlypis trichas sinuosa. Pp. 366–369 in P. Olofson (editor). Goals project 2000. Baylands ecosystem species and community profiles: life histories and environmental requirements of key plants, fish and wildlife. San Francisco Bay Area Wetlands Ecosystem Goals Project. San Francisco Bay Regional Water Quality Control Board, Oakland, CA.

TEWES, M. E. 1984. Opportunistic feeding by White-tailed Hawks at prescribed burns. Wilson Bulletin 96:135–136.

TEWKSBURY, J. J., A. E. BLACK, N. NUR, V. A. SAAB, B. D. LOGAN, AND D. S. DOBKIN. 2002. Effects of anthropogenic fragmentation and livestock grazing on western riparian bird communities. Studies in Avian Biology 25:158–202.

THAELER, C. J. 1961. Variation in some salt marsh populations of Microtus californicus. University of California Publications in Zoology 60:67–94.

THANNHEISER, D. 1981. The coastal vegetation of eastern Canada. The Munster Series of Geographical Studies 10:1–202.

THE CONSERVATION FOUNDATION. 1988. Protecting America's wetlands: an action agenda. National Wetlands Policy Forum, Conservation Foundation, Washington, DC.

THEODORAKIS, C. W., S. J. D'SURNEY, AND L. R. SHUGART. 1994. Detection of genotoxic insults as DNA strand breaks in fish blood cells by agarose gel electrophoresis. Environmental Toxicology and Chemistry 13:1023–1031.

THERON, E., K. HAWKINS, E. BERMINGHAM, R. E. RICKLEFS, AND N. I. MUNDY. 2001. The molecular basis of an avian plumage polymorphism in the wild: a melanocortin-1-receptor point mutation is perfectly associated with the melanic plumage morph of the Bananaquit, Coereba flaveola. Current Biology 11:550–557.

THOMAS, D. H., AND A. P. ROBIN. 1977. Comparative studies of thermoregulatory and osmoregulatory behaviour and physiology of five species of sandgrouse (Aves: Pterocliidae) in Morocco. Journal of Zoology 183:229–249.

THOMAS, E., T. GAPOTCHENKO, J. C. VAREKAMP, E. L. MECRAY, AND M. R. BUCHHOLTZ TEN BRINK. 2000. Benthic Foraminifera and environmental changes in Long Island Sound. Journal of Coastal Research 16:641–655.

THOMAS, M. L. H. 1983. Saltmarsh systems. Pp. 107–118 in M. L. H. Thomas (editor). Marine and coastal systems of the Quoddy Region, New Brunswick. Government of Canada-Fisheries and Oceans, Scientific Information and Publications Branch, Ottawa, ON, Canada.

THOMPSON, R. S., C. WHITLOCK, P. J. BARTLEIN, S. P. HARRISON, AND W. G. SPAAULDING. 1993. Climatic changes in the western United States since 18,000 yr. B.P. Pp. 468–513 in H. E. Wright, Jr., J. E. Kutzbach, T. Webb III, W. F. Ruddiman, F. A. Street-Perrott, and P. J. Bartlein (editors). Global climates since the last glacial maxima. University of Minnesota Press, Minneapolis, MN.

TILTON, M. A. 1987. Foraging habitats and behavior of the Red-winged Blackbird in a Rhode Island Spartina alterniflora marsh. M.S. thesis, University of Rhode Island, Kingston, RI.

TINER, R. W., JR. 1984. Wetlands of the United States: current status and recent trends. National Wetlands Inventory, USDI Fish and Wildlife Service, Washington, DC.

TITUS, J. G. (EDITOR). 1988. Greenhouse effects, sea level rise and coastal wetlands. Environmental Protection Agency 230-05-86-013. U.S. Government Printing Office, Washington, DC.

TITUS, J. G. 1991. Greenhouse effect and coastal wetland policy: how Americans could abandon an area the size of Massachusetts. Environmental Management 15:39–58.

TITUS, J. G., R. A. PARK, S. P. LEATHERMAN, J. R. WEGGEL, M. S. GREENE, P. W. MAUSEL, M. S. TREHAN, S. BROWN, C. GRANT, AND G. W. YOHE. 1991. Greenhouse effect and sea level rise: loss of land and the cost of holding back the sea. Coastal Management 19:171–204.

TOWNSEND, C. W. 1912. Notes on the summer birds of the St. John Valley, New Brunswick. Auk 29:16–23.

TRENHAILE, A. S. 1997. Coastal dynamics and landforms. Clarendon Press, Oxford, UK.

TROMBULAK, S. C., AND C. A. FRISSELL. 2000. Review of the ecological effects of roads on terrestrial and aquatic communities. Conservation Biology 14:18–30.

TRULIO, L. A., AND J. EVENS. 2000. California Black Rail Laterallus jamaicensis coturniculus. Pp. 341–345 in P. Olofson (editor). Goals project 2000. Baylands ecosystem species and community profiles: life histories and environmental requirements of key plants, fish and wildlife. San Francisco Bay Area Wetlands Ecosystem Goals Project. San Francisco Bay Regional Water Quality Control Board, Oakland, CA.

TUCKER, A. D., N. N. FITZSIMMONS, AND J. W. GIBBONS. 1995. Resource partitioning by the estuarine turtle *Malaclemys terrapin*: trophic, spatial, and temporal foraging constraints. Herpetologica 51:167–181.

TUCKER, A. D., J. W. GIBBONS, AND J. L. GREENE. 2003. Estimates of adult survival and migration for diamondback terrapins: conservation insight from local extirpation within a metapopulation. Canadian Journal of Zoology 79:2199–2209.

TUFTS, R. W. 1986. Birds of Nova Scotia. Nimbus Publishing and the Nova Scotia Museum, Halifax, NS, Canada.

TURNER, M. G. 1987. Effects of grazing by feral horses, clipping, trampling, and burning on a Georgia salt marsh. Estuaries 10:54–60.

U.S. ARMY CORPS OF ENGINEERS. 1916a. Tactical map, Cape Fortuna quadrangle, California, grid zone G. 1:62,500. Engineer Reproduction Plant, U.S. Army, Washington, DC.

U.S. ARMY CORPS OF ENGINEERS. 1916b. Tactical map, Rohnerville quadrangle, California. 1:62,500. Engineer Reproduction Plant, Fort Humphreys, Washington, DC.

U.S. ARMY CORPS OF ENGINEERS. 1987. Beneficial uses of dredged material. CECW-EH-D. Engineer Manual. EM 110-2-5026. Washington, DC.

U.S. CENSUS BUREAU. 2000. Census 2000. <http://www.census.gov> (6 June 2006).

U.S. COAST AND GEODETIC SURVEY. 1967a. Bathymetric map, coast and geodetic survey 1206N-15, Santa Barbara to Huntington Beach, California. 1:250,000. U.S. Department of Commerce, Washington DC.

U.S. COAST AND GEODETIC SURVEY. 1967b. Bathymetric map, coast and geodetic survey 1206N-16, Huntington Beach, California to Punta Sal Si Puedes, Mexico. 1:250,000. U.S. Department of Commerce, Washington DC.

U.S. COAST AND GEODETIC SURVEY. 1967c. Bathymetric map, coast and geodetic survey 1306N-19, Southwest of Santa Rosa Island, California. 1:250,000. U.S. Department of Commerce, Washington DC.

U.S. COAST AND GEODETIC SURVEY. 1967d. Bathymetric map, coast and geodetic survey 1306N-20 Cape San Martin to Point Concepcion, California. 1:250,000. U.S. Department of Commerce, Washington DC.

U.S. COAST AND GEODETIC SURVEY. 1969. Bathymetric map, coast and geodetic survey 1308N-12 Point Saint George, Oregon to Point Delgada, California. 1:250,000. U.S. Department of Commerce, Washington DC.

U.S. DEPARTMENT OF AGRICULTURE. 2003. Plant profile for *Salicornia virginica*. <http://plants.usda.gov/cgi_bin/plant_profile.cgi?symbol=SAVI> (6 June 2006).

U.S. ENVIRONMENTAL PROTECTION AGENCY. 1991. Registration eligibility decision: methoprene. U.S. Environmental Protection Agency, Office of Pesticides and Toxic Substances, Washington, DC.

U.S. ENVIRONMENTAL PROTECTION AGENCY. 1998. An SAB report: ecological impacts and evaluation criteria for the use of structures in marsh management. U.S. Environmental Protection Agency. EPA-SAB-EPEC-98-003. Washington, DC.

U.S. ENVIRONMENTAL PROTECTION AGENCY. 1998. Reregistration eligibility decision: *Bacillus thuringiensis*. U.S. Environmental Protection Agency, Office of Pesticide Programs, Washington, DC.

U.S. ENVIRONMENTAL PROTECTION AGENCY. 2003. Joint statement on mosquito control in the United States from the U.S. Environmental Protection Agency (EPA) and the U.S. Centers for Disease Control and Prevention (CDC). <http://www.epa.gov/pesticides/health/mosquitoes/mosquitojoint.htm> (6 June 2006).

U.S. GEOLOGICAL SURVEY. 1996. San Francisco Bay—1985 habitat. U.S. Geological Survey, Midcontinent Ecological Science Center, Fort Collins, CO.

U.S. GEOLOGICAL SURVEY. 2003. West Nile virus. U.S. Geological Survey, National Wildlife Health Center. <http://www.nwhc.usgs.gov/disease_information/west_nile_virus/index.jsp> (6 June 2006).

UETZ, P. 2005. The EMBL reptile data base. <http://www.embl-heidelberg.de/~uetz/LivingReptiles.html> (17 January 2006).

URIEN, C. M., L. R. MARTINIS, AND I. R. MARTINIS. 1980. Modelos desposicionales en la plataforma continental de Rio Grande du Sul, Uruguay, e Buenos Aires. Notas Tecnicas DECO-UFRGS 2:13–85.

USDI FISH AND WILDLIFE SERVICE. 1979. Light-footed Clapper Rail recovery plan. USDI Fish and Wildlife Service, Portland, OR.

USDI FISH AND WILDLIFE SERVICE. 1992. Status and trends report on wildlife of the San Francisco estuary. Prepared under EPA Cooperative Agreement CE-009519 by the USDI Fish and Wildlife Service, Sacramento Fish and Wildlife Enhancement Field Office, Sacramento, CA.

VALIELA, I., J. MCCLELLAND, J. HAUXWELL, P. J. BEHR, D. HERSH, AND K. FOREMAN. 1997. Macroalgal blooms in shallow estuaries: controls and ecophysiological and ecosystem consequences. Limnology and Oceanography 42:1105–1118.

VAN HORNE, B. 1983. Density as a misleading indicator of habitat quality. Journal of Wildlife Management 47:893–901.

VAN PROOSDIJ, D., J. OLLERHEAD, AND R. G. D. DAVIDSON-ARNOTT. 2000. Controls on suspended sediment deposition over single tidal cycles in a macrotidal saltmarsh, Bay of Fundy, Canada. Pp. 43–57 *in* K. Pye, and J. R. L. Allen (editors). Coastal and estuarine environments: sedimentology, geomorphology, and geoarcheology. Geological Society of London, London, UK.

VAN PROOSDIJ, D., J. OLLERHEAD, R. G. D. DAVIDSON-ARNOTT, AND L. SCHOSTAK. 1999. Allen Creek Marsh, Bay of Fundy: a macrotidal coastal saltmarsh. Canadian Landform Examples 37. Canadian Geographer 43:316–322.

VAN ZOOST, J. R. 1970. The ecology and waterfowl utilization of the John Lusby National Wildlife Area. M.S. thesis, Acadia University, Wolfville, NS, Canada.

VASQUEZ, E. A., E. P. GLENN, J. J. BROWN, G. R. GUNTENSPERGEN, AND S. G. NELSON. 2005. Growth characteristics and salinity tolerance underlying the cryptic invasion of North American salt marshes by an introduced haplotype of the common reed *Phragmites australis* (Poaceae). Marine Ecology Progress Series 298:1–8

VER PLANCK, W. E. 1958. Salt in California. California Division of Mines Bulletin 175. California Division of Mines, San Francisco, CA.

VERNER, J. 1965. Breeding biology of the Long-billed Marsh Wren. Condor 67:6–30.

VERNER, J., AND M. WILLSON. 1966. The influence of habitats on mating systems of North American passerine birds. Ecology. 47:143–147.

VICKERY, P. D. 1996. Grasshopper Sparrow (*Ammodramus savannarum*). *In* A. Poole, and F. Gill (editors). The birds of North America, No. 239. The Academy of Natural Sciences, Philadelphia, PA and The American Ornithologists' Union, Washington, DC.

VON BLOEKER, J. C., JR. 1932. Three new mammal species from salt marsh areas in southern California. Proceedings of the Biological Society of Washington 45:131–138.

VON BLOEKER, J. C., JR. 1937. Four new rodents from Monterey County, California. Proceedings of the Biological Society of Washington 50:153–157.

VON HAARTMAN, L. 1969. Nest-site and evolution of polygamy in European passerine birds. Ornis Fennica 46:1–12.

WAHRHAFTIG, C., AND J. H. BIRMAN. 1965. The Quaternary of the Pacific mountain system in California. Pp. 299–331 *in* H. E. Wright, Jr., and D. G. Frey (editors). The Quaternary of the United States. Princeton University Press, Princeton, NJ.

WALBECK, D. 1989. Use of open marsh water management ponds by ducks, wading bird, and shorebirds during late summer and fall. M.S. thesis, Frostburg State University, Frostburg, MD.

WALTER, H. S. 2004. The mismeasure of islands: implications for biogeographical theory and the conservation of nature. Journal of Biogeography 31:177–197

WALTERS, J. R. 1998. The ecological basis of avian sensitivity to habitat fragmentation. Pp. 181–192 *in* J. M. Marzluff, and R. Sallabanks (editors). Avian conservation. Island Press, Washington, DC.

WALTERS, J. R., S. R. BEISSINGER, J. W. FITZPATRICK, R. GREENBERG, J. D. NICHOLS, H. R. PULLIAM, AND D. W. WINKLER. 2000. The AOU Conservation Committee review of the biology, status, and management of Cape Sable Seaside Sparrows: final report. Auk 117:1093–1115.

WALTERS, M. J. 1992. A shadow and a song: the struggle to save an endangered species. Chelsea Green Publishing Company, Post Mills, VT.

WALTON, B. J. 1978. The status of the Salt Marsh Song Sparrows of the San Francisco Bay system, 1974–1976. M.A. thesis, San Jose State University, San Jose, CA.

WANG, X., S. VAN DER KAARS, P. KERSHAW, M. BIRD, AND F. JANSEN. 1999. A record of fire, vegetation and climate through the last three glacial cycles from Lombok Ridge Core G6-4, eastern Indian Ocean, Indonesia. Palaeogeography, Palaeoclimatology, Palaeoecology 147:241–256.

WANKET, J. A. 2002. Late Quaternary vegetation and climate of the Klamath Mountains. Ph.D. dissertation, University of California, Berkeley, CA.

WARD, D. W., AND J. BURGER. 1980. Survival of Herring Gull and domestic chicken embryos after simulated flooding. Condor 82:142–148.

WARNOCK, N., G. W. PAGE, T. D. RUHLEN, N. NUR, J. Y. TAKEKAWA, AND J. T. HANSON. 2002. Management and conservation of San Francisco Bay salt ponds: effects of pond salinity, area, tide, and season on Pacific Flyway waterbirds. Waterbirds 25 (Special Publication 2):79–92.

WARREN, R. S., P. E. FELL, R. ROZSA, A. H. BRAWLEY, A. C. ORSTED, E. T. OLSON, V. SWAMY, AND W. A. NIERING. 2002. Saltmarsh restoration in Connecticut: 20 years of science and management. Restoration Ecology 10:497–513.

WARREN, R. S., AND W. A. NIERING. 1993. Vegetation change on a northeast tidal marsh: interaction of sea-level rise and marsh accretion. Ecology 74: 96–103.

WARRICK, R. A., E. BARROW, AND T. M. L. WIGLEY (EDITORS). 1993. Climate and sea level change: observations, projections, and implications. Cambridge University Press, Cambridge, UK.

WASHINGTON STATE DEPARTMENT OF NATURAL RESOURCES. 1998. Our changing nature: natural resource trends in Washington State. Washington State Department of Natural Resources, Olympia, WA.

WASHINGTON STATE DEPARTMENT OF NATURAL RESOURCES. 2000. Changing our water ways: trends in Washington's water systems. Washington State Department of Natural Resources, Olympia, WA.

WATSON, E. B. 2002. Sediment accretion and tidal marsh formation at a Santa Clara valley tidal marsh. M.A. thesis, University of California. Berkeley, CA.

WEBER, L. M., AND S. M. HAIG. 1996. Shorebird use of South Carolina managed and natural coastal wetlands. Journal of Wildlife Management 60:73–82.

WEBSTER, C. G. 1964. Fall foods of Soras from habitats in Connecticut. Journal of Wildlife Management 28:163–165.

WEBSTER, W. D., AND J. K. JONES, JR. 1982. *Reithrodontomys megalotis*. Mammalian Species 167:1–5.

WEBSTER, W. M., J. F. PARNELL, AND W. C. BIGGS, JR. 1985. The mammals of the Carolinas, Virginia, and Maryland. University of North Carolina Press, Chapel Hill, NC.

WEINSTEIN, M. P., AND J. H. BALLETTO. 1999. Does the common reed, *Phragmites australis* affect essential fish habitat? Estuaries 22:63–72.

WEINSTEIN, M. P., J. R. KEOUGH, G. R. GUNTENSPERGEN, AND S. Y. LITVIN. 2003. *Phragmites australis*: a sheep in wolf's clothing. Proceedings from the technical forum and workshop. New Jersey Sea Grant Publication NJSG-03-516. Vineland, NJ.

WEINSTEIN, M. P., AND D. A. KREEGER (EDITORS). 2000. Concepts and controversies in tidal marsh ecology. Kluwer Academic Publishers, Dordecht, The Netherlands.

WELLER, M. W. 1994. Bird-habitat relationships in Texas estuarine marsh during summer. Wetlands 14:293–300.

WELLS, E. D., AND H. E. HIRVONEN. 1988. Wetlands of Atlantic Canada. Pp. 249–303 *in* National Wetlands Working Group. Wetlands of Canada. Sustainable

Development Branch, Environment Canada, and Polyscience Publications Inc., Ottawa, ON, Canada.

WELTER, W. A. 1935. The natural history of the Long-billed Marsh Wren. Wilson Bulletin 47:3–34.

WERNER, H. W. 1975. The biology of the Cape Sable sparrow. Report to USDI Fish and Wildlife Service, Everglades National Park, Homestead, FL.

WERNER, H. W., AND G. E. WOOLFENDEN. 1983. The Cape Sable sparrows: its habitat, habits and history. Pp. 55–75 in T. L. Quay, J. B. Funderburg, Jr., D. S. Lee, E. F. Potter, and C. S. Robbins (editors). The Seaside Sparrow: its biology and management. Occasional Papers of the North Carolina Biological Survey 1983-5. Raleigh, NC.

WETHERBEE, D. K. 1968. *Melospiza georgiana nigriscens* Coastal Plain Swamp Sparrow. Pp. 1490 in A. C. Bent (editor). Life Histories of North American cardinals, grosbeaks, buntings, towhees, finches, sparrows, and allies. Dover Publications, Inc., New York, NY.

WETMORE, A. 1926. Observations on the birds of Argentina, Paraguay, Uruguay and Chile. Bulletin of the U.S. National Museum of Natural History 133:1–448.

WHEELWRIGHT, N. T., AND J. D. RISING. 1993. Savannah Sparrow *(Passerculus sandwichensis)*. In A. Poole, and F. Gill (editors). The birds of North America, No. 45. The Academy of Natural Sciences, Philadelphia, PA and The American Ornithologists' Union, Washington, DC.

WHITE, G. C., AND K. P. BURNHAM. 1999. Program MARK: Survival estimation from populations of marked animals. Bird Study (Supplement) 46: 120–139.

WHITE, J. R., P. S. HOFMANN, K. A. F. URQUHART, D. HAMMOND, AND S. BAUMGARTNER. 1989. Selenium verification study, 1987–1988. A report to the California State Water Resources Control Board from California Department of Fish and Game, Sacramento, CA.

WHITT, M. B., H. H. PRINCE, AND R. R. COX, JR. 1999. Avian use of purple loosestrife dominated habitat relative to other vegetation types in a Lake Huron wetland complex. Wilson Bulletin 111:105–114.

WHITTINGHAM, L. A. 1989. An experimental study of paternal behavior in Red-winged Blackbirds. Behavioral Ecology and Sociobiology 25:73–80.

WHITTINGHAM, L. A., AND R. J. ROBERTSON. 1994. Food availability, parental care and male mating success in Red-winged Blackbirds *(Agelaius phoeniceus)*. Journal of Animal Ecology 63:139–150.

WIENER, J. G., C. C. GILMOUR, AND D. P. KRABBENHOFT. 2003. Mercury strategy for the bay-delta ecosystem: a unifying framework for science, adaptive management, and ecological restoration. Draft Final Report to CALFED. CALFED Bay-Delta Authority, Sacramento, CA.

WIENER, J. G., AND D. J. SPRY. 1996. Toxicological significance of mercury in freshwater fish. Pp. 297–339 in W. N. Beyer, G. H. Heinz, and A. W. Redmon-Norwood (editors). Environmental contaminants in wildlife: interpreting tissue concentrations. Lewis Boca Raton, FL.

WIENS, J. A. 1994. Habitat fragmentation: island v landscape perspectives on bird conservation. Ibis 137:S97–S104.

WIENS, J. A. 1996. Wildlife in patchy environments: metapopulations, mosaics, and management. Pp. 53–84 in D. R. McCullough (editor). Metapopulations and Wildlife Conservation. Island Press, Washington, DC.

WIENS, J. A., AND J. T. ROTENBERRY. 1981. Censusing and the evaluation of avian habitat occupancy. Studies in Avian Biology 6:522–532.

WILCOVE, D., C. MCLELLAN, AND A. DOBSON. 1986. Habitat fragmentation in the temperate zone. Pp. 237–256 in M. Soule (editor). Conservation biology: the science of scarcity and diversity. Sinauer, Sunderland, MA.

WILCOVE, D., S., D. ROTHSTEIN, J. DUBOW, A. PHILLIPS, AND E. LOSOS. 1998. Quantifying threats to imperiled species in the United States. BioScience 48: 607–615.

WILEY, R H. 1991. Associations of song properties with habitats for territorial oscine birds of eastern North America. American Naturalist 138:973-93.

WILLIAMS, D. F. 1979. Checklist of California mammals. Annals of the Carnegie Museum of Natural History 48:425–433.

WILLIAMS, D. F. 1986. Mammalian species of concern in California. Wildlife Management Division Administrative Report 86-1. Department of Fish and Game, Sacramento, CA.

WILLIAMS, J. B., AND B. DWINNEL. 1990. Field metabolism of free-living female Savannah Sparrows during incubation a study using doubly labeled water. Physiological Zoology 63:353–372.

WILLIAMS, J. B., AND K. A. NAGY. 1984. Daily energy expenditure of Savannah Sparrows: comparison of time-energy budget and doubly labeled water estimates. Auk 101:221–229.

WILLIAMS, J. B., AND K. A. NAGY. 1985. Daily energy expenditure by female Savannah Sparrows feeding nestlings. Auk 102:187–190.

WILLIAMS, J. B., W. R. SIEGFREID, S. J. MILTON, N. J. ADAMS, W. R. J. DEAN, M. A. DU PLESSIS, AND S. JACKSON. 1993. Field metabolism, water requirements, and foraging behavior of wild ostriches in the Namib. Ecology 74:390–404.

WILLIAMS, L. 1929. Notes on the feeding habits and behavior of the California Clapper Rail. Condor 31:52–56.

WILLIAMS, P. B., AND M. K. ORR. 2002. Physical evolution of restored breached levee salt marshes in the San Francisco Bay estuary. Restoration Ecology 10:527–542

WILLIAMSON, D. 1994. Use of *Spartina* and open intertidal flats by common birds of Willapa Bay. Pp. 32–44 in *Spartina* management program: integrated weed management for private lands in Willapa Bay, Washington. Spartina Task Force for the Noxious Weed Board and Pacific County Commissioners, Willapa Bay, WA.

WILLIAMSON, K. 1967. A bird community of accreting sand dune and salt marsh. British Birds 60:14–57.

WILLNER, G., G. A. FELDHAMMER, E. ZUCKER, AND J. A. CHAPMAN. 1980. *Ondatra zibethicus*. Mammalian Species 141:1–8.

WILSON, D. E., AND D. M. REEDER. 1993. Mammal species of the World. Smithsonian Institution Press, Washington, DC.

WILSON, D. E., AND S. RUFF. 1999. The Smithsonian book of North American mammals. Smithsonian Institution Press, Washington, DC.

WILSON, W. H., T. BRUSH, R. LENT, AND B. HARRINGTON. 1987. The effects of grid-ditching and open marsh water management on avian utilization of Massachusetts salt marshes. Pp. 333–349 *in* W. R. Whitman, and W. H. Meredith (editors). Waterfowl and wetlands symposium: proceedings of a symposium on waterfowl and wetlands management in the coastal zone of the Atlantic flyway. Delaware Coastal Management Program, Delaware Department of Natural Resources and Environmental Control, Dover, DE.

WINDHAM, L. 1999. Microscale spatial distribution of *Phragmites australis* (common reed) invasion into *Spartina patens* (salt hay)-dominated communities in brackish tidal marsh. Biological Invasions 1:137–148

WITHERS, P. C. 1992. Comparative animal physiology. Saunders College Publishing, New York, NY.

WOLF, C. M., T. GARLAND, JR., AND B. GRIFFITH. 1998. Avian and mammalian translocations: reanalysis with phylogenetically independent contrasts. Biological Conservation 86:243–255.

WOLFE, J. L. 1982. *Oryzymys palustris*. Mammalian Species 176:1–5.

WOLFE, M. F., S. SCHWARZBACH, AND R. A. SULAIMAN. 1998. Effects of mercury on wildlife: a comprehensive review. Environmental Toxicology and Chemistry 17:146–160.

WOLFE, R. A. 1996. Effects of open marsh water management on selected tidal marsh resources: a review. Journal of the American Mosquito Control Association 12:701–712.

WOOD, R. C. 1977. Evolution of the emydine turtles *Graptemys* and *Malaclemys* (Reptilia, Testudines, Emydidae). Journal of Herpetology 11:415–421.

WOOD, R. C., AND R. HERLANDS. 1997. Turtles and tires: the impact of roadkills on northern diamondback terrapin, *Malaclemys terrapin terrapin*. Pp. 46–53 *in* J. Van Abbema (editor). Proceedings: conservation, restoration and management of tortoises and turtles. New York Turtle and Tortoise Society, New York, NY.

WOODS, C. A., W. POST, AND C. W. KILPATRICK. 1982. *Microtus pennsylvanicus* (Rodentia: Muridae) in Florida: a Pleistocene relict in a coastal saltmarsh. Bulletin of the Florida State Museum Biological Sciences 28:25–52.

WOOLFENDEN, G. E. 1956. Comparative breeding behavior of *Ammospiza caudacuta* and *A. maritima*. University of Kansas Publications, Museum of Natural History 10:45–75.

WRAY, T., II. K. A, STRAIT, AND R. C. WHITMORE. 1982. Reproductive success of grassland bird on a reclaimed surface mine in West Virginia. Auk 99:157–164.

WRIGHT, S. A., AND D. H. SCHOELLHAMER. 2004. Trends in the sediment yield of the Sacramento River, California, 1957–2001. San Francisco Estuary and Watershed Science 2:2, Article 2. <http://repositories.cdlib.org/jmie/sfews/vol2/iss2/art2> (6 June 2006).

YANEV, K. P. 1980. Biogeography and distribution of three parapatric salamander species in coastal and borderland California. Pp. 531–549 *in* D. M. Power (editor). The California Islands: proceedings of a multidisciplinary symposium. Santa Barbara Museum of Natural History, Santa Barbara, CA.

YASUKAWA, K., F. LEANZA, AND C. D. KING. 1993. An observational and brood-exchange study of paternal provisioning in the Red-winged Blackbird, *Agelaius phoeniceus*. Behavioral Ecology 4:78–82.

YASUKAWA, K., J. L. MCCLURE, R. A. BOLEY, AND J. ZANOCCO. 1990. Provisioning of nestlings by male and female Red-winged Blackbirds, *Agelaius phoeniceus*. Animal Behaviour 40:153–166.

YASUKAWA, K., AND W. A. SEARCY. 1995. Red-winged Blackbird (*Agelaius phoeniceus*). *In* A. Poole, and F. Gill (editors). The birds of North America, No. 184. The Academy of Natural Sciences, Philadelphia, PA and The American Ornithologists' Union, Washington, DC.

YORIO, P. 1998. Anexo II: Capítulo 13. Costa Argentina (Patagonia). *In* P. Canevari, I. Davidson, D. E. Blanco, G. Castro, and E. H. Bucher (editors). Los humedales de América del Sur. Una agenda para la conservación de la biodiversidad y las políticas de desarrollo. Wetlands International. Wageningen, The Netherlands.

YOUNG, B. E. 1996. An experimental analysis of small clutch size in tropical House Wrens. Ecology 77:472–488.

YOUNG, K. E. 1987. The effect of Greater Snow Goose, *Anser caerulescens atlantica*, on grazing on a Delaware tidal marsh. Pp. 51–54 *in* W. R. Whitman, and W. H. Meredith (editors). Waterfowl and wetlands symposium: proceedings of a symposium on waterfowl and wetlands management in the coastal zone of the Atlantic flyway. Delaware Coastal Management Program, Delaware Department of Natural Resources and Environmental Control, Dover, DE.

YOUNG, K. R. 1994. Roads and the environmental degradation of tropical montane forests. Conservation Biology 8:972–976.

ZAR, J. H. 1999. Biostatistical analysis. Prentice Hall, Upper Saddle River, NJ.

ZEDLER, J. B. 1982. The ecology of southern California coastal salt marshes: a community profile. USDI Fish and Wildlife Service, Biological Services Program. FWS/OBS-81/54. Washington, DC.

ZEDLER, J. B. 1993. Canopy architecture of natural and planted cordgrass marshes: selecting habitat evaluation criteria. Ecological Applications 3:123–138.

ZEDLER, J. B. 1996. Coastal mitigation in southern California: the need for a regional restoration strategy. Ecological Applications 6:84–93.

ZEDLER, J. B. (EDITOR). 2001. Handbook for restoring tidal wetlands. *In* M. J. Kennish (editor). CRC Marine Science Series. CRC Press, Boca Raton, FL.

ZEDLER, J. B., J. C. CALLAWAY, AND G. SULLIVAN. 2001. Declining biodiversity: why species matter and how their functions might be restored in Californian tidal marshes. BioScience 51:1005–1017.

ZEDLER, J. B., J. COVIN, C. NORDBY, P. WILLIAMS, AND J. BOLAND. 1986. Catastrophic events reveal the dynamic nature of salt-marsh vegetation in southern California. Estuaries 9:75–80.

ZEMBAL, R., AND S. M. HOFFMAN. 2002. A survey of the Belding's Savannah Sparrow (*Passerculus sandwichensis beldingi*) in California, 2001. Habitat Conservation Planning Branch, Species Conservation and Recovery Program Report 2002-01, California Department of Fish and Game, Sacramento, CA.

ZEMBAL, R., S. M. HOFFMAN, AND J. R. BRADLEY. 1998. Light-footed Clapper Rail management and assessment, 1997. Habitat Conservation Planning Branch, Species Conservation and Recovery Program Report 1998-01, California Department of Fish and Game, Sacramento, CA.

ZEMBAL, R., J. KRAMER, R. J. BRANSFIELD, AND N. GILBERT. 1988. A survey of Belding's Savannah Sparrows in California. American Birds 42:1233–1236.

ZEMBAL, R., AND B. W. MASSEY. 1987. Seasonality of vocalizations by Light-footed Clapper Rails. Journal of Field Ornithology 58:41–48.

ZEMBAL, R., W. B. MASSEY, AND J. M. FANCHER. 1989. Movements and activity patterns of the light-footed Clapper Rail. Journal of Wildlife Management 53:39–42.

ZERVAS, C. 2001. Sea-level variations of the United States, 1854–1999. National Oceanic and Atmospheric Administration Technical Report NOS CO-OPS 36, Silver Spring, MD.

ZETTERQUIST, D. K. 1977. The salt marsh harvest mouse (*Reithrodontomys raviventris raviventris*) in marginal habitats. The Wasmann Journal of Biology 35:68–76.

ZIMMERMAN, T. D. 1992. Latitudal reproductive variation of the saltmarsh turtle, the diamondback terrapin (*Malaclemys terrapin*). M.S. thesis, University of Charleston, Charleston, SC.

ZINK, R. M., AND J. C. AVISE. 1990. Patterns of mitochondrial DNA and allozyme evolution in the avian genus *Ammodramus*. Systematic Zoology 39:148–161.

ZINK, R. M., AND R. C. BLACKWELL. 1996. Patterns of allozyme, mitochondrial DNA and morphometric variation in four sparrow genera. Auk 113:59–67.

ZINK, R. M., AND D. L. DITTMANN. 1993. Gene flow, refugia, and evolution of geographic variation in the Song Sparrow (*Melospiza melodia*). Evolution 47:717–729.

ZINK, R. M., D. L. DITTMANN, AND S. W. CARDIFF. 1991. Mitochondrial DNA variation and the taxonomic status of the Large-Billed Savannah Sparrow. Condor 93:1016–1019.

ZINK, R. M., AND H. W. KALE. 1995. Conservation genetics of the extinct Dusky Seaside Sparrow *Ammodramus maritimus nigrescens*. Biological Conservation 74:69–74.

ZINK, R. M., AND J. V. REMSEN, JR. 1986. Evolutionary processes and patterns of geographic variation in birds. Current Ornithology 4:1–69.

ZINK, R. M., J. D. RISING, S. MOCKFORD, A. G. HORN, J. M. WRIGHT, M. LEONARD, AND M. C. WESTBERG. In press. Mitochondrial DNA variation, species limits, and rapid evolution of plumage coloration and size in the Savannah Sparrow. Condor.

ZUCCA, J. J. 1954. A study of the California Clapper Rail. Wasmann Journal of Biology 12:135–153.

ZUG, D. A., AND W. A. DUNSON. 1979. Salinity preference in freshwater and estuarine snakes (*Nerodia sipedon* and *N. fasciata*). Florida Scientist 42:1–8.